W. Schnabel

Polymers and Light

1807–2007 Knowledge for Generations

Each generation has its unique needs and aspirations. When Charles Wiley first opened his small printing shop in lower Manhattan in 1807, it was a generation of boundless potential searching for an identity. And we were there, helping to define a new American literary tradition. Over half a century later, in the midst of the Second Industrial Revolution, it was a generation focused on building the future. Once again, we were there, supplying the critical scientific, technical, and engineering knowledge that helped frame the world. Throughout the 20th Century, and into the new millennium, nations began to reach out beyond their own borders and a new international community was born. Wiley was there, expanding its operations around the world to enable a global exchange of ideas, opinions, and know-how.

For 200 years, Wiley has been an integral part of each generation's journey, enabling the flow of information and understanding necessary to meet their needs and fulfill their aspirations. Today, bold new technologies are changing the way we live and learn. Wiley will be there, providing you the must-have knowledge you need to imagine new worlds, new possibilities, and new opportunities.

Generations come and go, but you can always count on Wiley to provide you the knowledge you need, when and where you need it!

William J. Pesce
President and Chief Executive Officer

Peter Booth Wiley
Chairman of the Board

W. Schnabel

Polymers and Light

Fundamentals and Technical Applications

WILEY-VCH Verlag GmbH & Co. KGaA

The Author

Prof. Dr. W. Schnabel
Divison of Solar Energy Research
Hahn-Meitner-Institut
Glienicker Str. 100
14109 Berlin
Germany

1st Edition 2007
  1st Reprint 2008

■ All books published by Wiley-VCH are carefully produced. Nevertheless, authors, editors, and publisher do not warrant the information contained in these books, including this book, to be free of errors. Readers are advised to keep in mind that statements, data, illustrations, procedural details or other items may inadvertently be inaccurate.

Library of Congress Card No.: applied for

British Library Cataloguing-in-Publication Data
A catalogue record for this book is available from the British Library.

Bibliographic information published by the Deutsche Nationalbibliothek
The Deutsche Nationalbibliothek lists this publication in the Deutsche Nationalbibliografie; detailed bibliographic data are available in the Internet at http://dnb.d-nb.de.

© 2007 WILEY-VCH Verlag GmbH & Co. KGaA, Weinheim

All rights reserved (including those of translation into other languages). No part of this book may be reproduced in any form – by photoprinting, microfilm, or any other means – nor transmitted or translated into a machine language without written permission from the publishers. Registered names, trademarks, etc. used in this book, even when not specifically marked as such, are not to be considered unprotected by law.

Composition K+V Fotosatz GmbH, Beerfelden
Printing betz-druck GmbH, Darmstadt
Bookbinding Litges & Dopf GmbH, Heppenheim
Cover Adam Design, Weinheim
Wiley Bicentennial Logo Richard J. Pacifico

Printed in the Federal Republic of Germany
Printed on acid-free paper

ISBN 978-3-527-31866-7

Contents

Preface *XIII*

Introduction *1*

Part I **Light-induced physical processes in polymers**

1 **Absorption of light and subsequent photophysical processes** *5*
1.1 Principal aspects *5*
1.2 The molecular orbital model *7*
1.3 The Jablonski diagram *10*
1.4 Absorption in non-conjugated polymers *10*
1.5 Absorption in conjugated polymers *12*
1.6 Deactivation of electronically excited states *13*
1.6.1 Intramolecular deactivation *13*
1.6.2 Intermolecular deactivation *14*
1.6.3 Energy migration and photon harvesting *16*
1.6.4 Deactivation by chemical reactions *21*
1.7 Absorption and emission of polarized light *22*
1.7.1 Absorption *22*
1.7.2 Absorption by chiral molecules *23*
1.7.3 Emission *26*
1.8 Applications *30*
1.8.1 Absorption spectroscopy *30*
1.8.1.1 UV/Vis spectroscopy *30*
1.8.1.2 Circular dichroism spectroscopy *32*
1.8.1.3 IR spectroscopy *35*
1.8.2 Luminescence *37*
1.8.3 Time-resolved spectroscopy *38*
1.8.3.1 General aspects *38*
1.8.3.2 Experimental techniques *39*
1.8.3.3 Applications of time-resolved techniques *41*
1.8.3.3.1 Optical absorption *41*

1.8.3.3.2	Luminescence 44	
	References 45	
2	**Photoconductivity** 49	
2.1	Introductory remarks 49	
2.2	Photogeneration of charge carriers 50	
2.2.1	General aspects 50	
2.2.2	The exciton model 52	
2.2.3	Chemical nature of charge carriers 54	
2.2.4	Kinetics of charge carrier generation 55	
2.2.5	Quantum yield of charge carrier generation 57	
2.3	Transport of charge carriers 60	
2.4	Mechanism of charge carrier transport in amorphous polymers 64	
2.5	Doping 66	
2.6	Photoconductive polymers produced by thermal or high-energy radiation treatment 69	
2.7	Photoconductive polymers produced by plasma polymerization or glow discharge 70	
	References 70	
3	**Electro-optic and nonlinear optical phenomena** 73	
3.1	Introductory remarks 73	
3.2	Fundamentals 74	
3.2.1	Electric field dependence of polarization and dipole moment 74	
3.2.2	Electric field dependence of the index of refraction 78	
3.3	Characterization techniques 79	
3.3.1	Second-order phenomena 79	
3.3.1.1	Determination of the hyperpolarizability β 79	
3.3.1.2	Determination of the susceptibility $\chi^{(2)}$ 81	
3.3.2	Third-order phenomena 82	
3.3.2.1	Third harmonic generation 83	
3.3.2.2	Self-focusing/defocusing 84	
3.3.2.3	Two-photon absorption (TPA) 85	
3.3.2.4	Degenerate four-wave mixing (DFWM) and optical phase conjugation 86	
3.4	Nonlinear optical materials 87	
3.4.1	General aspects 87	
3.4.2	Second-order NLO materials 89	
3.4.2.1	Guest-host systems and NLO polymers 89	
3.4.2.2	Orientation techniques 92	
3.4.3	Third-order NLO materials 93	
3.5	Applications of NLO polymers 96	
3.5.1	Applications relating to telecommunications 96	
3.5.2	Applications relating to optical data storage 99	

3.5.3	Additional applications *100*	
	References *101*	
4	**Photorefractivity** *103*	
4.1	The photorefractive effect *103*	
4.2	Photorefractive formulations *105*	
4.3	Orientational photorefractivity *107*	
4.4	Characterization of PR materials *108*	
4.5	Applications *110*	
	References *112*	
5	**Photochromism** *113*	
5.1	Introductory remarks *113*	
5.2	Conformational changes in linear polymers *115*	
5.2.1	Solutions *115*	
5.2.2	Membranes *122*	
5.3	Photocontrol of enzymatic activity *123*	
5.4	Photoinduced anisotropy (PIA) *123*	
5.5	Photoalignment of liquid-crystal systems *126*	
5.6	Photomechanical effects *130*	
5.6.1	Bulk materials *130*	
5.6.2	Monolayers *133*	
5.7	Light-induced activation of second-order NLO properties *134*	
5.8	Applicationss *136*	
5.8.1	Plastic photochromic eyewear *136*	
5.8.2	Data storage *137*	
	References *139*	
6	**Technical developments related to photophysical processes in polymers** *143*	
6.1	Electrophotography – Xerography *143*	
6.2	Polymeric light sources *146*	
6.2.1	Light-emitting diodes *147*	
6.2.1.1	General aspects *147*	
6.2.1.2	Mechanism *150*	
6.2.1.3	Polarized light from OLEDs *154*	
6.2.1.4	White-light OLEDs *155*	
6.2.2	Lasers *156*	
6.2.2.1	General aspects *156*	
6.2.2.2	Lasing mechanism *158*	
6.2.2.3	Optical resonator structures *159*	
6.2.2.4	Prospects for electrically pumped polymer lasers *162*	
6.3	Polymers in photovoltaic devices *162*	
6.4	Polymer optical waveguides *167*	
6.4.1	General aspects *167*	

6.4.2	Optical fibers 168
6.4.2.1	Polymer versus silica fibers 168
6.4.2.2	Compositions of polymer optical fibers (POFs) 169
6.4.2.3	Step-index and graded-index polymer optical fibers 170
6.4.3	Polymer planar waveguides 170
6.4.4	Polymer claddings 170
	References 171

Part II	Light-induced chemical processes in polymers
7	**Photoreactions in synthetic polymers** 177
7.1	Introductory remarks 177
7.1.1	Amplification effects 178
7.1.2	Multiplicity of photoproducts 178
7.1.3	Impurity chromophores 180
7.1.4	Photoreactions of carbonyl groups 182
7.2	Cross-linking 183
7.2.1	Cross-linking by cycloaddition of C=C bonds 184
7.2.2	Cross-linking by polymerization of reactive moieties in pendant groups 186
7.2.3	Cross-linking by photogenerated reactive species 188
7.2.4	Cross-linking by cleavage of phenolic OH groups 192
7.3	Simultaneous cross-linking and main-chain cleavage of linear polymers 193
7.4	Photodegradation of selected polymers 196
7.4.1	Poly(vinyl chloride) 196
7.4.2	Polysilanes 198
7.5	Oxidation 199
7.6	Singlet oxygen reactions 202
7.7	Rearrangements 202
	References 205

8	**Photoreactions in biopolymers** 207
8.1	Introductory remarks 207
8.2	Direct light effects 211
8 2.1	Photoreactions in deoxyribonucleic acids (DNA) 211
8.2.1.1	Dimeric photoproducts 212
8.2.1.2	Other DNA photoproducts 214
8.2.2	Photoreactions in proteins 214
8.2.2.1	Chemical alterations by UV light 215
8.2.2.2	Formation of stress proteins 216
8.2.2.3	Effects of visible light – photoreceptor action 217
8.2.2.4	Repair of lesions with the aid of DNA photolyases 219
8.2.3	Photoreactions in cellulose 221
8.2.4	Photoreactions in lignins and wood 221

8.3	Photosensitized reactions 222	
	References 228	
9	**Technical developments related to photochemical processes in polymers** 231	
9.1	Polymers in photolithography 231	
9.1.1	Introductory remarks 231	
9.1.2	Lithographic processes 231	
9.1.2.1	Projection optical lithography 233	
9.1.2.2	Maskless lithography 235	
9.1.3	Resists 236	
9.1.3.1	Classical polymeric resists – positive and negative resist systems 236	
9.1.3.2	Chemical amplification resists 239	
9.1.3.3	Resists for ArF (193 nm) lithography 242	
9.1.3.4	Resists for F_2 (157 nm) lithography 245	
9.1.4	The importance of photolithography for macro-, micro-, and nanofabrication 246	
9.2	Laser ablation of polymers 248	
9.2.1	General aspects 248	
9.2.1.1	Introductory remarks 248	
9.2.1.2	Phenomenological aspects 248	
9.2.1.3	Molecular mechanism 250	
9.2.2	Dopant-enhanced ablation 250	
9.2.3	Polymers designed for laser ablation 251	
9.2.4	Film deposition and synthesis of organic compounds by laser ablation 252	
9.2.5	Laser desorption mass spectrometry and matrix-assisted laser desorption/ionization (MALDI) 254	
9.2.6	Generation of periodic nanostructures in polymer surfaces 256	
9.2.7	Laser plasma thrusters 256	
9.3	Stabilization of commercial polymers 257	
9.3.1	Introductory remarks 257	
9.3.2	UV absorbers 258	
9.3.2.1	Phenolic and non-phenolic UV absorbers 258	
9.3.2.2	Mechanistic aspects 259	
9.3.3	Energy quenchers 260	
9.3.4	Chain terminators (radical scavengers) 262	
9.3.5	Hydroperoxide decomposers 265	
9.3.6	Stabilizer packages and synergism 266	
9.3.7	Sacrificial consumption and depletion of stabilizers 267	
	References 268	

Part III	Light-induced synthesis of polymers
10	**Photopolymerization** *275*
10.1	Introduction *275*
10.2	Photoinitiation of free radical polymerizations *276*
10.2.1	General remarks *276*
10.2.2	Generation of reactive free radicals *276*
10.2.2.1	Unimolecular fragmentation of type I photoinitiators *276*
10.2.2.2	Bimolecular reactions of type II photoinitiators *279*
10.2.2.3	Macromolecular photoinitiators *279*
10.2.2.4	Photoinitiators for visible light *281*
10.2.2.4.1	Metal-based initiators *282*
10.2.2.4.2	Dye/co-initiator systems *284*
10.2.2.4.3	Quinones and 1,2-diketones *285*
10.2.2.5	Inorganic photoinitiators *287*
10.3	Photoinitiation of ionic polymerizations *288*
10.3.1	Cationic polymerization *288*
10.3.1.1	General remarks *288*
10.3.1.2	Generation of reactive cations *290*
10.3.1.2.1	Direct photolysis of the initiator *290*
10.3.1.2.2	Sensitized photolysis of the initiator *291*
10.3.1.2.3	Free-radical-mediated generation of cations *292*
10.3.1.2.3.1	Oxidation of radicals *292*
10.3.1.2.3.2	Addition-fragmentation reactions *294*
10.3.2	Anionic polymerization *295*
10.3.2.1	General remarks *295*
10.3.2.2	Generation of reactive species *295*
10.3.2.2.1	Photo-release of reactive anions *295*
10.3.2.2.2	Photo-production of reactive organic bases *296*
10.4	Topochemical polymerizations *298*
10.4.1	General remarks *298*
10.4.2	Topochemical photopolymerization of diacetylenes *299*
10.4.3	Topochemical photopolymerization of dialkenes *301*
	References *302*
11	**Technical developments related to photopolymerization** *305*
11.1	General remarks *305*
11.2	Curing of coatings, sealants, and structural adhesives *307*
11.2.1	Free radical curing *307*
11.2.1.1	Solvent-free formulations *307*
11.2.1.2	Waterborn formulations *309*
11.2.2	Cationic curing *309*
11.2.3	Dual curing *310*
11.3	Curing of dental preventive and restorative systems *312*
11.4	Stereolithography – microfabrication *313*

11.5	Printing plates	*316*
11.5.1	Introductory remarks	*316*
11.5.2	Structure of polymer letterpress plates	*317*
11.5.3	Composition of the photosensitive layer	*317*
11.5.4	Generation of the relief structure	*317*
11.6	Curing of printing inks	*318*
11.7	Holography	*319*
11.7.1	Principal aspects	*319*
11.7.2	Mechanism of hologram formation	*321*
11.7.3	Multicolor holographic recording	*321*
11.7.4	Holographic materials	*322*
11.7.5	Holographic applications	*323*
11.8	Light-induced synthesis of block and graft copolymers	*324*
11.8.1	Principal aspects	*324*
11.8.2	Surface modification by photografting	*328*
	References	*329*

Part IV	**Miscellaneous technical developments**	
12	**Polymers in optical memories**	*337*
12.1	General aspects	*337*
12.2	Current optical data storage systems	*338*
12.2.1	Compact disk (CD) and digital versatile disk (DVD)	*338*
12.2.2	Blue-ray disks	*340*
12.3	Future optical data storage systems	*341*
12.3.1	General aspects	*341*
12.3.2	Volume holography	*342*
12.3.2.1	Storage mechanism	*342*
12.3.2.2	Storage materials	*343*
12.3.3	Photo-induced surface relief storing	*345*
	References	*345*

13	**Polymeric photosensors**	*347*
13.1	General aspects	*347*
13.2	Polymers as active chemical sensors	*349*
13.2.1	Conjugated polymers	*349*
13.2.1.1	Turn-off fluorescence detection	*350*
13.2.1.2	Turn-on fluorescence detection	*350*
13.2.1.3	ssDNA base sequence detection	*352*
13.2.1.4	Sensors for metal ions	*352*
13.2.1.5	Image sensors	*353*
13.2.2	Optical fiber sensors	*353*
13.2.3	Displacement sensors	*354*
13.3	Polymers as transducer supports	*355*
	References	*356*

14	**Polymeric photocatalysts** *359*	
14.1	General aspects *359*	
14.2	Polymers as active photocatalysts *359*	
14.2.1	Conjugated polymers *359*	
14.2.2	Linear polymers bearing pendant aromatic groups *361*	
14.3	Polymers as supports for inorganic photocatalysts *362*	
	References *364*	

Subject Index *365*

Preface

Light can do a lot of quite different things to polymers, and light is employed in various quite different technical applications related to polymers that have become beneficial to humans and are influencing the daily lives of many people. These applications include photocopying machines, computer chips, compact disks, polymer optical fiber systems in local area networks, and printing plates. There are many other very useful practical applications. Since these are commonly dealt with separately in monographs or review articles, the idea arose to comprehend and combine in a single book all important developments related to polymers and light that concern industrially employed practical applications or show potential for future applications. Actually, I first contemplated writing a book dealing with both physical and chemical aspects related to the interaction of light with polymers and to the synthesis of polymers with the aid of light while I was lecturing on certain topics of this field at the Technical University in Berlin and at Rika Daigaku (Science University) in Tokyo. However, I only started to immerse myself in this extensive project when I retired from active service some time ago. Upon retrieving and studying the salient literature, I became fascinated by the broadness of the field. The results of this project are presented here for the first time. In referring to the different topics, I have tried to deal with the fundamentals only to the extent necessary for an understanding of described effects. In attempting to be as concise as possible, descriptions of technical processes and tools have had to be restricted to a minimum in order to keep the extent of the book within reasonable limits. To somewhat compensate for this flaw, a rather comprehensive list of literature references also covering technical aspects is presented at the end of each chapter.

Writing a monograph implies that the author can both concentrate on the subject in a quiet office and rely on the cooperation of an effectively functioning library. Both were provided by the Hahn-Meitner-Institute, HMI, and I am very grateful to the management of this institute, especially to Prof. Dr. *M. Steiner*, Scientific Director, Chief Executive, for giving me the opportunity to work on this book after my transfer to emeritus status. Special thanks are due to Prof. Dr. *H. Tributsch*, head of the Solar Energy Research Division of HMI, for appreciating my intention to write this book and for providing a quiet room. The HMI library under the direction of Dr. *E. Kupfer* and his successor Dr. *W. Fritsch* has sub-

Polymers and Light. Fundamentals and Technical Applications. W. Schnabel
Copyright © 2007 WILEY-VCH Verlag GmbH & Co. KGaA, Weinheim
ISBN: 978-3-527-31866-7

stantially contributed to the preparation and completion of the manuscript by delivering necessary resources and executing many retrievals. The latter yielded most of the literature citations upon which this book is based. In this context, I wish to express my special gratitude to senior librarian Mr. *M. Wiencken*, who has performed an excellent job. Other people who proved very helpful in this project are Mr. *D. Gaßen*, who has kept the computer running, and Mrs. *P. Kampfenkel*, who has scanned various figures.

The personnel of the publisher, Wiley-VCH, worked carefully and rapidly on the editing of the manuscript after its completion in the summer of 2006. This is gratefully acknowledged.

Last but not least, credit has to be given to the efforts of the author's family. My wife *Hildegard* has accompanied the progress of the project with encouraging sympathy and moral support, and my two sons, Dr. *Ronald Schnabel* and Dr. *Rainer Florian Schnabel*, have given substantial advice. The latter has critically read all chapters of the manuscript.

Berlin, November 2006   *Wolfram Schnabel*

Introduction

The technological developments of the last decades have been essentially determined by trends to invent new materials and to establish new technical methods. These trends encompass the synthesis of novel polymeric materials and the employment of light in industrial processes. To an increasing extent, technical processes based on the interaction of light with polymers have become important for various applications. To mention a few examples, polymers are used as nonlinear optical materials, as core materials for optical wave guides, and as photoresists in the production of computer chips. Polymers serve as photoswitches and optical memories, and are employed in photocopying machines and in solar cells for the generation of energy. Moreover, certain polymeric materials can be utilized for the generation of light.

On the other hand, light serves also as a tool for the synthesis of polymers, i.e. for the initiation of the polymerization of small molecules, a method which is applied in technical processes involving the curing of coatings and adhesives and even by the dentist to cure tooth inlays.

Obviously, the field related to the topic *polymers and light* is a very broad one. A principle of order derived from the distinction of photophysical from photochemical processes may help to steer us through this wide field. Hence, photophysical and photochemical processes are addressed in separate parts of this book (Part I and Part II), where both fundamentals and related practical applications are dealt with. Regarding pure photophysical processes that are not combined with chemical alterations of the polymers (Part I), separate chapters are devoted to fundamentals concerning the interaction of light with polymers, photoconductivity, electro-optic and nonlinear phenomena, photorefractivity, and photochromism (Chapters 1–5, respectively). Important technical applications related to photophysical processes in polymers are dealt with in Chapter 6. These applications include xerography, light-emitting diodes (LEDs), lasers, solar cells, optical wave guides, and optical fibers.

In Part II, fundamentals of light-induced chemical processes are discussed by making a distinction between synthetic organic polymers (Chapter 7) and biopolymers (Chapter 8). Also in Part II, important technical applications related to photochemical processes in polymers are dealt with separately in Chapter 9. Here, important practical applications such as photolithography, which is a nec-

Polymers and Light. Fundamentals and Technical Applications. W. Schnabel
Copyright © 2007 WILEY-VCH Verlag GmbH & Co. KGaA, Weinheim
ISBN: 978-3-527-31866-7

essary tool for the production of computer chips, and laser ablation are covered. Moreover, one section of Chapter 9 is devoted to the stabilization of commercial polymers, a very important subject regarding the long-time stability of plastic materials.

The light-induced synthesis of polymers is the topic of Part III. While the various modes of photoinitiation of polymerization processes are discussed in Chapter 10, related technical applications are treated in Chapter 11. The latter include curing of coatings and dental systems, printing plates (used to print newspapers), holography (important for data storage), and the synthesis of block-and-graft copolymers.

Finally, Part IV reviews miscellaneous technical developments that do not fit neatly into the scheme of the preceding parts. These concern, in particular, the application of polymers in the field of optical memories, treated in Chapter 12, which refers also to currently important data storage systems (compact disks, digital versatile disks, and blue-ray disks). Moreover, the application potential of polymers in the fields of photosensors and photocatalysts is outlined in Chapters 13 and 14, respectively.

Part I
Light-induced physical processes in polymers

1
Absorption of light and subsequent photophysical processes

To open the way into the wide-ranging fields covered in this book, some elementary facts essential for an understanding of the material covered are outlined at the beginning. Since books [1–6] are available that comprehensively treat the principles of the interaction of light with matter, the aim here is to present the salient points in a very concise manner. Nevertheless, in citing typical cases, close adherence to the actual subject of the book has been sought by referring to polymers wherever possible.

1.1
Principal aspects

Photons are absorbed by matter on a time scale of about 10^{-15} s. During this very short time, the electronic structure of the absorbing molecule is altered, whereas the positions of the atomic nuclei in the molecule, vibrating on a time scale of 10^{-12} s, are not changed. There are two prerequisites for the absorption of a photon of energy $h\nu$ by a molecule: (1) the molecule must contain a chromophoric group with excitable energy states corresponding to the photon energy according to Eq. (1-1).

$$h\nu = E_n - E_0 \qquad (1\text{-}1)$$

E_n and E_0 denote the energies of the excited and the ground state, respectively. Typical chromophoric groups are listed in Table 1.1.

(2) The transition between the two energy states must cause a change in the charge distribution in the molecule, i.e. a change in the dipole moment. In terms of quantum mechanics: absorption of a photon is possible (allowed) if the transition moment M has a non-zero value. Since M is a vector composed of three components parallel to the three coordinates [Eq. (1-2)], at least one component must have a non-zero value.

$$M = M_x + M_y + M_z \qquad (1\text{-}2)$$

Polymers and Light. Fundamentals and Technical Applications. W. Schnabel
Copyright © 2007 WILEY-VCH Verlag GmbH & Co. KGaA, Weinheim
ISBN: 978-3-527-31866-7

1 Absorption of light and subsequent photophysical processes

Table 1.1 Typical chromophoric groups [4]

Chromophore	Typical compound	λ_{max} (nm) [a]	ε_{max} (L mol^{-1} cm^{-1}) [b]	Mode of electron transition
$\mathrm{C=C}$	Ethene	193	10^4	$\pi \rightarrow \pi^*$
$-\mathrm{C{\equiv}C}-$	Ethyne	173	6×10^3	$\pi \rightarrow \pi^*$
$\mathrm{C=O}$	Acetone	187	10^3	$\pi \rightarrow \pi^*$
		271	15	$n \rightarrow \pi^*$
$-\mathrm{N=N}-$	Azomethane	347	5	$n \rightarrow \pi^*$
$-\mathrm{N=O}$	t-Nitrosobutane	300	100	$\pi \rightarrow \pi^*$
		665	20	$n \rightarrow \pi^*$
$-\mathrm{O-N=O}$	Amyl nitrite	219	219	$\pi \rightarrow \pi^*$
		357	357	$n \rightarrow \pi^*$

a) Wavelength of maximum optical absorption.
b) Decadic molar extinction coefficient (log $I_0/I = \varepsilon c d$).

The higher the value of M, the more efficient is the absorption. As described by Eq. (1-3), M is composed of three integrals:

$$M = \int \psi_v^* \psi_v d\tau_v \int \psi_e^* \mu_{dp} \psi_e d\tau_e \int \psi_s^* \psi_s d\tau_s \qquad (1\text{-}3)$$

where ψ_v, ψ_e, and ψ_s are the vibronic, electronic, and electron-spin wave functions of the absorbing molecule, respectively. The asterisk denotes "excited state". μ_{dp} is the electronic dipole moment operator. $d\tau_v$, $d\tau_e$, and $d\tau_s$ refer to the three respective coordinates; $d\tau = dx \cdot dy \cdot dz$.

The three integrals in Eq. (1-3) are the basis of the so-called selection rules, which determine whether a transition is allowed or forbidden. $(\int \psi_v^* \psi_v d\tau_v)^2$ is the Franck-Condon factor and $\int \psi_s^* \psi_s d\tau_s$ applies to the spin properties of the excited and the ground states. If any of the three integrals in Eq. (1-3) is zero, the corresponding transition is forbidden, i.e. a final probability could only result from a second-order approximation. This applies, e.g., to the forbidden transitions between levels of the singlet and the triplet system. The magnitude of the Franck-Condon factor determines the probability of transitions with respect to molecular geometry. The rule states that the transition probability is highest if the geometries of the ground and excited states are equal. A more detailed treatment of these aspects is beyond the scope of this book and the reader is referred to relevant monographs [2–4].

The probability of the occurrence of an electronic transition is given by the (dimensionless) oscillator strength f, which is proportional to the square of the transition moment [Eq. (1-4)].

$$f = 8.75 \times 10^{-2} \Delta E |M|^2 \qquad (1\text{-}4)$$

Here, ΔE is equal to E_n-E_0 (given in eV). A large value of f corresponds to a strong absorption band and a short lifetime of the excited state. The maximum value is $f=1$.

Experimentally, the absorption of light is recorded as a function of the wavelength λ or the wave number $v' = \lambda^{-1}$ by measuring the change in the intensity of a light beam passing through a sample of unit path length (1 cm). For a homogeneous, isotropic medium containing an absorbing compound at concentration c (mol L^{-1}), the light absorption is described by Eq. (1-5), the Lambert-Beer law:

$$A = \lg_{10}(I_0/I) = \varepsilon c d \qquad (1\text{-}5)$$

where A is the absorbance (extinction, optical density), and I_0 and I denote the light intensity before and after absorption. Equivalent denotations for I_0 and I are *incident* and *transmitted radiant flux*, respectively. ε (L mol^{-1} cm^{-1}) is the decadic molar extinction coefficient at a given wavelength. The Lambert-Beer law does not hold at high light intensities, as experienced, e.g., with lasers. The oscillator strength f is related to the measured integrated extinction coefficient $\int \varepsilon dv'$ by Eq. (1-6), where ε and v' have to be given in units of L mol^{-1} cm^{-1} and cm^{-1}, respectively:

$$f = (2.3 \times 10^3 c^2 m / N e^2 \pi) F \int \varepsilon dv' = 4.32 \times 10^{-9} \, F \int \varepsilon dv' \qquad (1\text{-}6)$$

Here, c is the velocity of light, m and e are the mass and charge of an electron, respectively, and N is Avogadro's number. The factor F, which reflects solvent effects and depends on the refractive index of the absorbing medium, is close to unity. ε_{max}, the extinction coefficient at the maximum of an absorption band, is a measure of the intensity (magnitude) of the band and an indicator of the *allowedness* of the corresponding electronic transition.

1.2
The molecular orbital model

Changes in the electronic structure of a molecule can be visualized with the aid of the molecular orbital (MO) model [3, 4]. Molecular orbitals are thought to be formed by the linear combination of the valence shell orbitals of the atoms linked together in the molecule. The combination of two single orbitals of two adjacent atoms results in two molecular orbitals, one of lower and the other of higher energy than before combination. The low-energy orbital, denoted as the bonding orbital, is occupied by a pair of electrons of antiparallel spin. The high-energy molecular orbital is called an antibonding orbital. It is unoccupied in the ground state, but may be occupied by an electron upon electronic excitation of the molecule.

There are different kinds of molecular orbitals: bonding σ and π orbitals, non-bonding n orbitals, and antibonding σ^* and π^* orbitals. σ and σ^* orbitals are completely symmetrical about the internuclear axis, whereas π and π^* orbitals are antisymmetric about a plane including the internuclear axis. n orbitals, which are located on heteroatoms such as oxygen, nitrogen, or phosphorus, are nonbonding and are of almost the same energy as in the case of the isolated atom. A pair of electrons occupying an n orbital is regarded as a lone pair on the atom in question.

The simple MO model is based on several assumptions. For instance, σ and π orbitals are assumed not to interact. Moreover, molecules are described by localized orbitals each covering two nuclei only. Delocalized orbitals involving more than two nuclei are thought to exist only in the case of π-bonding in conjugated systems.

When a molecule in its ground state absorbs a photon, an electron occupying a σ, π or n orbital is promoted to a higher-energy σ^* or π^* orbital. In principle, the following transitions are possible: $\sigma \rightarrow \sigma^*$, $\pi \rightarrow \pi^*$, $n \rightarrow \pi^*$, and $n \rightarrow \sigma^*$. As

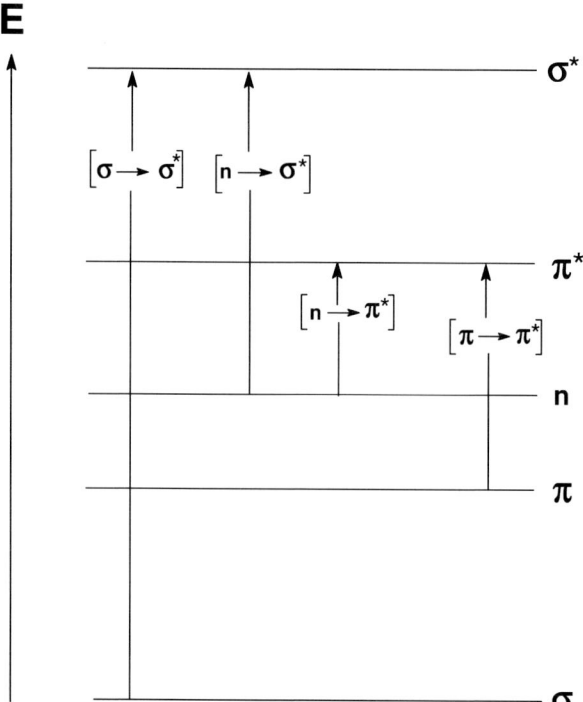

Fig. 1.1 Molecular orbitals (not to scale) and electronic transitions induced by the absorption of a photon.

Table 1.2 The correspondence of electron transition and optical absorption

Electron transition	Absorption region (nm)	Extinction coefficient (L mol^{-1} cm^{-1})
$\sigma \rightarrow \sigma^*$	100–200	10^3
n $\rightarrow \sigma^*$	150–250	10^2–10^3
$\pi \rightarrow \pi^*$		10^2–10^4
(Isolated π-bonds)	180–250	
(Conjugated π-bonds)	220–IR	
n $\rightarrow \pi^*$		1–400
(Isolated groups)	220–320	
(Conjugated segments)	250–IR	

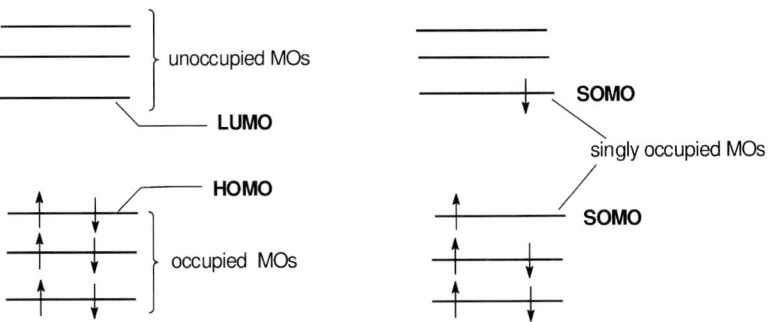

Fig. 1.2 Classification of molecular orbitals with respect to electron occupancy.

can be seen in Fig. 1.1, the orbital energy increases in the series $\sigma < \pi < n < \pi^* < \sigma^*$.

According to the differences in the orbital energies, the electron transitions indicated in Fig. 1.1 correspond to light absorption in different wavelength regions. This is illustrated in Table 1.2.

It follows that under conveniently practicable conditions ($\lambda > 200$ nm), photon absorption initiates transitions of n or π electrons rather than those of σ electrons.

Commonly, molecular orbitals are classified as *occupied* (doubly), *singly occupied*, and *unoccupied*. The acronyms HOMO and LUMO denote the frontier orbitals, i.e. the *H*ighest *O*ccupied and the *L*owest *U*noccupied *M*olecular *O*rbital, respectively. SOMO stands for *S*ingly *O*ccupied *M*olecular *O*rbital (see Fig. 1.2).

1.3
The Jablonski diagram

Photon-induced excitations of molecules also include vibrations of nuclei. This fact can be visualized with the aid of the Jablonski diagram (see Fig. 1.3).

The diagram shows the various energy states of a molecule, and further indicates the transitions related to the formation and deactivation of excited states. Here, photon absorption leads to electron transitions from the ground state S_0 to the excited states S_1, S_2, etc. Electron release occurs when the photon energy exceeds the ionization energy E_I. This is not the case within the wavelength range of UV and visible light, i.e. $\lambda = 200–800$ nm ($h\nu = 6.2–1.6$ eV).

1.4
Absorption in non-conjugated polymers

Figure 1.4 shows absorption spectra of the typical unconjugated linear polymers presented in Chart 1.1.

Due to the fact that electronic excitations also involve vibronic and rotational sublevels (the latter are not shown in Fig. 1.3), the absorption spectra of molecules consist of bands rather than single lines. It is notable that the maxima of the absorption spectra shown in Fig. 1.4 are located in the UV region. They reflect spin-state-conserving electronic transitions, i.e. transitions in the singlet manifold: upon photon absorption, molecules in the singlet ground state S_0 are

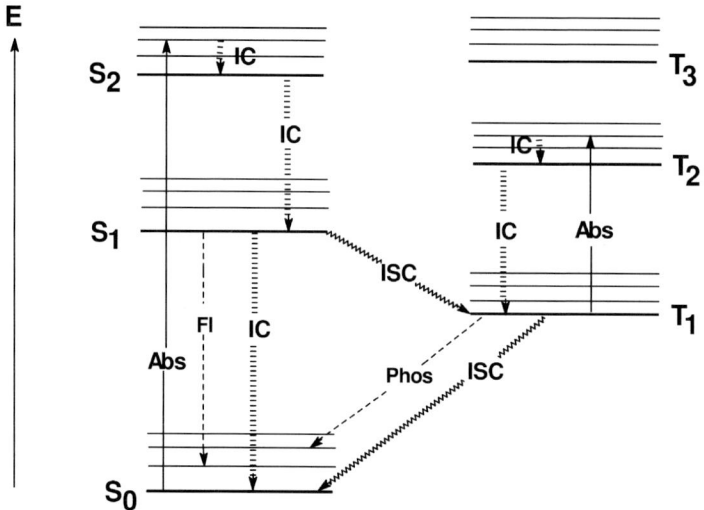

Fig. 1.3 Jablonski-type diagram. Abbreviations and acronyms:
Abs: absorption, Fl: fluorescence, Phos: phosphorescence,
IC: internal conversion, ISC: intersystem crossing.

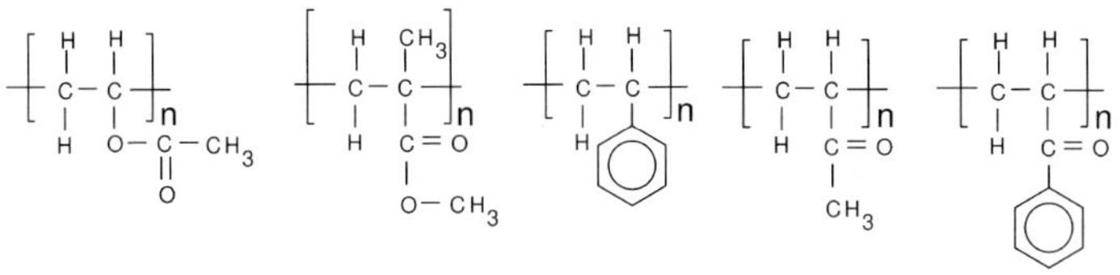

1.4 Absorption in non-conjugated polymers

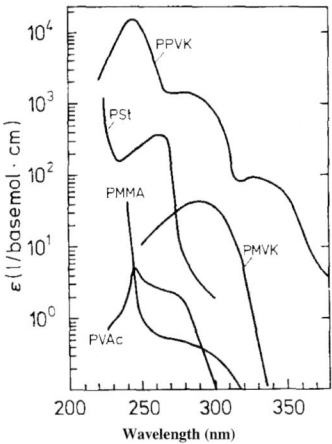

Chart 1.1 Chemical structures of poly(vinyl acetate), PVAc; poly(methyl methacrylate), PMMA; polystyrene, PSt; poly-(methyl vinyl ketone), PMVK; poly(phenyl vinyl ketone), PPVK.

Fig. 1.4 Absorption spectra of non-conjugated polymers. Adapted from Schnabel [7] with permission from Carl Hanser.

converted into molecules in an excited singlet state S_n. At long wavelengths (low photon energies), photon absorption generates S_1 states. At shorter wavelengths, S_2 and higher states are excited. In the case of polymers containing carbonyl groups, the absorption bands located at long wavelengths correspond to $n \rightarrow \pi^*$ transitions with low extinction coefficients, i.e. low values of the transition moment. At shorter wavelengths, $\pi \rightarrow \pi^*$ transitions with larger transition moments are excited. In this connection, the reader's attention is directed to Table 1.2, which indicates the relative orders of magnitude of the extinction coefficients of the different electron transitions.

1.5
Absorption in conjugated polymers

In recent years, various aromatic polymers with conjugated double bonds, so-called conjugated polymers, have been synthesized and thoroughly investigated with regard to applications in the fields of electroluminescence (organic light-emitting diodes) and photovoltaics (energy conversion of sunlight). Figure 1.5 presents typical absorption spectra of conjugated polymers (see Chart 1.2).

The maxima of the absorption spectra of conjugated polymers are located in the visible wavelength region.

Certain phenomena observed with conjugated polymers cannot be rationalized in terms of the model described in Section 1.1. This concerns, above all, the generation of charge carriers with the aid of UV and visible light, and the conduction of photogenerated charge carriers. A rationale for these phenomena is provided by the *exciton model*, which was originally developed for inorganic semiconductors and dielectrics [9–11]. According to this model, the absorption

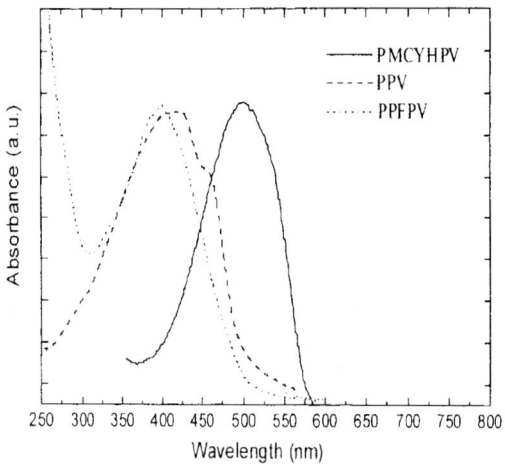

Fig. 1.5 Absorption spectra of conjugated polymers. Adapted from Shim et al. [8] with permission from Springer.

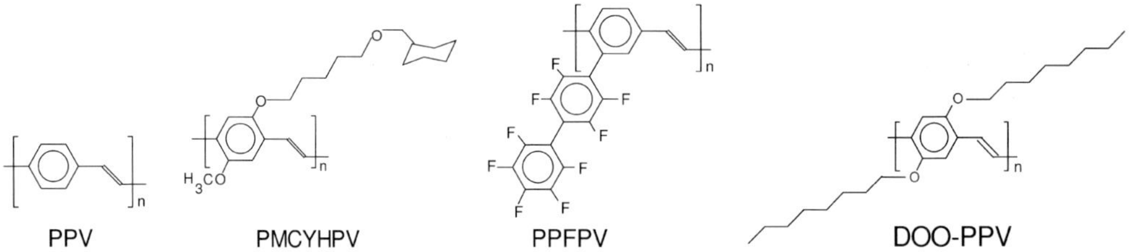

Chart 1.2 Chemical structures of poly(1,4-phenylene vinylene), PPV, and three PPV derivatives.

of a photon by a conjugated polymer promotes an electron from the ground state to an upper electronically excited state, which takes on the quality of a quasi-particle resembling a hydrogen-like system and can be considered as an electron/hole pair. The electron and hole are bound together, i.e. they cannot move independently of one another in the medium. Significantly, however, excitons are considered to be able to diffuse and, under certain circumstances, to dissociate into free charge carriers. This aspect is also treated in Section 2.2.2.

1.6
Deactivation of electronically excited states

1.6.1
Intramolecular deactivation

In condensed media, vibrational relaxation (internal conversion) is usually so fast that molecules excited to vibronically excited states S_{1v}, S_{2v}, etc. relax to the lowest excited singlet state, S_1, before they can undergo other processes. Further intramolecular deactivation processes of S_1 states (see the Jablonski diagram in Fig. 1.3) may be radiative or non-radiative. There is one radiative deactivation path resulting in photon emission, termed *fluorescence*, and two non-radiative processes competing with fluorescence: internal conversion (IC) to the ground state and intersystem crossing (ISC) to the triplet manifold. The latter process involves a change in electron spin, i.e. a molecule excited to the singlet state having solely pairs of electrons with antiparallel spins is converted into a molecule in an excited triplet state possessing one pair of electrons with parallel spins. Triplet states are commonly formed via this route. The direct formation of triplet states from the ground state through photon uptake is strongly spin-forbidden. In other words, $S_0 \rightarrow T_1$ transitions are very unlikely, i.e., the respective extinction coefficients are very low. In analogy, $T_1 \rightarrow S_0$ transitions are also spin-forbidden, which implies that the lifetime of triplet states is quite long and significantly exceeds that of S_1 states. Triplet states can deactivate radiatively. The emission of photons from triplet states is termed *phosphorescence*. Both luminescence processes, fluorescence and phosphorescence, cover a variety of transitions to the various vibronic levels of the S_0 state (see Fig. 1.6) and, therefore, yield emission spectra with several bands instead of a single line, as would be expected for the sole occurrence of 0-0 transitions. Figure 1.7 presents as a typical example the emission spectrum of poly(2,5-dioctyloxy-*p*-phenylene vinylene), DOO-PPV (see Chart 1.2) [12].

Since fluorescence is emitted from the non-vibronically excited S_1 state (see Fig. 1.6) and absorption involves higher, i.e., vibronically excited S_1 states, the maximum of the fluorescence spectrum is shifted to lower energy (higher wavelengths) relative to the absorption maximum (Stokes shift). The maximum of the phosphorescence spectrum is located at even higher wavelengths, since phosphorescence originates from the non-vibronically excited T_1 state, which is of lower energy than the corresponding S_1 state (see Fig. 1.3). The emission spectrum

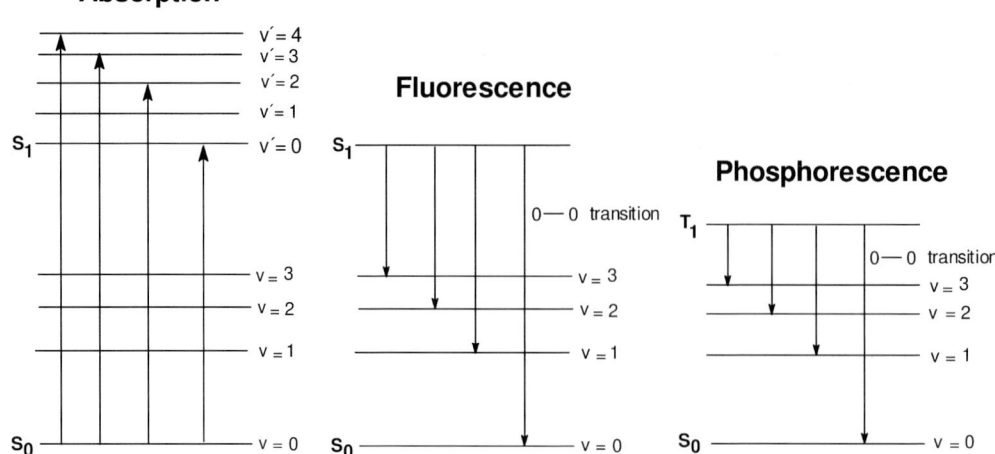

Fig. 1.6 Schematic depiction of transitions occurring during absorption, fluorescence, and phosphorescence.

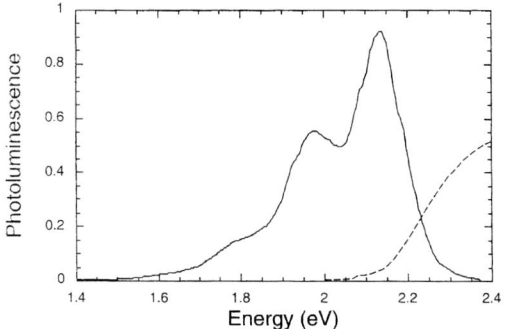

Fig. 1.7 Emission spectrum (full curve) and part of the absorption spectrum (dotted curve) of DOO-PPV. Adapted from Lane et al. [12] with permission from Wiley-VCH.

presented in Fig. 1.7 features three bands at 2.15 eV (577 nm), 1.98 eV (626 nm), and 1.8 eV (689 nm), which may be attributed to the zero-phonon (0-0), the one-phonon (1-0), and the two-phonon (2-0) transitions, respectively.

1.6.2
Intermolecular deactivation

Energy transfer from electronically excited molecules to ground-state molecules of different chemical composition represents a highly important intermolecular deactivation path. In general terms, energy transfer occurs according to Eq. (1-7) from a donor to an acceptor, the latter frequently being referred to as a quencher.

$$D^* + A \rightarrow D + A^* \tag{1-7}$$

This process is energetically favorable in the case of exothermicity, i.e., if the excitation energy of D^* exceeds that of A^*: $E(D^*) > E(A^*)$. A typical case concerns the stabilization of polymeric plastics. If an electronically excited macromolecule P^* transfers its excitation energy to an additive A according to Eq. (1-8), hydrogen abstraction [Eq. (1-9)] is inhibited and the macromolecule remains intact.

$$P^* + A \rightarrow P + A^* \tag{1-8}$$

$$P^* + RH \rightarrow {}^\bullet PH + R^\bullet \tag{1-9}$$

There are two major mechanisms by which energy transfer can occur: (1) The dipole-dipole (coulombic) mechanism, also denoted as the *Förster mechanism*, operating through mutual repulsion of the electrons in the two molecules. It is characterized by relatively large interaction distances ranging up to a molecular separation of 5 nm. (2) The exchange mechanism, also denoted as the *Dexter mechanism*, according to which a transient complex is formed on close approach of the partner molecules.

The dependence of the rate constant, k_{ET}, of intermolecular energy-transfer processes on the distance, R, is given by Eqs. (1-10) and (1-11) [13].

Long-range interaction: $\quad k_{ET} = k_D^0 (R_0/R)^6 \tag{1-10}$

Short-range interaction: $\quad k_{ET} = k_D^0 \exp(-aR) \tag{1-11}$

Here, k_D^0 is the unimolecular decay rate constant of the excited donor, and R_0 is the critical distance between D^* and A, at which the probabilities of spontaneous deactivation and of energy transfer are equal. Typical R_0 values are listed in Table 1.3, which also includes values for self-transfer [14]. The latter process is of relevance for down-chain energy transfer (energy migration), which is referred to below.

In principle, energy-transfer processes from both singlet and triplet excited donors to ground-state acceptors are possible [see Eqs. (1-12) and (1-13), respectively].

Table 1.3 Typical R_0 values (in Å) for aromatic chromophores [14].

	Naphthalene	Phenanthrene	Pyrene	Anthracene
Naphthalene	7.35	13.16	28.97	23.16
Phenanthrene		8.77	14.43	21.72
Pyrene			10.03	21.30
Anthracene				21.81

$$D^*(S_1) + A(S_0) \rightarrow D(S_0) + A^*(S_1) \quad (1\text{-}12)$$

$$D^*(T_1) + A(S_0) \rightarrow D(S_0) + A^*(T_1) \quad (1\text{-}13)$$

Commonly, singlet energy transfer takes place by the dipole-dipole mechanism, whereas triplet energy transfer occurs by the exchange mechanism since the dipole-dipole mechanism is spin-forbidden in this case.

If electronically excited, chemically identical species are generated at a high concentration, for example at high absorbed dose rates or during the simultaneous excitation of various chromophores attached to the same polymer chain, annihilation processes according to Eq. (1-14) can become important.

$$M^* + M^* \rightarrow M^{**} + M \quad (1\text{-}14)$$

M^{**} denotes a highly excited species that can emit a photon differing in energy to that emitted by M^*, or can undergo ionization or bond breakage. Annihilation is a self-reaction of excited species that may be singlets or triplets.

1.6.3
Energy migration and photon harvesting

A polymer-specific mode of energy transfer concerns *energy migration* in linear homopolymers, i.e. in macromolecules composed of identical repeating units. Since all of the repeating units contain identical chromophores, excitation energy can travel down the chain, provided that the geometrical conditions are appropriate (large R_0 for self-transfer) and the lifetime of the excited state τ_{exc} is longer than the energy-hopping time τ_h, i.e., $\tau_{exc} > \tau_h$. There are various pathways that may ensue, following the absorption of a photon by a certain chromophoric group. Figure 1.8 shows, besides the energy migration process, energy transfer to an external acceptor molecule and light emission.

Actually, *monomer emission* needs to be distinguished from *excimer emission*. The latter process originates from a transient complex formed, e.g., in the case of aromatic compounds, by the interaction of an excited molecule with a non-excited chemically identical molecule leading to an excited dimer, denoted as an *excimer* (see Scheme 1.1). In linear macromolecules bearing pendant aromatic groups, this process corresponds to the interaction between neighboring repeating units, as demonstrated in Scheme 1.1.

Excimers can usually be detected by a shift of the fluorescence emission maximum to a wavelength longer than in the case of monomer emission.

After down-chain energy migration in linear polymers had been evidenced by triplet-triplet annihilation and enhanced phosphorescence quenching [15–17], the idea arose to guide electronic excitation energy along the chain to defined sites, where it might serve to initiate chemical or physical processes. Obviously, such a mechanism is relevant to *photon harvesting* processes employed by nature in photosynthetic systems operating on the following principle, which is also re-

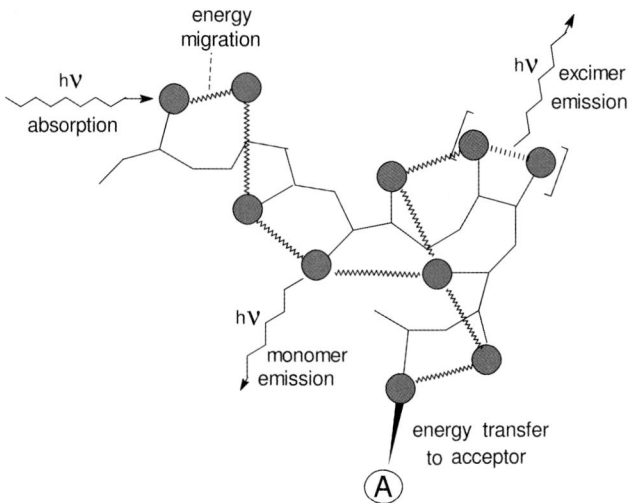

Fig. 1.8 Pathways of excitation energy in a linear macromolecule.

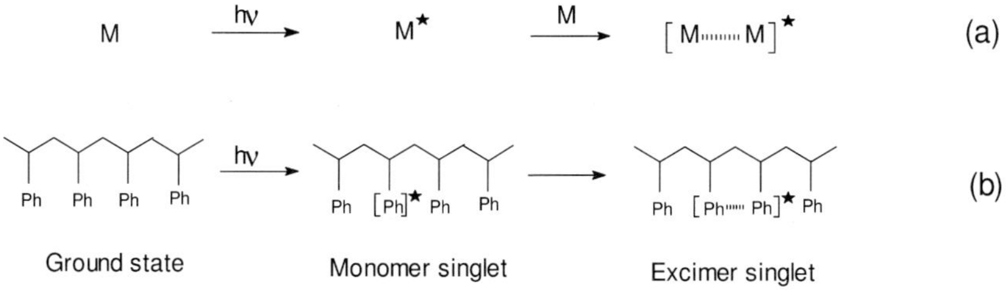

$$M \xrightarrow{h\nu} M^* \xrightarrow{M} [M\cdots M]^* \quad (a)$$

Ground state — Monomer singlet — Excimer singlet (b)

Scheme 1.1 Excimer formation: (a) general description, (b) in polystyrene.

ferred to as the *antenna effect* [18]: a large number of chromophores collect photons and guide the absorbed energy to one reaction center. As regards synthetic polymers, early studies on photon harvesting were devoted to linear polymers composed overwhelmingly of repeating units bearing the same donor chromophore (naphthalene) and to a very small extent the acceptor chromophore (anthracene) acting as an energy trap [15, 19]. Relevant work concerning linear polymers has been thoroughly reviewed by Webber [13]. Very interesting recent studies concerning multiporphyrin systems of various nonlinear structures have been reviewed by Choi et al. [20] and are considered below. In the case of the linear polymers mentioned above, practically all photons are absorbed by naphthalene moieties upon exposure to light in the wavelength range 290–320 nm. As illustrated in Scheme 1.2, excitation energy taken up by a naphthalene chromophore migrates down the chain and eventually reaches an anthracene trap.

This process is evidenced by the anthracene fluorescence, which is quite distinct from that of naphthalene. The quantum yield of anthracene sensitization

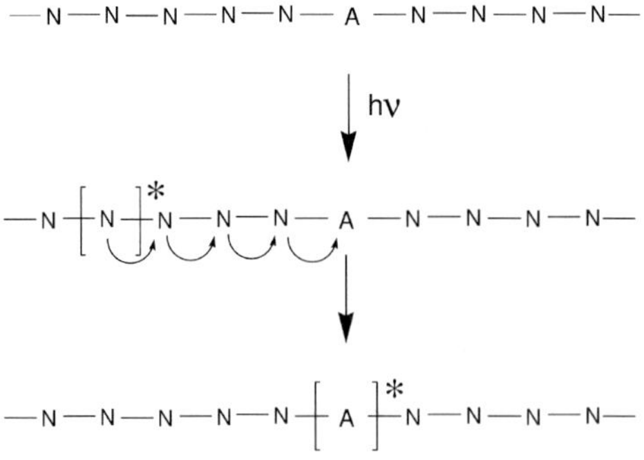

1 Absorption of light and subsequent photophysical processes

—N—N—N—N—N— A —N—N—N—N—

↓ hν

—N—[N—N—N—N—N]*—A —N—N—N—N—

↓

—N—N—N—N—N—[A]*—N—N—N—N—

Scheme 1.2 Mechanism of photon harvesting. Illustration of the transport of excitation energy by self-transfer through donor moieties (naphthalene) to an acceptor trap (anthracene).

Φ_S, i.e., the number of sensitized acceptors per directly excited donor, can be obtained from Eq. (1-15).

$$\Phi_S = 1 - [I(D)/I(D)^0] \quad (1\text{-}15)$$

Here, $I(D)^0$ and $I(D)$ are the donor fluorescence intensities in the absence and in the presence of the acceptor, respectively. Φ_S values varying between 0.1 and 0.7 have been found by examining, in aqueous or organic solvents, a variety of polymers having naphthalene and anthracene groups attached to the main

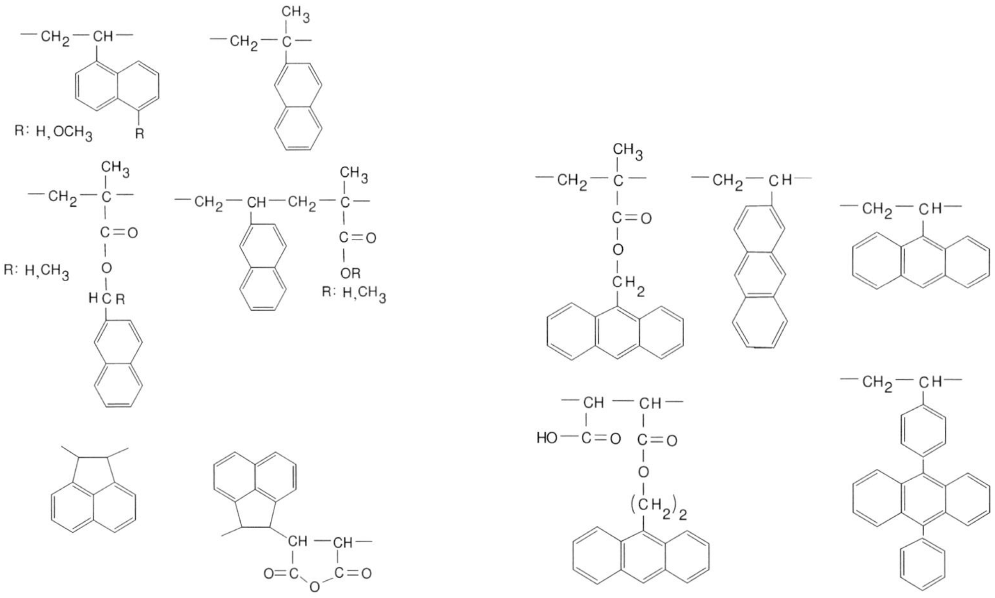

Chart 1.3 Chemical structures of repeating units bearing naphthalene and anthracene groups contained in copolymers employed in photon-harvesting studies [13].

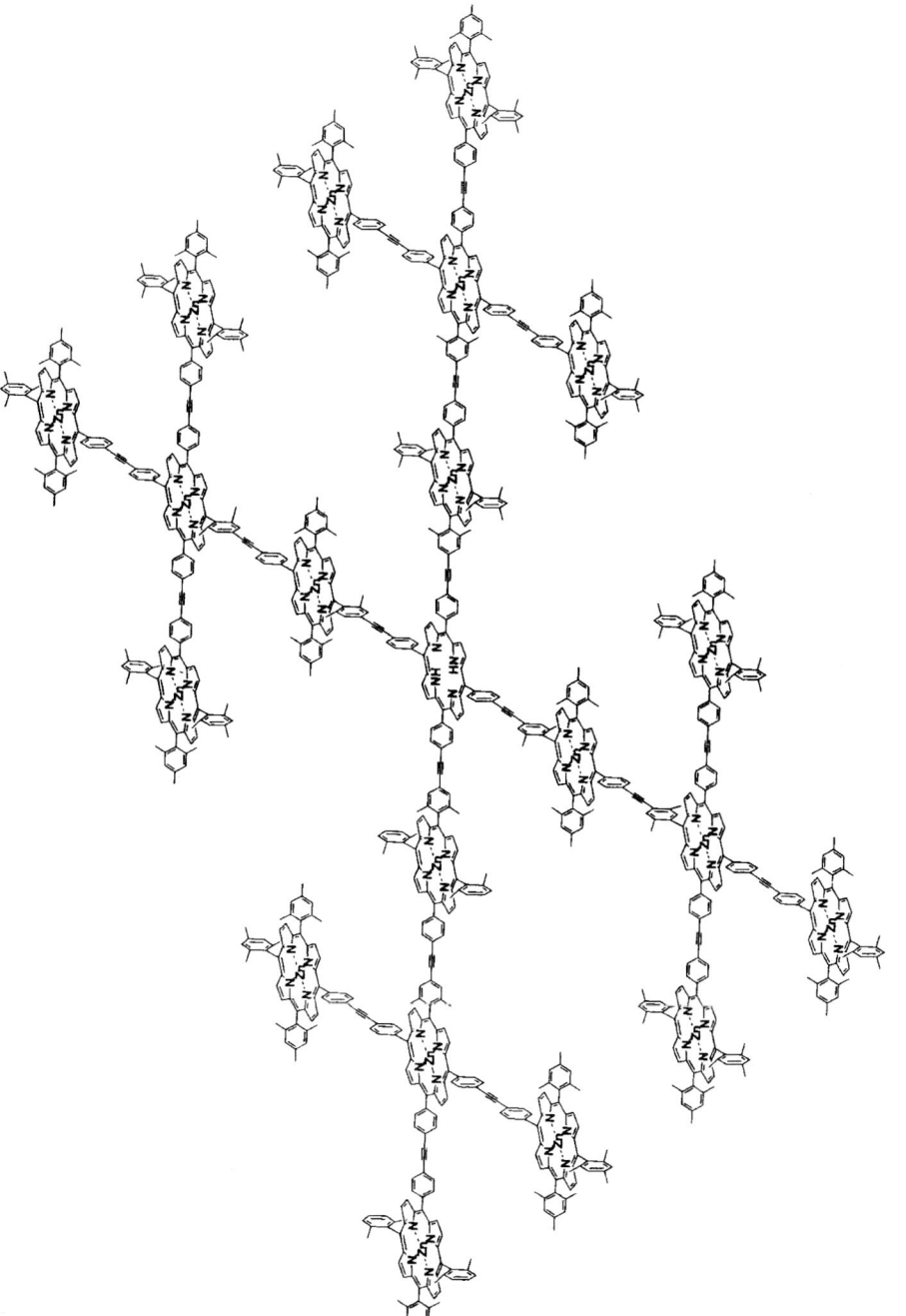

Chart 1.4 Chemical structure of a dendritic 21-porphyrin array consisting of 20 Zn porphyrin units attached to a Zn-free porphyrin focal core [21, 22].

chain in different modes (see Chart 1.3). The largest Φ_S values were found in cases in which excimer formation was unlikely [13].

Obviously, excimer formation represents a serious obstacle to energy migration, since the excimer site itself functions as a trap, and, after excitation, is mostly deactivated by emission of a photon rather than by energy transfer to a neighboring donor moiety ($\tau_{exc} < \tau_h$). Moreover, any effect on coil density exerted by the choice of temperature or solvent can dramatically effect the efficiency of energy trapping.

Chart 1.5 Chemical structure of a dendritic multiporphyrin array consisting of four wedges of a Zn porphyrin heptamer anchored to a Zn-free porphyrin focal core [22].

The light-harvesting multiporphyrin arrays synthesized in recent years seem to mimic natural photosynthetic systems much more closely than the linear polymers of the early studies. As outlined in the review by Choi [20], strategies for the synthesis of multiporphyrin arrays of various architectures have been developed. These comprise, besides ring-, star-, and windmill-shaped structures, also dendritic arrays. With the aim of a high photon-harvesting efficiency combined with vectorial energy transfer over a long distance to a designated point, dendritic light-harvesting antennae have proved to be most promising. A typical example is the system shown in Chart 1.4. It consists of a total of 21 porphyrin units, i.e. 20 P_{Zn} Zn-complexing porphyrin moieties, which are connected via diarylethyne linkers to one centrally located P_{free} unit, i.e. a non-complexing porphyrin moiety. The quantum yield for the energy transfer $P_{Zn} \rightarrow P_{free}$ is $\Phi_{ET}=0.92$ [21].

The structure of another large dendritic system is depicted in Chart 1.5. It consists of four heptameric Zn-porphyrin segments acting as energy donors. They are anchored to a central P_{free} moiety acting as the acceptor [22]. Photon absorption by the P_{Zn} moieties, at $\lambda=589$ nm or 637 nm, results in very effective $P_{Zn} \rightarrow P_{free}$ energy transfer ($\Phi_{ET}=0.71$, $k_{ET}=1.04\times10^9$ s^{-1}), as indicated by a strongly increased light emission from the P_{free} moieties.

1.6.4
Deactivation by chemical reactions

Triplet excited molecules formed in condensed media are liable to undergo bimolecular chemical reactions, since their long lifetimes permit a large number of encounters between the reaction partners. The hydrogen abstraction reaction, Eq. (1-16), of triplet excited carbonyl groups is a typical example.

$$\diagup\!\!\!\!C=O^* + RH \rightarrow \diagup\!\!\!\!{}^{\bullet}C-OH + R^{\bullet} \tag{1-16}$$

Singlet excited molecules are usually relatively short-lived and, therefore, are not very likely to undergo bimolecular reactions. In many cases, however, chemical bond cleavage competes with physical monomolecular deactivation paths. For example, singlet excited carbonyl groups contained in a polyethylene chain can undergo the Norrish type I reaction, resulting in a free radical couple [see Eq. (1-17)].

$$-CH_2-CH_2-CH_2-\left(\begin{array}{c}C\\||\\O\end{array}\right)^*CH_2-CH_2- \rightarrow -CH_2-CH_2-CH_2-C^{\bullet}+{}^{\bullet}CH_2-CH_2 \atop ||\ O \tag{1-17}$$

More details of chemical deactivation processes are provided in Chapter 7.

1.7
Absorption and emission of polarized light

1.7.1
Absorption

The absorption of linearly polarized light is characterized by the fact that only those chromophores with a component of the absorption transition moment located in the same direction as the electric (polarization) vector of the incident light can be excited. No light will be absorbed if the direction of the transition moment is perpendicular to the electric vector of the incident light. This dichroic behavior is exhibited by anisotropic organic materials in the solid state, such as single crystals of certain substances in which the transition moments of all molecules are fixed in a parallel orientation. In the case of linear polymers, it is possible to generate some degree of optical anisotropy in highly viscous or rigid samples by aligning the macromolecules in a specific direction. Various methods have been employed to achieve orientation, such as mechanical alignment, Langmuir-Blodgett (LB) film deposition, liquid-crystalline self-organization, and alignment on specific substrates. As a typical example, Fig. 1.9 shows absorption spectra recorded from an LB film placed on the surface of a fused silica substrate and consisting of 100 monolayers of DPOPP (see Chart 1.6) [23].

Electron microscopy revealed that the LB film had a liquid-crystalline-like structure. This means that many polymer chains were oriented parallel to the substrate plane and exhibited a preferential orientation of their backbones along the dipping direction. Absorption spectra recorded with the incident light polarized either parallel or perpendicular to the dipping direction show a maximum at 330 nm (3.76 eV) in both cases, but $A_\parallel$ and $A_\perp$, the absorbances parallel and

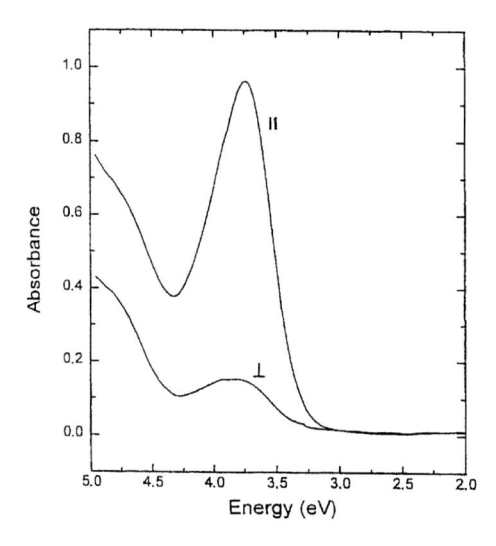

Fig. 1.9 Absorption spectra of an LB film consisting of 100 monolayers of DPOPP recorded with linearly polarized incident light ($\parallel$ and $\perp$: parallel and perpendicular to the dipping direction, respectively). Adapted from Cimrova et al. [23] with permission from Wiley-VCH.

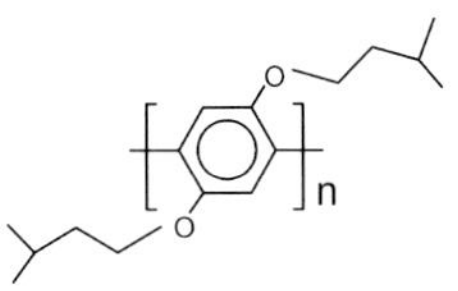

Chart 1.6 Chemical structure of poly(2,5-di-isopentyloxy-p-phenylene), DPOPP.

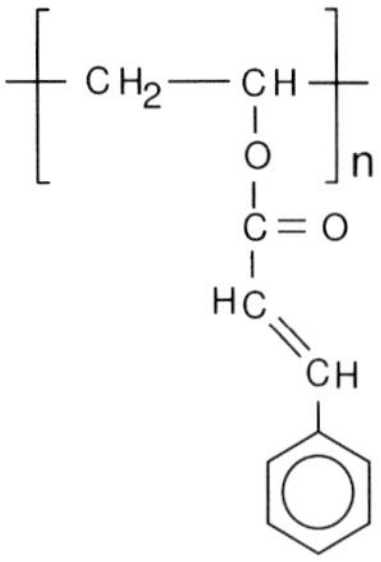

Chart 1.7 Chemical structure of poly(vinyl cinnamate).

perpendicular to the dipping direction, respectively, differ by a factor of about five, the in-plane order parameter, $S = (A_\| - A_\perp)/(A_\| + A_\perp)$, being 0.67.

It might be noted that, in principle, it is possible to create anisotropy upon irradiating an ensemble of randomly oriented photochromic chromophores with linearly polarized light, since photons are only absorbed by chromophores with transition moments parallel to the electric vector of the incident light. This applies, e.g., to thin films of poly(vinyl cinnamate) (see Chart 1.7) and its derivatives. Exposure to linearly polarized light induces the preferential orientation of liquid-crystal molecules in contact with the film surface [24]. The photoalignment is likely to be caused by the *trans-cis* isomerization of the cinnamoyl groups, a separate process to cross-linking through [2+2] addition, which is a major photoreaction of this polymer.

The creation of anisotropy is treated in some detail in Section 4.4, which deals with the *trans-cis* isomerization of azobenzene compounds.

1.7.2
Absorption by chiral molecules

A chiral molecule is one that is not superimposable on its mirror image. It contains one or more elements of asymmetry, which can be, for example, carbon atoms bearing four different substituent groups. In principle, chiral molecules can exist in either of two mirror-image forms, which are not identical and are called *enantiomers*. Chiral molecules have the property of rotating the plane of polarization of traversing linearly polarized monochromatic light, a phenomenon called *optical activity*. Linearly polarized light can be viewed as the result of the superposition of opposite circularly polarized light waves of equal amplitude and phase. The two circularly polarized components traverse a medium containing chiral molecules with different velocities. Thereby, the wave remains plane-polarized, but its plane of polarization is rotated through a certain angle a, the *optical rotation, OR*. In other words, optical activity stems from the fact that n_r and n_l, the refractive indices for the two circularly polarized components of linearly polarized light, are different, a phenomenon referred to as *circular birefringence*.

Optically active compounds are commonly characterized by their specific rotation $[a]$, measured in solution [see Eq. (1-18)].

$$[a] = 100a/cd \quad (\deg \text{cm}^3 \text{ dm}^{-1} \text{ g}^{-1}) \tag{1-18}$$

where c is the concentration in units of g/100 cm^3, and d is the path length of the light in dm. $[a]$ depends on λ, the wavelength of the light, and the temperature. Actually, $[a]$ is proportional to the difference in the refractive indices n_r and n_l: $[a] \propto n_r - n_l$. Since n_r and n_l have different dependences on λ, $[a]$ also depends on λ. A plot of $[a]$ vs. λ yields the *optical rotary dispersion (ORD)* curve of the substance. In many cases, ORD curves exhibit, at wavelengths of light absorption, a sine-wave form, which is referred to as the *Cotton effect* (see Fig. 1.10) [25]. The inversion point of the S-shaped curve (c) in Fig. 1.10 corresponds to λ_{max}, the wavelength of the absorption maximum, at which n_r is equal to n_l.

In addition to their optical activity, chiral molecules are characterized by the property of absorbing the two components of incident linearly polarized light, i.e. left- and right-circularly polarized light, to different extents. This phenomenon, called *circular dichroism*, CD, can be quantified by the difference in molar extinction coefficients $\Delta\varepsilon = \varepsilon_l - \varepsilon_r$. CD is characterized by the fact that a linearly polarized light wave passing through an optically active medium is transformed into an elliptically polarized light wave. With the aid of commercially available instruments, the actual absorbance A of each circularly polarized light compo-

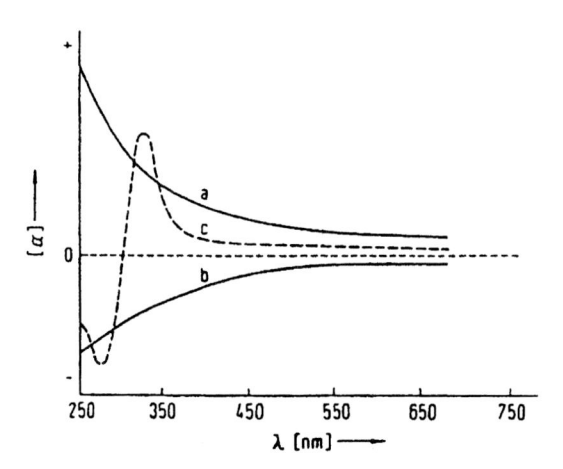

Fig. 1.10 Schematic depiction of optical rotary dispersion (ORD) curves for positive and negative rotation, (a) and (b), respectively, for wavelength regions without absorption. The S-shaped curve (c) is typical of the Cotton effect, reflecting light absorption. Adapted from Perkampus [25] with permission from Wiley-VCH.

nent is measured, yielding the difference $A_l - A_r$. The latter is related to the ellipticity Θ, given either in degrees (deg) or radians (rad), according to Eqs. (1-19) and (1-20), respectively.

$$\Theta = 2.303(A_l - A_r)180/4\pi \quad \text{(deg)} \tag{1-19}$$

$$\Theta = 2.303/4(A_l - A_r) \quad \text{(rad)} \tag{1-20}$$

Commonly, for the sake of comparison, the molar ellipticity $[\Theta]=100\ \Theta/cd$ in units of $\text{deg cm}^2\ \text{dmol}^{-1}$ is recorded, where c is the concentration in mol L^{-1} and d is the optical path length. If, in the case of polymers such as proteins, the molar concentration is related to the molar mass of the residue, i.e. to the repeating (base) unit, the *mean residue weight* ellipticity $[\Theta]_{MRW}$ is obtained.

In recent years, circular dichroism spectroscopy has been widely applied in investigations concerning the molecular structure of chiral polymers. It is a powerful tool for revealing the secondary structures of biological macromolecules, for instance of polypeptides, proteins, and nucleic acids in solution. An

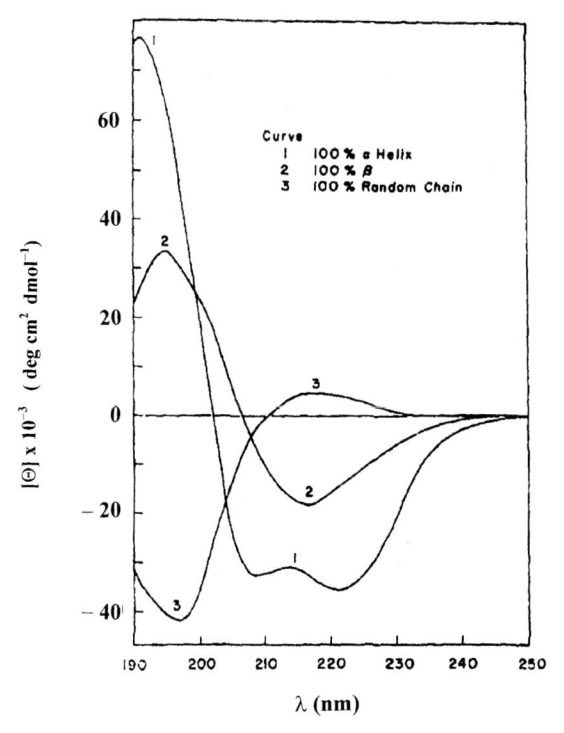

Fig. 1.11 Circular dichroism spectra of poly(L-lysine) in its α-helical, β-sheet, and random coil conformations. Adapted from Greenfield et al. [26] with permission from the American Chemical Society.

important feature is the possibility of monitoring conformational alterations of optically active macromolecules by CD measurements. Typical data are presented in Fig. 1.11, which shows CD spectra of poly(L-lysine) in three different conformations [26]. Poly(L-lysine) adopts three different conformations, depending on the pH and temperature: random coil at pH 7.0, α-helix at pH 10.8, and β-sheet at pH 11.1 (after heating to 52 °C and cooling to room temperature once more). These conformational transitions are due to changes in the long-range order of the amide chromophores. For detailed information on circular dichroism of chiral polymers, the reader is referred to relevant publications [27–30].

1.7.3
Emission

Provided that the transition moment does not change direction during the lifetime of an excited state, fluorescent light is polarized parallel to the incident light. For linearly polarized incident light, this implies that the direction of the electric vector of both the incident and the emitted light is the same. Therefore, in the case of oriented polymers, fluorescence can only be generated with linearly polarized light if the components of the absorption transition moments of the chromophores are aligned parallel to the electric vector of the incident light. If the alignment of the macromolecules is not perfect, the emitted light is not perfectly polarized. This is commonly characterized by the degree of polarization P, defined by Eq. (1-21).

$$P = \frac{I_\| - I_\perp}{I_\| + I_\perp} \tag{1-21}$$

Here, $I_\|$ and $I_\perp$ are the intensities of the fluorescence polarized parallel and perpendicular to the electric vector of the incident light. Usually, set-ups with the geometry shown in Fig. 1.12 are employed for fluorescence measurements. The

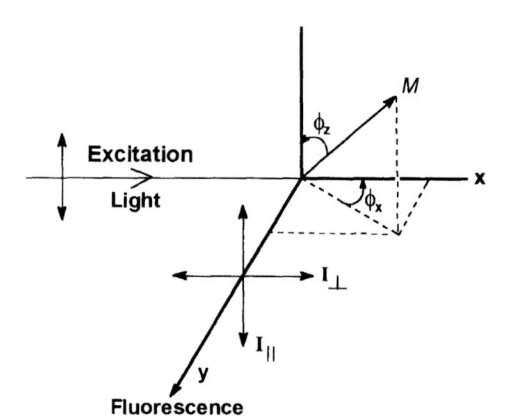

Fig. 1.12 Geometry of experimental set-ups employed in fluorescence depolarization measurements.

sample is excited with light incident along the x-axis and the fluorescence is monitored along the y-axis. M denotes the transition dipole moment.

As a typical example, Fig. 1.13 shows fluorescence spectra recorded from an LB film of DPOPP (for the absorption spectra, see Fig. 1.9). The exciting light was polarized parallel to the dipping direction.

In accordance with the conclusion derived from the absorption spectra, the emission spectra also reveal the partially ordered structure of the film. As in the case of absorption, $I_{\parallel}$ and $I_{\perp}$, the fluorescence intensities parallel and perpendicular to the dipping direction, respectively, differ appreciably, in this case by a factor of three to four. Much higher dichroic ratios have been found with other oriented systems, e.g. with highly aligned films consisting of blends of polyethylene with 1 wt.% MEH-PPV (see Chart 1.8) [31, 32]. The films, fabricated by tensile drawing over a hot pin at 110–120 °C, proved to be highly anisotropic (dichroic ratio >60), with the preferred direction parallel to the draw axis.

In principle, oriented polymeric systems capable of generating linearly polarized light have the potential to be used as backlights for conventional liquid-crystal displays (LCDs), a subject reviewed by Grell and Bradley [33]. In this connection, systems generating circularly polarized (CP) light also became attractive. CP light can be utilized for backlighting LCDs either directly with the aid of appropriate systems or after transformation into linearly polarized light with the aid of a suitable $\lambda/4$ plate [33]. CP light has been generated, for example, with a highly ordered polythiophene bearing chiral pendant groups,

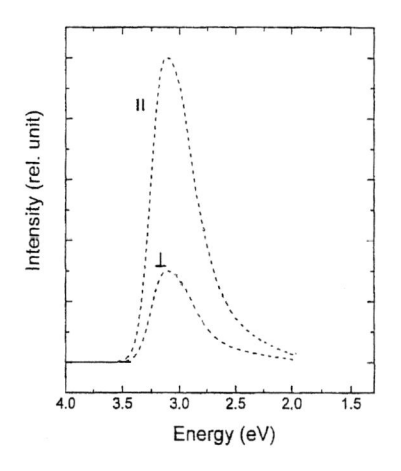

Fig. 1.13 Fluorescence spectra of a DPOPP film prepared by the LB technique. $I_{\parallel}$ and $I_{\perp}$: fluorescence intensities parallel and perpendicular to the dipping direction. Exciting light: λ_{exc}=320 nm, polarized parallel to the dipping direction. Adapted from Cimrova et al. [23] with permission from Wiley-VCH.

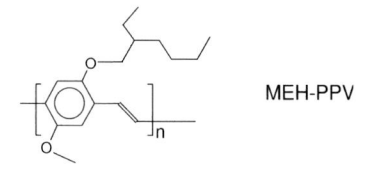

MEH-PPV

Chart 1.8 Chemical structure of poly[2-methoxy-5-(2'-ethylhexyloxy)-p-phenylenevinylene], MEH-PPV.

poly{3,4-di[(S)-2-methylbutoxy]thiophene} (see Chart 1.12) [34]. In this case, however, the dissymmetry factor g_e was low; g_e is defined as $2(I_L-I_R)/(I_L+I_R)$, and $|g_e|$ is equal to two for pure, single-handed circularly polarized light. I_L and I_R denote the left- and right-handed emissions, respectively. Circularly polarized light is produced quite efficiently when a conventional luminophore is embedded within a chiral nematic matrix consisting of a mixture of compounds A and B (see Chart 1.9) [35]. When this system was exposed to unpolarized light of $\lambda=370$ nm, the dissymmetry factor g_e approached −2 in the 400–420 nm wavelength range.

Another aspect, also considered in Subsection 1.8.3.3.2, concerns fundamental time-resolved fluorescence studies. Here, the emphasis is placed on fluorescence depolarization measurements, which are very helpful in following rotational and segmental motions and for studying the flexibility of macromolecules. If the polymer under investigation does not contain intrinsically fluorescent probes (e.g., certain amino acid moieties in proteins), then the macromolecules have to be labeled with fluorescent markers. Information concerning the rate of rotation or segmental motion then becomes available, provided that the emission rate is on a similar time scale. Only when this condition is met can the rate of depolarization be measured. If the emission rate is much faster, there is no depolarization, whereas if it is much slower, the depolarization will be total.

Commonly, the emission anisotropy r(t) is determined as a function of time. r(t) is defined by Eq. (1-22).

$$r(t) = \frac{I_{||}(t) - I_{\perp}(t)}{I_{||}(t) + 2I_{\perp}(t)} \tag{1-22}$$

By irradiating a sample with a short pulse of linearly polarized light and separately recording $I_{||}$ and $I_{\perp}$ as a function of time t after the pulse, the sum $S(t) = I_{||} + 2I_{\perp}$ and the difference $D(t) = I_{||} - I_{\perp}$ may be obtained. The application of an appropriate correlation function to $r(t) = D(t)/S(t)$ yields the relaxation time τ. In

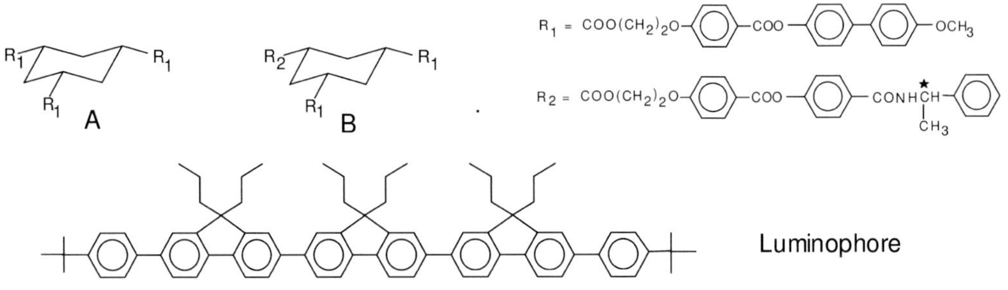

Chart 1.9 Chemical structures of compounds A and B, forming a chiral nematic matrix, and of an oligomeric luminophore.

general, the time dependence of r(t) is rather complex, i.e. the decay of r(t) does not follow a single exponential decay function. Theories have been developed to analyze the experimentally observed decay functions. However, it is beyond the scope of this book to deal with the relevant theoretical work, which has been thoroughly reviewed elsewhere as part of the overall subject of fluorescence depolarization [36, 37]. In simple cases, r(t) decays according to a single exponential decay law. Provided that this applies to the rotational motion of macromolecules, the rotational relaxation time τ_r can be evaluated by assuming spherically shaped macromolecules. For a rotating spherical body, r(t) is expressed by Eq. (1-23).

$$r(t) = (2/5)\exp(-6D_r t) \qquad (1\text{-}23)$$

The rotational diffusion constant D_r is given by Eq. (1-24), the Einstein law.

$$D_r = 1/\tau_r = kT/V\eta \qquad (1\text{-}24)$$

Here, V is the volume of the sphere and η is the viscosity of the solvent.

As can be seen in Table 1.4, the τ_r values of proteins such as bovine serum albumin and trypsin in aqueous solution lie in the ns range and become larger with increasing molar mass. The proteins were labeled with fluorescent markers such as 1-dimethylamino-5-sulfonyl-naphthalene groups (see Chart 1.10) [38].

Segmental motions and molecular flexibility have been studied for various polymers, such as polystyrene and the Y-shaped immunoglobulins IgA and IgG. Relaxation times in the range of 10–100 ns were found. In these studies, the

Table 1.4 Rotational correlation times τ_r of proteins in aqueous solution at 25 °C determined by time-resolved fluorescence depolarization measurements [37].

Protein	Molar mass (g mol^{-1})	τ_r (ns)
Apomyoglobin	17 000	8.3
Trypsin	25 000	12.9
Chymotrypsin	25 000	15.1
β-Lactoglobulin	36 000	20.3
Apoperoxidase	40 000	25.2
Serum albumin	66 000	41.7

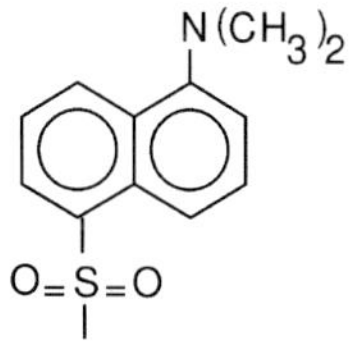

Chart 1.10 Chemical structure of the 1-dimethylamino-5-sulfonyl-naphthalene group.

polymers were labeled with small amounts of appropriate fluorescent markers, such as anthracene in the case of PSt [39].

Again, it is a prerequisite for such measurements that the fluorescence decays at a rate similar to that of the motion under investigation. Measurable rotational relaxation times are in the range 1 ns to 1 µs, corresponding to the rotation of species with molar masses up to 10^6 g mol^{-1} in aqueous solution.

1.8
Applications

1.8.1
Absorption spectroscopy

1.8.1.1 UV/Vis spectroscopy
There are numerous applications reliant upon the ultraviolet and visible (UV/Vis) wavelength range. For example, absorption spectroscopy is applied to analyze and identify polymers and copolymers containing chromophores that absorb in this wavelength range, such as aromatic or carbonyl groups. In this context, the investigation of photochemical reactions, for instance of reactions occurring in degradation processes, is noteworthy. Moreover, absorption measurements allow the monitoring of alterations in the tertiary structure of macromolecular systems, for instance, in the case of the denaturation of biomacromolecules, especially proteins and nucleic acids. Figure 1.14 demonstrates the increase in the optical absorption observed upon heating an aqueous solu-

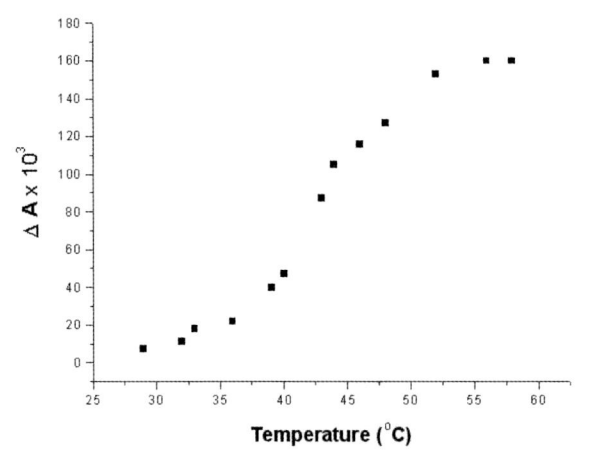

Fig. 1.14 Thermal denaturation of lysozyme in aqueous solution. Differential absorption vs. temperature. [lysozyme]: 10 g L^{-1}, pH: 1.45, [KCl]: 0.2 M. Adapted from Nicolai et al. [40] with permission from John Wiley & Sons, Inc.

$$\begin{array}{c}\diagdown\\N-H\cdots\cdots O=C\\\diagup\end{array}\begin{array}{c}\diagup\\\diagdown\end{array}\longrightarrow\begin{array}{c}\diagdown\\N-H+O=C\\\diagup\end{array}\begin{array}{c}\diagup\\\diagdown\end{array}$$

Scheme 1.3 Destruction of hydrogen bonds.

tion of lysozyme, a globular protein that acts as an enzyme in the cleavage of certain polysaccharides [40]. The absorption change reflects the unfolding of the polypeptide chains due to the destruction of intramolecular interactions such as hydrogen bonds (see Scheme 1.3).

The thermal denaturation of other superstructures, such as those of collagen and deoxyribonucleic acid (DNA), may also be monitored by following the increase in the optical absorption. Collagen is the most abundant protein in connective tissues and constitutes a major part of the matrix of bones. In its native state, it adopts a three-stranded helical structure. Dissociation of the three chains at temperatures above 40 °C is accompanied by an increase in optical absorption. DNA, the carrier of genetic information and an essential constituent of the nuclei of biological cells, contains the bases adenine, guanine, cytosine, and thymine, and hence absorbs UV light. The intensity of its absorption spectrum (λ_{max}=260 nm) is reduced by about 30% when single strands combine to form the double-stranded helix. Conversely, the optical absorption increases upon denaturation [41]. This is illustrated in Fig. 1.15.

Generally, changes in optical absorption related to molecular alterations not involving chemical bond breakage are denoted by the terms *hypochromicity* (also *hypochromy*) and *hyperchromicity* (also *hyperchromy*), depending on whether the optical absorption decreases or increases, respectively. As regards nucleic acids in solution, hypochromicity applies to a decrease in optical absorbance when single-stranded nucleic acids combine to form double-stranded helices. The hypochromic effect is not restricted to nucleic acids, proteins, and other polymers, but has also been observed with aggregates of dyes and clusters of aromatic compounds. In interpreting this effect, it has been assumed that the electron clouds of chromophores brought into close proximity are strongly interacting. The resulting alteration in the electron density causes changes in the absorption spectrum. The hypochromicity phenomenon and relevant theories are discussed in detail in a recent monograph [42].

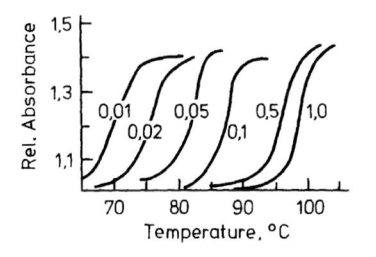

Fig. 1.15 Thermal denaturation of DNA (E. coli). Relative absorbance at 260 nm vs. temperature at various concentrations of KCl (given in the graph in units of mol L^{-1}). Adapted from Marmur et al. [41] with permission from Elsevier.

1.8.1.2 Circular dichroism spectroscopy

Circular dichroism (CD) spectroscopy is a form of absorption spectroscopy based on measuring the difference in the absorbances of right- and left-circularly polarized light by a substance (see Section 1.7.2). Regarding polypeptides, proteins, and nucleic acids, it is a powerful tool for analyzing secondary and tertiary structures and for monitoring conformational changes. In the case of proteins, it allows the discrimination of different structural types, such as a-helix, parallel and antiparallel β-pleated sheets, and β-turns, and, moreover, allows estimation of the relative contents of these structures. Details are given in review articles [43–45].

Since appropriate instruments have become commercially available, CD spectroscopy has developed into a routine method for the characterization of the chirality of newly synthesized polymers. As a typical example, the rather high chiro-optical activity of the ladder-type poly(p-phenylene) of the structure shown in Chart 1.11 was revealed CD spectroscopically: molar ellipticity $[\Theta] = 2.2 \times 10^6$ rad cm^2 mol^{-1} (at $\lambda_{max} = 461$ nm), corresponding to an anisotropy factor of $g = \Delta\varepsilon/\varepsilon = 0.003$ [46].

The following three examples serve to demonstrate the general importance of CD spectroscopy. (1) Consider first the case of optically active polythiophene derivatives. They belong to the class of polymers of which the optical activity is based on the enantioselective induction of main-chain chirality by the presence of enantiomerically pure side groups. In the case of PDMBT (Chart 1.12), CD spectroscopy permits the detection of a pronounced thermochromic effect. When dichloromethane solutions that do not exhibit chiro-optical activity related to the $\pi \rightarrow \pi^*$ transition at $\lambda = 438$ nm at 20 °C are cooled to –30 °C, the onset of absorption is significantly red-shifted. Moreover, a CD spectrum exhibiting a strong bisignate Cotton effect (see Fig. 1.16) is recorded. The chiro-optical activity, which is observed for n-decanol solutions even at room temperature ($g = \Delta\varepsilon/\varepsilon = 0.02$), is ascribed to highly ordered packing of the polythiophene chains in chiral aggregates [34].

(2) In the case of thin films of PMBET (see Chart 1.13), another optically active polythiophene derivative, CD spectroscopy reveals stereomutation of the main chain. As can be seen in Fig. 1.17, a CD spectrum that is the mirror im-

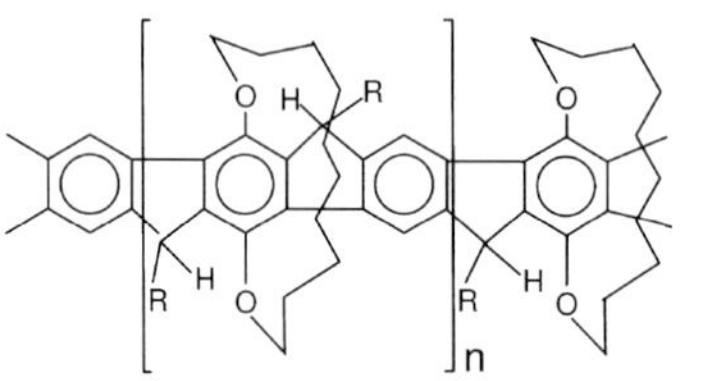

Chart 1.11 Chemical structure of a ladder-type poly(p-phenylene).

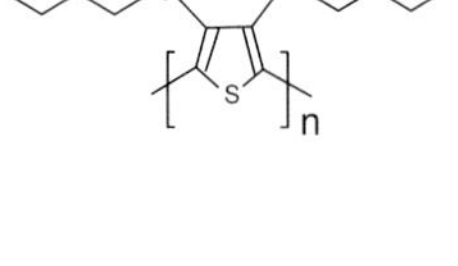

Chart 1.12 Chemical structure of poly{3,4-di[(S)-2-methylbutoxy]thiophene}, PDMBT.

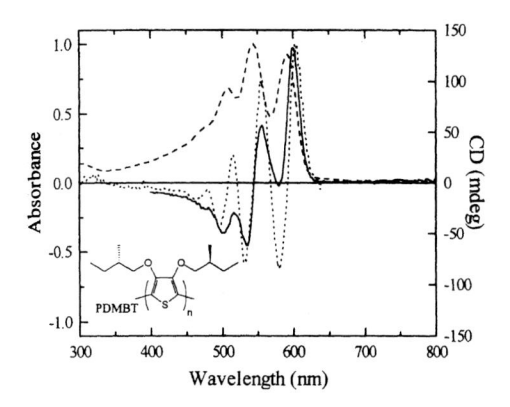

Fig. 1.16 Normalized absorption spectrum (dashed line) and CD spectrum (solid line) of PDMBT, recorded in dichloromethane solution at -30 °C. Dotted line: first derivative of the absorption spectrum. Adapted from Langeveld-Voss et al. [34] with permission the American Chemical Society.

Chart 1.13 Chemical structure of poly(3-{2-[(S)-2-methylbutoxy]ethyl}thiophene), PMBET.

age of the original spectrum is recorded when PMBET is rapidly cooled from the disordered melt to the crystalline state. Apparently, by rapid cooling of the melt, a metastable chiral associated form of the polymer that exhibits the mirror-image main-chain chirality is frozen-in [47].

(3) A final example demonstrating the usefulness of CD spectroscopy concerns the detection of light-induced switching of the helical sense in polyisocyanates bearing chiral pendant groups [48]. Polyisocyanates (see Chart 1.14) exist as stiff helices comprising equal populations of dynamically interconverting right- and left-handed helical segments. The relative population of these segments is extraordinarily sensitive to chiral perturbations. This is demonstrated by the CD spectra shown in Fig. 1.18. They were recorded from polyisocyanate, PICS (see Chart 1.14), that had been irradiated with circularly polarized light (CPL) of opposite handedness. Initially, the pendant groups consist of a racemic mixture of the two enantiomers and a CD spectrum is not observed. Absorption

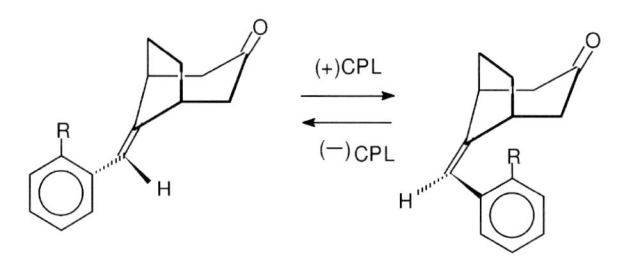

Scheme 1.4 Isomerization of the pendant groups of PICS.

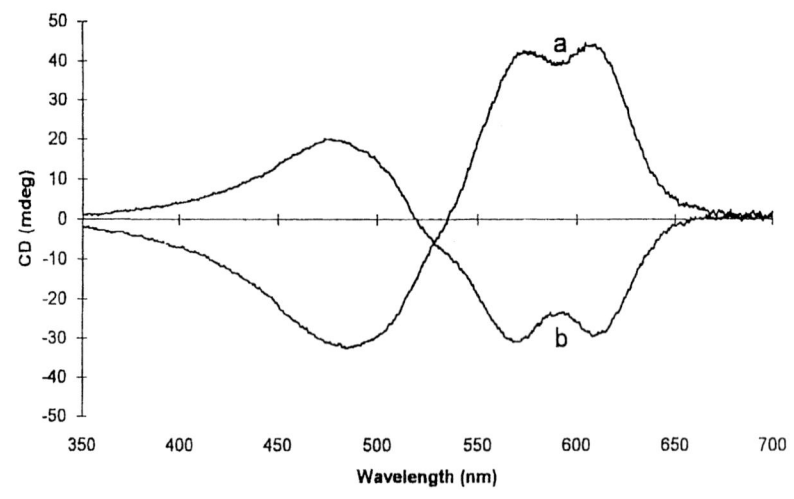

Fig. 1.17 CD spectra of PMBET recorded at room temperature from thin films spin-coated onto glass plates after fast (a) and slow (b) cooling from 200 °C to 20 °C. Adapted from Bouman et al. [47] with permission from Wiley-VCH.

Chart 1.14 Chemical structure of polyisocyanates. General structure, left. PICS, right.

of light induces isomerization at the C-C double bond (see Scheme 1.4). Thus, irradiation with circularly polarized light, which is absorbed by the two enantiomers to different extents, results in an optically active partially resolved mixture, and the CD spectra shown in Fig. 1.18 are observed. Remarkably, an enantiomeric excess of just a few percent, i.e. close to the racemic state, converts the polymer into one having a disproportionate excess of one helical sense. In other words, chiral amplification takes place since the minor enantiomeric group takes on the helical sense of the major enantiomeric group.

Interestingly, the helical sense of the polymer may be reversibly switched by alternating irradiation with (+)- or (−)-CPL, or returned to the racemic state by irradiation with unpolarized light.

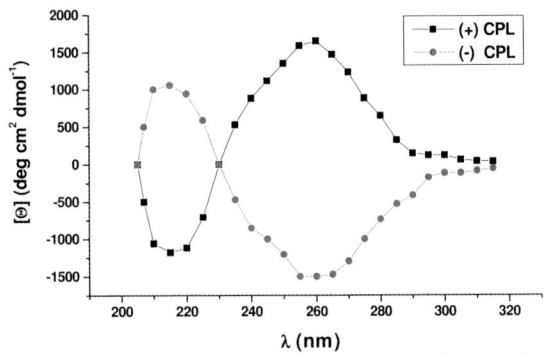

Fig. 1.18 CD spectra of polyisocyanate, PICS, irradiated with circularly polarized light (CPL) of opposite handedness at $\lambda > 305$ nm. The spectra were recorded in dichloromethane/tetrahydrofuran (1:1) solution. Adapted from Li et al. [48] with permission from the American Chemical Society.

1.8.1.3 IR spectroscopy

Infrared (IR) spectroscopy has become a very powerful chemical-analytical tool in the analysis and identification of polymers. It also plays a prominent role in tests related to chemical alterations generated by extrinsic forces and serves, for example, in the monitoring of polymer degradation. The wavelength regime of importance ranges from about 2.5 to 50 µm (4000 to 200 cm^{-1}). This corresponds to the energies required to excite vibrations of atoms in molecules. Precisely speaking, the full spectrum of infrared radiation covers the wavelength range from 0.75 to 10^3 µm, i.e. besides the aforementioned mid-IR region there is the near-IR region (0.75 to 2.5 µm) and the far-IR region (50 to 10^3 µm).

IR light is absorbed when the oscillating dipole moment corresponding to a molecular vibration interacts with the oscillating vector of the IR beam. The absorption spectra recorded with the aid of IR spectrometers consist of bands attributable to different kinds of vibrations of atom groups in a molecule, especially valence and deformation (bending) vibrations, as can be seen in Fig. 1.19.

Figure 1.20 presents a typical example of the application of IR spectroscopy. Here, the UV radiation-induced chemical modification of a polyester containing in-chain cinnamoyl groups (see Chart 1.15) is illustrated [49].

As can be seen in Fig. 1.20, the FTIR spectrum of the unirradiated polymer features absorption bands at 1630, 1725, and 1761 cm^{-1}, which may be assigned

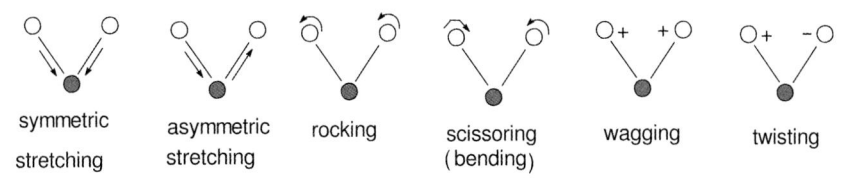

Fig. 1.19 Notation of group vibrations.

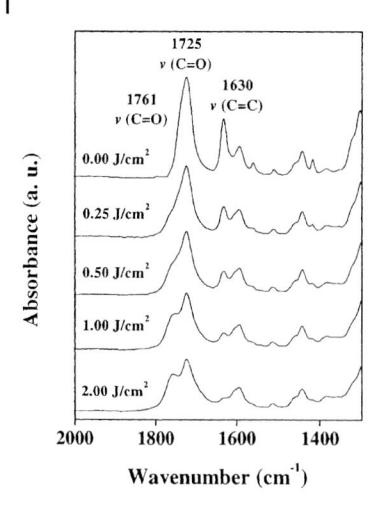

Fig. 1.20 FTIR spectra of a Cn-polyester recorded before and after irradiation with UV light (260–380 nm) to different absorbed doses. Adapted from Chae et al. [49] with permission from Elsevier.

to the stretching vibrations of vinylene double bonds and conjugated and non-conjugated carbonyl bonds, respectively. Upon irradiation, the intensities of the vinylene and the conjugated carbonyl bands decrease, whereas the band due to the non-conjugated carbonyl groups intensifies with increasing absorbed dose. This behavior may be explained in terms of simultaneously occurring *trans-cis* isomerizations and [2+2] cycloadditions (dimerizations). The band at 1630 cm^{-1} decreases since the extinction coefficient of *cis* C=C bonds is lower than that of *trans* C=C bonds. The growth in the intensity of the band at 1761 cm^{-1} indicates the occurrence of dimerizations.

Modern commercial IR spectrometers operating with the aid of a Michelson interferometer produce interferograms, which, upon mathematical decoding by means of the Fourier transformation, deliver absorption spectra commonly referred to as Fourier-transform infrared (FTIR) spectra [50]. Comprehensive collections of IR spectra of polymers, monomers, and additives are available [51]. Moreover, the reader's attention is directed to several books [52–58].

Chart 1.15 Chemical structure of the polyester referred to in Fig. 1.20.

1.8.2
Luminescence

Many problems in the physics and chemistry of polymers have been investigated by means of fluorescence techniques. Within the scope of this book, it is merely possible to point out the high versatility of these techniques rather than to discuss the innumerable publications. Among the features of luminescence that account for the variety of its applications is the fact that emission spectra can be recorded at extremely low chromophore concentrations. Thus, a polymer may be labeled with such a small amount of luminophore that the labeling does not perturb the properties of the system. As regards linear polymers in solution, it is possible to derive information on the conformational state and the behavior of the macromolecules. This concerns such topics as the interpenetration of polymer chains, the microheterogeneity of polymer solutions, conformational transitions of polymer chains, and the structures of polymer associates. Relevant work has been reviewed by Morawetz [59]. Here, only one typical example is described, which concerns the kinetics of HCl transfer from aromatic amino moieties to much more basic aliphatic amino groups attached to discrete macromolecules, in this case poly(methyl methacrylate)s (see Scheme 1.5).

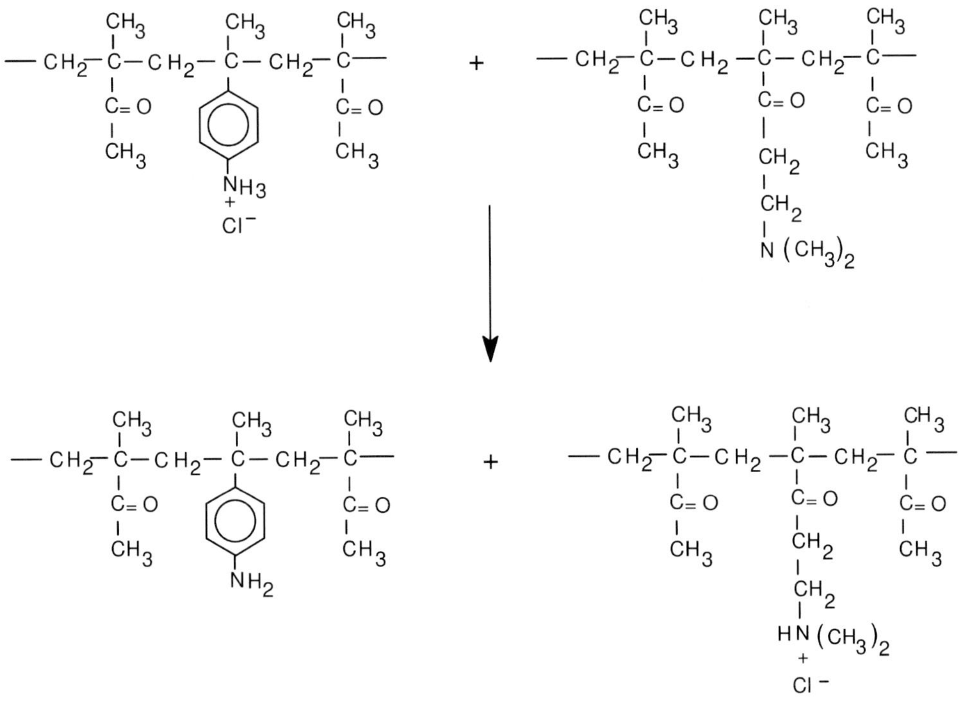

Scheme 1.5 HCl transfer from aromatic to aliphatic amino groups.

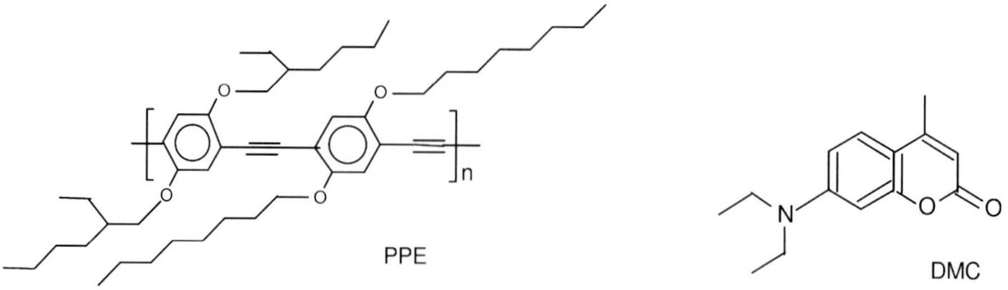

Chart 1.16 Chemical structures of a poly(2,5-dialkoxy-p-phenylene ethynylene), PPE, and 7-diethylamino-4-methyl-coumarin, DMC.

The release of HCl from the aminostyrene groups increases the fluorescence intensity, since protonation prevents light emission. Thus, the rate of HCl transfer between the different macromolecules can be measured in a stopped-flow experiment. It was found that the rate constant of the reaction decreased with increasing chain length of the interacting polymers [60]. This result may be interpreted in terms of the excluded volume effect: flexible polymer chains in good solvent media strongly resist mutual interpenetration, a phenomenon that becomes more pronounced with increasing chain length.

Another quite different kind of luminescence application pertains to the generation of polarized light with the aid of aligned systems. Here, the concept of polarizing excitonic energy transfer, EET, comes to prominence. Thus, in appropriate systems, randomly oriented sensitizer molecules harvest the incoming unpolarized light by isotropic absorption and subsequently transfer the energy to a uniaxially oriented polymer. The latter emits light with a high degree of linear polarization. According to this concept, all incident light can be funnelled into the same polarization. The incorporation of the polarizing EET process into colored liquid-crystal displays (LCDs) would imply that dichroic polarizers are no longer required for the generation of polarized backlights in conventional LCDs. A system functioning in this way consists of a ternary blend of high molar mass (4×10^6 g mol^{-1}) polyethylene, 2 wt% of a derivative of PPE, and 2 wt% of the sensitizer DMC (see Chart 1.16) [61]. Blend films prepared by solution-casting from xylene are uniaxially drawn at 120 °C to a draw ratio of about 80.

1.8.3
Time-resolved spectroscopy

1.8.3.1 General aspects
With the advent of powerful lasers capable of generating short light pulses, a new era of research commenced [62–64]. Notably, the new light sources permit the measurement of lifetimes of excited states and the detection of short-lived intermediates such as free radicals and ions. The concomitant development of sophisticated detection methods has also brought about continuous progress during the

last decades in the fields of polymer physics and chemistry [9, 65–68]. While researchers were initially fascinated by studying processes on the microsecond (1 µs = 10^{-6} s) and nanosecond (1 ns = 10^{-9} s) time scale, more recent research has concentrated on the picosecond (1 ps = 10^{-12} s) and femtosecond (1 fs = 10^{-15} s) time region. In this way, a wealth of information has become available that allows the identification of extremely short-lived intermediates and elucidates the mechanisms of many photophysical and photochemical processes. The aim here is not to review work on the technical development of pulsed lasers and on the invention of highly sensitive detection methods. In a more general way, information is given on the wide-ranging potential of time-resolved measurements and their benefits in the fields of polymer photophysics and photochemistry.

Time-resolved measurements were initiated both by physicists, who were principally interested in photophysical processes that left the chemical structures of the molecules intact, and by chemists, who were mainly interested in the chemical alterations of the irradiated molecules, but also in the associated photophysical steps. The parallel development of these two lines of research is reflected in the terminology. For example, the term *flash photolysis*, as used by chemists, applies to time-resolved measurements of physical property changes caused by chemical processes induced by the absorption of a light flash (pulse). Flash photolysis serves to identify short-lived intermediates generated by bond breakage, such as free radicals and radical ions. Moreover, it allows the determination of rate constants of reactions of intermediates. Therefore, this method is appropriate for elucidating reaction mechanisms.

1.8.3.2 Experimental techniques

For pico- and femtosecond studies, time-resolved measurements require powerful pulsed laser systems operated in conjunction with effective detection techniques. Relevant commercially available laser systems are based on Ti:sapphire oscillators, tunable between 720 and 930 nm (optimum laser power around 800 nm). For nanosecond work, Nd^{3+}:YAG (neodymium-doped yttrium-aluminum-garnet) (1064 nm) and ruby (694.3 nm) laser systems are commonly employed. For many applications, light pulses of lower wavelength are produced with the aid of appropriate nonlinear crystals through second, third, or fourth harmonic generation. For example, short pulses of λ = 532, 355, and 266 nm are generated in this way by means of Nd^{3+}:YAG systems. Moreover, systems based

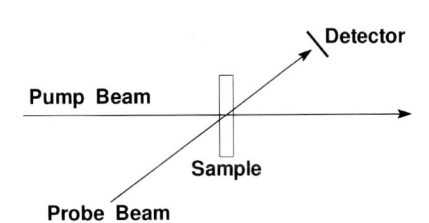

Fig. 1.21 Schematic depiction of a set-up for time-resolved optical absorption measurements.

on mode-locked dye lasers have occasionally been employed for ultrafast measurements in the fs and ps time domain [12].

Principally, the pump and probe technique depicted in Fig. 1.21 is applied in time-resolved transient absorption experiments. A pump beam, directed onto the sample, generates excited species or reactive intermediates such as free radicals. The formation and decay of these species can be monitored with the aid of an analyzing (probe) light beam that passes through the sample perpendicular to the direction of the pump beam. In principle, a set-up of this kind is also suitable for recording luminescence, if it is operated without the probe beam.

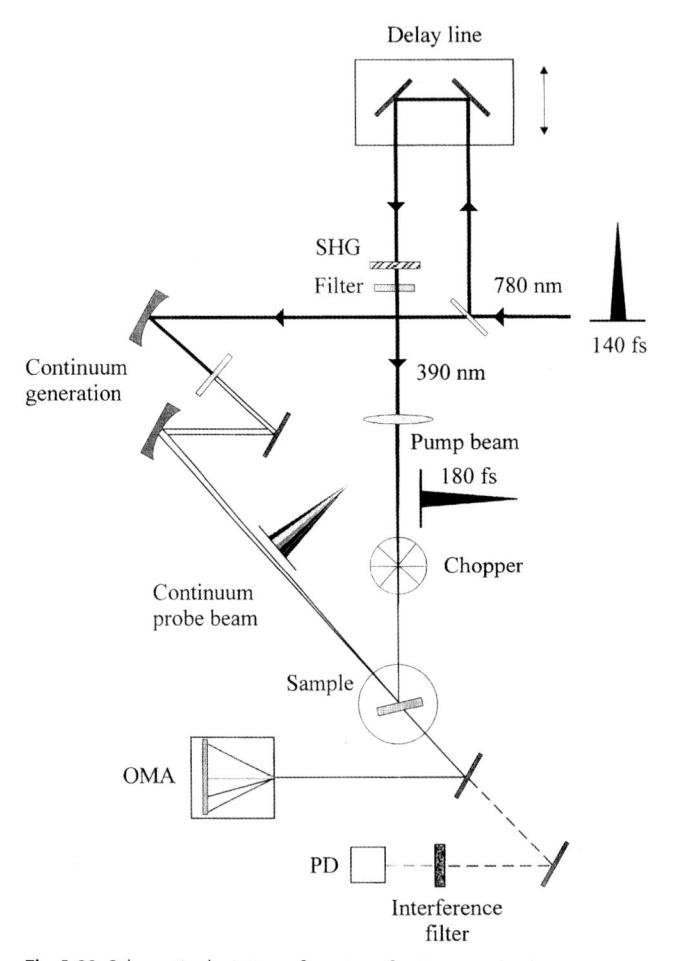

Fig. 1.22 Schematic depiction of a set-up for time-resolved optical absorption measurements in the femtosecond time domain. SHG: second harmonic generation crystal; PD: photodiode; OMA: optical multichannel analyzer. Adapted from Lanzani et al. [68] with permission from Wiley-VCH.

A typical set-up employed for time-resolved measurements in the femtosecond time domain is presented in Fig. 1.22 [68]. Here, a Ti:sapphire system operated in conjunction with a LiB_3O_5 crystal functioning as a frequency doubler provides the pump pulse ($\lambda = 390$ nm, repetition rate 1 kHz). The pulse intensity (excitation density) can be varied between 0.3 and 12 mJ cm^{-2}. For the generation of the analyzing white light, a fraction of the pump pulse is split off and focused through a thin sapphire plate. The resulting supercontinuum, which extends from 450 to 1100 nm, is passed through the sample prior to hitting the detector. Through mechanical operation of the delay line, transient absorption spectra are recorded at various times after the pump pulse by averaging over 100 to 1000 laser pulses.

Modern detection systems are based on the charge-coupled device (CCD) technique, which is not indicated in the schematic of Fig. 1.22.

Prior to the advent of powerful lasers, high-speed flash techniques were employed as light sources in time-resolved studies. Research was focused mainly on luminescence studies, aimed at determining fluorescence and phosphorescence lifetimes. In this connection, the development and successful application of sophisticated methods such as the single-photon time-correlation method and high-speed photography methods (streak camera) are worthy of note. Detailed technical information on these topics is available in a book by Rabek [69]. The physical principles of lifetime determinations have been described by Birks [70].

1.8.3.3 Applications of time-resolved techniques

1.8.3.3.1 Optical absorption

Optical absorption measurements are much more difficult to perform than emission measurements. This applies, for instance, to the detection of species having a low extinction coefficient at the relevant wavelengths. The surrounding molecules should be transparent, which is important in the case of solutions. Moreover, it has to be taken into account that invariably one has to measure an absorbance difference and not an absolute quantity as in the case of luminescence. In principle, molecules that have been promoted to an excited state of sufficiently long lifetime can absorb photons. Provided that the absorption coefficients are large enough, the absorption spectrum can permit identification of the excited state, and from its decay the lifetime of the excited state is obtained. In the relevant literature, this kind of absorption is frequently denoted by the acronyms PIA or PA, referring to *photoinduced absorption*. In many cases, excited triplet states are relatively long-lived and can easily be detected by light absorption measurements. As a typical result, Fig. 1.23 shows the T-T absorption spectrum, i.e. the spectrum of excited triplet states, of the polymer PPVK (see Chart 1.17), generated by irradiation in benzene solution at room temperature with a 15 ns pulse of 347 nm light. The triplet lifetime amounts to several microseconds in this case [71].

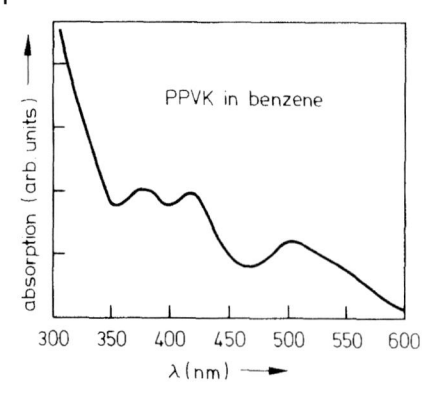

Fig. 1.23 Triplet-triplet absorption spectrum of poly(phenyl vinyl ketone) in benzene solution at room temperature. Recorded at the end of a 15 ns pulse of 347 nm light.

Commonly, excited singlet states have very short lifetimes and can only be detected by means of femtosecond absorption spectroscopy. A typical case is illustrated in Fig. 1.24, which shows the differential transmission spectrum of MEH-DSB (see Chart 1.18).

The differential transmission is defined as $\Delta T/T = (T-T_0)/T_0$, where T and T_0 are the transmissions in the presence and the absence of the pump beam, respectively. It may be recalled that $T = (I/I_0) = e^{-ad}$, where I_0 and I denote the light intensities before and after the sample; a and d are the absorption coefficient and the sample thickness, respectively. The absorbance A is equal to ad. In the small signal limit, commonly 10^{-5} to 10^{-3}, i.e. $(\Delta T/T) \ll 1$, $\Delta T/T$ is proportional to Δa, the change in the absorption coefficient: $(\Delta T/T) \approx -\Delta ad$. Negative values of $\Delta T/T$ correspond to photoinduced absorption (PIA). Thus, in Fig. 1.24, the band between 600 and 1100 nm with a peak at about 900 nm reflects the absorption of singlet intrachain excitons [72]. Positive values of $\Delta T/T$ correspond to bleaching or stimulated emission, SE. Thus, in Fig. 1.24, the band between 450 and 500 nm is assigned to bleaching due to depopulation of ground-state electrons and the band at around 535 nm, coinciding with the photoluminescence (PL) spectrum, is ascribed to SE [72]. The spectral features shown by the solid line in Fig. 1.24 are similar to those reported for many poly(arylene vinylene)s. The phenomenon of stimulated emission is dealt with in more detail in Section 6.2.2. Also typical of poly(arylene vinylene)s, Fig. 1.25 presents differential transmission kinetic traces recorded at 800 nm at varying pulse intensities for a thin film of poly[2-methoxy-5-(2′-ethylhexyloxy)-p-phenylene vinylene], MEH-PPV. The absorption decays on the ps time scale and the decay dynamics depends on the excitation density. The higher the pulse intensity, the faster is

Chart 1.17 Chemical structure of poly(phenyl vinyl ketone).

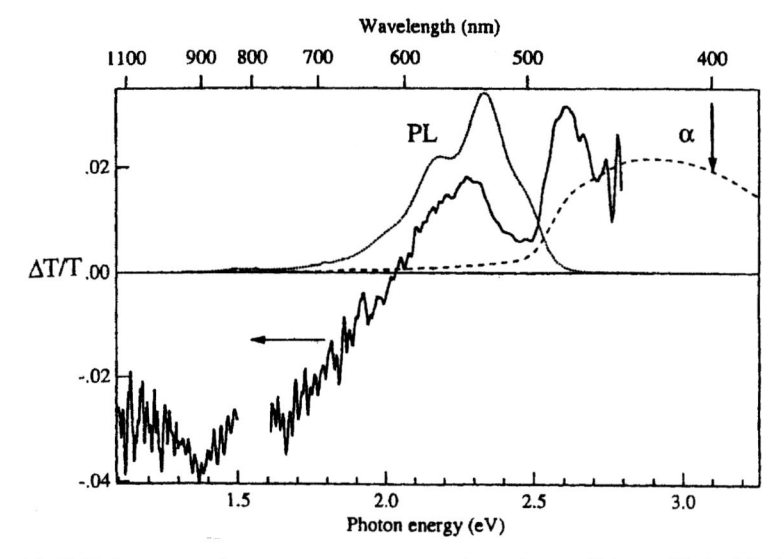

Fig. 1.24 Femtosecond spectroscopy at $\lambda_{exc}=400$ nm, pulse length: 150 fs, pulse energy: 1 mJ, pulse repetition rate: 1 kHz. Differential transmission spectrum of a thin film of MEH-DSB (solid line) recorded at the end of the pulse. Also shown: ground-state absorption coefficient α (dashed line) and photoluminescence spectrum, PL (dotted line). Adapted from Maniloff et al. [72] with permission from the American Physical Society.

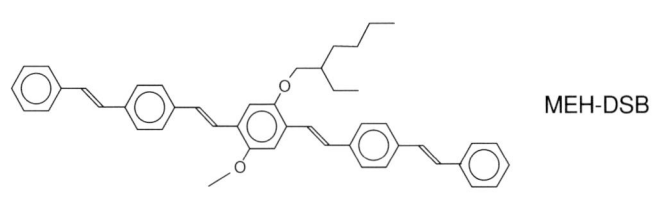

Chart 1.18 Chemical structure of a phenylene vinylene oligomer.

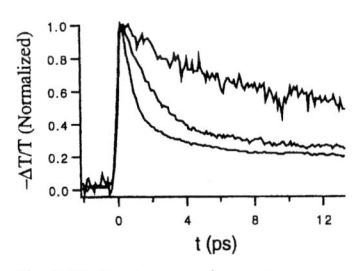

Fig. 1.25 Femtosecond spectroscopy. Differential transmission traces recorded at $\lambda_{rec}=850$ nm from thin films of poly[2-methoxy-5-(2'-ethylhexyloxy)-p-phenylene vinylene], MEH-PPV, irradiated as indicated in the legend of Fig. 1.24 at varying photon fluences, from upper to lower curves: 1.0×10^{13}, 3.1×10^{14}, and 9.3×10^{14} cm^{-2}, respectively. Adapted from Maniloff et al. [72] with permission from the American Physical Society.

the decay. Since the decay dynamics of the PIA band at around 800 nm and of the SE band at 535 nm are correlated, it is concluded that both bands arise from the same species, namely intrachain excitons. The intensity-dependent decay dynamics may be interpreted in terms of exciton-exciton annihilation, a process involving interaction of nearby excitons and resulting in non-radiative relaxation to the ground state [72].

1.8.3.3.2 Luminescence

During the past decades, time-resolved fluorescence measurements have helped to address many problems in the polymer field. A typical example concerns the determination of the rate of rotational and segmental motions of macromolecules in solutions as dealt with in Section 1.7.3. Moreover, time-resolved fluorescence measurements permit the investigation of energy migration and excimer formation in linear polymers. Down-chain energy migration in a linear polymer bearing overwhelmingly naphthalene plus a few anthracene pendant groups was evidenced by a decrease in the naphthalene fluorescence and a concomitant increase in anthracene fluorescence [17]. Similarly, the decay of the monomer emission was found to be correlated with the build-up of the excimer fluorescence in the case of polystyrene in dilute solution in dichloromethane [73]. This is illustrated in Fig. 1.26.

The remainder of this section focuses on the phenomenon of spectral or gain narrowing, which has been discovered in more recent fluorescence studies. As can be seen in Fig. 1.27, the shape of the spectrum of light emitted from BuEH-PPV (see Chart 1.19) changes drastically when the intensity of the exciting light pulse is increased beyond a threshold value. The broad emission spectrum extending over a wavelength range of about 200 nm recorded at low incident light intensity is transformed into a narrow band with $\Delta\lambda \approx 10$ nm at high light intensity [74].

The phenomenon of spectral narrowing is attributed to a cooperative effect in light emission, the so-called *amplified spontaneous emission* effect, which involves the coherent coupling of a large number of emitting sites in a polymer matrix.

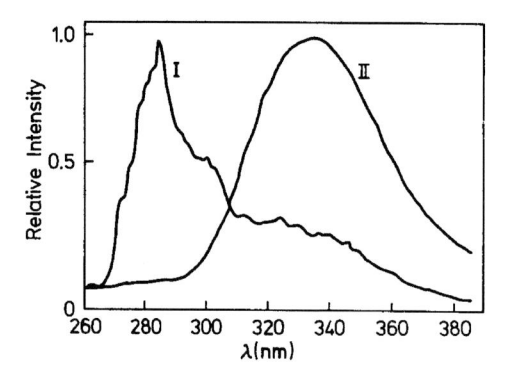

Fig. 1.26 Fluorescence spectra of polystyrene in oxygen-free CH_2Cl_2 solution (1 g L^{-1}). I: Monomer emission, recorded at the end of a 10 ns flash ($\lambda_{exc} = 257$ nm). II: Excimer emission, recorded 45 ns after the flash. Adapted from Beavan et al. [73] with permission from John Wiley & Sons, Inc.

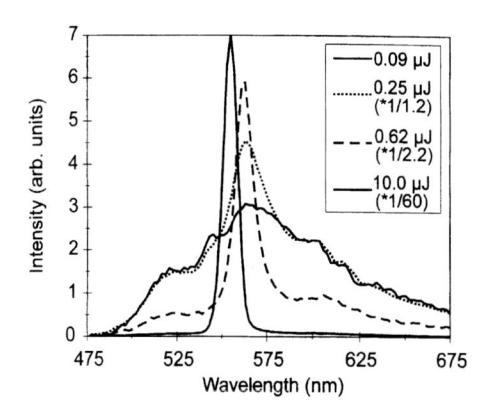

Fig. 1.27 Spectral narrowing in the case of BuEH-PPV. Emission spectra recorded at different excitation pulse energies. Pulse duration: 10 ns, λ_{exc} = 532 nm. Film thickness: 210 nm [74]. Adapted from Lemmer et al. [75] with permission from Wiley-VCH.

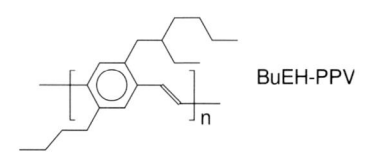

Chart 1.19 Chemical structure of poly[2-butyl-5-(2'-ethylhexyl)-1,4-phenylene vinylene], BuEH-PPV.

Spectral narrowing has been observed for thin polymer films (200–300 nm thick) on planar glass substrates. The films act as wave guides since the refractive index of the polymer is larger than that of the surrounding air or the glass substrate. Immediately after absorption of a light pulse, some photons are spontaneously emitted from certain excited sites. These photons are coupled into the guided-wave mode and stimulate radiative deactivation processes of other excited sites upon propagation through the film, a process denoted as amplified spontaneous emission. The phenomenon of spectral narrowing is explained by the fact that the emission of photons with the highest net gain coefficient is favored [75].

References

1 J.D. Coyle, *Introduction to Organic Photochemistry*, Wiley, Chichester (1986).
2 H.H. Jaffe, M. Orchin, *Theory and Applications of Ultraviolet Spectroscopy*, Wiley, New York (1962).
3 G.M. Barrow, *Introduction to Molecular Spectroscopy*, McGraw-Hill Kogakusha, Tokyo (1962).
4 H.G.O. Becker (ed.), *Einführung in die Photochemie*, Thieme, Stuttgart (1983).
5 J. Kopecky, *Organic Photochemistry: A Visual Approach*, VCH, Weinheim (1992).
6 M. Pope, C.E. Swenberg, *Electronic Processes in Organic Crystals and Polymers*, 2nd Edition, Oxford University Press, New York (1999).
7 W. Schnabel, *Polymer Degradation, Principles and Practical Applications*, Hanser, München (1981).
8 H.-K. Shim, J.-I. Jin, *Light-Emitting Characteristics of Conjugated Polymers*, in K.-S. Lee (ed.), *Polymers for Photonics Applications I*, Springer, Berlin, Adv. Polym. Sci. 158 (2002) 193.

9 N. S. Sariciftci (ed.), *Primary Photoexcitations in Conjugated Polymers: Molecular Exciton versus Semiconductor Band Model*, World Scientific, Singapore (1997).

10 J. Cornil, D. A. dos Santos, D. Beljonne, Z. Shuai, J.-L. Bredas, *Gas Phase to Solid State Evolution of the Electronic and Optical Properties of Conjugated Chains: A Theoretical Investigation*, in G. Hadziioannou, P. F. van Hutten (eds.), *Semiconducting Polymers*, Wiley-VCH, Weinheim (2000), p. 235.

11 K. Pichler, D. Halliday, D. C. Bradley, P. L. Burn, R. H. Friend, A. B. Holmes, J. Phys. Cond. Matter 5 (1993) 7155.

12 P. A. Lane, S. V. Frolov, Z. V. Vardeny, *Spectroscopy of Photoexcitations in Conjugated Polymers*, in G. Hadziioannou, P. F. van Hutten (eds.), *Semiconducting Polymers*, Wiley-VCH, Weinheim (2000), p. 189.

13 S. E. Webber, Chem. Rev. 90 (1990) 1469.

14 I. B. Berlman, *Energy Transfer Parameters of Aromatic Compounds*, Academic Press, New York (1973).

15 R. F. Cozzens, R. B. Fox, J. Chem. Phys. 50 (1969) 1532.

16 C. David, M. Lempereur, G. Geuskens, Eur. Polym. J. 8 (1972) 417.

17 J. W. Longworths, M. D. Battista, Photochem. Photobiol. 11 (1970) 207.

18 J. E. Guillet, *Polymer Photophysics and Photochemistry*, Cambridge University Press, Cambridge, U.K. (1985).

19 J. S. Aspler, C. E. Hoyle, J. E. Guillet, Macromolecules 11 (1978) 925.

20 M. S. Choi, T. Yamazaki, I. Yamazaki, T. Aida, Angew. Chem. Int. Ed. 43 (2004) 150.

21 M. R. Benites, E. T. Johnson, S. Weghorn, L. Yu, P. D. Rao, J. R. Diers, S. I. Yang, C. Kirmaier, D. J. Bocian, D. Holten, J. S. Lindsey, J. Mater. Chem. 12 (2002) 65.

22 M. S. Choi, T. Aida, T. Yamazaki, I. Yamazaki, T. Aida, Angew. Chem. Int. Ed. 40 (2001) 3194.

23 V. Cimrova, M. Remmers, D. Neher, G. Wegner, Adv. Mater. 8 (1996) 146.

24 K. Ichimura, Y. Akita, H. Akiyama, K. Kudo, Y. Hayashi, Macromolecules 30 (1997) 903.

25 H.-H. Perkampus, *Encyclopedia of Spectroscopy*, VCH, Weinheim (1995).

26 N. J. Greenfield, G. D. Fasman, *Computed Circular Dichroism Spectra for the Evaluation of Protein Conformation*, Biochemistry 8 (1969) 4108.

27 A. Rodger, B. Norden, *Circular Dichroism and Linear Dichroism*, Oxford University Press, Oxford (1997).

28 G. D. Fasman (ed.), *Circular Dichroism and the Conformational Analysis of Biomolecules*, Plenum Press, New York (1996).

29 K. Nakanishi, N. Berova, R. W. Woody (eds.), *Circular Dichroism: Principles and Applications*, VCH Publishers, Weinheim (1994).

30 R. W. Woody, *Circular Dichroism of Peptides*, in E. Gross, J. Meienhofer (eds.), *The Peptides: Analysis, Synthesis, Biology*, Academic Press, New York (1985), Vol. 7, p. 14.

31 T. W. Hagler, K. Pakbaz, J. Moulton, F. Wudl, P. Smith, A. J. Heeger, Polym. Commun. 32 (1991) 339.

32 T. W. Hagler, K. Pakbaz, K. F. Voss, A. J. Heeger, Polym. Commun. Phys. Rev. B 44 (1991) 8652.

33 M. Grell, D. D. C. Bradley, Adv. Mater. 11 (1999) 895.

34 B. M. W. Langeveld-Voss, R. A. J. Janssen, M. P. T. Christiaans, S. C. J. Meskers, H. P. J. M. Dekkers, E. W. Meijer, J. Am. Chem. Soc. 118 (1996) 4908.

35 S. H. Chen, D. Katsis, A. W. Schmid, J. C. Mastrangelo, T. Tsutsui, N. T. Blanton, Nature 397 (1999) 506.

36 E. A. Anufrieva, Yu. Ya. Gotlib, *Investigation of Polymers in Solution by Polarized Luminescence*, Adv. Polym. Sci. 40, Springer, Berlin (1981), p. 1.

37 K. P. Ghiggino, A. Roberts, D. Phillips, *Time-Resolved Fluorescence Techniques in Polymer and Biopolymer Studies*, Adv. Polym. Sci. 40, Springer, Berlin (1981), p. 69.

38 P. Wahl, C. R. Acad. Sci. 263 (1966) 1525.

39 See literature cited in [37].

40 D. F. Nicolai, G. B. Benedek, Biopolymers 15 (1976) 2421.

41 J. Marmur, P. Doty, J. Mol. Biol. 5 (1962) 109.

42 N. L. Veksin, *Photonics of Biopolymers*, Springer, Berlin (2002).

43 R.W. Woody, *Circular Dichroism*, Methods Enzymol. 246 (1995) 34.
44 W.C. Johnson Jr., Methods Enzymol. 210 (1992) 426.
45 W.C. Johnson Jr., Proteins 7 (1990) 205.
46 R. Fiesel, J. Huber, U. Scherf, Angew. Chem. 108 (1996) 2233.
47 M.M. Bouman, E.W. Meijer, Adv. Mater. 7 (1995) 385.
48 J. Li, G.B. Schuster, K.-S. Cheon, M.M. Green, J.V. Selinger, J. Am. Chem. Soc. 122 (2000) 2603.
49 B. Chae, S.W. Lee, M. Ree, S.B. Kim, Vibrational Spectrosc. 29 (2002) 69.
50 W. Klöpffer, *Introduction to Polymer Spectroscopy*, Springer, Berlin (1984).
51 D.O. Hummel, *Atlas of Polymer and Plastics Analysis*, 3rd Edition, Wiley-VCH, Weinheim (2005).
52 A. Elliott, *Infrared Spectra and Structure of Organic Long-Chain Polymers*, Arnold, London (1969).
53 M. Claybourn, *Infrared Reflectance Spectroscopy of Polymers. Analysis of Films, Surfaces and Interfaces*, Adhesion Society, Blacksburg, VA (1998).
54 R.A. Meyers (ed.), *Encyclopedia of Analytical Chemistry: Application, Theory and Instrumentation*, Wiley, Chichester (2000).
55 J.M. Chalmers, P.R. Griffiths (eds.), *Handbook of Vibrational Spectroscopy*, Wiley, Chichester (2002).
56 H.W. Siesler, Y. Ozaki, S. Kawata, H.M. Heise, *Near-Infrared Spectroscopy*, Wiley-VCH, Weinheim (2002).
57 J. Workman Jr., *Handbook of Organic Compounds, NIR, IR, Raman and UV-Vis Spectra Featuring Polymers and Surfactants*, Academic Press, San Diego (2000).
58 H.M. Mantsch, D. Chapman, *Infrared Spectroscopy of Biomolecules*, Wiley, New York (1996).
59 H. Morawetz, J. Polym. Sci.: Part A: Polym. Chem. 37 (1999) 1725.
60 Y. Wang, H. Morawetz, Macromolecules 23 (1990) 1753.
61 A. Montali, C. Bastiaansen, P. Smith, C. Weder, Nature 392 (1998) 261.
62 R.R. Alfano, *Semiconductors Probed by Ultrafast Laser Spectroscopy*, Academic Press, New York (1984).
63 J.L. Martin, A. Mignus, G.A. Mourou, A.H. Zewail (eds.), *Ultrafast Phenomena*, Springer Series in Chemical Physics, Vol. 55, Springer, Berlin (1992).
64 G. Porter, *Flash Photolysis into the Femtosecond – A Race against Time*, in J. Manz, L. Wöste (eds.), *Femtosecond Chemistry*, Wiley-VCH, Weinheim (1995).
65 F.C. DeSchryver, S. De Feyter, G. Schweitzer (eds.), *Femtochemistry*, Wiley-VCH, Weinheim (2001).
66 D.W. McBranch, M.B. Sinclair, *Ultrafast Photo-Induced Absorption in Nondegenerate Ground State Conjugated Polymers: Signatures of Excited States*, in [9], p. 587.
67 J.-Y. Bigot, T. Barisien, *Excited-State Dynamics of Conjugated Polymers and Oligomers*, in F.C. DeSchryver, S. De Feyter, G. Schweitzer (eds.), *Femtochemistry*, Wiley-VCH, Weinheim (2001).
68 G. Lanzani, S. De Silvestre, G. Cerullo, S. Stagira, M. Nisoli, W. Graupner, G. Leising, U. Scherf, K. Müllen, *Photophysics of Methyl-Substituted Poly(para-Phenylene)-Type Ladder Polymers*, in G. Hadziioannou, P.F. van Hutten (eds.), *Semiconducting Polymers*, Wiley-VCH, Weinheim (2000), p. 235.
69 J.F. Rabek, *Experimental Methods in Photochemistry and Photophysics*, Wiley, Chichester (1982).
70 J.B. Birks, *Photophysics of Aromatic Molecules*, Wiley-Interscience, London (1970), p. 94.
71 W. Schnabel, J. Kiwi, *Photodegradation*, in H.H.G. Jellinek (ed.), *Aspects of Degradation and Stabilization of Polymers*, Elsevier Scientific Publ., Amsterdam (1978), p. 195.
72 E.S. Maniloff, V.I. Klimov, D.W. McBranch, Phys. Rev. B 56 (1997) 1876.
73 S.W. Beavan, J.S. Hargreaves, D. Phillips, *Photoluminescence in Polymer Science*, Adv. Photochem. 11 (1978) 207.
74 F. Hide, M.A. Diaz-Garcia, B.J. Schartz, M.R. Anderson, P. Qining, A.J. Heeger, Science 273 (1996) 1833.
75 U. Lemmer, A. Haugeneder, C. Kallinger, J. Feldmann, *Lasing in Conjugated Polymers*, in G. Hadziioannou, P.F. van Hutten (eds.), *Semiconducting Polymers*, Wiley-VCH, Weinheim (2000), p. 309.

2
Photoconductivity

2.1
Introductory remarks

A photoconductive solid material is characterized by the fact that an electric current flows through it under the influence of an external electric field when it absorbs UV or visible light. There are two essential requirements for photoconductivity: (1) the absorbed photons must induce the formation of charge carriers, and (2) the charge carriers must be mobile, i.e. they must be able to move independently under the influence of an external electric field. Photoconductivity was first detected in inorganic materials, for example in crystals of alkali metal halides containing color centers (trapped electrons in anion vacancies) or in materials possessing atomic disorder such as amorphous silicon or selenium. As regards organic materials, dye crystals and, more recently, also various polymeric systems have been found to exhibit photoconductivity. Two groups of photoconducting polymeric systems may be distinguished: (a) solid solutions of active compounds of low molar mass in inert polymeric matrices, also denoted as molecularly doped polymers, and (b) polymers possessing active centers in the main chain or in pendant groups. Examples belonging to group (a) are polycarbonate and polystyrene molecularly doped with derivatives of triphenylamine, hydrazone, pyrazoline or certain dyes (see Table 2.1). Molecularly doped polymers are widely used as transport layers in the photoreceptor assemblies of photocopying machines.

Typical examples of photoconductive polymers (group (b)) are listed in Table 2.2. Concerning the field of conducting polymers, including photoconducting polymers, the reader is referred to various books and reviews [1–21].

Table 2.1 Typical dyes applied as dopants in photoconducting polymeric systems.

Chemical structure	Denotation
	Perylene dye
	Azo dye
	Quinone dye
	Squaraine dye
M : Cd, Zn, TiO etc.	Phthalocyanine dye

2.2
Photogeneration of charge carriers

2.2.1
General aspects

Regarding inorganic semiconductors, the photogeneration of charge carriers has been explained in terms of the so-called band model, according to which the nuclei of atoms are situated at fixed sites in a lattice [22]. Since the charges of the nuclei are largely compensated by their inner-shell electrons, an average constant potential is attributed to the outer-shell electrons, denoted as valence electrons. The energy levels of the valence electrons differ only slightly and are, therefore, considered as being located in the so-called valence band (see Fig. 2.1).

At $T=0$, the absolute zero temperature, all valence electrons reside in the valence band; at higher temperatures, some electrons are promoted to the so-called conduction band. The probability of an electron being in a quantum state of energy E is given by Eq. (2-1).

Table 2.2 Chemical structures of typical photoconducting polymers.

Chemical structure	Acronym	Denotation
(N-vinyl carbazole structure)	PVC	Poly(N-vinyl carbazole)
(polyacetylene structure)	PAC	trans-Polyacetylene
(thiophene structure)	PT	Polythiophene
(fluorene structure)	PFO	Poly(dialkyl fluorene)
(phenylene vinylene structure)	PPV	Poly(p-phenylene vinylene)
(phenylene structure)	PPP	Poly(p-phenylene)
(ladder-type structure) X = CH₃, R₁ = C₆H₁₃, R = –⌬–C₁₀H₂₁	m-LPPP	Methyl-substituted ladder-type poly(p-phenylene)
(silylene structure) R₁ and R₂: alkyl or aryl groups		Polysilylene
(aniline structure)	PANI	Polyaniline

$$f(E - E_F) = \frac{\exp[-\beta'(E - E_F)]}{1 + \exp[-\beta'(E - E_F)]} \qquad (2\text{-}1)$$

Here, $f(E - E_F)$ is the Fermi distribution function, β' is equal to $(kT)^{-1}$, where k is the Boltzmann constant, T is the absolute temperature, and E_F is the Fermi energy.

The Fermi level of inorganic semiconductors lies between the valence band and the conduction band, in contrast to metals, for which the Fermi level lies within the valence band. According to this model, the phenomenon of dark conductivity is feasible. Photoconductivity implies that, upon irradiation, electrons

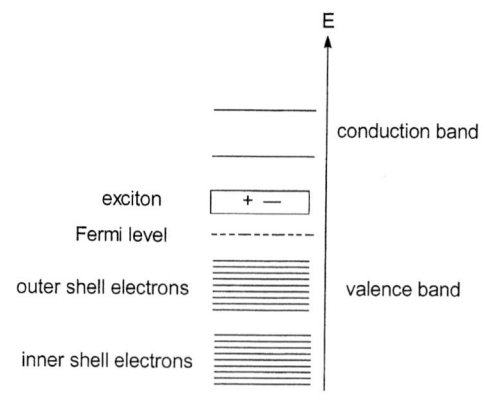

Fig. 2.1 Energy levels of a semiconductor. Also shown: energy level of an exciton state, as generated upon photon absorption.

are promoted from the valence band to the conduction band. Thus, the total electrical conductivity κ is composed of two terms representing the dark conductivity κ_d and the photoconductivity κ_p.

$$\kappa = \kappa_d + \kappa_p \qquad (2\text{-}2)$$

Band-to-band transitions of electrons require photon energies exceeding the energy of the band gap. Since the energy states of the conduction band are not localized, i.e. not attributable to specific atomic nuclei, electrons transferred to the conduction band lose their local binding and become mobile. Regarding polymeric systems, this aspect is at variance with recent experimental and theoretical work, which overwhelmingly led to the conclusion that in such systems localized states are involved both in the photogeneration of charges and in the carrier transport, and that the theoretical model developed for inorganic semiconductors is not applicable for polymeric systems. At present, the generation of charge carriers is explained in terms of the *exciton concept* and a generally accepted carrier transport mechanism presumes *charge hopping* among discrete sites, as will be described in the following subsections.

2.2.2
The exciton model

The exciton model is based on the fact that, in organic photoconductors, the light-induced transition of an electron to an excited state causes a pronounced polarization of the chromophoric group. Because of the relatively high stability of this state, it is considered to be an entity of special nature. This entity, called an exciton, is an excited state of quasi-particle character located above the valence band. It resembles a hydrogen-like system with a certain binding energy, which can, besides other non-radiative or radiative deactivation routes, also give rise to the formation of a geminate electron/hole pair. Under certain conditions, the latter can dissociate and thus give rise to the generation of free, i.e. independent, charge carriers:

$$\text{exciton} \rightarrow [h^+/e^-] \rightarrow h^+ + e^- \tag{2-3}$$

It is generally accepted that the dissociation of electron/hole pairs is induced, or at least strongly assisted, by an external electric field. Whether electron/hole pair dissociation generally also occurs intrinsically, i.e. in the absence of an external electric field, has not yet been fully established. In certain cases, such as in m-LPPP [23] or in PPV [24], this process has been evidenced. However, in these and similar cases, electron/hole pair dissociation is likely to be due to the presence of impurities such as molecular oxygen and/or structural defects in the macromolecular system such as conformational kinks or chain twists that function as dissociation sites. The existence of these sites and the capability of excitons to approach them are presumably prerequisites for dissociation. In this connection, it is notable that excitons are conjectured to diffuse over certain distances. It has been suggested that charge generation, i.e. the formation of free charge carriers, occurs preferentially at specially structured sites on the surface of the sample.

In view of the highly variable nature of photoconducting materials, different types of exciton states have been postulated. For instance, an exciton state with a radius of the order of 100 Å, a so-called Wannier exciton, is assumed to be formed in amorphous silicon, in which the wave function spreads over the electronic orbitals of many Si atoms. In contrast, in conjugated polymers such as poly(phenyl vinylene) or polysilanes (see Table 2.2), the formation of less extended, so-called Frenkel excitons, with radii of the order of 10 Å is assumed. In this case, the polymer system is considered to be an ensemble of short molecular segments that are characterized by localized wave functions and discrete energy levels, and an exciton generated by the absorption of a photon exists within the intra-chain delocalization length. For systems permitting the formation of charge-transfer (CT) states, the existence of charge-transfer or quasi-Wannier excitons, having radii exceeding those of Frenkel excitons, is postulated. This applies, for example, to poly(methyl phenyl silylene) [25]. In this case, the absorption of photons in main-chain segments generates Frenkel excitons, which are converted to CT excitons through intramolecular interaction with pendant phenyl groups (see Scheme 2.1).

Moreover, CT excitons are thought to be formed by intermolecular interaction in certain polymeric systems containing small molecules. A typical example is poly(N-vinyl carbazole) doped with trinitrofluorenone (TNF), a system which played a major role in early photoconductive studies on polymeric systems (see Chart 2.1).

As regards the nature of the so-called dissociation sites referred to above, it may be noted that generally any kind of disorder-induced kink may play an activating

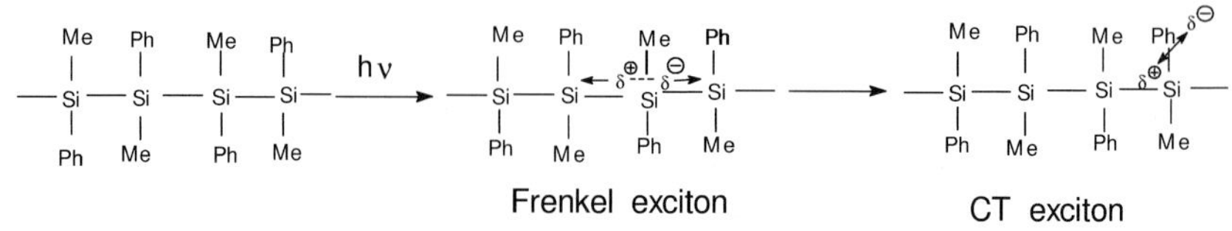

Scheme 2.1 Generation of charge-transfer excitons in poly(methyl phenyl silylene) [25].

Chart 2.1 Chemical structures of poly(N-vinyl carbazole) and trinitrofluorenone.

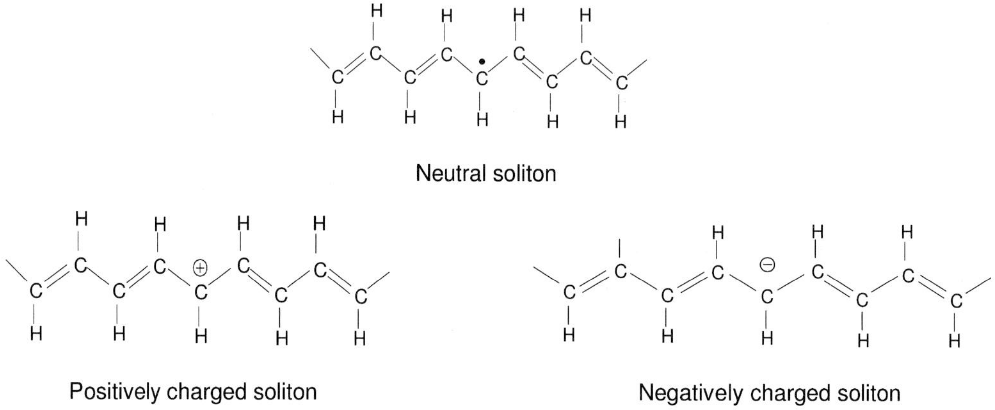

Chart 2.2 Chemical structures of solitons formed in *trans*-polyacetylene.

role in the dissociation of electron/hole pairs. In the case of *trans*-polyacetylene, which has been examined quite extensively, so-called neutral solitons (see Chart 2.2) resulting from incomplete *cis-trans* isomerization are postulated to function as dissociation sites. Neutral solitons are characterized by a free spin and are, therefore, detectable by electron-spin resonance (ESR) measurements [26].

2.2.3
Chemical nature of charge carriers

In the earlier literature, charge carriers generated in polymers are frequently denoted as polarons and bipolarons and it is assumed that these charged species are formed instantaneously upon optical excitation [27]. The fundamental and often quite controversial debate on the nature of the primary photoexcitations in π-conjugated polymers has attracted much attention in the scientific community and has resulted in a series of articles being compiled in a book edited by Sariciftci [9]. This book is wholeheartedly recommended for further reading. The currently accepted notion that optical absorption generates primarily neutral excitations (excitons) rather than charged species was adopted in Section 2.2.2. The earlier model is based to some extent on the assignment of transient optical absorption bands at around 0.6 and 1.6 eV, recorded with PPV-type polymers, to bipolarons. However, this assignment was contradicted by unambiguous experimental evidence for an attribution of these transient absorption bands to singly-charged ions [28]. The definition of the term polaron, which can sometimes be rather elusive in older work, has been subject to alterations and many authors now denote the products of the dissociation of electron/hole pairs as negative and positive polarons. However, by doing so, the difficulty of precisely describing the chemical nature of the charge carriers is merely circumvented. As a matter of fact, the release of an electron should lead to a radical cation and the capture of an electron to a radical anion. Actually, relatively little work has hitherto been dedicated to clarifying the nature of photogenerated charge carriers. Time-resolved spectroscopy has helped to evidence the existence of radical cations acting as charge carriers in certain polymeric systems. In this case, radical cations were generated by hole injection from an indium tin oxide (ITO) electrode by applying an external electric field to polysulfone systems containing tris(stilbene) amine derivatives [29]. Moreover, the formation of radical cations in poly(methyl phenyl silylene) with $\Phi_{CC} \approx 1 \times 10^{-3}$ was evidenced by means of transient optical absorption measurements (absorption bands at around 375 and 460 nm formed upon irradiation with 20 ns laser pulses, $\lambda = 347$ nm) [25]. In the case of *m*-LPPP irradiated with 380 nm laser pulses, a transient optical absorption band at around 691 nm (1.91 eV) attributed to *positive polarons* was detected (see below) [23]. Obviously, quite different charge carriers will be produced depending on the chemical nature of the polymer. For example, in the case of *trans*-polyacetylene, the dissociation of electron/hole pairs at neutral solitons is considered to give rise to positively and negatively charged solitons (scc Chart 2.2) [30].

2.2.4
Kinetics of charge carrier generation

The research concerning the mechanism and kinetics of the photogeneration of charge carriers has focused on conjugated polymers since these are of great importance for applications in light-emitting diodes and organic photovoltaic cells (see Sections 6.2.1 and 6.3). Typical work performed with *m*-LPPP (see Table 2.2) revealed that charge carriers are generated within a few hundred femtoseconds in a very small yield in the absence of an external electric field [23]. The polymer was irradiated with 180 fs pulses of 380 nm light at 77 K. Transmission difference spectra plotted as ΔT/T exhibited, besides the emission and absorption bands of excitons, an absorption band at 1.9 eV (650 nm) attributable to individ-

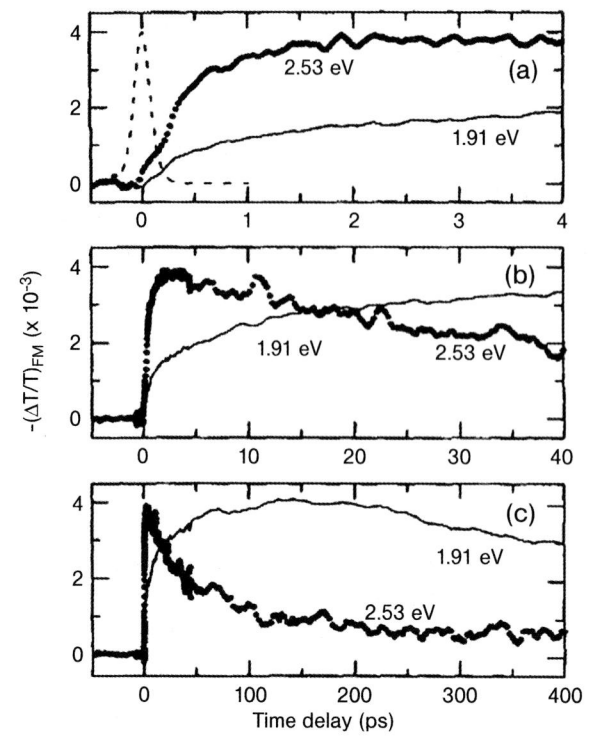

Fig. 2.2 Dissociation of excitons into charge carriers in *m*-LPPP under the influence of an external electric field (13 V). Kinetic traces, on different time scales, demonstrating changes in the field-induced differential transmission $(\Delta T/T)_{FM}$ at 1.91 eV (hole absorption) and 2.53 eV (exciton emission) following irradiation of a 100 nm thick polymer film at 77 K with 180 fs pulses of 380 nm light. Trace (a) also shows the pulse profile (dashed line). Adapted from Lanzani et al. [23] with permission from Wiley-VCH.

ual positive polarons (holes). This band was formed within the duration of the pulse. When an external electric field was applied, the yield of charge carriers was significantly increased. As can be seen from the kinetic traces shown in Fig. 2.2, the formation of the polaron absorption corresponds to the decay of the exciton emission, thus demonstrating that excitons dissociate into charge carriers.

Upon applying a field modulation technique, it was possible to record directly field-induced changes in the $\Delta T/T$ spectra. Therefore, the kinetic traces in Fig. 2.2 reflect the time dependence of the field-induced differential transmission $(\Delta T/T)_{FM}$, which is the difference between $\Delta T/T$ recorded in the presence and absence of the electric field: $(\Delta T/T)_{FM} = (\Delta T/T)_F - (\Delta T/T)_{F=0}$.

2.2.5
Quantum yield of charge carrier generation

It has been pointed out above that the deactivation of excitons may result in the formation of geminate electron/hole pairs that can eventually form free charge carriers. This process proceeds with strong competition from charge recombination and can be affected by an external electric field. According to the Onsager theory [31], the probability P_r of recombination can be estimated with the aid of Eq. (2-4).

$$P_r \propto \exp\left(-\frac{r_c}{r}\right) \exp\left(-\frac{eFr}{2kT}[1 - \cos\Theta]\right) \quad (2\text{-}4)$$

Here, e is the elementary charge, F is the electric field strength, k is the Boltzmann constant, T is the temperature, and Θ is the angle between the vector connecting the charges and the direction of the electric field.

The Onsager theory considers two potentials determining the fate of an electron/hole pair: the Coulomb potential $e^2/\varepsilon r$ (ε = dielectric constant) and the thermal energy kT. Pairs having a radial distance r larger than r_c will escape recombination. At the critical radial distance r_c, the thermal energy is equal to the Coulomb potential [see Eq. (2-5)].

$$kT = \frac{e^2}{\varepsilon r_c} \quad (2\text{-}5)$$

According to Eq. (2-4), the recombination probability decreases with increasing field strength, i.e. the escape probability $P_e = 1 - P_r$ increases. Therefore, the quantum yield for charge carrier generation Φ_{cc} should increase with increasing field strength. Figure 2.3 shows a double logarithmic plot of the dependence of Φ_{cc} on the electric field strength, measured at $T = 295$ K for three polysilylenes [32].

The quantum yield increases dramatically by about three orders of magnitude in the cases of the polysilylenes PBMSi and PMPSi, having aromatic substituents, whereas the fully aliphatic polysilane PDHeSi is quite ineffective in charge carrier production, presumably because CT excitons cannot be formed in this case. Interestingly, Φ_{cc} is markedly higher for the biphenyl-substituted polysilane than for the phenyl-substituted one, which might be due to a larger initial electron/hole distance in the former case. The curves in Fig. 2.3 were obtained with the aid of Eq. (2-6) [33], which is based on calculations by Mozumder [34].

$$\Phi_{cc} = \Phi_{cc,0} \int 4\pi r^2 f(r, F, T) g(r) dr \quad (2\text{-}6)$$

Here, $\Phi_{cc,0}$ denotes the primary quantum yield, $f(r, F, T)$ is the dissociation probability of pairs at radial distance r, and $g(r)$ is the initial spatial distribution

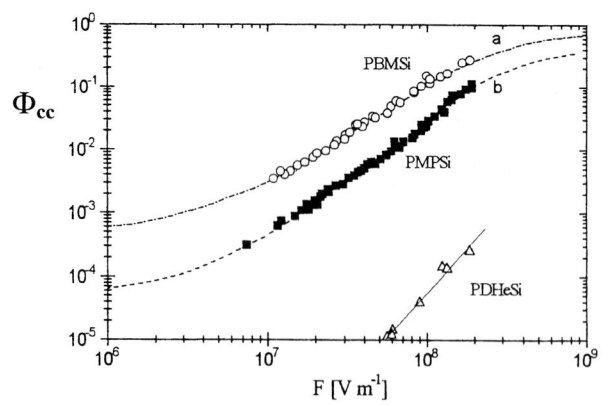

Fig. 2.3 Quantum yield for charge carrier generation as a function of the electric field strength determined at 295 K for three polysilylenes: poly(biphenyl methyl silylene), PBMSi; poly(methyl phenyl silylene), PMPSi; and poly(dihexyl silylene), PDHeSi. Adapted from Eckhardt [32] with permission from the author.

of electron/hole pairs. Satisfactory data fits were obtained by applying a Gaussian distribution function for electron/hole pair distances [see Eq. (2-7)].

$$g(r) = (\pi^{-3/2} a^{-3}) \exp\left(-\frac{r^2}{a^2}\right) \qquad (2\text{-}7)$$

Here, a is a material parameter.

Regarding the curves in Fig. 2.3, data fitting was performed with $\Phi_{cc,0}=0.85$ and $a=1.6$ nm in the case of PBMSi and $\Phi_{cc,0}=0.45$ and $a=1.3$ nm in the case of PMPSi. These data are in accordance with the assumption that $\Phi_{cc,0}$ increases with increasing initial electron/hole radial distance r_0, since statistically a is a measure of r_0.

Φ_{cc} values are most accurately determined by the xerographic (electrophotographic) discharge method, which is based on the determination of the light-induced change in the surface potential, $U=Q/C$, generated by a corona process. Q and C denote the surface charge density and the capacitance per unit area, respectively. ΔU is recorded at a given sampling frequency and the discharge quantum yield is obtained with the aid of Eq. (2-8)

$$\Phi_{cc} = \left(\frac{1}{efI}\right)\left(\frac{\Delta Q}{\Delta t}\right)_{t \to t_0} = \left(\frac{C}{efI}\right)\left(\frac{\Delta U}{\Delta t}\right)_{t \to t_0} = \left(\frac{\varepsilon \varepsilon_0}{edfI}\right)\left(\frac{\Delta U}{\Delta t}\right)_{t \to t_0} \qquad (2\text{-}8)$$

with the following denotations: dielectric constant ε (dimensionless), vacuum dielectric constant $\varepsilon_0 = 8.85 \times 10^{-14}$ A s V^{-1} cm^{-1}, elementary charge $e = 1.6022 \times 10^{-19}$ A s, sample thickness d [cm], light intensity I [photons cm^{-2} s^{-1}], surface potential U [V], and fraction of absorbed light f. Figure 2.4 shows a schematic depiction of a typical experimental set-up, which includes a rotating metal disk

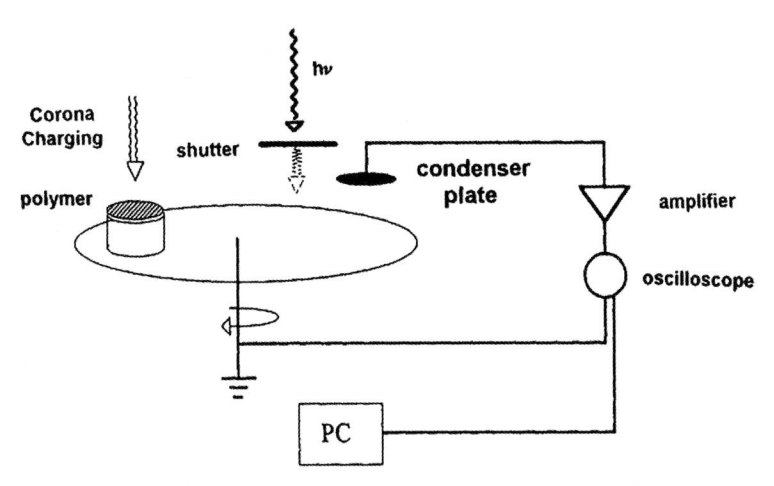

Fig. 2.4 Schematic illustration of a set-up used to determine Φ_{cc} by means of the xerographic discharge method. Adapted from Eckhardt [32] with permission from the author.

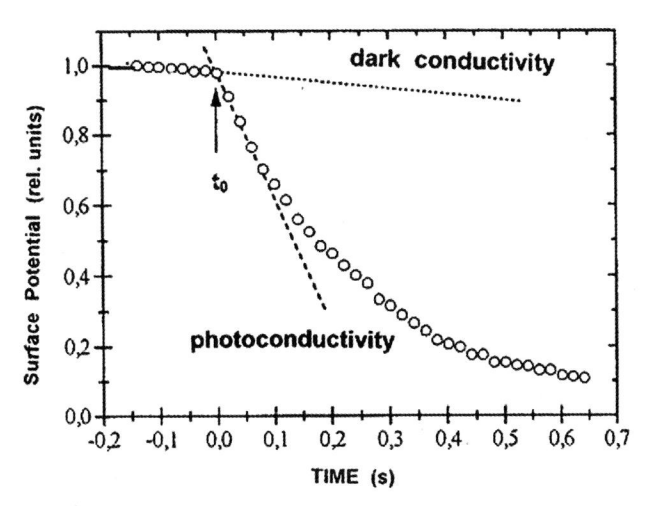

Fig. 2.5 Light-induced decrease in the surface potential recorded for poly(methyl phenyl silylene) at $\lambda_{exc} = 337$ nm; t_0 = onset of irradiation. Adapted from Eckhardt [32] with permission from the author.

carrying the sample. Upon rotation (600–2400 rpm), the sample passes a continuous light beam and a condenser plate for determination of the change in the surface potential.

A typical result obtained upon irradiation of poly(methyl phenyl silylene) at $\lambda_{exc} = 337$ nm is shown in Fig. 2.5 [32].

2.3
Transport of charge carriers

The transport of charge carriers through a solid is characterized by the drift mobility μ, which is defined as the hole or electron velocity per unit electric field strength. μ, frequently given in units of cm^2 V^{-1} s^{-1}, can be obtained with the aid of Eq. (2-9) by measuring the transit time t_r, which is the time required for charge carriers to pass a sample of thickness d when an external electric field of strength F is applied.

$$\mu = \frac{d}{t_r F} \tag{2-9}$$

Commonly, the so-called time-of-flight (TOF) method is applied to determine μ. Figure 2.6 shows a schematic depiction of a typical set-up.

A sandwich-type sample consisting of a semi-transparent ITO electrode, a polymer layer, and a metal (usually aluminum) electrode (see Fig. 2.7a) is irradiated with a short laser flash through the ITO electrode. During the light flash, which is totally absorbed by a very thin sheet at the surface of the polymer layer, charge carriers are generated and start to drift towards the metal electrode under the influence of an external electric field. The photocurrent is recorded as a function of time after the flash. Notably, the transport of both sorts of charge carriers cannot be recorded simultaneously. In the case of a negatively polarized metal electrode, hole migration can be observed, while electron migration can be followed with a positively polarized metal electrode. For mobility measurements in thin samples or materials inappropriate for photochemical charge car-

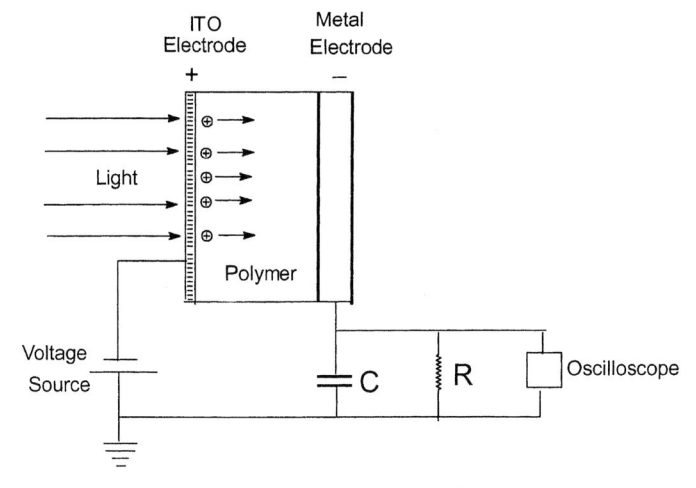

Fig. 2.6 Schematic illustration of a typical time-of-flight (TOF) set-up used for the determination of the mobility μ.

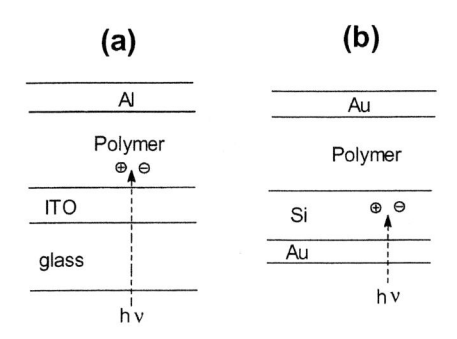

Fig. 2.7 Sandwich-type assemblies applied in time-of-flight determinations of charge carrier mobility: (a) carrier generation in the polymer layer, (b) carrier generation in the silicon substrate.

rier generation (low absorption coefficient, low quantum yield Φ_{cc}), a sandwich-type arrangement consisting of gold/silicon/polymer/gold layers (see Fig. 2.7b) is used [35]. Here, after passing through the lower gold layer, the light is totally absorbed by the silicon substrate, thus generating charges that are injected into the polymer layer.

Usually, only one sort of charge carrier is capable of migrating through the polymer film. In the cases of carbon-catenated π-conjugated and silicon-catenated σ-conjugated polymers, the photoconductivity is due to hole conduction. On the other hand, electrical conductivity due to electron conduction has been observed with low molar mass compounds such as tris(8-oxyquinolato)aluminum, Alq$_3$, dispersed in polymethacrylates bearing special pendant groups (see Chart 2.3 and also Table 6.3 in Section 6.2.1.2).

Figure 2.8 shows a typical result obtained for conjugated polymers [36]. Here, charge carriers are generated in a poly(methyl phenyl silylene) sample by a 15 ns flash of 347 nm light. The photocurrent is formed during the flash and a fraction decays very rapidly until a plateau is reached. In the subsequent phase, the current decreases slowly. The initial phase after the flash is characterized by the rapid formation of charge carriers and the rapid recombination of a fraction of them. The plateau corresponds to the migration of the holes, which drift at different velocities through the sample, and the end of the plateau corresponds to the time at which the fastest holes arrive at the metal electrode.

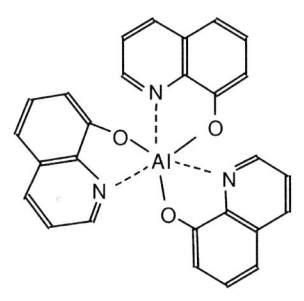

Chart 2.3 Chemical structure of tris(8-oxyquinolato)-aluminum, Alq$_3$.

Table 2.3 Hole mobilities at T=295 K and F≈10^5 V cm^{-1}.

Polymer	μ (cm^2 V^{-1} s^{-1})	References
Crystals of low molar mass organic compounds	10^{-1}–10^0	[5, 28]
Amorphous silicon	10^{-1}	[5]
m-LPPP	10^{-3}	[37]
Poly(9,9-dioctylfluorene)	10^{-4}	[38]
Poly(methyl phenyl silylene)	10^{-4}	[32]
Poly(p-phenylene vinylene)	10^{-5}	[39]
Polythiophene	10^{-5}	[40]
Poly(N-vinyl carbazole)	10^{-7}–10^{-6}	[41]

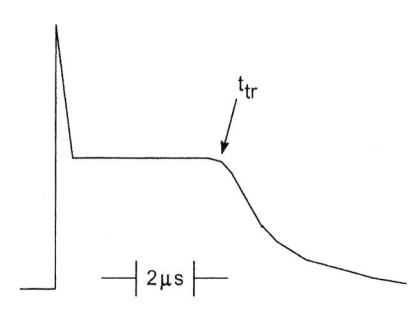

Fig. 2.8 Time-of-flight experiment performed with poly(methyl phenyl silylene). Photocurrent trace recorded with a positively biased ITO electrode at F=2.5×10^7 V m^{-1}, d=2 µm, λ_{exc}=347 nm, flash duration: 20 ns. Adapted from Eckhardt et al. [36] with permission from Taylor & Francis Ltd.

From Table 2.3, which lists typical μ values, it can be seen that the hole mobility in conjugated polymers is lower than that in organic crystals and amorphous silicon, but much larger than that in undoped poly(N-vinyl carbazole). Therefore, conjugated polymers have potential for applications in conducting opto-electronic and photonic devices. In principle, this also applies to liquid-crystal systems that can exhibit enhanced molecular order due to their self-organizing ability, as has been pointed out in a progress report [42].

The fact that liquid crystallinity enhances carrier transport as compared to non-ordered systems was convincingly demonstrated in the case of poly(9,9-dioctylfluorene). A relatively high hole mobility of 9×10^{-3} cm^2 V^{-1} s^{-1} was obtained when the polymer was examined as a uniformly aligned nematic glass. This value is significantly larger than the μ=4×10^{-4} cm^2 V^{-1} s^{-1} measured for an isotropic film of the same polymer [43]. Although significant progress has been made in developing materials with improved charge carrier mobilities, it seems that future applications will require materials possessing much further improved transport properties. Apparently, interchain interactions and morphological complexities strongly control charge carrier transport in bulk polymeric systems. Taking this into account, recent work on hole transport has led to quite high mobility values. For example, high mobilities were measured for very thin films (70–100 nm) of poly(3-hexylthiophene), P3HT, having a regioregularity of 96% [40]. (*Regioregularity* denotes the percentage of stereoregular head-to-tail attachments of thiophene rings bearing hexyl groups in the 3-position.) The films consisted of large amounts of microcrystalline domains embedded in an amorphous matrix. During film processing, the macromolecules arranged by self-organization into a lamellar structure composed of two-dimensional conjugated sheets. For a lamellae orientation parallel to the substrate, hole mobility values as high as 0.1 $cm^2\,V^{-1}\,s^{-1}$ were found. In this context, work with isolated linear polymer chains (*molecular wires*) is also noteworthy [44]. It revealed that the hole transport mobility along isolated polymer chains can exceed 0.1 $cm^2\,V^{-1}\,s^{-1}$, as can be seen in Table 2.4. Here, μ values were obtained from a pulse radiolysis study on dilute polymer solutions. Holes were generated by charge transfer from benzene radical cations to the polymer. By means of a time-resolved microwave conductivity method, it was shown that the conductivity of the solution increased significantly after the holes were produced, indicating that the mobility of holes in the polymer chains is considerably higher than the mobility of the initially formed benzene radical cations.

Interestingly, electron transport has been observed with a diene compound of the structure shown in Chart 2.4.

Table 2.4 Hole mobility μ in linear polymers in dilute solution in benzene [44].

Chemical structure	Acronym	μ (cm^2 V^{-1} s^{-1})
	DEH-PF	0.74
	MEH-PPV	0.43
	m-LPPP	0.16
	P3HT	0.02
	PAPS6	0.23

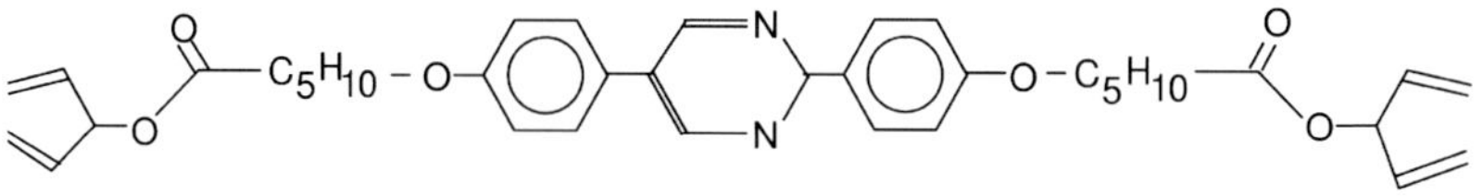

Chart 2.4 Chemical structure of a diene compound amenable to electron transport [45].

For this compound, which forms a smectic C phase at room temperature, an electron mobility of 1.5×10^{-5} cm^2 V^{-1} s^{-1} was reported. By virtue of its reactive groups, this diene compound can be photopolymerized to form a polymeric network [45].

2.4
Mechanism of charge carrier transport in amorphous polymers

At present, a hopping mechanism is generally accepted for the transport of charge carriers through amorphous polymeric media under the influence of an external electric field [23, 46]. After separation of electron/hole pairs, the independent charge carriers are temporarily trapped at certain sites. The latter have the quality of potential wells formed by single molecules or segments of polymer chains. Assisted by an external electric field, the carriers are removed from these sites by thermal activation and move until recaptured by other sites. With regard to this model, Gill has formulated an empirical relationship [Eq. (2-10)] for the dependence of the mobility μ on electric field strength and temperature [47].

$$\mu(F,T) = \mu_0 \exp\left(-E_{a,0} + \frac{\beta F^{1/2}}{kT_{\text{eff}}}\right) \quad (2\text{-}10)$$

Here, $E_{a,0}$ is the average activation energy, $\beta = (e^3/\pi \varepsilon \varepsilon_0)^{1/2}$ is the Poole-Frenkel factor, and T_{eff} is an effective temperature, where $T_{\text{eff}}^{-1} = T^{-1} - T_0^{-1}$. T_0 is the temperature at which Arrhenius plots of μ with varying F intersect, and $\mu_0 = \mu$ (T=T$_0$).

More recently, a relationship for the dependence of μ on F and T was derived by Bässler [21, 28] on the basis of the so-called disorder concept. The latter takes into account that carrier hopping in amorphous polymers is determined by the energy state of the transport sites and by the geometrical localization of the sites. The values of the energy states of the sites vary within a certain distribution, the so-called density of states (DOS) distribution, which is referred to as diagonal disorder. The width of this distribution is characterized by a parameter σ. Regarding the geometrical localization of the sites, it is taken into account that they are randomly distributed within the three-dimensional system, which is referred to as off-diagonal disorder. The width of this distribution is characterized by the geometrical disorder parameter Σ. The two distributions can be il-

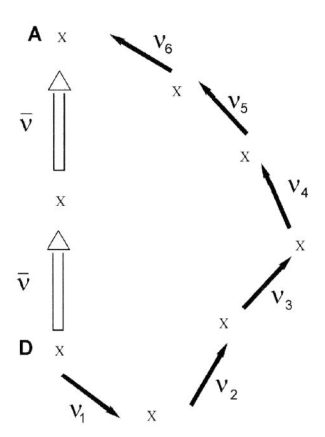

Fig. 2.9 Schematic depiction of a carrier trajectory in a polymeric matrix reflecting the geometrical (off-diagonal) disorder. The electric field acts along the D–A direction; v: jump rate. Adapted from Bässler et al. [21] with permission from Wiley-VCH.

lustrated as follows. Diagonal disorder: transport sites are traps of varying depths; off-diagonal disorder: the trajectories of carriers do not follow lines parallel to the field direction but show significant deviations therefrom, especially at low electric field strengths, as is demonstrated in Fig. 2.9.

In conclusion, charge transport in amorphous polymers occurs by way of carrier hopping within a positionally random and energetically disordered system of localized states [48]. The dependence of the carrier mobility on diagonal and off-diagonal disorder is taken into account by Eq. (2-11):

$$\mu(F, T) = \mu_0 \exp\left[-\frac{4\sigma^{*2}}{9}\right] \exp\left[C(\sigma^{*2} - \Sigma^2)F^{1/2}\right] \quad (2\text{-}11)$$

Here, $\sigma^* = \sigma/kT$, with σ being the width of the Gaussian distribution of energy states, C is an empirical constant, and μ_0 is a material constant.

According to Eq. (2-11), $\ln \mu$ is proportional to $F^{1/2}$ and $1/T^2$. Regarding the field strength dependence of μ, typical results obtained with poly(methyl phenyl silylene) are presented in Fig. 2.10 [32].

Note that the square-root dependence does not hold for the entire field regime, which is in accordance with findings for other polymers [28]. Note also that Eq. (2-11) predicts that the field dependence changes sign if $\Sigma > \sigma/kT$ and that the phenomenologically defined Gill temperature T_0 is related to the disorder parameter σ of the system, $T_0 = \sigma/k\Sigma$. For example, T_0 is equal to 387 K for $\Sigma = 3$ and $\sigma = 0.1$ eV [28]. The applicability of the model described above was scrutinized by Bässler [28] and is still being examined, as indicated by recent publications [49–51]. It has been pointed out, for instance, that in the case of m-LPPP the dependence of μ on electric field strength and temperature resembles that of molecular crystals, except that μ is two orders of magnitude lower, a behavior at variance with the present version of the disorder model. Attempts to modify the disorder model have to some extent been focussed on the interaction of charge carriers with the surrounding matrix, i.e. on the so-called polaronic effect. The latter implies that a localized

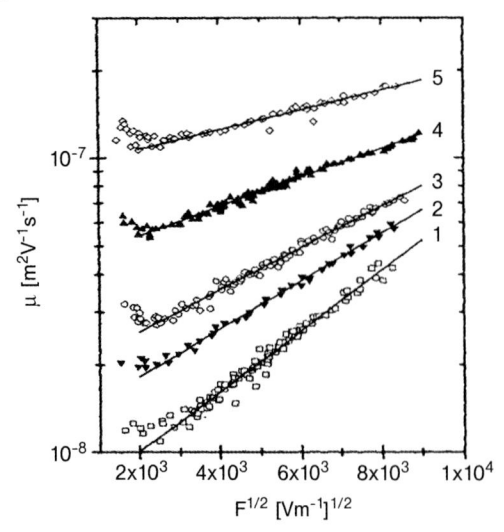

Fig. 2.10 Electric field dependence of the mobility, μ, of holes in poly-(methyl phenyl silylene) at various temperatures: (1) 295 K, (2) 312 K, (3) 325 K, (4) 355 K, (5) 385 K. Adapted from Eckhardt [32] with permission from the author.

carrier is strongly coupled either to local polarization or to vibrations and/or rotations of the molecule at which it resides. Since the coupling is induced by the charge carrier itself, the process is referred to as self-trapping and gives rise to the denotation of charge carriers as *polarons*. When a polaron moves, it carries along the associated structural deformation. As regards the hopping model, polaronic effects can be taken into account by considering that the activation energy for the mobility in a random hopping system is composed of two components, a polaronic component $E_a^{(p)}$ and a disorder component $E_a^{(d)}$ [see Eq. (2-12)].

$$E_a = E_a^{(p)} + E_a^{(d)} \tag{2-12}$$

Therefore, the dependence of the charge carrier mobility on electric field strength and temperature can be described by Eq. (2-13):

$$\mu(F, T) = \mu_0 \exp\left[-\left(\frac{E_p}{2kT} + \frac{4\sigma^{*2}}{9}\right)\right] \exp\left[C\left(\sigma^{*2} - \Sigma^2\right) F^{1/2}\right] \tag{2-13}$$

Here, E_p denotes the polaron binding energy.

2.5
Doping

It is possible to make inert polymers photoconductive and to improve the photoconduction performance of conducting polymers by doping, i.e. by the addition of appropriate low molar mass substances to the polymers. Relevant work has been reviewed by Mylnikov [3]. Early studies with inert polymers such as poly-

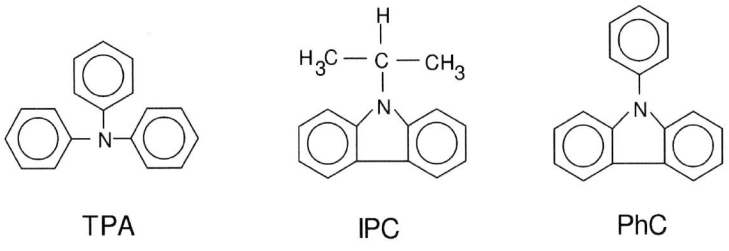

Chart 2.5 Chemical structures of electron-donating compounds: triphenylamine (TPA), isopropylcarbazole (IPC), and phenylcarbazole (PhC).

carbonate, polystyrene, and poly(vinyl chloride) revealed that μ, the hole mobility, and Φ_{cc}, the quantum yield of charge carrier generation, were increased when electron-donating compounds such as those presented in Chart 2.5 were incorporated as dopants. Actually, large amounts of dopants have to be applied to accomplish significant variations in Φ_{cc} and μ.

Figure 2.11 depicts the increase in Φ_{cc} with increasing triphenylamine content in commercial bisphenol A polycarbonate (see Chart 2.6) [52], and Fig. 2.12 shows a plot of log μ vs. 1/T. It can be seen that the hole mobility may be varied over several orders of magnitude by changing the TPA concentration [53]. Here, irradiations were performed at wavelengths of λ_{exc} = 300 and 337 nm, respec-

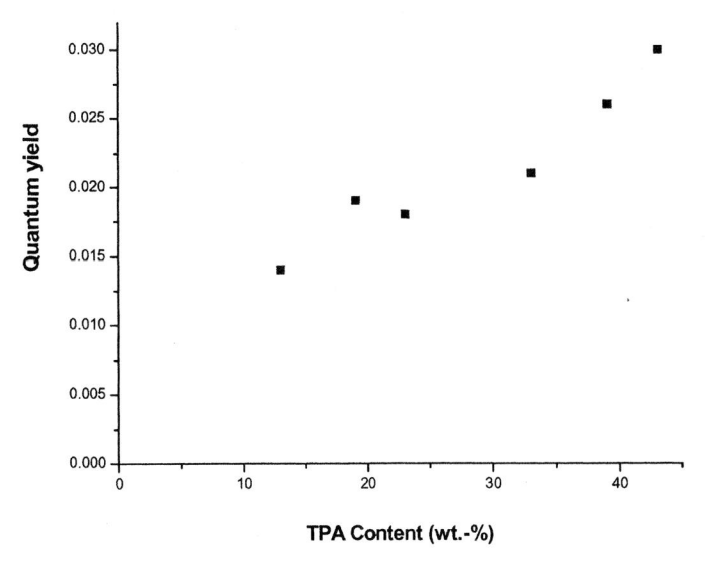

Fig. 2.11 Doping of an inert polymer, bisphenol A polycarbonate, with triphenylamine (TPA). The quantum yield of charge carrier formation Φ_{cc} as a function of the TPA content. λ_{exc} = 300 nm. Adapted from Borsenberger et al. [52] with permission from the American Institute of Physics.

Chart 2.6 Chemical structure of bisphenol A polycarbonate, poly(oxycarbonyloxy-1,4-phenylene-isopropylidene-1,4-phenylene).

tively, at which the polycarbonate is transparent and the light is absorbed solely by TPA.

As regards photoconducting polymers, typical work has been carried out with poly(N-vinylcarbazole), PVK, and polysilylenes. The first commercial photoconductor was based on a 1:1 charge-transfer (CT) complex between PVK and trinitrofluorenone (TNF) [11]. Similar photoconductor properties were found with a 1:1 CT complex of TNF with poly[bis(2-naphthoxy)phosphazene] (see Chart 2.7), which is an insulator if dopant-free [54].

Results obtained with poly(methyl phenyl silylene) are presented in Table 2.5, which demonstrate that, at low concentration (3 mol%), electron-accepting dopants having zero dipole moment are capable of increasing both μ and Φ_{cc}. The increase in Φ_{cc} is more pronounced the higher the value of the electron affinity,

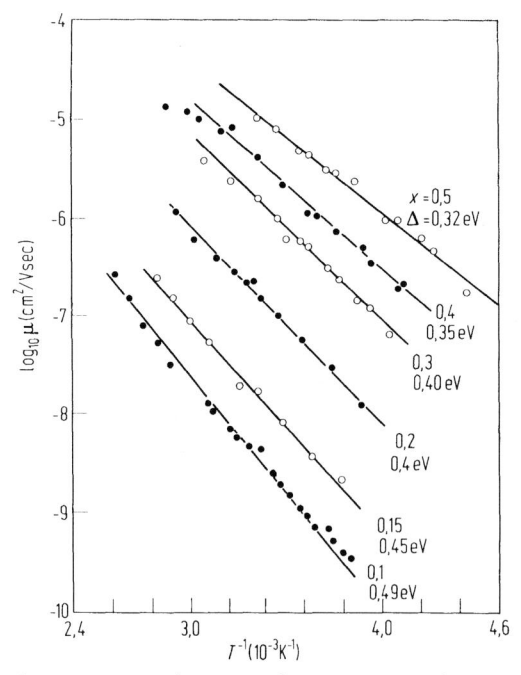

Fig. 2.12 Doping of an inert polymer, bisphenol A polycarbonate, with triphenylamine (TPA). Temperature dependence of the hole mobility. Plot of log μ vs. 1/T for various TPA contents denoted as weight fraction x. $\lambda_{exc} = 337$ nm, $F = 7 \times 10^5$ V cm^{-1}. Δ denotes the activation energy. Adapted from Pfister [53] with permission from the American Physical Society.

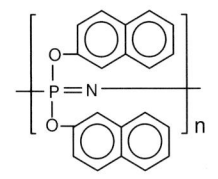

Chart 2.7 Chemical structure of poly[bis(2-naphthoxy)phosphazene], P2NP.

P2NP

Table 2.5 The photoconduction performance of poly(methyl phenyl silylene) containing electron-acceptor-type dopants [55].

Additive (3 mol%)	EA [a] (eV)	Dipole moment (Debye)	μ [b] (cm² V⁻¹ s⁻¹)	Φ_{cc} [c]
None			2.28×10⁻⁴	1.9×10⁻² [d]
o-DNB [g]	0.0	6.0	5.02×10⁻⁵	2.3×10⁻² [d]
m-DNB	0.3	3.8	1.42×10⁻⁴	2.3×10⁻² [d]
p-DNB	0.7	0.0	3.10×10⁻⁴	3.4×10⁻² [d]
Tetracene	1.0	0.0	3.06×10⁻⁴	9.6×10⁻² [e]
Chloranil	1.3	0.0	4.12×10⁻⁴	12.5×10⁻² [e]
TCNQ [f]	1.7	0.0	5.71×10⁻⁴	10.0×10⁻² [e]

a) Electron affinity.
b) Hole mobility.
c) Quantum yield of charge carrier formation.
d) λ_{exc} = 355 nm.
e) λ_{exc} = 339 nm.
f) TCNQ: tetracyanoquinone.
g) DNB: dinitrobenzene.

EA. Polar dopants also cause an increase in the quantum yield, but the hole mobility is concomitantly decreased [55].

Fullerene, C_{60}, is quite an effective dopant. It is an excellent electron acceptor, capable of accepting up to six electrons. Photoinduced electron transfer from conducting polymers such as poly(3-octylthiophene), P3OT, and poly[2-methoxy-5-(2'-ethylhexyloxy)-p-phenylene vinylene], MEH-PPV, to fullerene C_{60} occurs on a timescale of less than 1 ps. A C_{60} content of a few percent is sufficient to enhance Φ_{cc} in the ps time domain by more than an order of magnitude [56].

2.6
Photoconductive polymers produced by thermal or high-energy radiation treatment

Certain polymers become photoconductive upon exposure to heat or high-energy radiation, an aspect that has been reviewed by Mylnikov [3]. For example, polyacrylonitrile (maximum sensitivity at λ = 420 nm) or polypyrrole (maximum sensitivity at λ = 500–600 nm) exhibit photoconductivity after heat treatment, which is thought to be due to the formation of conjugated double bonds. High-energy electron irradiation, on the other hand, renders polyethylene photoconductive with maximum sensitivity in the near-infrared region. This phenomenon was postulated as being due to radiation-generated donor- and acceptor-type traps.

2.7
Photoconductive polymers produced by plasma polymerization or glow discharge

Various polymeric materials prepared by plasma polymerization or glow discharge become conductive when exposed to UV light. This applies, for example, to a polymer obtained by plasma polymerization of styrene. The polymer was examined as a thin sheet coated with gold layers on both sides [57]. Also, thin polymer layers deposited by glow discharge of tetramethylsilane, tetramethylgermanium or tetramethyltin on conducting substrates were found to be photoconductive in the wavelength region 200–350 nm [58].

References

1 D. Mort, D. Pai (eds.), *Photoconductivity and Related Phenomena*, Elsevier, Amsterdam (1976).
2 D. Mort, N. Pfister (eds.), *Electronic Properties of Polymers*, Wiley-Interscience, New York (1982).
3 V. Mylnikov, *Photoconducting Polymers*, Adv. Polym. Sci, 115 (1994) 1.
4 D. Haarer, *Photoconductive Polymers: A Comparison with Inorganic Materials*, Adv. Solid State Phys. 30 (1990) 157.
5 D. Haarer, Angew. Makromol. Chem. 183 (1990) 197.
6 T.A. Skotheim (ed.), *Handbook of Conducting Polymers*, Marcel Dekker, New York (1986).
7 T.A. Skotheim, R.L. Elsenbaumer, J.R. Reynolds (eds.), *Handbook of Conducting Polymers*, 2nd Edition, Marcel Dekker, New York (1997).
8 G. Zerbi, *Organic Materials for Photonics*, Elsevier Science, Amsterdam (1993).
9 N.S. Sariciftci (ed.), *Primary Photoexcitations in Conjugated Polymers: Molecular Exciton versus Semiconductor Band Model*, World Scientific, Singapore (1997).
10 K.Y. Law, Chem. Rev. 93 (1993) 449.
11 P.M. Borsenberger, D.S. Weiss, *Organic Photoreceptors for Xerography*, Marcel Dekker, New York (1998).
12 P.M. Borsenberger, D.S. Weiss, *Organic Photoreceptors for Imaging Systems*, Marcel Dekker, New York (1993).
13 N.V. Joshi, *Photoconductivity*, Marcel Dekker, New York (1990).
14 H.S. Nalwa (ed.), *Handbook of Organic Conductive Molecules and Polymers*, Vol. 3, Wiley, New York (1997).
15 H.S. Nalwa (ed.), *Handbook of Advanced Electronic and Photonic Materials and Devices*, Academic Press, San Diego (2001).
16 G. Hadziioannou, P.F. van Hutten (eds.), *Semiconducting Polymers*, Wiley-VCH, Weinheim (2000).
17 M. Pope, C.E. Swenberg, *Electronic Processes in Organic Crystals and Polymers*, 2nd ed., University Press, Oxford (1999).
18 D. Fichou (ed.), *Handbook of Oligo- and Polythiophenes*, Wiley-VCH, Weinheim (1998).
19 A. Pron, P. Rannou, *Processible Conjugated Polymers: From Organic Semiconductors to Organic Metals and Superconductors*, Prog. Polym. Sci. 27 (2002) 135.
20 H. Kies, *Conjugated Conducting Polymers*, Springer, Berlin (1992).
21 H. Bässler, Phys. Stat. Sol. B. 175 (1993) 15.

22 G. von Bünau, T. Wolff, *Photochemie, Grundlagen, Methoden, Anwendungen*, VCH, Weinheim (1987).
23 G. Lanzani, S. de Sylvestre, G. Cerullo, S. Stagira, M. Nisoli, W. Graupner, G. Leising, U. Scherf, K. Müllen, *Photophysics of Methyl-Substituted Poly(paraphenylene)-Type Ladder Polymers*, in [16], p. 235.
24 K. Pichler, D. Halliday, D. C. Bradley, P. L. Burn, R. H. Friend, A. B. Holmes, J. Phys. Cond. Matter 5 (1993) 7155.
25 S. Nespurek, V. Herden, W. Schnabel, A. Eckhardt, Czechoslovak J. Phys. 48 (1998) 477.
26 J. Knoester, M. Mostovoy, *Disorder and Solitons in trans-Polyacetylene*, in [16], p. 63.
27 R. H. Friend, D. D. C. Bradley, P. D. Townsend, J. Phys. D: Appl. Phys. 20 (1987) 1367.
28 H. Bässler, *Charge Transport in Random Organic Semiconductors*, in [16], p. 365.
29 M. Redecker, H. Bässler, H. H. Hörhold, J. Phys. Chem. 101 (1997) 7398.
30 M. Lögdlund, W. R. Salaneck, *Electronic Structure of Surfaces and Interfaces in Conjugated Polymers*, in [16], p. 115.
31 L. Onsager, Phys. Rev. 54 (1938) 554.
32 A. Eckhardt, Ph.D. Thesis, Technical University Berlin (1995).
33 V. Cimrova, I. Kminek, S. Nespurek, W. Schnabel, Synth. Metals 64 (1994) 271.
34 A. Mozumder, J. Chem. Phys. 60 (1974) 4300.
35 B. J. Chen, C. S. Lee, S. T. Lee, P. Webb, Y. C. Chan, W. Gambling, H. Tian, W. H. Zhu, Jpn. Appl. Phys. 39 (2000) 1190.
36 A. Eckhardt, V. Herden, S. Nespurek, W. Schnabel, Phil. Mag. B 71 (1995) 239.
37 D. Hertel, U. Scherf, H. Bässler, Adv. Mat. 10 (1998) 1119.
38 M. Redecker, D. D. C. Bradley, M. Inbasekaran, E. P. Woo, Appl. Phys. Lett. 73 (1998) 1565.
39 E. Lebedev, T. Dittrich, V. Petrova-Koch, S. Karg, W. Brütting, Appl. Phys. Lett. 71 (1997) 2686.
40 H. Sirringhaus, P. J. Brown, R. H. Friend, M. M. Nielsen, K. Bechgaard, B. M. W. Langeveld-Voss, A. I. Spiering, R. A. J. Janssen, E. W. Meijer, D. M. de Leeuw, Nature 401 (1999) 685.
41 E. Müller-Horsche, D. Haarer, H. Scher, Phys. Rev. B 35 (1987) 1273.
42 M. O'Neill, S. M. Kelly, Adv. Mater. 15 (2003) 1135.
43 M. Redecker, D. D. C. Bradley, M. Inbasekaran, E. P. Woo, Appl. Phys. Lett. 74 (1998) 1400.
44 F. C. Grozema, L. D. A. Siebbeles, J. M. Warman, S. Seki, S. Tagawa, U. Scherf, Adv. Mater. 14 (2002) 228.
45 P. Vlachos, S. M. Kelly, B. Mansoor, M. O'Neill, Chem. Commun. (2002) 874.
46 M. Abkowitz, H. Bässler, M. Stolka, Phil. Mag. B 63 (1991) 201.
47 W. D. Gill, J. Appl. Phys. 43 (1972) 5033.
48 V. I. Arkhipov, P. Heremans, E. V. Emelianova, G. J. Andriaenssens, H. Bässler, Appl. Phys. Lett. 82 (2003) 3245.
49 S. Nespurek, Macromol. Symp. 104 (1996) 285.
50 V. I. Arkhipov, J. Reynaert, Y. D. Jin, P. Heremans, E. V. Emelianova, G. J. Andriaenssens, H. Bässler, Synth. Met. 138 (2003) 209.
51 V. I. Arkhipov, P. Heremans, E. V. Emelianova, G. J. Andriaenssens, H. Bässler, Chem. Phys. 288 (2003) 51.
52 P. Borsenberger, G. Contois, A. Ateya, J. Appl. Phys. 50 (1979) 914.
53 G. Pfister, Phys. Rev. B 16 (1977) 3676.
54 P. G. Di Marco, G. M. Gleria, S. Lora, Thin Solid Films 135 (1986) 157.
55 A. Eckhardt, V. Herden, W. Schnabel, *Photoconductivity in Polysilylenes: Doping with Electron Acceptors*, in N. Auner, J. Weis (eds.), *Organosilicon Chemistry III*, Wiley-VCH, Weinheim (1997), p. 617.
56 B. Kraabel, C. H. Lee, D. McBranch, D. Moses, N. S. Sariciftci, A. J. Heeger, Chem. Phys. Lett. 213 (1993) 389.
57 S. Morita, M. Shen, J. Polym. Sci. Phys. Ed. 15 (1977) 981.
58 N. Inagaki, M. Mitsuuchi, Polym. Sci. Lett. Ed. 22 (1978) 301.

3
Electro-optic and nonlinear optical phenomena

3.1
Introductory remarks

Electro-optic (EO) phenomena are related to the interaction of an electric field with an optical process. The classical electro-optic effects, the Pockels and the Kerr effect, discovered in 1893 and 1875 with quartz and carbon disulfide, respectively, refer to the induction of birefringence in certain materials under the influence of an external electric field. Application of an electric field to the sample causes a change in the refractive index. In the case of the Pockels effect, Δn is linearly proportional to E, the strength of the applied electric field [see Eq. (3-1)]. Hence, it is also called the linear electro-optic effect. In contrast, Δn is proportional to E^2 in the case of the Kerr effect [see Eq. (3-2)].

Linear electro-optical effect – *Pockels effect*: $\Delta n = rE$ (3-1)

Quadratic electro-optical effect – *Kerr effect*: $\Delta n = q\lambda E^2$ (3-2)

r (m V^{-1}) and q (m V^{-2}) are the Pockels and the Kerr constants, respectively, E is the electric field strength (V m^{-1}), and λ (m) is the wavelength of the light.

Pockels cells containing an appropriate crystal, such as potassium dihydrogen phosphate, and Kerr cells containing an appropriate liquid, e.g. nitrobenzene, are used as light shutters (in conjunction with polarizers) and intensity modulators of linearly polarized laser light beams. Actually, the technical importance of EO effects is increasing because of various applications in optical communication devices, particularly concerning EO modulators, that are used in fiber-optic communication links. In the search for novel EO materials, organic compounds and particularly polymeric systems have also been explored. While polymers are cheap and easily processable, many of them are inferior to inorganic crystals because of their low thermal stabilities. Therefore, the application potential of polymeric systems is limited. Nevertheless, a large volume of research has been devoted to the use of polymers in photonic devices based on EO effects. Some highlights regarding the achievements in this field are reported in this chapter.

It should be emphasized that the Kerr effect refers to a quadratic, i.e. a nonlinear dependence of the refractive index on the strength of the externally ap-

Polymers and Light. Fundamentals and Technical Applications. W. Schnabel
Copyright © 2007 WILEY-VCH Verlag GmbH & Co. KGaA, Weinheim
ISBN: 978-3-527-31866-7

plied electric field. In this respect, the Kerr effect is the first nonlinear optical phenomenon that has gained both fundamental and practical importance. The interest in nonlinear phenomena flourished after the construction of the first ruby laser in 1960 by T.H. Maiman [1] and the observation of second harmonic generation (SHG), i.e. frequency doubling of laser light, in 1961 [2]. Since then, the field of nonlinear optics has developed very rapidly, as demonstrated by a plethora of articles and books. To a large extent, these also cover research on organic materials, including polymers [3–14].

3.2
Fundamentals

3.2.1
Electric field dependence of polarization and dipole moment

Electric field-induced changes in refractive index can be explained with the aid of the following model: under the influence of the electric field, the charge distribution in the molecules is perturbed and the molecules are polarized. The dipole moment p_i induced by an electric field along the molecular axis can be expressed by an expansion [see Eq. (3-3)] [15].

$$p_i = \mu_0 + \sum_j a_{ij} E_j + \sum_{jk} \beta_{ijk} E_j E_k + \sum_{jkl} \gamma_{ijkl} E_j E_k E_l + \ldots \qquad (3\text{-}3)$$

Here, μ_0 denotes the permanent dipole moment. The coefficients are tensors termed as linear polarizability, a_{ij}, and first and second molecular hyperpolarizabilities, β_{ijk} and γ_{ijkl}, respectively. The indices refer to the tensor elements expressed in the frame of the molecule using Cartesian coordinates. E_j, E_k, and E_l denote the applied electric field strength components. Commonly, the response time ranges from picoseconds to femtoseconds. Therefore, if an alternating electric field with a frequency of less than 10^{12} Hz is applied, the direction of the polarization alternates with the oscillations of the applied field.

The polarization induced at the molecular level can cause a polarization in the bulk of the sample and lead to macroscopically detectable property changes, for instance in the refractive index. The macroscopic polarization P_I induced by the electric field can be expressed by the expansion given by Eq. (3-4).

$$P_I = P_0 + \sum_J \chi^{(1)}_{IJ} E_J + \sum_{JK} \chi^{(2)}_{IJK} E_J E_K + \sum_{JKL} \chi^{(3)}_{IJKL} E_J E_K E_L + \ldots \qquad (3\text{-}4)$$

Here, P_0 is the permanent polarization, and $\chi^{(2)}$ and $\chi^{(3)}$ denote the second- and third-order nonlinear optical three-dimensional susceptibility tensors. The indices attached to the χ tensors refer to the tensor elements, and the indices associated with the E values refer to the components of the electric field strength, here expressed in the laboratory frame.

In the case of weak applied fields, the higher terms in Eq. (3-4) can be neglected and, if the sample is not permanently polarized, Eq. (3-4) reduces to Eq. (3-5).

$$P_{linear} = \chi^{(1)} E \tag{3-5}$$

If the medium is isotropic, $\chi^{(1)}$ is a scalar, i.e. the relationship between E and P_{linear} is independent of the direction of the field vector E and the polarization is parallel to E. Many polymers possess amorphous structures and their optical properties are isotropic. However, electro-optic polymeric systems containing polar moieties can be made anisotropic by orienting these moieties, for example, by electric field-induced or optical alignment. In this case, the polarization is not necessarily parallel to the direction of E and its component in one direction is related to the field components in all three directions:

$$\begin{aligned} P_X &= \chi_{11} E_X + \chi_{12} E_Y + \chi_{13} E_Z \\ P_Y &= \chi_{21} E_X + \chi_{22} E_Y + \chi_{23} E_Z \quad \Leftrightarrow \quad P_I = \sum_J \chi_{IJ} E_J \\ P_Z &= \chi_{31} E_X + \chi_{32} E_Y + \chi_{33} E_Z \end{aligned} \tag{3-6}$$

Note that the indices X, Y, and Z, expressed in upper-case letters, represent the coordinates of the macroscopic laboratory frame. As indicated in Fig. 3.1, lower-case letters are used to denote the coordinates of the molecular frame.

The susceptibility of an anisotropic medium is represented by a tensor. Tensors are composed of 3^{a+1} elements, where a is the number of interacting vectors and $a+1$ denotes the rank. With $a=1$, $\chi^{(1)}$ is a second-rank tensor with $3^2 = 9$ elements, which can be expressed by the matrix given in Eq. (3-7).

$$\chi^{(1)} = \begin{pmatrix} \chi_{11} & \chi_{12} & \chi_{13} \\ \chi_{21} & \chi_{22} & \chi_{23} \\ \chi_{31} & \chi_{32} & \chi_{33} \end{pmatrix} \tag{3-7}$$

Polarization can be induced in matter not only by an externally applied electric field, but also by the electric field of a passing light beam. This kind of interaction does not lead to a loss of intensity of the beam, in contrast to absorption, which reduces the intensity. The overall situation, taking into account both

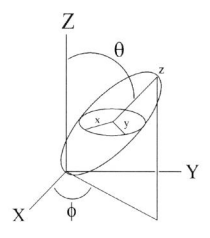

Fig. 3.1 The macroscopic laboratory frame (X, Y, Z) and the molecular frame (x, y, z). Adapted from Kippelen et al. [15] with permission from Springer.

kinds of interaction, i.e. polarization and absorption, can be described on the basis of complex and frequency-dependent entities consisting of a real and an imaginary part. This concerns the dielectric constant, the optical susceptibility, and the refractive index. For example, the complex refractive index n' [see Eqs. (3-8) and (3-9)] is given by the sum of the real part n and the imaginary part ik', the latter corresponding to light absorption [15]:

$$n' = n + ik' \tag{3-8}$$

$$a' = \frac{2k'\omega}{c} \tag{3-9}$$

Here, a' (cm^{-1}) is the linear absorption coefficient, ω (s^{-1}) is the frequency of the optical field, and c (cm s^{-1}) is the speed of light.

When a high-intensity laser beam impinges on material, its electromagnetic field induces electrical polarization that gives rise to a variety of nonlinear optical properties, because in this case the higher terms in Eq. (3-4) are not negligible. The determination of the coefficients $\chi^{(2)}$ and $\chi^{(3)}$ that serve to characterize the nonlinear properties is complicated by the fact that they are composed of many elements. With a being equal to two and three, $\chi^{(2)}$ and $\chi^{(3)}$ are composed of $3^{a+1} = 27$ and 81 elements, respectively. Fortunately, these tensors possess symmetry properties that can be invoked to reduce the number of independent elements, for instance, when the optical frequencies involved in the nonlinear interaction are far away from resonance (absorption) [15].

In the case of second harmonic generation, for example, the second-order susceptibility tensor elements are symmetrical in their last two indices. Therefore, the number of independent tensor elements is reduced from 27 to 18. Moreover, the tensor elements $\chi^{(2)}_{IJK}$ can be expressed in contracted form $\chi^{(2)}_{IL}$. The index I takes the value 1, 2 or 3, corresponding to the three Cartesian coordinates, and the index L varies from 1 to 6. The values of L refer to the six different combinations of the indices J and K according to the following convention [15]:

L:	1	2	3	4	5	6
J,K	1,1	2,2	3,3	2,3 or 3,2	1,3 or 3,1	1,2 or 2,1

Therefore, $\chi^{(2)}$ can be expressed by the matrix given by Eq. (3-10).

$$\bar{\chi}^{(2)} = \begin{pmatrix} \chi^{(2)}_{11} & \chi^{(2)}_{12} & \chi^{(2)}_{13} & \chi^{(2)}_{14} & \chi^{(2)}_{15} & \chi^{(2)}_{16} \\ \chi^{(2)}_{21} & \chi^{(2)}_{22} & \chi^{(2)}_{23} & \chi^{(2)}_{24} & \chi^{(2)}_{25} & \chi^{(2)}_{26} \\ \chi^{(2)}_{31} & \chi^{(2)}_{32} & \chi^{(2)}_{33} & \chi^{(2)}_{34} & \chi^{(2)}_{35} & \chi^{(2)}_{36} \end{pmatrix} \tag{3-10}$$

For poled polymers that belong to the ∞mm symmetry group, some of the tensor elements vanish and the $\chi^{(2)}$ tensor reduces to Eq. (3-11) [15].

$$\bar{\chi}^{(2)} = \begin{pmatrix} 0 & 0 & 0 & 0 & \chi^{(2)}_{15} & 0 \\ 0 & 0 & 0 & \chi^{(2)}_{15} & 0 & 0 \\ \chi^{(2)}_{31} & \chi^{(2)}_{32} & \chi^{(2)}_{33} & 0 & 0 & 0 \end{pmatrix} \qquad (3\text{-}11)$$

When Kleinman symmetry $\left(\chi^{(2)}_{ijk} = \chi^{(2)}_{ikj} = \chi^{(2)}_{jkl} = \chi^{(2)}_{jik} = \chi^{(2)}_{kij} = \chi^{(2)}_{kji}\right)$ is valid [16], $\chi^{(2)}_{15}$ is equal to $\chi^{(2)}_{31}$. Therefore, only two independent tensor elements, namely $\chi^{(2)}_{31}$ and $\chi^{(2)}_{33}$, remain. Methods that are commonly applied to determine macroscopic susceptibilities are based on geometrical arrangements permitting the usage of these simplifications. Regarding the relationship between the macroscopic susceptibilities and the molecular hyperpolarizabilities, equations have been derived for the practically very important case of rigid, polar moieties containing polymeric systems that have been or are subject to an alignment process [15]. It is beyond the scope of this book to treat this subject in detail. A typical result concerning the relation of $\chi^{(2)}$ to β is given by Eqs. (3-12) and (3-13) [17]. In this case, it was assumed that the macroscopic susceptibility of a given volume is the sum of all corresponding molecular contributions in this volume and that each molecular component is mapped onto the corresponding macroscopic vector.

$$\chi^{(2)}_{ZZZ} = NF\beta_{zzz}\langle \cos^3 \Theta \rangle \qquad (3\text{-}12)$$

$$\chi^{(2)}_{XXZ} = \chi^{(2)}_{YYZ} = \chi^{(2)}_{XZY} = \chi^{(2)}_{YZY} = \chi^{(2)}_{ZXX} = \chi^{(2)}_{ZYY} = \frac{1}{2} NF\beta_{zzz}\langle \cos \Theta \sin^2 \Theta \rangle \qquad (3\text{-}13)$$

Here, N is the number of hyperpolarizable groups per unit volume (number density), F is a factor correcting for local field effects, and Θ is the angle between the permanent dipole μ_0 of the molecule (z direction) and the direction of the poling field (Z direction). The brackets indicate an averaging over all molecular orientations weighted by an orientational distribution function.

The importance of the hyperpolarizability and susceptibility values relates to the fact that, provided these values are sufficiently large, a material exposed to a high-intensity laser beam exhibits nonlinear optical (NLO) properties. Remarkably, the optical properties of the material are altered by the light itself, although neither physical nor chemical alterations remain after the light is switched off. The quality of nonlinear optical effects is crucially determined by symmetry parameters. With respect to the electric field dependence of the vector P given by Eq. (3-4), second- and third-order NLO processes may be discriminated, depending on whether $\chi^{(2)}$ or $\chi^{(3)}$ determines the process. The discrimination between second- and third-order effects stems from the fact that second-order NLO processes are forbidden in centrosymmetric materials, a restriction that does not hold for third-order NLO processes. In the case of centrosymmetric materials, $\chi^{(2)}$ is equal to zero, and the nonlinear dependence of the vector P is solely determined by $\chi^{(3)}$. Consequently, third-order NLO processes can occur with all materials, whereas second-order optical nonlinearity requires non-centrosymmetric materials.

The significances of the susceptibilities $\chi^{(1)}$, $\chi^{(2)}$, and $\chi^{(3)}$ are related to specific phenomena. $\chi^{(1)}$ relates to optical refraction and absorption. Common effects related to $\chi^{(2)}$ are frequency doubling (second harmonic generation, SHG) and the linear electro-optic effect (Pockels effect). Typical effects connected with $\chi^{(3)}$ are frequency tripling (third harmonic generation, THG), sum and difference frequency mixing, two-photon absorption, and degenerate four-wave mixing.

3.2.2
Electric field dependence of the index of refraction

Regarding light frequencies in the non-resonant regime, electro-optic (EO) activity relates to the control of the index of refraction of a material by application of an external electric field. Either DC or AC (ranging from 1 Hz to more than 100 GHz) voltages are applied. The index of refraction, n, corresponds to the speed of light, c, in the material ($n = c_0/c$, with c_0 being the speed of light in vacuo). Therefore, the electro-optic activity relates to a voltage-controlled phase shift of the light. The change in the refractive index of a non-centrosymmetric material in a modulating electric field E can be represented by the expansion given by Eq. (3-14) [18].

$$\Delta n_{IJ} = \frac{1}{2} n_{IJ}^3 r_{IJK} E_K + \frac{1}{2} n_{IJ}^3 p_{IJKK} E_K^2 + \ldots \quad (3\text{-}14)$$

Provided that higher terms are negligible, Eq. (3-14) reduces to Eq. (3-15), which relates to the Pockels effect.

$$\Delta n_{IJ} = \frac{1}{2} n_{IJ}^3 r_{IJK} E_K \quad (3\text{-}15)$$

The susceptibility tensor $\chi_{IJK}^{(2)}$ is related to the Pockels tensor r_{IJK} [Eq. (3-16)] [19].

$$\chi_{IJK}^{(2)} = \frac{1}{2} n_I^4 r_{IJK} \quad (3\text{-}16)$$

$\chi_{IJK}^{(2)}$ is invariant under permutation of the first two indices. Therefore, a condensed notation, resulting in only two indices, L and K, can be used. The first index L represents the combination IJ and may have the value 1=XX, 2=YY, 3=ZZ, 4=YZ, 5=ZX, or 6=XY=YX, and the second index K may have the value 1=X, 2=Y or 3=Z [17].

Technical applications based on the Pockels effect require systems that are non-centrosymmetric on a macroscopic level. This relates particularly to polymeric systems containing physically admixed or chemically incorporated components with permanent dipoles. In such cases, macroscopic second-order nonlinearity can be accomplished by poling, i.e. by aligning the permanent dipole moments of the components with the aid of an external electric field that is applied at temperatures in the vicinity of the polymer's glass transition temperature, T_g. The order thus obtained is frozen-in by cooling to a low temperature $T \ll T_g$. The refractive index of the uniformly poled polymer is uniaxial, with a long axis n_e in the poling direction (direction 3) and a short axis n_0 perpendicular to the poling direction (directions 1 and 2). If a modulating electric field is applied in the poling direction, the two Pockels coefficients r_{33} and r_{31} can be discriminated. They are described by Eqs. (3-17) and (3-18) in relation to the susceptibilities $\chi^{(2)}_{333}$ and $\chi^{(2)}_{311}$, and are related to the hyperpolarizability β through Eqs. (3-12) and (3-13).

$$\Delta n = \frac{1}{2} n_e^3 r_{33} E_{mod} = \frac{\chi^{(2)}_{333} E_{mod}}{n_e} = \frac{NF\beta \langle \cos^3 \Theta \rangle E_{mod}}{n_e} \tag{3-17}$$

$$\Delta n = \frac{1}{2} n_0^3 r_{13} E_{mod} = \frac{\chi^{(2)}_{113} E_{mod}}{n_0} = \frac{NF\beta \langle \frac{1}{2} \cos \Theta \sin^2 \Theta \rangle E_{mod}}{n_0} \tag{3-18}$$

Here, N is the number density of hyperpolarizable groups, Θ is the angle between the permanent dipole μ_0 of the molecule (z direction) and the direction of the poling field (Z direction), and F is a local field factor. Commonly, $\langle \cos^3 \Theta \rangle$ is larger than $\langle 0.5 \cos \Theta \sin^2 \Theta \rangle$. Therefore, the most efficient EO modulation is achieved if r_{33} is used, rather than r_{13} [17].

3.3
Characterization techniques

3.3.1
Second-order phenomena

3.3.1.1 Determination of the hyperpolarizability β

Commonly, two methods are employed to determine the hyperpolarizability β: (1) electric field-induced second harmonic generation, EFISH and (2) hyper-Raleigh scattering, HRS. HRS is applicable to both nonpolar and polar molecules as well as ions, but EFISH applies only to polar, non-ionic molecules. While in the EFISH method only the component of β parallel to the dipole moment is measured, HRS yields several of the tensor components. In the case of EFISH, one measures $I_{2\omega}$, the intensity of light at frequency 2ω emitted from a solution of the sample that is submitted to an external electric field E_0 and simultaneously irradiated with laser light of frequency ω. Provided that the external electric field is applied along the Z-axis in the laboratory frame and the laser light is polarized along the same axis, the macroscopic polarization $P(2\omega)$ induced in the solution by the electric field of the incident laser wave E_ω is given by Eq. (3-19).

$$P_Z(2\omega) = \chi^{(3)}_{ZZZZ} E_0 E_Z(\omega) E_Z(\omega) \tag{3-19}$$

Here, $\chi^{(3)}_{ZZZZ}$ is the macroscopic third-order susceptibility, which is related to the first and second molecular hyperpolarizabilities, β and γ, by Eq. (3-20) [20]:

$$\chi^{(3)}_{ZZZZ} = NF_{2\omega}F^2_{\omega}F_0\left(\gamma + \frac{\mu\beta_z}{5kT}\right) \qquad (3\text{-}20)$$

N is the number density of chromophoric groups; $F_{2\omega}$, F_ω, and F_0 are local field factors at frequencies 2ω, ω, and zero; μ is the ground-state dipole moment, and β_z is the vectorial component of β along the ground-state dipole moment, taken to be oriented along the z-axis in the molecular framework ($\beta_z = \beta_{zxx} + \beta_{zyy} + \beta_{zzz}$). In the case of π-conjugated chromophores, γ is negligibly small in comparison with $\mu\beta_z/5kT$. Therefore, according to Eq. (3-20), the product $\mu\beta_z$ is directly available from $\chi^{(3)}_{ZZZZ}$, obtained by measuring the intensity $I_{2\omega}$ of the second harmonic generated by sample solutions. $I_{2\omega}$ is proportional to $\chi^{(3)}_{ZZZZ}$ [see Eq. (3-21)].

$$I_{2\omega} \propto \chi^{(3)}_{ZZZZ} I^2_\omega E^2_0 \qquad (3\text{-}21)$$

Commonly, the evaluation of the susceptibility $\chi^{(3)}_{ZZZZ}$ is related to a reference standard. A detailed description of both experimental techniques and data evaluation is given in the article by Singer et al. [20].

In contrast to the EFISH method, the hyperpolarizability β can be measured directly by means of the HRS method developed by Clays and Persoons [21, 22]. This method involves measuring the intensity of the incoherently scattered, frequency-doubled light from isotropic solutions. As shown in Fig. 3.2, an infrared laser beam is focused on the center of a cell containing a solution of the NLO-active compound.

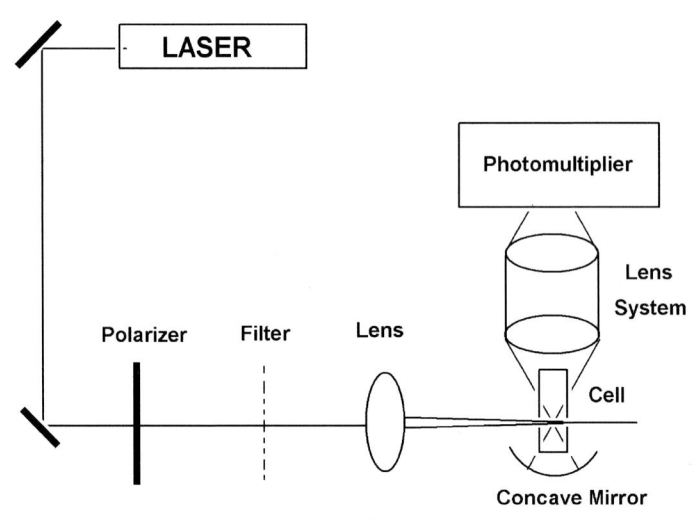

Fig. 3.2 Schematic depiction of a set-up for measuring second-order hyperpolarizability by means of the hyper-Rayleigh scattering method.

The intensity of the scattered light, $I_{2\omega}$, is proportional to the square of the intensity of the incident light, I_ω, as given by Eq. (3-22).

$$I_{2\omega} = g\left(N_1 \left\langle \beta^2_{IJKsolvent} \right\rangle + N_2 \left\langle \beta^2_{IJKsolute} \right\rangle \right) I_\omega^2 \qquad (3\text{-}22)$$

Here, g is a set-up dependent factor; N_1 and N_2 are the number densities of solvent and solute molecules, respectively; and $\left\langle \beta^2_{IJK} \right\rangle$ is the mean value of the square of hyperpolarizability tensor components in the laboratory framework [23]. It must be noted that the HRS process is extremely inefficient. Typically, the number of scattered photons is 10^{-14} times the number of incident photons [20]. In principle, a low output intensity would be expected for an isotropic solution, where the fields emitted from the individual NLO molecules interfere destructively. That a measurable amount of incoherently scattered harmonic light can be generated may be rationalized by assuming that fluctuations in orientation can produce regions of alignment [22]. The rather low intensity of the scattered light requires the application of powerful lasers, such as an Nd-YAG system producing 1064 nm light pulses, in conjunction with a sampling technique involving more than 100 pulses.

3.3.1.2 Determination of the susceptibility $\chi^{(2)}$

Several techniques have been developed for determining the second-order susceptibility $\chi^{(2)}$ [24]. Of practical importance are methods that may be employed for aligned polymeric systems containing polar moieties [4, 8]. Methods making use of the Pockels or linear electro-optic (EO) effect are based on the measurement of the variation in the refractive index of thin polymer films induced by an external electric field. In this way, values of the electro-optic coefficients r_{33} and r_{13} are obtained, which are related to the corresponding $\chi^{(2)}$ values through Eq. (3.16).

A quite direct method for measuring $\chi^{(2)}$ is based on second harmonic generation, SHG. Figure 3.3 depicts a typical set-up used to determine the SHG coefficients d_{31} and d_{33}, defined as $d = \chi^{(2)}/2$, by way of SHG measurements.

A polarized laser beam of frequency ω passes through the polymer sample and an IR-blocking filter. The SHG signal is selected by means of an interference filter operating at the frequency 2ω, and is measured using a photomultiplier tube connected to a boxcar integrator. The intensity, $I_{2\omega}$, is proportional to the square of the SHG coefficient, d, and to the square of the intensity of the fundamental laser beam [see Eqs. (3-23) and (3-24)] [8].

$$I_{2\omega} = K d^2 I_\omega^2 \qquad (3\text{-}23)$$

$$K = \frac{512\pi t_\omega^4 T_{2\omega} t_0^2 p^2 \sin^2 \Psi(\Theta)}{A(n_\omega^2 - n_{2\omega}^2)} \qquad (3\text{-}24)$$

Here, A is the area of the laser beam; Θ is the incident angle; t_0, t_ω, and $T_{2\omega}$ are transmission factors; p is a projection factor; $\Psi(\Theta)$ is an angular factor re-

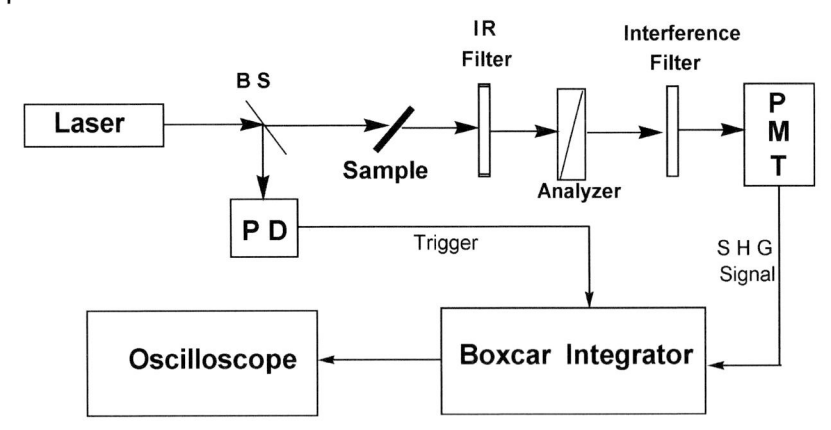

Fig. 3.3 Schematic depiction of a set-up for measuring second harmonic generation (SHG). BS: beam splitter, PD: photodiode, PMT: photomultiplier tube. Adapted from Jerphagnon et al. [25] with permission from the American Institute of Physics.

lated to the sample thickness, the fundamental wavelength, and the refraction angles; and n_ω and $n_{2\omega}$ are the refractive indices of the sample at ω and 2ω. The coefficient, d, of the polymer is obtained by comparing the $I_{2\omega}$ value with that measured for a standard reference sample, commonly Y-cut quartz with $d_{11}=0.49$ pm V^{-1}, at $\lambda=1064$ μm.

3.3.2
Third-order phenomena

Several measuring techniques giving evidence of third-order nonlinear behavior are listed in Table 3.1 [26, 27].

It is difficult to compare the third-order susceptibilities of systems examined using different measuring techniques. Since they are based on fundamentally different origins, they do not yield identical $\chi^{(3)}$ values. Different nonlinear mechanisms contribute in a specific manner to $\chi^{(3)}$, and values measured for the same material by different techniques may differ by several orders of magnitude. This applies, for instance, to the case of the combined resonant and non-resonant interaction of light with matter. A full expression of $\chi^{(3)}$ reflects non-resonant and resonant contributions [see Eq. (3-25)].

$$\chi^{(3)} = \chi^{(3)}_{\text{nonresonant}} + \chi^{(3)}_{\text{resonant}} \tag{3-25}$$

Resonance occurs at wavelengths around that of the absorption band. Moreover, the strong frequency (wavelength) dependence of $\chi^{(3)}$ and the influence of repetition frequency and pulse duration of the laser on $\chi^{(3)}$ have to be taken into account. It is beyond the scope of this book to describe the various measuring

3.3 Characterization techniques

Table 3.1 Measuring techniques for third-order susceptibilities.

Method	Acronym	Denotation of process
Third harmonic generation	THG	$\chi^{(3)}(3\omega;\omega,\omega,\omega)$
Z-scan		$\chi^{(3)}(-\omega;\omega,-\omega,\omega)$
Two-photon absorption	TPA	$\chi^{(3)}(-\omega;\omega,-\omega,\omega)$
Degenerate four-wave mixing	DFWM	$\chi^{(3)}(-\omega;\omega,\omega,-\omega)$
Electric field-induced second harmonic generation	EFISH	$\chi^{(3)}(-2\omega;\omega,\omega,0)$
Optical Kerr gate	OKG	$\chi^{(3)}(-\omega;\omega,\omega,-\omega)$

techniques. However, some of the most widely used methods are briefly considered below, with the additional aim of providing some insight into the fascinating field of third-order nonlinear effects.

3.3.2.1 Third harmonic generation

The term *third harmonic generation*, THG, refers to the generation of a light beam that consists of photons having three times the energy of the photons of the input beam. THG can be easily detected and is, therefore, widely employed in the third-order nonlinear characterization of newly developed materials [28]. THG is a four-photon process, in which three incident photons with angular frequency ω create a photon with frequency 3ω. The off-resonant THG process can be represented by a transition between virtual excited states, as shown by the dashed lines in Fig. 3.4.

In the case of THG, the third-order susceptibility corresponds to a nonlinear polarization component, which oscillates at the third harmonic frequency of the incident laser beam. Regarding the simplified case of an isotropic solution, only the element $\chi^{(3)}_{XXXX}(-3\omega;\omega,\omega,\omega)$ of the third-order susceptibility tensor creates a polarization at 3ω, which is parallel to the incident electrical field E_ω, assumed to be parallel to the X-axis [see Eq. (3-26)].

$$P_{3\omega} = \frac{1}{4}\chi^{(3)}_{XXXX}(-3\omega;\omega,\omega,\omega)E_\omega^3 \qquad (3\text{-}26)$$

For THG measurements, pulsed laser systems operating at infrared wavelengths (typically 1064, 1850, 1907 or 2100 nm) are used. Most commonly, $\chi^{(3)}_{XXXX}$ is obtained by relating the third-harmonic signal of the sample to that measured si-

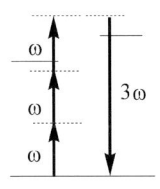

Fig. 3.4 Energy level diagram illustrating third harmonic generation. Arrows denote photon energies, horizontal solid lines represent energy states of the medium, and dashed lines represent virtual excited states.

multaneously with a fused silica plate serving as a reference. The incident beam is focused on the sample in a vacuum chamber and a water filter removes the fundamental frequency from the output beam, which is further attenuated so that it lies within the linear range of the photomultiplier.

3.3.2.2 Self-focusing/defocusing

Thin polymer sheets allowing unhindered passage of a low-intensity light beam of a given non-resonant wavelength can act as lenses if a high-intensity beam is passed through them. This is a consequence of the intensity dependence of the refractive index n [see Eq. (3-27)].

$$n = n_0 + n_2 I \tag{3-27}$$

Here, n_0 denotes the linear refractive index (at low intensity I) and n_2 is the nonlinear refractive index, which can be measured by means of a Z-scan experiment [29, 30]. A typical set-up is schematically depicted in Fig. 3.5 a.

The incoming beam is split into two equal parts: one part is guided to the detector D_1, while the other is passed through the sample and an aperture prior to reaching the detector D_2. Provided that the sample is nonlinearly active, the phenomena outlined below will be observed if the sample is moved through the focused laser beam along the optical axis. Thus, the transmission through the aperture is reduced if the sample is moved to the left of the original focus z_0, because the beam is defocused. On the other hand, if the sample is placed to the

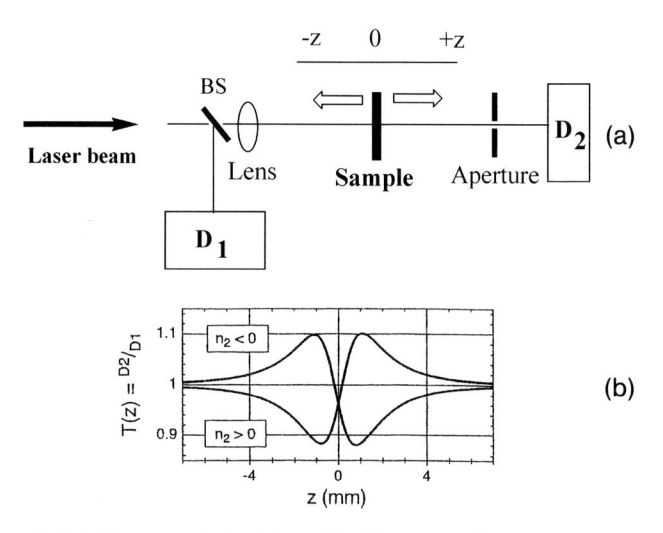

Fig. 3.5 (a) Schematic depiction of the Z-scan experiment. BS: beam splitter. (b) Typical Z-scan curves for $n_2>0$ and $n_2<0$. Adapted from Gubler et al. [30] with permission from Springer.

right of z_0, the beam is focused on the aperture and the transmission through it is increased. This applies in the case of $n_2>0$. The opposite behavior is observed if $n_2<0$. Both cases are shown schematically in Fig. 3.5b, in which the signal ratio D_2/D_1 is plotted against the distance z. The nonlinear refractive index n_2 can be obtained from the z-scan in the following way: ΔT_{pv}, the difference in the transmittance between peak and valley, is proportional to the phase distortion, $\Delta \Phi_0$, according to the empirical relationship $\Delta T_{pv} = k \Delta \Phi_0$, where k is a constant determined by the lay-out of the apparatus. With $\Delta \Phi_0 = (2\pi/\lambda) n_2 I_0 L$, one obtains Eq. (3-28) [29].

$$n_2 = \frac{\Delta T_{pv} \lambda}{2\pi k I_0 L} \tag{3-28}$$

Here, I_0 and L denote the light intensity and the thickness of the sample, respectively. The third-order susceptibility $\chi^{(3)}$ can then be obtained by using Eq. (3-29) [26].

$$n_2 = \frac{12\pi^2}{cn_0} \chi^{(3)} \tag{3-29}$$

This applies when esu units are used for both n_2 and $\chi^{(3)}$. It is interesting to note that the set-up shown in Fig. 3.5a can also be used to determine the two-photon absorption coefficient a_2. In this case, the Z-scan experiment is performed without the aperture.

3.3.2.3 Two-photon absorption (TPA)

The simultaneous absorption of two photons of equal energy can occur if a laser beam (ps or fs pulses) is focused within a material [31, 32]. The process, depicted schematically in Fig. 3.6, is related to the excitation of a molecule to an energy level $h\nu_1 = 2\ h\nu_2$ by the simultaneous absorption of two photons of energy $h\nu_2$ ($\nu = \omega/2\pi$).

Two-photon absorption is possible, provided that both photons are spatially and temporally coincident. It occurs with a probability proportional to the square of the light intensity.

TPA can be measured by the transmission method or by the Z-scan technique. Moreover, two-photon fluorescence can serve to measure TPA absorption cross-sections, provided that a fluorescent excited state is reached by TPA. In nonlinear

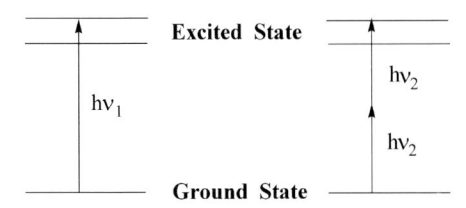

Fig. 3.6 Energy level diagram depicting single-photon and two-photon absorptions.

transmission experiments, the transmission of the sample, T_r, is measured as a function of the input intensity, I_0. At high incident intensities, TPA is proportional to I_0^2 and there is a linear relationship between $1/T_r$ and I_0 [see Eq. (3-30)].

$$\frac{1}{T_r} = \frac{I_0}{I} = 1 + a_2 I_0 L \tag{3-30}$$

Here, L is the sample thickness and a_2 is the absorption coefficient for the pure two-photon absorption process.

3.3.2.4 Degenerate four-wave mixing (DFWM) and optical phase conjugation

Degenerate four-wave mixing (DFWM) is frequently employed to measure $\chi^{(3)}$ values and response times of polymeric systems. The DFWM technique is based on the interaction between three spatially distinguishable light beams of equal frequency ω. The interaction results in the generation of a fourth beam of the same frequency ω. Figure 3.7 shows the commonly used backward-wave geometry, with three incident beams spatially overlapping in the sample.

The pump beams 1 and 3 are counterpropagating. The signal beam 4 is emitted in the direction opposite to the probe beam 2. Its intensity depends on $\chi^{(3)}$ and on the intensities of beams 1, 2, and 3 according to Eq. (3-31) [27].

$$I_4 = \frac{\omega^2}{4c^2 n^2} (\chi^{(3)})^2 L^2 I_1 I_3 I_2 \tag{3-31}$$

Here, c, n, and L denote the velocity of light in vacuo, the refractive index of the sample, and the pathlength in the medium, respectively. Equation (3-31) holds in the case of there being no linear or nonlinear light absorption. The retracement of the probe beam is characteristic of the phenomenon of *optical phase conjugation*, OPC [33]. This refers to the property of materials to act as mirrors and to reflect an incident light beam exactly in phase with its former phase. Unlike a conventional mirror, whereby rays are redirected according to the ordinary law of reflection, a phase conjugate mirror, also called a phase conjugator, retroreflects all incoming rays back to their origin. Figure 3.8 illustrates the difference between a conventional and a phase conjugate mirror.

At a conventional mirror, only the wave vector component normal to the surface changes sign, while the tangential components remain unchanged. The propagation direction of the reflected ray depends on the angle between the surface normal

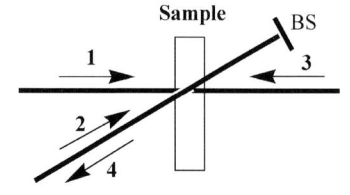

Fig. 3.7 Degenerating four-wave mixing with counterpropagating pump beams 1 and 3. BS: beam stopper.

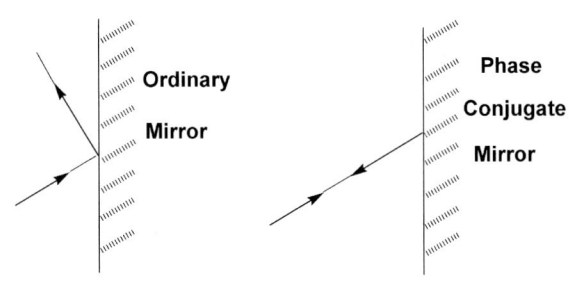

Fig. 3.8 The reflection of a ray of light off an ordinary mirror and off a phase conjugate mirror.

and the incident ray. A phase conjugate mirror, on the other hand, changes the sign of the complex wave vector so that the reflected ray is antiparallel to the incident ray. Phase conjugation by degenerate four-wave mixing may result in reflectivities $R = I_4/I_2$ exceeding 100%. For example, using picosecond pulses, $R = 2.5$ has been found for poly(methyl methacrylate) doped with 5×10^{-4} mol L^{-1} rhodamine 6G [34]. For detailed information concerning the DFWM technique and additional techniques not dealt with here, the reader is referred to the literature [26, 27].

3.4
Nonlinear optical materials

3.4.1
General aspects

Second-order NLO materials. Originally, second-order nonlinear optics was developed with the aid of inorganic crystals such as lithium niobate, LiNbO$_3$, and potassium dihydrogen phosphate, KH$_2$PO$_4$ (KDP). The nonlinear optical behavior of these crystals is due to light-induced displacement of the ions in the lattice. Certain organic substances having a non-centrosymmetric structure and containing delocalized π-electrons behave similarly. They undergo very fast, light-induced, intramolecular perturbations of their charge distributions. In other words, irradiation with light at non-resonant wavelengths causes an almost instantaneous shift in the π-electron density over the molecule, which accounts for the large and fast polarization. 2-Methyl-4-nitroaniline, MNA, and 4-dimethylamino-4'-nitrostilbene, DANS, are typical organic compounds exhibiting second-order NLO activity (see Chart 3.1).

These compounds are so-called charge-transfer molecules having the general structure shown in Chart 3.2.

Here, an electron-donating and an electron-accepting moiety are connected by an extended π-electron system. In such compounds, the electron displacement occurs on a subpicosecond time scale and can be much more pronounced than in inorganic crystals. Polymeric organic systems are of practical importance.

MNA **DANS**

Chart 3.1 Chemical structures of 2-methyl-4-nitroaniline, MNA, and 4-dimethylamino-4'-nitrostilbene, DANS.

ACCEPTOR ——[π-conjugated system]—— DONOR

Chart 3.2 General structure of charge-transfer molecules (AπD molecules).

They consist either of polymers containing admixed AπD compounds or of polymers with AπD moieties chemically incorporated into the main chain or in pendant groups. As pointed out above, for an organic material to undergo a significant change in its dipole moment upon exposure to an intense light beam, it needs to have a non-centrosymmetric molecular structure. This requirement also pertains to the macroscopic level. In other words, both a large hyperpolarizability β of the molecular constituents and a large macroscopic susceptibility $\chi^{(2)}$ are required. Macroscopic non-centrosymmetry can be attained by aligning the assemblies so that the individual tensor components of β add constructively.

Third-order NLO materials. Unlike for second-order NLO activities, there are no molecular symmetry restrictions for the third-order nonlinear response of materials. In principle, all materials, including air, are capable of exhibiting third-order NLO activity. Generally, for most centrosymmetric compounds, the hyperpolarizability γ is very small. This does not apply, however, for organic π-conjugated compounds. It is the almost instantaneous shift in π-electron density over the whole molecule or extended parts of it that occurs upon irradiation which accounts for the large susceptibilities $\chi^{(3)}$ of conjugated compounds. As regards the field of macromolecules, π-conjugated polymers such as polyacetylenes or polydiacetylenes (see Chart 3.3) exhibit pronounced third-order NLO activities. $\chi^{(3)}$ values of non-conjugated polymers such as poly(methyl methacrylate) are several orders of magnitude lower than those of conjugated polymers.

Polyacetylene   Polydiacetylene   Polysilane

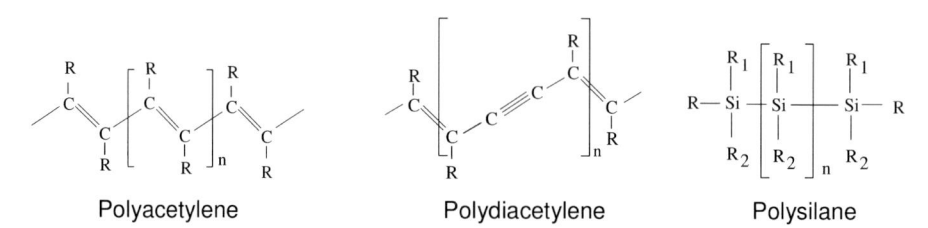

Chart 3.3 Polymers exhibiting third-order NLO activities. R, R_1, and R_2 denote aliphatic or aromatic groups.

Interestingly, σ-conjugated polymers such as polysilanes (see Chart 3.3) also exhibit remarkably large third-order susceptibilities $\chi^{(3)}$.

3.4.2
Second-order NLO materials

3.4.2.1 Guest-host systems and NLO polymers
Fundamentally, there are two categories of second-order NLO polymeric systems, commonly also referred to as electro-optically active polymeric systems [4, 35]: (1) guest-host systems consisting of rigid solutions of small AπD compounds in polymeric matrices, and (2) systems consisting of polymers with AπD moieties incorporated into either the main chain or side groups [36]. In the latter case, the rigidity of the polymeric matrix can be improved by chemical crosslinking. General structures of such polymers are depicted in Fig. 3.9.

In this context, research concerning non-centrosymmetric structures with supramolecular helical organization is interesting. In the case of thermally stable (up to 400 °C) polyesters containing π-conjugated donor-acceptor segments (see Chart 3.4), the hyperpolarizability values, β, turned out to be much larger than those of the respective monomeric chromophores.

At a chiral unit content of 50%, the second harmonic generation (SHG) efficiency of the polymer (at $\lambda=532$ nm) is 4.8 times that of the monomer, and β is equal to 20.7×10^{-30} esu. This enhancement may be rationalized in terms of the directional orientation of dipole segments in the polymer as a consequence of the chiral organization of the polymer chains [37].

Typical low molar mass AπD compounds and polymers containing AπD moieties are listed in Table 3.2 [38] and Table 3.3 [39, 40]. In this context, it is noticeable that electro-optically active compounds have been tabulated [7].

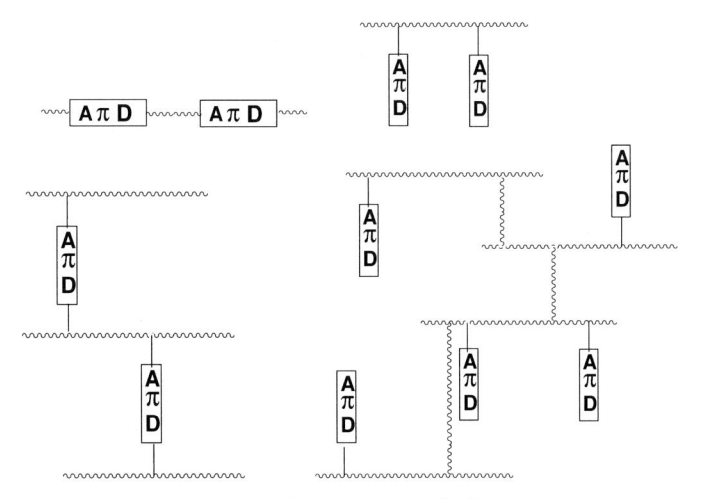

Fig. 3.9 Schematic depiction of the structures of polymeric matrices containing AπD moieties.

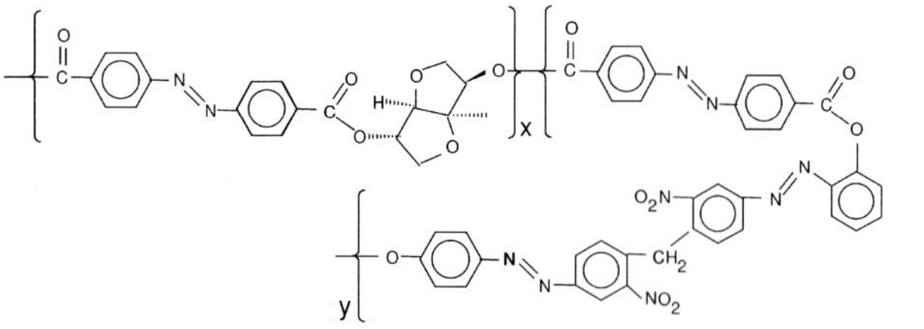

3 Electro-optic and nonlinear optical phenomena

Polyester

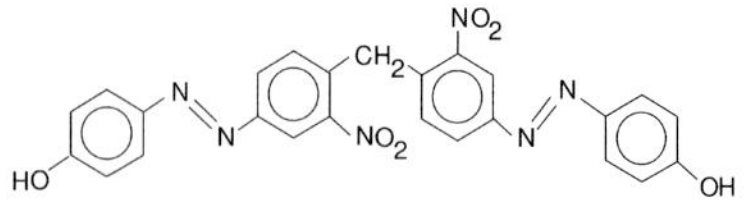

Monomer

Chart 3.4 Chemical structures of an electro-optically active polyester and a chemically related monomer.

Table 3.2 Characteristics of electro-optically active chromophores determined in chloroform solution. Adapted from Swalen and Moylan [38].

Denotation	Structure	λ_{max} [a] (nm)	μ [b] (Debye)	β_0 [c] (10^{-30} esu)
I	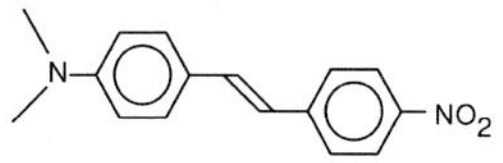	438	6.7	81.3
II	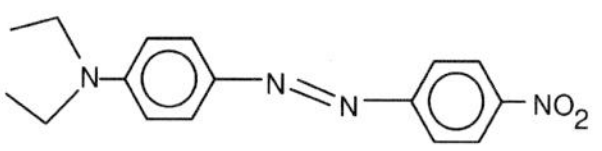	494	8.0	95.2
III	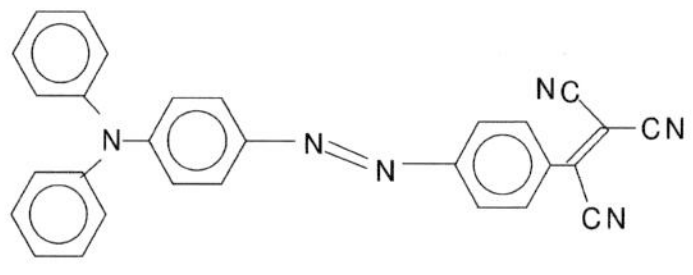	602	7.1	259
IV	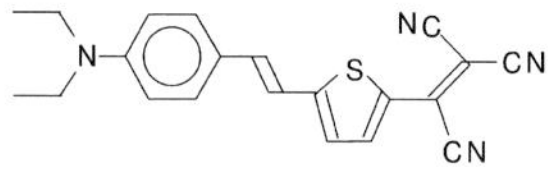	698	10.4	359
V	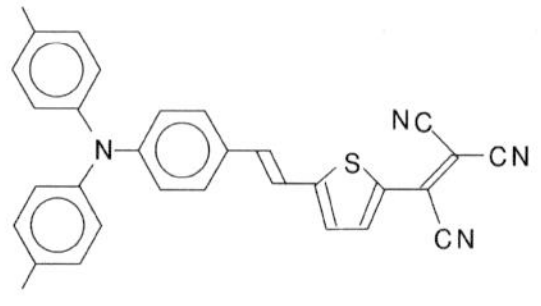	680	8.3	479

a) Wavelength of maximum.
b) Dipole moment.
c) Off-resonance hyperpolarizability.

Table 3.3 Characteristics of electro-optically active poled polymer films. Adapted from Bertram et al. [39] and Lipscomb et al. [40].

Chemical structure	Acronym	f_{ac} [a]	T_{Pol} [b] (°C)	λ [c] (µm)	r_{33} [d] (pm V^{-1})
	Ber-1		100	1.55	42
	3RDCYXY	15 mol%	140	1.3	30
	GT-P3	62 wt%	180	1.541	12
	ROI-4	17 mol%	215	1.3	16

a) Fraction of active compound.
b) Poling temperature.
c) Wavelength.
d) Component of the Pockels coefficient tensor directed parallel to the applied electric field.

At present, various compounds are commercially available [41]. Typical examples are given in Table 3.4.

Second-order NLO polymers have potential for technical applications (see Section 3.5 below), for example, in electro-optic modulation and switching or frequency doubling. A large body of compounds has hitherto been explored and, at present, relevant research is mainly focused on optimizing secondary properties such as thermal stability, adhesion, thermal expansion, etc.

Table 3.4 Commercially available NLO polymers [41].

Denotation	Chemical Structure
Poly[4-(2,2-dicyanovinyl)-N-bis(hydroxyethyl)aniline-alt-(isophoronediisocyanate)]urethane	
Poly{4-(2,2-dicyanovinyl)-N-bis(hydroxyethyl)aniline-alt-[4,4′-methylenebis(phenyl)isocyanate]}urethane	
Poly[4-(2,2-dicyanovinyl)-N-bis(hydroxyethyl)aniline-alt-p-phenylenediacrylate]	
Poly[1-methoxy-4-(0-disperse red 1)-2,5-bis(2-methoxyethyl)-benzene]	
Poly[1-methoxy-4-(0-disperse red 1)-2,5-phenylenevinylene]	

3.4.2.2 Orientation techniques

Practical applications demand optimum alignment of the AπD moieties in the sample in a non-centrosymmetric fashion. To this end, the most common approach involves electric field-induced alignment of glassy, i.e. amorphous, polymer films, a process commonly referred to as poling. Thereby, a net orientation of the molecular dipole moments along a polar axis of the macroscopic sample is attained. Poling is carried out at a temperature close to the glass transition temperature of the polymer matrix, at which the molecules are relatively mobile. Electric field-induced alignment can be achieved either by sandwiching the polymer samples between electrodes, which is referred to as electrode poling, or by corona poling. Figure 3.10 shows a schematic diagram of a corona poling set-up with wire-to-plane configuration.

A corona discharge is induced upon application of an electric potential of several kV across the electrodes. Ionized molecules from the air are forced by the electric field to move to the surface of the sample. The deposited ions induce image charges on the earthed electrode. Thereby, a static electric field of about

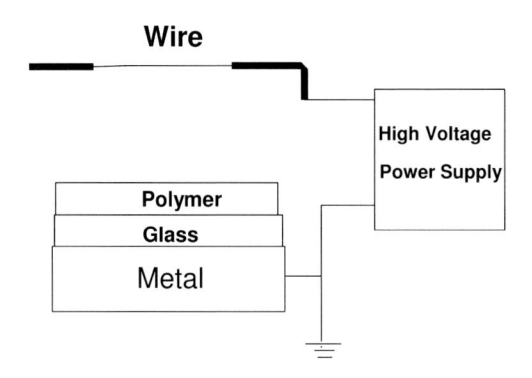

Fig. 3.10 Schematic diagram showing a corona poling set-up with wire-to-plane configuration. The tungsten wire is placed above and parallel to the sample. Adapted from Eich et al. [42] with permission from the Optical Society of America.

10^6 V cm^{-1} is generated across the sample, which induces alignment of the NLO moieties with respect to the direction of the electric field. Poled samples are represented by $C_{\infty v}$ symmetry. Alternative alignment methods are based on the Langmuir-Blodgett (LB) and self-assembly techniques, both of which are difficult to perform.

In the case of polymer systems containing photochromic chromophores, e.g. azo groups, alignment can be achieved upon exposure to light instead of a static electric field. This method is referred to as *optical poling* (see also Section 5.5). With such systems, optimum results have been obtained by applying a combined electro-optical poling method. As can be seen in Table 3.3, Pockels coefficients exceeding 10 pm V^{-1} have been measured for appropriate polymers poled by the combined electro-optical method. More detailed information concerning the various alignment techniques can be obtained from review articles [4, 8, 43, 44].

3.4.3
Third-order NLO materials

Table 3.5 presents a selection of $\chi^{(3)}$ values of various conjugated polymers, determined by THG measurements, while Table 3.6 shows $\chi^{(3)}$ values of some full-ladder and semi-ladder polymers, determined by means of the DFWM technique.

It must be noted that the $\chi^{(3)}$ values reported in the literature vary over broad ranges. Therefore, the values listed here reflect only the general behavior of several classes of compounds. It can be seen in Table 3.5 that *trans*-polyacetylenes (PAs) and polydiacetylenes (PDAs) exhibit the largest third-order NLO susceptibilities. The $\chi^{(3)}$ value of *cis*-PA (not shown) is more than an order of magnitude smaller than that of *trans*-PA. Derivatives of poly-*p*-phenylene, poly(phenylene vinylene), and polythiophene also exhibit NLO activity, but to a much lesser extent than PAs and PDAs. As pointed out above, polysilanes also possess quite large $\chi^{(3)}$ values. This is explained by the σ-conjugation of the silicon chain, which implies a pronounced delocalization of σ-electrons. A very large $\chi^{(3)}$ value

Table 3.5 Third-order susceptibilities, $\chi^{(3)}$, obtained by third harmonic generation measurements. Adapted from Kajzar [28] and Nalwa [45].

Polymer	Acronym [c]	$\chi^{(3)}$ (esu)	λ [a] (nm)	Remarks
	trans-PA	5.6×10^{-9}	1907	Isotropic film
	trans-PA	2.7×10^{-8}	1907	Oriented film
R: $-(CH_2)_4-O-CO-NH-(CH_2)_3-CH_3$	PDA-C$_4$UC$_4$	2.9×10^{-10}	1907	Oriented film
	PDA-CH	1.0×10^{-10}	1907	
	PPV	1.4×10^{-10}	1450	Isotropic film
	PBT	2.9×10^{-11}	1907	Spun film
	PTV	3.2×10^{-11}	1850	
	PTT	2×10^{-11}	1907	Isotropic film
R: C≡CH	PDES	3.0×10^{-9} [b]	620	

Table 3.5 (continued)

Polymer	Acronym [c]	$\chi^{(3)}$ (esu)	λ [a] (nm)	Remarks
[Si(R)(R)]$_n$, R: (CH$_2$)$_5$CH$_3$	PDHS	1×10^{-11}	1064	
[C(CH$_3$)(C$_6$H$_4$CH$_3$)–CH$_2$]$_n$	PVT	3×10^{-14}	1907	

a) Fundamental wavelength.
b) Determined by the DFWM method.
c) Abbreviations: *trans*-PA: *trans*-polyacetylene; PDA-C$_4$UC$_4$: poly[5,7-dodecadiyne-1,12-diol-bis(*n*-butoxycarbonyl methylurethane)]; PDA-CH: poly[1,6-di-(*N*-carbazoyl)-2,4-hexadiyne]; PPV: poly(*p*-phenylene vinylene); PBT: poly(3-butylthiophene); PTV: poly(2,5-thienylene vinylene); PTT: poly(thieno-3,2-bithiophene); PDES: poly(diethynylsilane); PDHS: poly(di-*n*-hexylsilane); PVT: poly(vinyltoluene).

Table 3.6 Third-order susceptibilities, $\chi^{(3)}$, obtained by the DFWM method. Adapted from Wijekoon et al. [46].

Polymer	Acronym	$\chi^{(3)}$ (esu)	λ [a] (nm)
benzobisthiazole polymer	PBT	1.0×10^{-10}	602
benzobisoxazole polymer	PBO	1.0×10^{-10}	602
polyimide structure	LARC-TPI	2.0×10^{-12}	602
BBL structure	BBL	1.5×10^{-11}	1064
BBB structure	BBB	5.5×10^{-12}	1064

a) Fundamental wavelength.

(3×10^{-9} esu) has been found for poly(diethynylsilane), PDES. In this case, a response time of 135 fs was measured [47]. Compared with those of conjugated polymers, the $\chi^{(3)}$ values of non-conjugated polymers are very low. For example, $\chi^{(3)}$ values of 4.0 and 3.4×10^{-14} esu have been measured for poly(methyl methacrylate) and poly(vinyltoluene), respectively. As regards the polymers listed in Table 3.6, it is notable that some of them, for instance BBL and BBB, are soluble and film-forming, in spite of their quasi-two-dimensional structures. For practical applications, materials with large $\chi^{(3)}$ values, low optical losses a, and ultrafast response times t_{resp} are desired. Ideal targets set for device applications are $\chi^{(3)} \geq 10^{-7}$ esu, $a \leq 10^2$ cm^{-1}, and $t_{resp} \leq 1$ ps. Therefore, appropriate materials should possess a figure of merit, $\chi^{(3)}/a$, of 10^{-9} esu cm. Although most polymeric materials exhibit much lower $\chi^{(3)}/a$ values, various promising devices have been proposed and fabricated [45]. For detailed information concerning third-order NLO properties of polymers and other compounds the reader is referred to the literature [28, 45, 46].

3.5
Applications of NLO polymers

The application potential of the effects dealt with in this chapter covers a broad field extending from specific electro-optical devices to the all-optical computer. For many applications, polymeric materials have proven appropriate and equivalent to inorganic materials. This section is focused on two aspects, the electro-optical (EO) or Pockels effect and two-photon absorption, which have been exploited extensively. Technical developments relating to polymeric modulators operating on the basis of the Pockels effect have reached the stage of commercialization [5].

3.5.1
Applications relating to telecommunications

With the advent of optical fibers in telecommunications in the late 1970s, practical applications for nonlinear optical devices operating on the basis of the EO effect became a serious goal. Besides inorganic materials, which were used exclusively in the early days, more recently polymeric electro-optic materials have also found use in a variety of device configurations. They can function as tunable Bragg wavelength filters, ultra-high bandwidth signal modulators for telecommunications, fast modulators for optical 3D sensing, electrical-to-optical signal transducers, switches at nodes in optical networks, and controllers of the phase of radiofrequency, optical signals, etc. [5]. Typical configurations, the Mach-Zehnder (MZ) interferometer and the birefringent modulator, are depicted schematically in Fig. 3.11.

In the case of the MZ interferometer (Fig. 3.11a), application of an electric field to one arm results in a phase retardation relative to the signal traversing

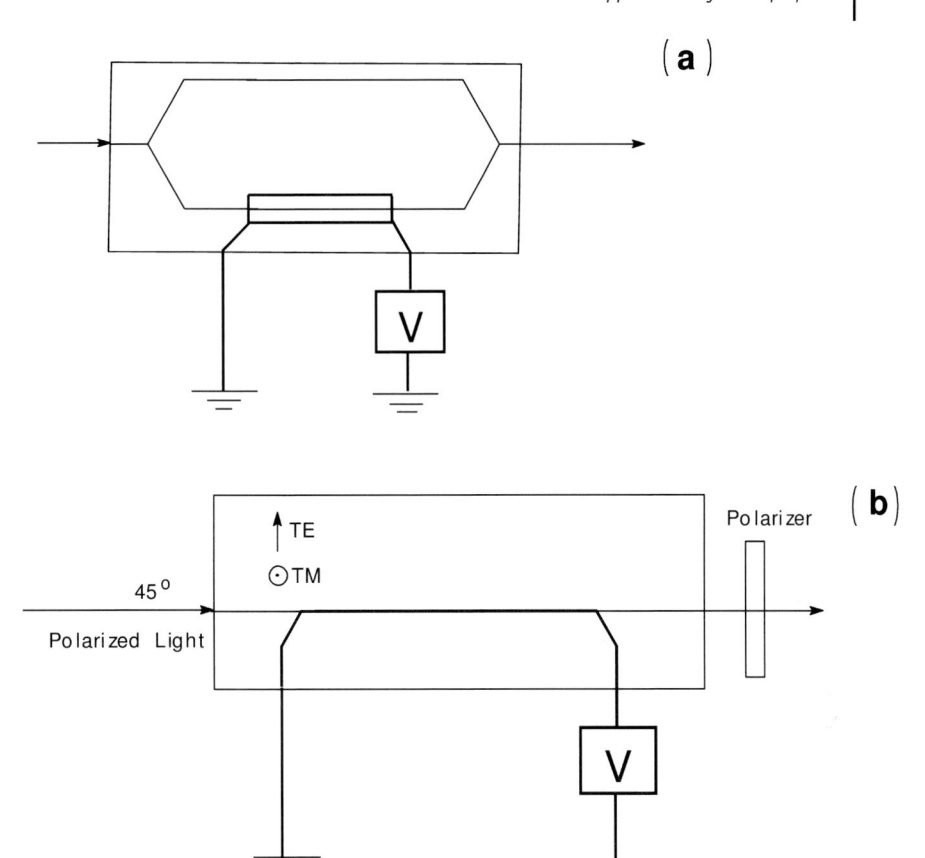

Fig. 3.11 Electro-optic device configurations. (a) Mach-Zehnder interferometer, (b) birefringent modulator. TM and TE denote transverse magnetic and transverse electric polarization, respectively.

the second arm and in destructive interference at the output. The phase retardation, $\Delta\varphi$, of light traversing the material of optical path length L under the influence of an electric field E is proportional to Δn, the change in the index of refraction [see Eq. (3-32)].

$$\Delta\varphi = \frac{2\pi\Delta nL}{\lambda} = \frac{\pi n^3 ErL}{\lambda} \tag{3-32}$$

As a consequence of the voltage-controlled destructive interference, the applied electrical signal is transduced onto the optical beam as an amplitude modulation. The birefringent modulator depicted in Fig. 3.11b functions as an electrical-to-optical signal transducer. Here, both TM and TE optical modes traverse the EO material. The application of an electric field produces a voltage-dependent birefringence, which is turned into amplitude modulation with the aid of a polarizer positioned at the output of the device.

The drive voltage, V_D, required to achieve full-wave modulation is inversely proportional to the EO coefficient of the material. Since drive voltages of the order of 1 V or less are required for lossless communication links, materials with large EO coefficients are desirable. V_D depends on the device configuration. For example, V_D for the birefringent modulator exceeds that for the MZ-type modulator by a factor of 1.5 [5]. It should be noted that the change in the refractive index ($\Delta n = 0.5\ n^3 rE$) is rather small. For example, if $n^3 = 5$, $r = 5 \times 10^{-12}$ m V^{-1}, and $E = 10^6$ V m^{-1}, Δn is equal to 1.25×10^{-5}.

Very successful efforts in employing polymeric materials as modulators have been made with the guest/host systems shown in Table 3.7. The guest compounds are characterized by the cyanofuran moiety. A thermally rather stable host matrix, denoted as APC, is a copolymer: poly[bisphenol A carbonate-co-4,4'-(3,3,5-trimethylcyclohexylidene)diphenol]. The systems shown in Table 3.7 are employed in commercially available modulators; the relevant industrial companies are cited in Dalton's review article [5]. These polymeric systems are

Table 3.7 Characteristics of electro-optically active chromophores in a PMMA matrix. Adapted from Dalton [5].

Denotation	Chemical Structure	μ [a] (Debye)	r [b] (pm V^{-1})
FTC		12.19	50
CLD		13.42	70
GLD		13.88	105

a) Dipole moment obtained by quantum mechanical calculation.
b) Pockels coefficient at a number density of about 1.5×10^{20} molecules cm^{-3}, measured at $\lambda = 1.3$ μm.

Table 3.8 Comparison of lithium niobate and polymeric EO materials. Adapted from Dalton [5].

Property	LiNbO$_3$	EO Polymer
Pockels coefficient, r (pm V^{-1}) at $\lambda=1.3$ μm	31	>70
Dielectric constant, ε	28	2.5–4
Refractive index, n	2.2	1.6–1.7
Figure of merit ($n^3 r/\varepsilon$)	12	>100
Optical loss (dB cm^{-1}) at $\lambda=1.3$ μm	0.2	0.2–1.1
Maximum optical power (mW)	250	250
Bandwidth×length product[a] $\Delta f\, L$ (GHz cm)	10	>100

a) Δf: Bandwidth in a device of Mach-Zehnder configuration.
L: Interaction length of light with the modulating electric field.

superior to lithium niobate with respect to various important properties, as can be seen in Table 3.8.

Pockels coefficients measured at the technologically important wavelengths 1.3 and 1.55 μm are higher than in the case of lithium niobate. Moreover, the difference in the dielectric constants is important: $\varepsilon=28$ (LiNbO$_3$) and $\varepsilon=2.5$–4 (EO polymer). The lower ε value corresponds to a decreased device power consumption and an enhanced speed of operation.

3.5.2
Applications relating to optical data storage

Potential applications of polymeric materials with large $\chi^{(3)}$ values concern photonic devices in various fields, such as optical fiber communication, optical computing, imaging, dynamic holography, optical switching, and optical data storage. Two-photon absorption, a third-order nonlinear effect (see Section 3.3.2.3), has gained importance for optical data storage [48]. Two-photon absorption is possible, provided that both photons are spatially and temporally coincident. As this requirement has to be fulfilled, optical sectioning can be accomplished, i.e. absorption events can be directed to selected layers. In other words, information can be recorded in previously defined layers of a film, and thereby three-dimensional bit optical data storage within the volume of a recording medium is possible. Photochemical free radical polymerization (see Section 10.2) can be employed to achieve optical data storage at a density as high as 0.4 Tb cm^{-3} with a bit spacing of 1 μm and a layer spacing of 3 μm [49, 50]. For this technique, a recording medium consisting of a monomer solution containing a photoinitiator is typically used. Since the initiation is restricted to two-photon absorption, the polymerization is confined to the region of the focus spot. To prevent distortion of the recorded planes through shrinkage or flow, gelation of the system by UV pre-irradiation is carried out. Polymerization at the recorded bit changes the refractive index. The pattern of recorded bits can thus be read by producing a phase/intensity map by means of differential interference contrast microscopy [51].

3.5.3
Additional applications

Additional potential applications based on other nonlinear phenomena, such as second harmonic generation (frequency doubling of laser light), phase conjugation, and optical bistability, may be envisaged. *Phase conjugation* (see Section 3.3.2.4) allows the distortionless transmission of images, because, upon retracement, the beam reflected from a phase conjugator corrects every distortion of the probe beam. *Optical bistability* is the basis for the transphasor, the optical transistor, a device switching light with light without the aid of an electrical current. This can be achieved by focusing two laser beams, a strong constant beam and a weak variable probe beam, onto the front face of a Fabry-Perot interferometer containing a substance having a nonlinear refractive index. Since the latter depends on the light intensity, constructive interference sets in at a certain intensity of the probe beam, and the transmittance increases to a high level, as shown in Fig. 3.12. The term bistability refers to the existence of two quasi-stable levels.

Another potential application relates to *optical limiters*, i.e. materials that can be used for the protection of eyes and sensors from intense light pulses, and generally, for devices that are required to have a high transmittance at low intensities and a low transmittance at high intensities [52, 53]. Appropriate substances contain chromophores that exhibit nonlinear light absorption, termed reverse saturable absorption. Such chromophores become more strongly absorbing as the incident light intensity is increased. The nonlinear response may be exhibited when chromophores absorb weakly in the ground state and strongly in the excited state. Optical limiting may also be due to two-photon (or, more generally, multi-photon) absorption (see Section 3.3.2.3).

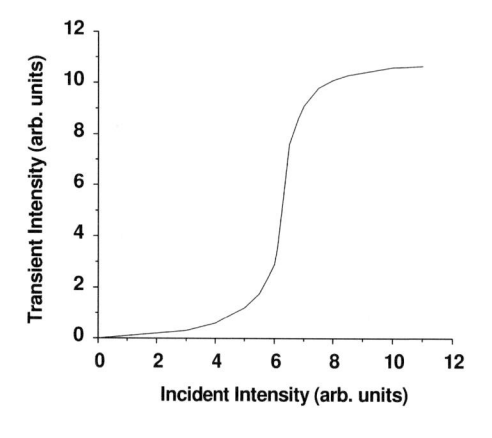

Fig. 3.12 The transmittance behavior of a transphasor (optical transistor). Plot of the transmitted intensity as a function of the incident intensity.

References

1 H.-H. Perkampus, *Encyclopedia of Spectroscopy*, VCH, Weinheim (1995).
2 P. A. Franken, L. E. Hill, C. W. Peters, G. Weinreich, Phys. Rev. Lett. 7 (1961) 118.
3 S. K. Yesodha, C. K. S. Pillai, N. Tsutsumi, *Stable Polymeric Materials for Nonlinear Optics: A Review Based on Azobenzene Systems*, Prog. Polym. Sci. 29 (2004) 45.
4 F. Kajzar, K.-S. Lee, A. K.-Y. Jen, *Polymeric Materials and their Orientation Techniques for Second-Order Nonlinear Optics*, Adv. Polym. Sci. 161 (2003) 1.
5 L. Dalton, *Nonlinear Optical Polymeric Materials: From Chromophore Design to Commercial Applications*, Adv. Polym. Sci. 158 (2002) 1.
6 Z. Sekkat, W. Knoll (eds.), *Photoreactive Organic Thin Films*, Academic Press, Amsterdam (2002).
7 M. G. Kuzyk, C. W. Dirk (eds.), *Characterization Techniques and Tabulations for Organic Nonlinear Optical Materials*, Marcel Dekker, New York (1998).
8 J. I. Chen, S. Marturunkakul, L. Li, S. T. Tripathy, *Second-Order Nonlinear Optical Materials*, in T. A. Skotheim, R. L. Elsenbaumer, J. R. Reynolds (eds.), *Handbook of Conducting Polymers*, 2nd Edition, Marcel Dekker, New York (1998), p. 727.
9 S. Bauer-Gogonea, R. Gerhard-Multhaupt, *Nonlinear Optical Electrets*, in R. Gerhard-Multhaupt (ed.), *Electrets*, 3rd Edition, Vol. 2, Laplacian Press, Morgan Hill, CA (1999), p. 260.
10 H. S. Nalwa, S. Miyata (eds.), *Nonlinear Optics of Organic Molecules and Polymers*, CRC Press, Boca Raton, FL, USA (1997).
11 D. M. Burland, R. D. Miller, C. A. Walsh, *Second-Order Nonlinearity in Poled Polymer Systems*, Chem. Rev. 94 (1994) 31.
12 N. P. Prasad, D. J. Williams, *Introduction to Nonlinear Optical Effects in Molecules and Polymers*, Wiley, New York (1991).
13 B. S. Wherrett, in C. Flytzanis, J. L. Oudar (eds.), *Nonlinear Optics: Materials and Devices*, Springer, Berlin (1986).
14 M. Canva, G. I. Stegeman, *Quadratic Parametric Interactions in Organic Waveguides*, Adv. Polym. Sci. 158 (2002) 87.
15 B. Kippelen, N. Peyghambarian, *Photorefractive Polymers and their Applications*, Springer, Berlin, Adv. Polym. Sci. 161 (2003) 87.
16 D. A. Kleinmann, Phys. Rev. 126 (1962) 1977.
17 G. R. Möhlmann, C. P. J. M. van der Vorst, R. A. Huijts, C. T. J. Wreesmann, Proc. SPIE 971 (1988) 252.
18 E. Cavicchi, J. Kumar, S. Tripathy, *Nonlinear Optical Spectroscopy of Polymers*, in H. Bässler (ed.), *Optical Techniques to Characterize Polymer Systems*, Elsevier, Amsterdam (1989), p. 325.
19 C. P. J. M. van der Vorst, D. J. Picken, *Electric Field Poling of Nonlinear Optical Side Chain Polymers*, in V. P. Shibaev (ed.), *Polymers as Electrooptical and Photooptical Active Media*, Springer, Berlin (1996).
20 K. D. Singer, S. F. Hubbard, A. Schober, L. M. Hayden, K. Johnson, *Second Harmonic Generation*, in [7], p. 311.
21 K. Clays, A. Persoons, Phys. Rev. Lett. 66 (1991) 2980; Rev. Sci. Instrum. 63 (1992) 3285.
22 K. Clays, A. Persoons, L. De Mayer, *Modern Linear Optics, Part 3*, Adv. Chem. Phys., Wiley, New York (1993).
23 J. A. Delaire, E. Ishov, K. Nakatani, *Photoassisted Poling and Photoswitching of NLO Properties of Spiropyrans and other Photochromic Molecules in Polymers and Crystals*, in Z. Sekkat, W. Knoll (eds.), *Photoreactive Organic Thin Films*, Academic Press, Amsterdam (2002).
24 T. Watanabe, H. S. Nalwa, S. Miyata, *Measurement Techniques for Refractive Index and Second-Order Optical Nonlinearities*, Chapter 3 in [10].
25 J. Jerphagnon, S. K. Kurtz, J. Appl. Phys. 41 (1970) 1667.
26 H. S. Nalwa, *Measurement Techniques for Third-Order Optical Nonlinearities*, Chapter 10 in [10].
27 J. L. Bredas, C. Adant, P. Tackx, A. Persoons, *Third-Order Optical Response in Organic Materials: Theoretical and Experimental Aspects*, Chem. Rev. 94 (1994) 243.
28 F. Kajzar, *Third Harmonic Generation*, in [7].

29 E.W. Van Stryland, M. Sheik-Bahae, *Z-Scan*, Chapter 8 in [7].
30 U. Gubler, C. Bosshard, *Molecular Design for Third-Order Optics*, Adv. Polym. Sci. 158 (2002) 125.
31 T.-C. Lin, S.-J. Chung, K.-S. Kim, X. Wang, G.S. He, J. Swiatkiewicz, H.E. Pudavar, P.N. Prasal, *Organics and Polymers with High Two-Photon Activities and their Applications*, Springer, Berlin, Adv. Polym. Sci. 161 (2003) 157.
32 S. Kershaw, *Two-Photon Absorption*, Chapter 7 in [7].
33 M. Gower, D. Proch (eds.), *Optical Phase Conjugation*, Springer, Berlin (1994).
34 K. Abe, M. Amano, T. Omatsu, Optics Express 12 (2004) 1243.
35 H.S. Nalwa, T. Watanabe, S. Miyata, *Organic Materials for Second-Order Nonlinear Optics*, Chapter 4 in [10].
36 N. Pereda, J. Extebarria, C.L. Focia, J. Ortega, C. Artal, M.R. Ros, J.C. Serano, J. Appl. Phys. 87 (2000) 217.
37 B. Philip, K. Sreekumar, J. Polym. Sci. Part A: Polym. Chem. 40 (2002) 2868.
38 J.D. Swalen, C.R. Moylan, *Linear Optical Properties*, Chapter 4 in [7].
39 R.P. Bertram, E. Soergel, H. Blank, N. Benter, K. Buse, R. Hagen, S.G. Kostromine, J. Appl. Phys. 94 (2003) 6208.
40 G.F. Lipscomb, J.I. Thackara, R. Lytel, *Electro-Optic Effect*, in [7].
41 Aldrich, ChemFiles 4 (2004) 4.
42 M. Eich, H. Looser, D. Yoon, R. Twieg, G. Bjorklund, J. Baumert, J. Opt. Soc. Am. B 6 (1989) 1590.
43 F. Kajzar, J.M. Nunzi, *Molecular Orientation Techniques*, in F. Kajzar, R. Reinisch (eds.), *Beam Shaping Control with Nonlinear Optics*, Plenum Press, New York (1998), p. 101.
44 S. Bauer, Appl. Phys. Rev. 80 (1996) 5531.
45 H.S. Nalwa, *Organic Materials for Third-Order Nonlinear Optics*, Chapter 11 in [10].
46 W.M.K.P. Wijekoon, P.N. Prasad, *Nonlinear Optical Properties of Polymers*, in J.E. Mark (ed.), *Physical Properties of Polymers Handbook*, AIP Press, Woodbury, NY (1995) Chapter 38.
47 K.S. Wong, S.G. Han, Z.V. Vardeny, J. Shinar, Y. Pang, I. Maghsoodi, T.J. Barton, S. Grigoras, B. Parbhoo, Appl. Phys. Lett. 58 (1991) 1695.
48 P. Boffi, D. Piccinin, M.C. Ubaldi (eds.), *Infrared Holography for Optical Communications: Techniques, Materials, and Devices*, Topics in Applied Physics 86, Springer, Berlin (2003).
49 B.H. Cumpton, S.P. Ananthavel, S. Barlow, D. Dyer, J.E. Ehrlich, L.L. Erskine, A.A. Heikal, S.M. Kuebler, I.Y.S. Lee, D. McCord-Maughon, J. Qin, H. Röckel, M. Rumi, X.L. Wu, S.R. Marder, J.W. Perry, Nature 398 (1999) 51.
50 H.B. Sun, T. Matsuo, H. Misawa, Appl. Phys. Lett. 74 (1999) 786.
51 D. Day, M. Gu, A. Smallridge, *Review of Optical Data Storage*, in [48], p. 1.
52 J.W. Perry, *Organic and Metal-Containing Reverse Saturable Absorbers for Optical Limiters*, Chapter 13 in [10].
53 E.W. Van Stryland, D.J. Hagan, T. Xia, A.A. Said, *Application of Nonlinear Optics to Passive Optical Limiting*, Chapter 14 in [10].

4
Photorefractivity

4.1
The photorefractive effect

The photorefractive (PR) effect refers to the spatial modulation of the index of refraction in an electro-optically active material that is non-uniformly irradiated. Notably, the refractive index of an electro-optically active material is electric field dependent. The PR effect is based on the light-induced generation and subsequent migration of charge carriers and, therefore, is strongly connected to the phenomena of photogeneration and conduction of charge carriers in polymeric systems dealt with in Chapter 2. The PR effect was first observed in inorganic materials such as $LiNbO_3$, $BaTiO_3$, $InP:Fe$, and GaAs [1–9] and later also in organic materials. Work related to polymers has been reviewed [10–12]. Materials exhibiting the PR effect should be capable of forming charge carriers, i.e. pairs of positively and negatively charged ions, in a sufficiently high quantum yield upon exposure to light, and these charge carriers should migrate with a sufficiently high mobility. A prerequisite for the occurrence of the PR effect is separation of the charges, which is commonly accomplished if only one type of charge carrier is mobile and the material contains traps where the migrating carriers are captured. A non-uniform irradiation of polymeric materials can be accomplished by placing foils in the interference region of two coherent light waves. In this way, a fringe pattern of brighter and darker regions, i.e. of strongly and weakly or not at all irradiated regions, is produced. Notably, the charge separation due to the exclusive migration of charge carriers of the same sign from the irradiated to the non-irradiated regions results in the build-up of a space-charge field, i.e. of an internal electric field between the irradiated and unirradiated regions, which allows the linear electro-optic effect (Pockels effect, see Section 3.1) to become operative. In other words, the formation of the space-charge field gives rise to a change in the refractive index, and in this way a refractive index fringe pattern is generated. The magnitude of the refractive index modulation Δn, frequently also referred to as the *dynamic range*, depends on the space-charge field strength E_{SC} according to Eq. (4-1).

$$\Delta n = -\frac{n^3 r_e E_{SC}}{2} \qquad (4\text{-}1)$$

Polymers and Light. Fundamentals and Technical Applications. W. Schnabel
Copyright © 2007 WILEY-VCH Verlag GmbH & Co. KGaA, Weinheim
ISBN: 978-3-527-31866-7

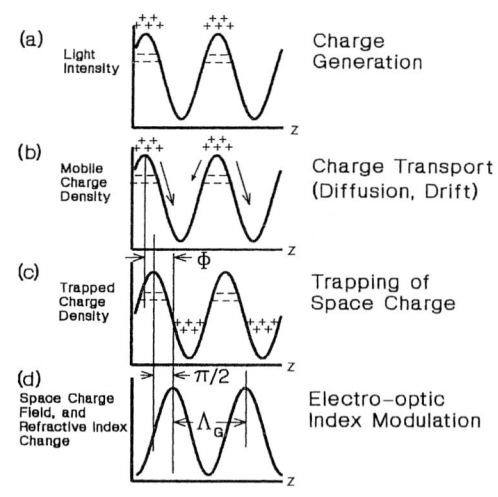

Fig. 4.1 The photorefractive effect. One-dimensional illustration of the charge generation by non-uniform irradiation of a polymer film and the subsequent generation of a refractive index grating through transport and trapping of the mobile holes. Adapted from Valley and Klein [13] and Moerner and Silence [12] with permission from the American Chemical Society.

Here, r_e is the electro-optic (or Pockels) coefficient for a given geometry and n is the refractive index.

Commonly, holes are the mobile charge carriers in photorefractive polymers. Since the migration of holes by diffusion is a rather slow process, a drift is enforced by the application of an external electric field. The latter not only promotes hole migration, but also provides essential assistance during the photo-

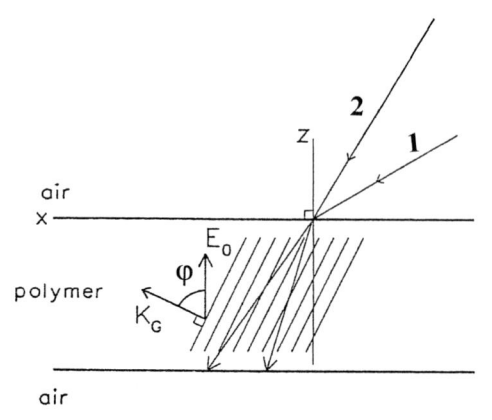

Fig. 4.2 Schematic depiction of the experimental geometry employed for writing a refractive index grating in a PR polymer. Adapted from Moerner and Silence [12] with permission from the American Chemical Society.

generation process (see Section 2.2). Significantly, there is a phase shift between the irradiation pattern and the refractive index pattern, as can be seen in Fig. 4.1, which illustrates the mechanism of grating formation.

A schematic depiction of the formation of a grating in a polymer film located in an external electric field is shown in Fig. 4.2.

The grating is written by beams 1 and 2, which enter the film at angles of incidence a_1 and a_2 with respect to the sample normal. The grating is written at a wave vector K_G at an angle φ with respect to the external electric field E_0. The spatial periodicity Λ_G of the grating is given by Eq. (4-2).

$$\Lambda_G = \frac{\lambda_0}{2n \sin[(a_1 - a_2)/2]} \qquad (4\text{-}2)$$

Here, n is the refractive index and λ_0 is the wavelength of the light in vacuo.

4.2
Photorefractive formulations

An organic photorefractive system has to contain different functional groups providing for the generation, transport, and trapping of charge carriers. Moreover, a plasticizing function is required for certain formulations. Apart from the latter, these requirements may, in principle, be met by fully functionalized polymers, i.e. by polymers containing, in their main chain and side chains, the various requisite functional groups. However, since this approach is rather difficult to implement, research activities have concentrated mostly on the so-called host/guest approach, which is based on formulations consisting of a host polymer and various low molar mass guest compounds. Typical polymers and low molar mass compounds used for formulations exhibiting a photorefractive effect are shown in Chart 4.1 and Chart 4.2, respectively.

The system PMMA-PNA:DEH:TNF is a typical photorefractive formulation, with PMMA-PNA acting as the host polymer and DEH (30 wt.%) and TNF (0.1 wt.%) as charge-transporting agent and charge-generating sensitizer, respectively. In order to ensure bulk transport of the photogenerated holes by the hopping mechanism, the concentration of the transporting agent has to be rather high. Typical examples of fully functionalized polymers are also presented in Chart 4.1 (polymers VI [14] and VII [15]). In the case of polymer VI, photoexcitation of the chromophores MHB^+Br^- at $\lambda=647$ nm induces electron transfer from the aromatic amino groups (Am) according to reaction (b) in Scheme 4.1. In this way, trapped electrons, $MHB^\bullet Br^-$, and mobile radical cations, $Am^{\bullet+}$, are formed. The hole transport according to reaction (c) is a multiple successive electron-hopping process from neutral Am groups to neighboring radical cations.

Polymer VII belongs to a group of conjugated polymers containing porphyrin or phthalocyanine complexes synthesized by Lu et al. [16]. Here, the polymer backbone consists of phenylene vinylene moieties, which facilitate hole trans-

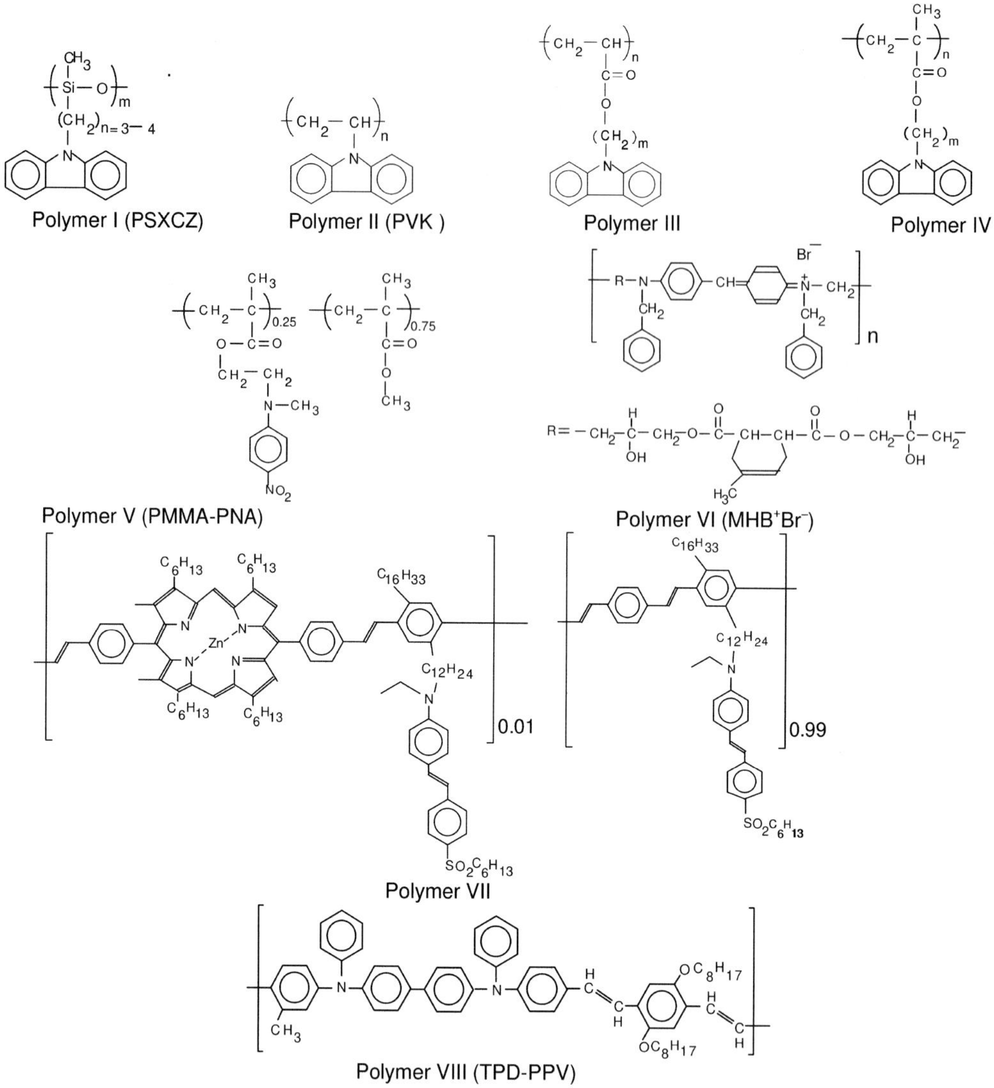

Chart 4.1 Polymers employed in photorefractive formulations.

port through intramolecular migration and interchain hopping. Charge carriers are formed as a result of the selective absorption of near-infrared light (e.g., He-Ne laser light, $\lambda = 632.8$ nm) by the porphyrin or phthalocyanine complexes, and trapping might occur at the side groups.

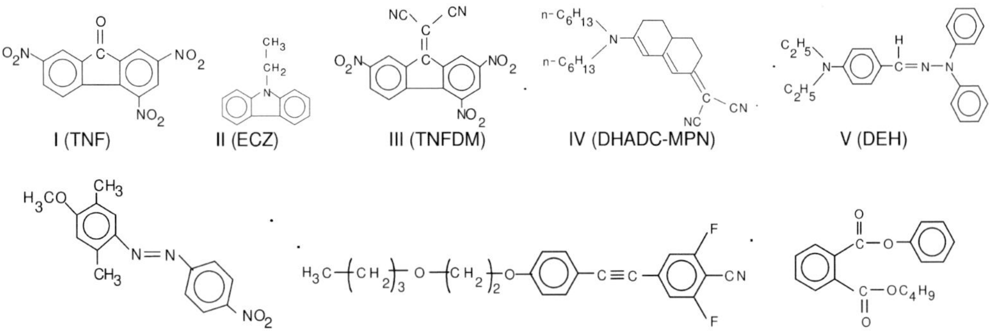

I (TNF) II (ECZ) III (TNFDM) IV (DHADC-MPN) V (DEH)

VI (DMNPAA) VII (FTCN) VIII (BBP)

Chart 4.2 Low molar mass compounds employed in photorefractive formulations.

$$\begin{array}{lcccll}
MHB^+Br^- & + & h\nu & \rightarrow & ((MHB^+Br^-)^* & \text{(a)} \\
(MHB^+Br^-)^* & + & Am & \rightarrow & MHB^\bullet Br^- + Am^{\bullet+} & \text{(b)} \\
Am^{\bullet+} & + & Am & \rightarrow & Am + Am^{\bullet+} \rightarrow \text{etc.} & \text{(c)}
\end{array}$$

Scheme 4.1 Generation and transport of charge carriers in polymer VI.

4.3
Orientational photorefractivity

During the development of new photorefractive materials, the employment of chromophoric compounds with a permanent dipole moment turned out to lead to unexpectedly high Δn values, provided that the glass transition temperature of the formulation was close to ambient temperature, such that the chromophores were mobile and could become oriented under the influence of an electric field, a process referred to as *poling*. Poling-induced orientation of the chromophoric molecules leads to macroscopic electro-optical properties and especially to birefringence. Notably, the total effective electric field in a photorefractivity experiment results from a superposition of the internal space-charge field and the externally applied electric field. Consequently, the spatial refractive index modulation is controlled not only by the space-charge field but also by a strong contribution from the orientational birefringence, a fact referred to by the term orientational photorefractivity. Notably, in this case, the refractive index change has a quadratic dependence on the total electric field, which is a superposition of the internal space-charge field and the externally applied field, and to a rough approximation the dependence of the dynamic range Δn on the field strength E is given by Eq. (4-3).

$$\Delta n = \mathrm{pE}^2 = \frac{\mathrm{pV}^2}{d^2} \qquad (4\text{-}3)$$

Here, p is a material parameter, V is the applied voltage, and d is the sample thickness.

DMNPAA and DHADC-MPN (see Chart 4.2) are typical optically anisotropic compounds with permanent dipole moments, which can be oriented in an electric field at room temperature in formulations plasticized with ECZ and, therefore, have low T_g values. Typical values reported in the literature are $p = 86$ cm^2 V^{-2} for the system DMNPAA:PVK:ECZ:TNF and $p = 333$ cm^2 V^{-2} for the system DHADC-MPN:PVK:ECZ:TNFDM [10].

4.4
Characterization of PR materials

Commonly, the PR properties of materials are characterized and tested by two-beam coupling and four-wave mixing experiments. *Two-beam coupling* (2BC) refers to the energy exchange between the two interfering laser beams employed to write the grating. During the formation of the grating, the two writing beams diffract from the forming grating, i.e. each writing beam is partially diffracted in the direction of the other beam by the forming grating. In a 2BC experiment, the change in the transmitted intensity of either of the write beams is recorded as the other write beam is switched on and the grating is formed. This can be seen in Fig. 4.3, which shows beam intensity as a function of time, as recorded in two experiments in which the intensities of the two writing beams (before the sample) were kept equal [14].

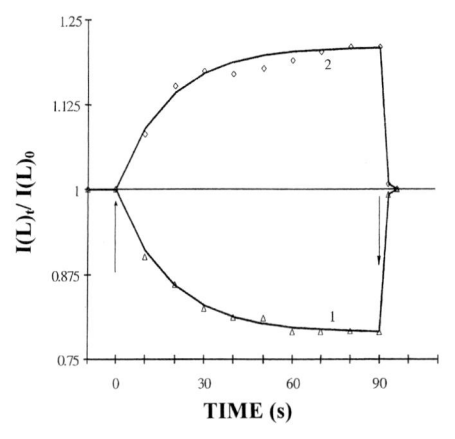

Fig. 4.3 Two-beam coupling experiments yielding evidence for the occurrence of the PR effect in a film consisting of polymer VI (MHB$^+$Br$^-$). The intensity of beam 1 was monitored as beam 2 was switched on at $t=0$ and switched off at $t=90$ s, and the intensity of beam 2 was monitored as beam 1 was switched on at $t=0$ and switched off at $t=90$ s. $\lambda = 647$ nm, $E = 26$ V μm^{-1}, and $d = 19.4$ μm; I_0 (1) $= I_0$ (2) $= 78$ mW cm^{-2}. Adapted from Vannikov et al. [14] with permission from Elsevier.

In the first experiment, in which beam 2 was switched on and off and beam 1 was monitored, the intensity of the latter decreased. Conversely, when beam 1 was switched on and off and beam 2 was monitored, the intensity of the latter increased. The occurrence of such asymmetric energy transfer unambiguously confirms the PR nature of the optical encoding and allows a distinction to be made between a grating based on the PR effect and other types of gratings.

From plots of the type shown in Fig. 4.3, the beam coupling ratio γ_0, as defined by Eq. (4-4), can be determined.

$$\gamma_0 = \frac{I(L)_{sat}}{I(L)_0} \tag{4-4}$$

Here, $I(L)_{sat}$ and $I(L)_0$ denote the intensity at saturation and at time $t=0$, respectively, of the writing beam under consideration, measured *after* passage through the sample. The beam coupling gain coefficient, Γ, is given by Eq. (4-5).

$$\Gamma = \frac{1}{L}[\ln(\gamma_0 \beta) - \ln(\beta + 1 - \gamma_0)] \tag{4-5}$$

Here, β is the ratio of the intensities of the two beams before the sample and L is the optical path length given by Eq. (4-6).

$$L = \frac{d}{\cos a} \tag{4-6}$$

Here, d is the sample thickness and a is the angle of incidence of the beam with respect to the sample normal.

The total refraction index modulation, Δn, is given by Eq. (4-7).

$$\Delta n = \frac{\Gamma \lambda}{4\pi} \tag{4-7}$$

Typical results obtained with polymer VI at $I_1(0) = 720$ mW cm^{-2}, $\beta = 22$, $E = 8$ V μm^{-1}, $\lambda = 647$ nm, and $d = 7.4$ μm are: $\Gamma = 313$ cm^{-1}, $\Delta n = 1.6 \times 10^{-3}$, $\eta = \sin^2(\Gamma L/2) = 2.1\%$, and $\tau = 4$ s (grating build-up or response time) [14].

The *four-wave mixing* technique serves to measure the diffraction efficiency η during the writing process as a function of time and as a function of the strength of the external electric field. Figure 4.4 shows a schematic representation of a typical set-up employed in four-wave mixing experiments.

Notably, a reading beam is used in addition to the two writing beams. Commonly, the reading beam is of the same wavelength as the two writing beams, but of a much lower intensity, and it is counterpropagating one of the writing beams. η is defined according to Eq. (4-8) as the ratio of the intensities of the diffracted beam, I_d, and of the incoming reading beam, I_0.

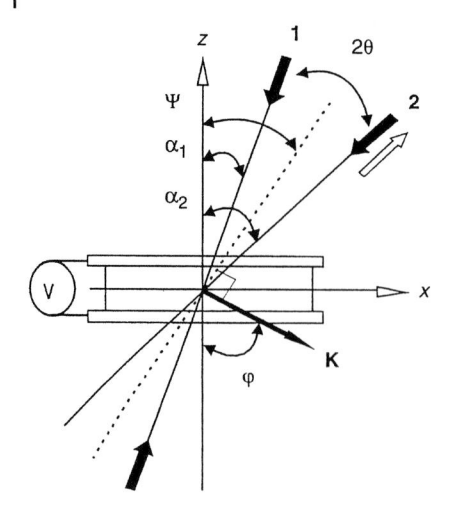

Fig. 4.4 Schematic depiction of a set-up for a four-wave mixing experiment, as employed to measure diffraction efficiency as a function of the strength of an external electric field. Reading beam counterpropagating with writing beam (1). Diffracted beam counterpropagating with writing beam (2). Adapted from Kippelen et al. [11] with permission from the International Society for Optical Engineering.

$$\eta = \frac{I_d}{I_0} \tag{4-8}$$

Usually, the electric field is applied to the sample by sandwiching the polymer between two transparent electrodes, such as ITO (indium tin oxide)-coated glass slides. The diffraction efficiency, η, can be obtained from Kogelnik's coupled-wave theory for thick holograms with the aid of Eq. (4-9) [17].

$$\eta \propto \sin^2\left(\frac{f_g \pi d \Delta n}{\lambda}\right) \tag{4-9}$$

Here, f_g is a geometrical factor dependent on the polarization of the beams and the experimental geometry, and λ is the wavelength of the light of the reading beam.

4.5
Applications

Photorefractive polymeric systems can be used to record, in real-time and with a high storage density, optically encoded information with low-power lasers such as semiconductor diode lasers. They are appropriate for recording *holograms*. The storage of a large number of holograms at a single spot in the storage medium (*multiplexing*, see Section 12.3) is possible. Therefore, there is a significant application potential. Actually, applications concerning dynamic holographic interferometry, holographic storage, and real-time processing have been demonstrated and future technical applications seem likely [18–22]. With respect to commercial applications, it is noteworthy that the PR effect is reversible, i.e.

previously recorded holograms can be erased by irradiation with a spatially uniform light beam. Moreover, holograms can be overwritten.

There is a long list of technical requirements for holographic materials, such as optical quality, near-IR sensitivity, large refractive index modulation, short response time, self-processing, inertness and long shelf-life, non-destructive readout, and low cost. Successful technical applications depend on the availability of materials that fulfil all or most of these requirements. Interesting proposals have been made to overcome still existing technical problems, such as that concerning destructive readout. To retrieve information from holograms with good fidelity, the reading and writing beams have to be of the same wavelength. However, since the material is photosensitive at the relevant wavelength, the readout process partially erases the stored information. According to Kippelen et al., this problem can be overcome with the aid of a photorefractive system containing a substituted diphenylacetylene (compound VII in Chart 4.2) that is sensitive to two-photon absorption [23]. In a system of the composition FTCN/PVK/BBP/ECZ (25:55:10:10 wt.%), charge carriers are generated exclusively by two-photon processes and holographic recording is achieved with high-intensity writing beams ($\lambda=650$ nm, 0.25 mW each). For readout, a low-intensity beam ($\lambda=650$ nm, 0.25 µW), which does not affect the photorefractive system, is sufficient.

The requirements of high near-IR sensitivity and short response time are largely fulfilled by applying a pre-irradiation method, denoted as *time-gated holographic imaging* [24]. Pre-irradiation provides for charge carriers before the writing starts and thus affords a significant reduction in response time. According to Mechner et al. [24], pre-irradiation at $\lambda=633$ nm prior to holographic recording at $\lambda=830$ nm improved the response time by a factor of 40 ($\tau \approx 30$ ms) in investigations with a formulation containing TPD-PPV (polymer VIII in Chart 4.2) (see Table 4.1).

Note that holograms can also be generated in polymeric media by other methods, for instance by photopolymerization of appropriate monomers contained in special formulations (see Section 11.7).

Table 4.1 Composition of a photorefractive material suitable for holographic recording by means of time-gated holographic imaging [24].

Components	Content (wt%)	Function
Polymer VIII (TPD-PPV)	56	Conductive host matrix
1:1 Mixture of 2,5-dimethyl-(4-*p*-nitrophenylazo)-anisole and 3-methoxy-(4-*p*-nitrophenylazo)-anisole	30	Electro-optical material
Diphenyl phthalate	13	Plasticizer
[6,6]-Phenyl-C_{61}-butyric acid methyl ester	1	Sensitizer

References

1. F. S. Chen, J. Appl. Phys. 38 (1967) 3418.
2. P. Günter, *Holography, Coherent Light Amplification and Optical Phase Conjugation with Photorefractive Materials*, Phys. Rep. 93 (1982) 199.
3. T. J. Hall, R. Jaura, L. M. Conners, P. D. Foote, *The Photorefractive Effect – A Review*, Prog. Quant. Electron. 10 (1985) 77.
4. J. Feinberg, *Photorefractive Nonlinear Optics*, Phys. Today 41 (1988) 46.
5. P. Günter, J.-P. Huignard, *Photorefractive Materials and Their Applications I and II*, in Topics in Applied Physics 61, Springer, Berlin (1988).
6. M. P. Petrov, S. L. Stepanov, A. V. Khomenko, *Photorefractive Crystals in Coherent Optical Systems*, Springer, Berlin (1991).
7. M. Gower, D. Proch (eds.), *Optical Phase Conjugation*, Springer, Berlin (1994).
8. P. Yeh, *Introduction to Photorefractive Nonlinear Optics*, Wiley, New York (1993).
9. D. D. Nolte (ed.), *Photorefractive Effects and Materials*, Kluwer Academic Publ., Boston (1995).
10. B. Kippelen, *Overview of Photorefractive Polymers for Holographic Data Storage*, in J. Coufal, D. Psaltis, G. T. Sincerbox (eds.), *Holographic Data Storage*, Springer, Berlin, Series in Optical Sciences 76 (2000) 159.
11. B. Kippelen, N. Peyghambarian, *Current Status and Future of Photorefractive Polymers for Photonic Applications*, Crit. Rev. Opt. Sci. Technol., CR 68 (1997) 343.
12. W. E. Moerner, S. M. Silence, *Polymeric Photorefractive Materials*, Chem. Rev. 94 (1994) 127.
13. G. C. Valley, M. B. Klein, Opt. Eng. 22 (1983) 704.
14. A. V. Vannikov, A. D. Grishina, L. Ya. Pereshivko, T. V. Krivenko, V. V. Savelyev, L. I. Kostenko, R. W. Rychwalski, J. Photochem. Photobiol. A: Chem. 150 (2002) 187.
15. L. Lu, J. Polym. Sci., Part A: Polym. Chem. 39 (2001) 2557.
16. L. Q. Wang, M. Wang, L. Lu, Adv. Mater. 12 (2000) 974.
17. H. Kogelnik, Bell Syst. Tech. J. 48 (1969) 2909.
18. R. Bittner, K. Meerholz, G. Steckman, D. Psaltis, Appl. Phys. Lett. 81 (2002) 211.
19. C. Poga, P. M. Lundquist, V. Lee, R. M. Shelby, R. J. Twieg, D. M. Burland, Appl. Phys. Lett. 69 (1996) 1047.
20. P. M. Lundquist, R. Wortmann, C. Geletneky, R. J. Twieg, M. Jurich, V. Y. Lee, C. R. Moylan, D. M. Burland, Science 274 (1996) 1182.
21. B. L. Volodin, Sandalphon, K. Meerholz, B. Kippelen, N. Kukhtarev, N. Peyghambarian, Opt. Eng. 34 (1995) 2213.
22. B. L. Volodin, B. Kippelen, K. Meerholz, B. Jaridi, N. Peyghambarian, Nature 383 (1996) 58.
23. B. Kippelen, P.-A. Blanche, A. Schülzgen, C. Fuentes-Hernandez, G. Ramos-Ortiz, J. F. Wang, N. Peyghambarian, S. R. Marder, A. Leclercq, D. Beljonne, J.-L. Bredas, Adv. Funct. Mater. 12 (2002) 615.
24. E. Mechner, F. Gallego-Gomez, H. Tillmann, H.-H. Hörhold, J. C. Hummelen, K. Meerholz, Nature 418 (2002) 959.

5
Photochromism

5.1
Introductory remarks

There are substances that are transformed from form A into form B, having a different absorption spectrum, upon the absorption of light of wavelength λ_1 and that return to the initial state A either thermally or by the absorption of light of wavelength λ_2 (see Scheme 5.1).

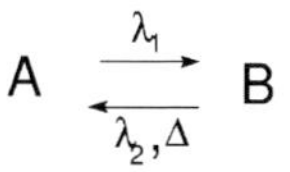

$$A \underset{\lambda_2, \Delta}{\overset{\lambda_1}{\rightleftarrows}} B$$

Scheme 5.1 Photochromic transformation of molecules.

Substances capable of undergoing color changes in this way are denoted as photochromic and the corresponding phenomenon is termed *photochromism*. As can be seen from Table 5.1, in which typical photochromic systems are presented, photochromism can be based on various chemical processes.

trans-cis (E/Z) Isomerization occurs in azobenzene compounds (example (a)) and also in the cases of azines, stilbenes, and certain biological receptors in living systems. Pericyclic reactions (electrocyclizations) occur in the cases of spiropyrans and spirooxazines (examples (b) and (c)), and also with diarylethenes (example (d)) and fulgides (example (e)). Heterolytic bond cleavage resulting in ionic dissociation occurs in the case of triphenylmethanes (example (f)). Concise information on organic photochromism, including details of the various families of photochromic compounds and the chemical processes involved in photochromic transformations, is given in an IUPAC Technical Report [1]. Moreover, this subject has been dealt with in various review articles and books that emphasize its importance and potential for applications in the fields of molecular switches and information storage [2–9]. With respect to the present book, various publications focusing on polymers have to be pointed out [10–21].

The transformations presented in Table 5.1 are always accompanied by changes in physical properties. Besides the color changes, there are also changes in dipole moment and in the geometrical structure at the molecular level. Regarding bulk properties, there are changes in the refractive index, which give rise to photo-induced birefringence and dichroism.

Polymers and Light. Fundamentals and Technical Applications. W. Schnabel
Copyright © 2007 WILEY-VCH Verlag GmbH & Co. KGaA, Weinheim
ISBN: 978-3-527-31866-7

5 Photochromism

Table 5.1 Typical photochromic processes.

trans-cis Isomerization
(a) Azobenzene:

trans ⇌ (λ₁ forward, Δ or λ₂ reverse) cis

Pericyclic reactions
(b) Spiropyrans:

Closed Form ⇌ (λ₁ forward, Δ or λ₂ reverse) Open Form

(c) Spirooxazines:

Closed Form ⇌ (λ₁ forward, Δ or λ₂ reverse) Open Form

(d) Diarylethenes:

Open Form ⇌ (λ₁ forward, λ₂ reverse) Closed Form

(e) Fulgides and fulgimides:
(X = O) (X = NR)

Open Form ⇌ (λ₁ forward, λ₂ reverse) Closed Form

Heterolytic bond cleavage
(f) Triarylmethanes:

⇌ (λ forward, Δ reverse)

With respect to polymeric systems containing photochromic groups, special aspects have to be addressed. For instance, in linear macromolecules, not only the chromophoric moieties but also neighboring units of the polymer chain or surrounding molecules may be affected upon the absorption of photons by the chromophoric groups. Conformational changes in linear polymers in solution induced in this way may lead to a change in viscosity or even to phase separation. For instance, in liquid-crystalline polymeric systems, phase transitions can be generated. In the case of rigid polymer matrices, photomechanical effects are induced, i.e. photoisomerization causes shrinkage or expansion. Interestingly, stable relief surface gratings can be generated in polymer foils containing photochromic moieties. Notably, the photostimulated conformational change in polymers may result in an enormous amplification effect, i.e. the absorption of a single photon affects not only one moiety but also several neighboring ones or even the whole macromolecule.

Potential applications of photochromic transformations relate to the reversible control of the properties of appropriate materials. In this connection, polymers offer the advantage of easy fabrication and, therefore, a plethora of studies has been devoted to polymers containing photochromic groups or to polymers with admixed photochromic compounds. Apparently, among the various photochromic polymeric systems dealt with in the literature, those containing azobenzene groups [19, 20] have attracted the main interest, although it seems that others, particularly those containing diarylethenes [5] and furyl fulgides [6], deserve special attention because of their excellent performance. Light-induced coloration/discoloration cycles could be repeated more than 10^4 times with certain diarylethenes, thus proving their extraordinary resistance to fatigue [5]. Thermal irreversibility and fatigue resistance are prerequisites for applications related to data storage and switching of photonic devices [21], which are considered in Chapter 12 of this book.

5.2
Conformational changes in linear polymers

5.2.1
Solutions

Photochromic transformations may induce conformational changes in linear macromolecules containing appropriate chromophoric groups. Commonly, the transformation of these groups is accompanied by a change in polarity. This change is most pronounced if the transformation generates electrically charged groups, e.g. in the cases of triphenylmethane or spiropyran groups. However, azobenzene groups also undergo a drastic change in polarity. The change in the geometry of the azobenzene group from the planar (*trans* or *E*-form) to the non-planar (*cis* or *Z*-form) leads to a decrease in the distance between the *para* carbon atoms of the benzene rings from 9.9 to 5.5 Å and to an increase in the di-

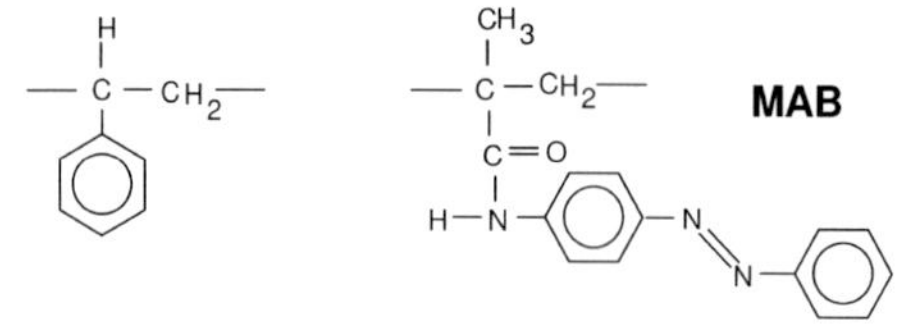

Chart 5.1 Chemical structures of co-monomer moieties: styrene (left) and 4-(methylacryloylamino)azobenzene (right).

pole moment from 0.5 to 5.5 D. Regarding linear polymers containing pendant photochromic groups, the change in polarity affects not only the intermolecular interaction between the chromophore and surrounding solvent molecules, but also the intramolecular interaction between pendant groups. As a consequence, random coil macromolecules undergo conformational alterations leading to expansion or shrinkage. For example, a copolymer with pendant azobenzene groups consisting of styrene and 4–6 mol% 4-(methylacryloylamino)azobenzene, MAB (see Chart 5.1), precipitates in dilute cyclohexane solution at temperatures above the critical miscibility temperature upon irradiation with UV light. This phenomenon is explained in terms of *cis*-azobenzene groups having, in contrast to *trans*-azobenzene groups, the capability of interacting rather strongly with styrene moieties. Therefore, immediately after *trans-cis* isomerization, *cis*-azobenzene groups interact preferentially with neighboring styrene moieties, thus causing a contraction of the coil. Interactions of the *cis*-azobenzene groups with styrene moieties of other macromolecules result in aggregation, a process that ultimately leads to precipitation [22, 23]. This is illustrated schematically in Fig. 5.1.

In solution, coil expansion and contraction is readily reflected by changes in viscosity and in the intensity of scattered light. As can be seen in Fig. 5.2, the optical absorption at 620 nm and the reduced viscosity, η_{spec}/c, increase simultaneously when a poly(N,N-dimethylacrylamide) sample containing 9.1 mol% pendant triphenylmethane leucohydroxide groups is irradiated in dilute methanol solution with UV light ($\lambda > 270$ nm). In the dark, the reduced viscosity returns to the initial value. The development of a green color in conjunction with the increase in the viscosity indicates the formation of triphenylmethyl cations. Obviously, the polymer coils become expanded due to electrostatic repulsion of ionized pendant groups formed according to Scheme 5.2 [24].

In the case of an azobenzene-modified poly(arylether ketone amide) (see Chart 5.2), a pronounced volume contraction due to photo-induced *trans-cis* isomerization of the azobenzene groups was evidenced by means of size-exclusion chromatography (SEC) [25]. When irradiated in dilute N,N-diethylacetamide solution, this polymer underwent a reduction in its hydrodynamic radius by a factor of 2.7, corresponding to a contraction of the hydrodynamic volume by a factor of about 20. This pronounced shrinkage effect is believed to be due to a large number of conformationally restricted backbone segments, because other more flexible polyamides and polyurea polymers exhibit much weaker contraction effects.

Intramolecular Interaction

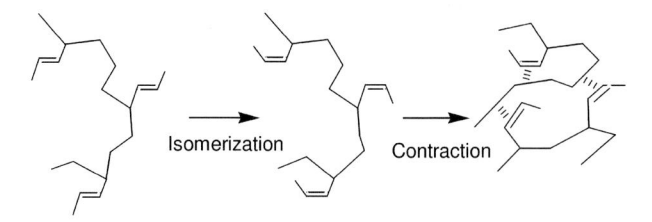

Intermolecular Interaction

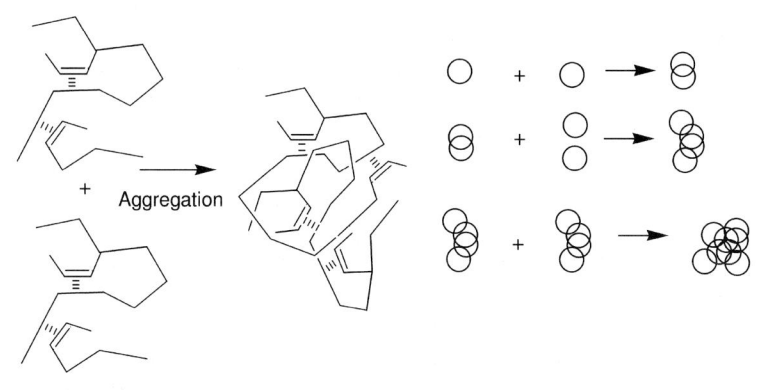

Fig. 5.1 Coil contraction and precipitation of polystyrene bearing pendant azobenzene groups.

The dynamics of conformational changes can be measured by following the change in the light-scattering intensity. Relevant studies relate to a polyamide containing in-chain azobenzene groups (see Chart 5.3) that was brought into the compact form through *trans-cis* isomerization by continuous UV irradiation in N,N-dimethylacetamide solution and subsequently exposed to a 20 ns flash of 532 nm light. On recording the changes in the optical absorption and in the light-scattering intensity, both at $\lambda=514$ nm, as a function of time, it turned out that the *cis-trans* isomerization was completed within the 20 ns flash and that the polymer chains unfolded on the ms time scale. Obviously, after isomerization, the polymer chains maintain the initial compact conformation, and the strain energy built-up in this way causes coil expansion [26]. The whole process is shown schematically in Scheme 5.3.

The possibility of photo-inducing geometrical alteration in polymers in solution has attracted special interest with regard to various polypeptides (see Chart 5.4).

Besides unordered random coil structures, polypeptides are capable of assuming stable geometrically ordered structures, namely α-helix and β-structures. As shown in Fig. 5.3, these structures can be conveniently discriminated by recording circular dichroism (CD) spectra [14].

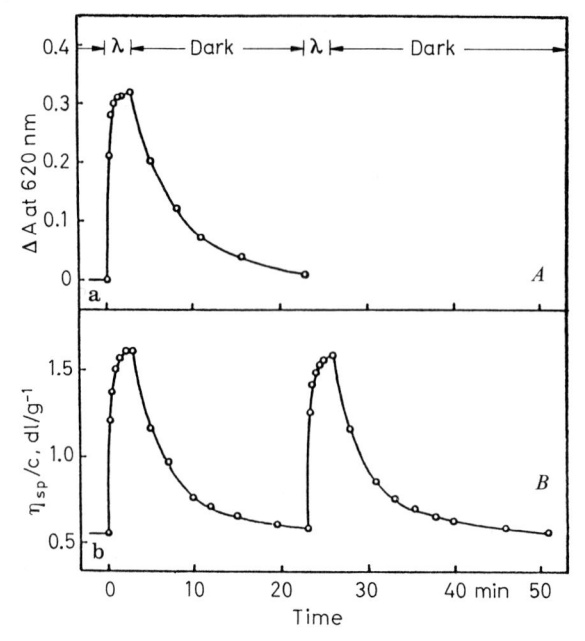

Fig. 5.2 Coil expansion of poly(N,N-dimethylacrylamide) containing pendant triphenylmethane leucohydroxide (9.1 mol%) in methanol upon exposure to UV light ($\lambda > 270$ nm). (a) Optical absorption at $\lambda = 620$ nm, (b) reduced viscosity η_{spec}/c ($\eta_{spec} = (\eta_{solution}/\eta_{solvent}) - 1$). Adapted from Irie [11] with permission from Springer.

Scheme 5.2 Photogeneration of triphenylmethyl cations in poly(N,N-dimethylacrylamide) containing pendant triphenylmethane leucohydroxide groups.

Chart 5.2 Chemical structure of an azobenzene-modified poly(arylether ketone amide).

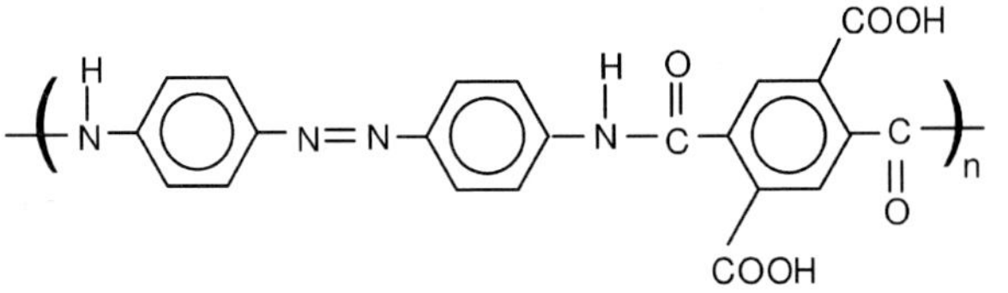

5.2 Conformational changes in linear polymers

Chart 5.3 Chemical structure of a polyamide containing in-chain azobenzene groups.

Compact Conformation

10^{-7} s | cis-trans isomerization

Compact Conformation

10^{-3} s | unfolding

Extended Conformation

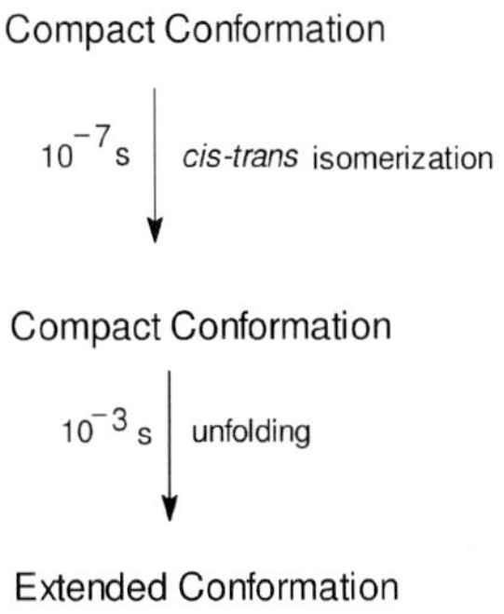

Scheme 5.3 Conformational change of a polyamide containing in-chain azobenzene groups due to cis-trans isomerization.

poly(L-lysine)   poly(L-glutamic acid)

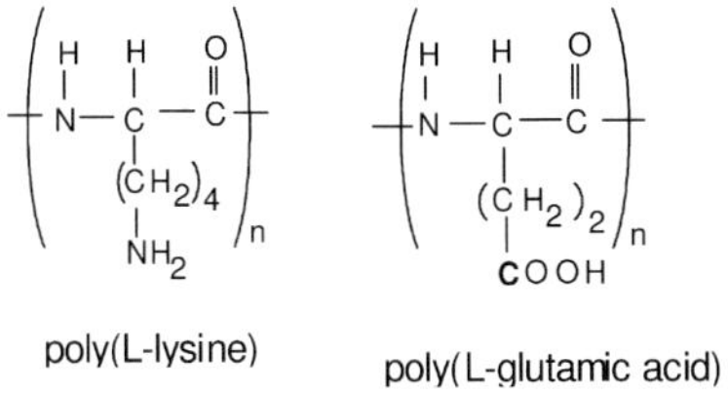

Chart 5.4 Chemical structures of poly(L-lysine) and poly(L-glutamic acid).

Light-induced transformations from one structure to another have been studied with many modified polypeptides [13, 14] bearing pendant photochromic groups such as azobenzene or spiropyran groups. Typical examples are the modified poly(L-glutamic acids) PGA-1 and PGA-2 presented in Chart 5.5.

The spiropyran-modified poly(L-glutamic acid) PGA-2 undergoes a coil→helix transition upon exposure to visible light in hexafluoro-2-propanol solution. In the dark, the polypeptide, containing 30–80 mol% chromophore units in the open charged form, adopts a random coil conformation. Irradiation causes isomerization in the side chains, as indicated by complete bleaching of the colored solution (see Scheme 5.4). The formation of the colorless and uncharged spiropyran form induces spiralization of the polypeptide chain. The coil→helix transition can be followed with the aid of CD spectra, as shown in Fig. 5.4.

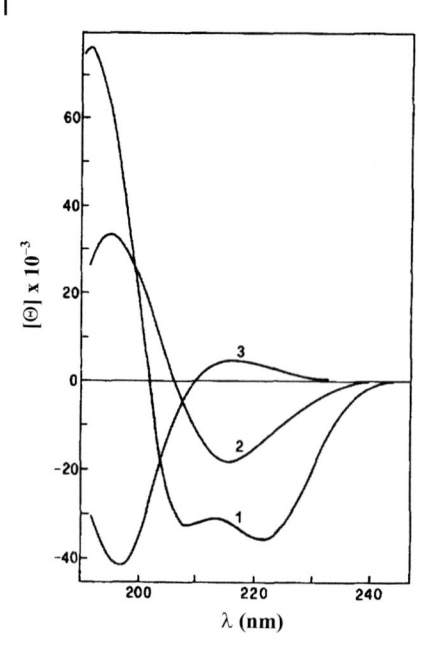

Fig. 5.3 Standard circular dichroism (CD) spectra of common polypeptide structures: (1) α-helix, (2) β-structure, and (3) random coil. Adapted from Pieroni et al. [14] with permission from Elsevier.

The coil → helix transition proceeds rapidly within seconds, whereas the back reaction requires several hours for full conversion. Notably, in this case, the photochromic behavior of the spiropyran groups is opposite to that observed in other solvents (see example (b) in Table 5-1). The reverse photochromism is due to the high polarity of hexafluoro-2-propanol, which stabilizes the charged merocyanine form better than the uncharged spiropyran form.

Chart 5.5 Chemical structures of modified poly(L-glutamic acids).

Scheme 5.4 Isomerization of the spiropyran-modified poly(L-glutamic acid) PGA-2.

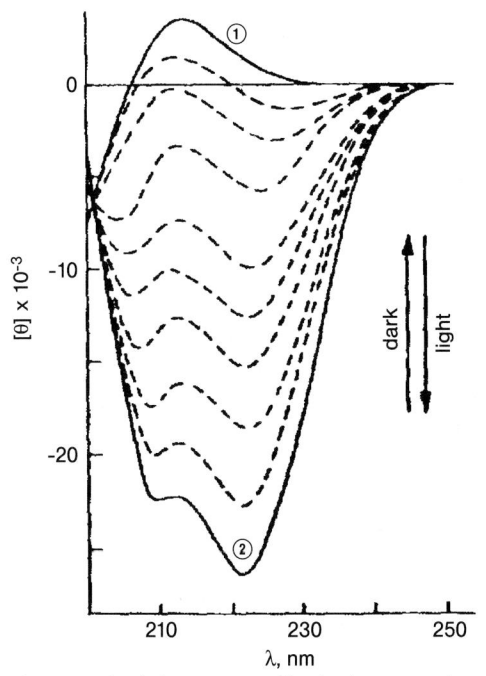

Fig. 5.4 Coil → helix transition of poly(glutamic acid) PGA-2 containing 80 mol% spiropyran units in the side chains. CD spectra recorded in hexafluoro-2-propanol solution in the dark (1) and after exposure to sunlight (2). Adapted from Pieroni et al. [14] with permission from Elsevier.

5.2.2
Membranes

As an extension of the work described in the previous section, one goal was the development of artificial membranes, the physical properties of which, such as permeability, electrical conductivity, and membrane potential, could be controlled in response to light. Typically, in the case of membranes consisting of poly(L-glutamic acid) bearing azo groups in the side chains, the water content increases upon light exposure. Concomitantly, the dissociation of acid groups is accelerated and augmented and the potential across the membrane and the cross-membrane conductance are enhanced [15]. Typical results are presented in Fig. 5.5.

Moreover, a low molar mass spiropyran compound entrapped in a membrane consisting of plasticized poly(vinyl chloride) rendered the latter photoresponsive. A membrane potential change of more than 100 mV was induced by irradiation with light [27]. For further details and additional references, the reader is referred to the relevant reviews [11, 28].

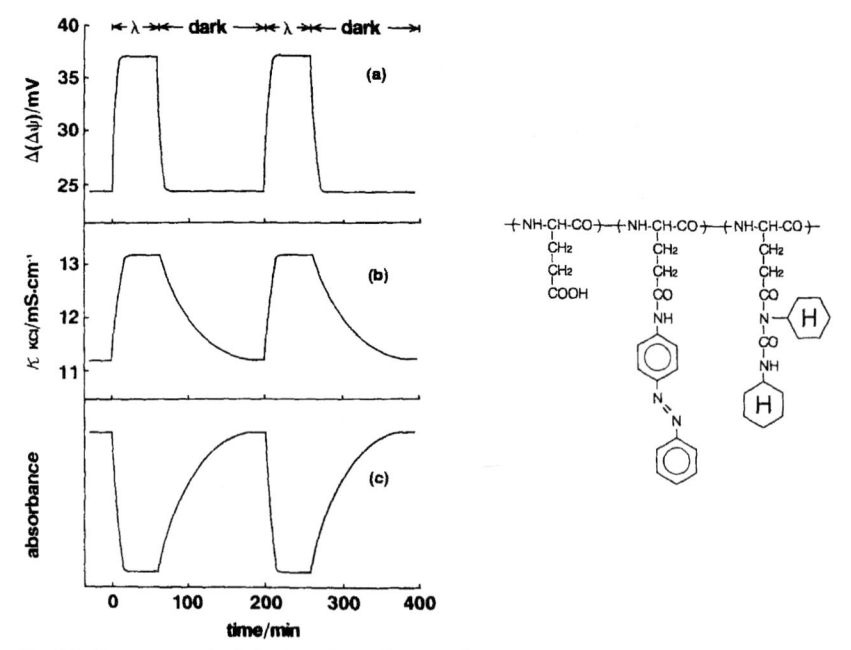

Fig. 5.5 Photoresponsive behavior of membranes of an azo-modified poly(L-glutamic acid) containing 12–14 mol% azobenzene groups at 60 °C. (a) Membrane potential, (b) conductance, and (c) absorbance at 350 nm. Adapted from Kinoshita [15] with permission from Elsevier.

5.3
Photocontrol of enzymatic activity

Photochromic groups covalently attached to enzymes are, in certain cases, capable of affecting the tertiary protein structure upon light-induced isomerization. As a consequence, the biocatalytic activity of the enzymes can be switched on and off [29]. For example, the catalytic activity of papain is inhibited when 4-carboxy-*trans*-azobenzene groups covalently linked to the lysine moieties of the enzyme undergo *trans-cis* isomerization (see Scheme 5.5). At a loading of five units per enzyme molecule, 80% of the catalytic activity is retained.

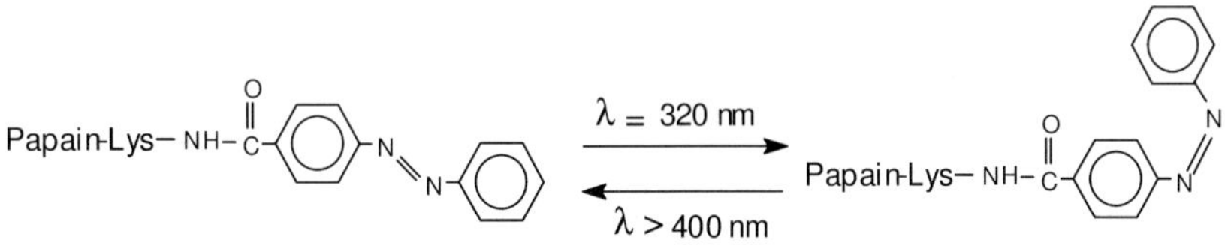

Scheme 5.5 Photoisomerization of azobenzene groups covalently linked to the lysine moieties of papain.

The inactivity of enzyme molecules bearing *cis*-azobenzene groups is explained by their incapability of binding to the reaction substrate. Similarly, the binding of α-D-manopyranose to concanavalin A is photocontrollable, provided that the enzyme is modified by the attachment of thiophenefulgide or nitrospiropyran. However, the general applicability of this method has to be subject to scrutiny because the photoswitching behavior is quite sensitive to the level of loading. Low loadings may result in a low switching efficiency and high loadings often deactivate the biomaterials in both isomeric forms.

5.4
Photoinduced anisotropy (PIA)

Exposure of polymer films bearing azobenzene groups to linearly polarized laser light induces optical dichroism and birefringence. This is due to the fact that during exposure a major fraction of the chromophores becomes oriented perpendicular to the polarization direction of the light. Photons of linearly polarized light are preferentially absorbed by molecules with a transition moment parallel to the polarization plane of the light. The absorbed photons induce *trans-cis* isomerizations in conjunction with rotational diffusion. The relaxation of the *cis* molecules results in *trans* molecules with a new orientation distribution, i.e. the fraction of *trans* molecules with a transition moment parallel to the polarization plane of the incident light becomes smaller. Continuous repetition of this cycle steadily reduces this fraction and makes the system more transparent to the incident light as the *trans* molecules can no longer be excited.

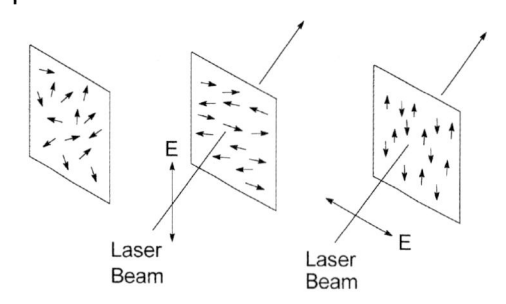

Fig. 5.6 Schematic illustration of the generation of anisotropy upon irradiation of a film containing photochromic entities with linearly polarized light.

To sum up, during the irradiation, azobenzene groups with transition moments that are not initially perpendicular to the polarization direction of the laser light undergo a series of *trans-cis-trans* isomerization cycles accompanied by a change in orientation until they finally line up in directions approximately perpendicular to the polarization direction of the laser light (see Fig. 5.6).

In this way, an orientation distribution with an excess of azobenzene groups oriented in the direction perpendicular to the polarization plane of the laser light is attained. The resulting birefringence can be detected with the aid of another laser beam that is not absorbed by the photochromic compound. Notably, the anisotropy can be erased if the sample is irradiated with circularly polarized laser light or is heated to a temperature in excess of the glass transition temperature. This behavior is demonstrated for a typical case in Fig. 5.7. Here, it can be seen that the birefringence (monitored at 633 nm) of a 400–500 nm thick film of pMNAP polymer (see Chart 5.6) is built up upon irradiation with a linearly polarized laser beam (λ=488 nm) [30]. The birefringence relaxes down to a certain level when the writing beam is turned off and is completely eliminated upon turning on a circularly polarized light beam (λ=488 nm).

Photo-induced anisotropy (PIA) is quantitatively described by Eqs. (5-1) and (5-2), by Δn in terms of the induced birefringence and by the parameter S in terms of light absorption behavior.

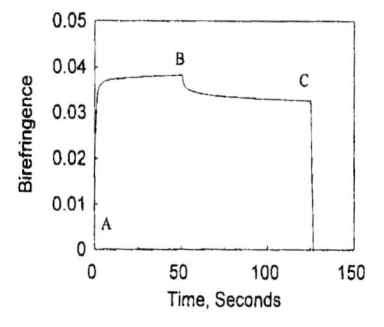

$$\Delta n = n_{\parallel} - n_{\perp} \tag{5-1}$$

Fig. 5.7 Generation of birefringence upon irradiation of pMNAP polymer with linearly polarized light (λ=488 nm). A: light turned on; B: light turned off; C: circularly polarized light turned on. Adapted from Meng et al. [30] with permission from John Wiley & Sons, Inc.

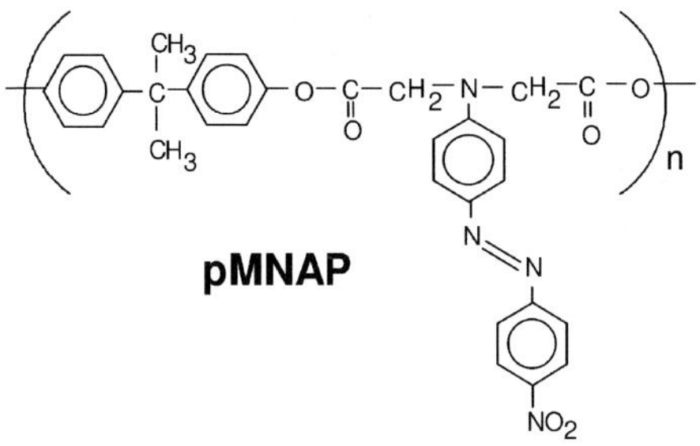

Chart 5.6 Chemical structure of pMNAP polymer used for the photo-generation of birefringence (see Fig. 5.7).

pMNAP

$$S = (A_\| - A_\perp)/(A_\| + 2A_\perp) \tag{5-2}$$

Here, $A_\|$ and $A_\perp$ and $n_\|$ and $n_\perp$ denote the absorbances and the refractive indices at orientations parallel and perpendicular to the polarization plane of the exciting probe light, respectively.

In recent years, optical dichroism and birefringence based on photo-induced *trans-cis-trans* isomerization of azobenzene groups has been observed with pre-oriented liquid-crystalline polymers [31-35] at temperatures above the glass transition temperature, and also with various amorphous polymers at temperatures well below the glass transition temperature. In the case of a polyimide (see Chart 5.7), a quasi-permanent orientation can be induced [36–38]. Here, the azobenzene groups are rather rigidly attached to the backbone and photoisomerization occurs at room temperature, i.e. 325 °C below the glass transition temperature, $T_g = 350$ °C. This behavior is in accordance with the fact that the isomerization quantum yields of azobenzene compounds are very similar in solution and in polymer matrices: $\Phi(trans \rightarrow cis) \approx 0.1$ and $\Phi(cis \rightarrow trans) \approx 0.5$.

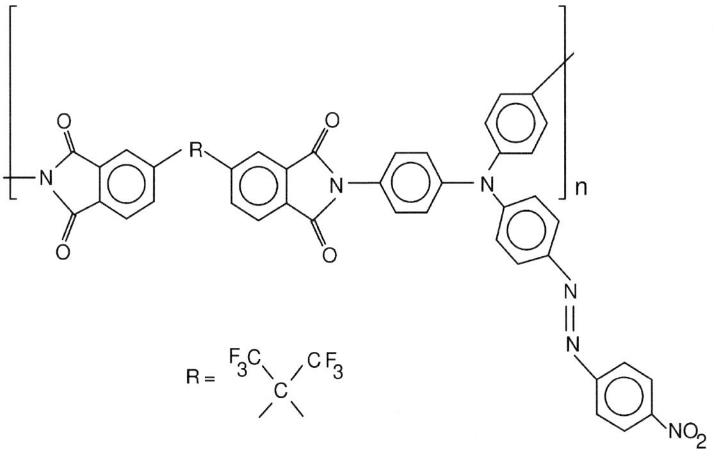

Chart 5.7 Chemical structure of a polyimide bearing pendant azobenzene groups.

Because of the importance of the PIA phenomenon for applications in optical data storage systems, a large variety of homopolymers and copolymers has been studied and the reader is referred to the literature cited in a relevant review article [39]. In this connection, it is also worthwhile to cite work performed with cyclic siloxane oligomers bearing pendant photochromic groups. Compounds of this family possessing relatively high glass transition temperatures and capable of forming cholesteric liquid-crystalline phases have been examined as potential optical recording materials [40].

5.5
Photoalignment of liquid-crystal systems

It has been shown in Section 5.4 that linearly polarized laser light induces a change in the orientation of azobenzene groups contained in polymers. Interestingly, this change in orientation can be greatly amplified if the azobenzene groups are contained in liquid-crystalline polymers. This phenomenon, which has been the subject of extensive investigations [16, 41–44], is described here in some detail for the case of a methacrylate-based copolymer consisting mainly of non-photosensitive mesogenic side groups and a small fraction of azobenzene-containing side groups (see Chart 5.8) [45].

Initially, this copolymer is an isotropic (polydomain) liquid-crystalline polymer with a glass transition temperature of $T_g = 45\,°C$ and a clearing temperature (transition from nematic to isotropic phase) of $T_{N-I} = 112\,°C$. Irradiation with linearly polarized light at $\lambda = 366$ nm (2.8 mW cm^{-2}) and $T = 106\,°C$, i.e. just below T_{N-I}, induces anisotropy. By repetitive *trans-cis-trans* isomerization, the optical axis of the azobenzene groups becomes aligned perpendicular to the electric vector of the incident light. In this way, a cooperative motion of the neighboring photoinactive mesogenic groups is triggered. Thus, the entire assembly of mesogenic side groups becomes aligned in one direction and forms a monodomain

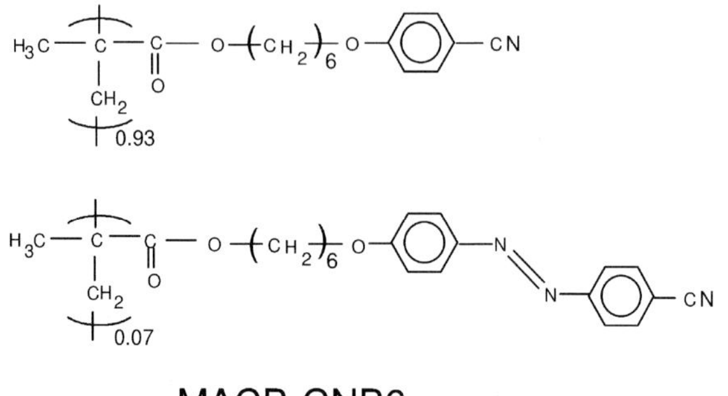

MACB-CNB6

Chart 5.8 Chemical structures of the components of a liquid-crystalline copolymer exhibiting amplified photoalignment (see Fig. 5.8).

nematic phase. This was evidenced by measuring the transmittance of an irradiated (λ_{exc}=633 nm) copolymer film placed between a pair of crossed polarizers at various rotation angles. As can be seen in Fig. 5.8, the transmittance has maxima at 45°, 135°, 225°, and 315°, and minima at 0°, 90°, 180°, and 270°.

Materials such as the LC copolymer considered here possess an application potential for image storage. This is demonstrated in Fig. 5.9, which shows (a) the transmittance response of the copolymer during alternating irradiation with polarized and unpolarized light, and (b) a one-year-old stored image, which was generated by irradiation of a copolymer film through a standard photo mask [45].

The field of liquid-crystalline polymers is still growing and a significant number of the relevant papers deal with subjects related to photochemical and photophysical problems, as has been documented in several reviews [46–48]. The progress in research is demonstrated here by referring to an interesting development concerning the photochromic amplification effect based on the surface-assisted alignment of liquid-crystalline compounds in cells possessing so-called *command surfaces* [16, 41–43]. The latter consist of silica glass plates or polymer films bearing attached photochromic groups at an area density of about one unit per nm^2. The light-induced isomerization of the photochromic moieties triggers reversible alignment alterations of the low molar mass liquid-crystalline compounds contained in the cell. Chemical structures of appropriate compounds forming nematic crystalline phases are shown in Chart 5.9.

It should be noted that the intermolecular interaction between surface azobenzene units and liquid-crystal molecules is strongly determined by their chemical nature, an aspect that has been thoroughly investigated [43] but is not elaborated here. It is estimated that the amplification involves up to 10^4 liquid-

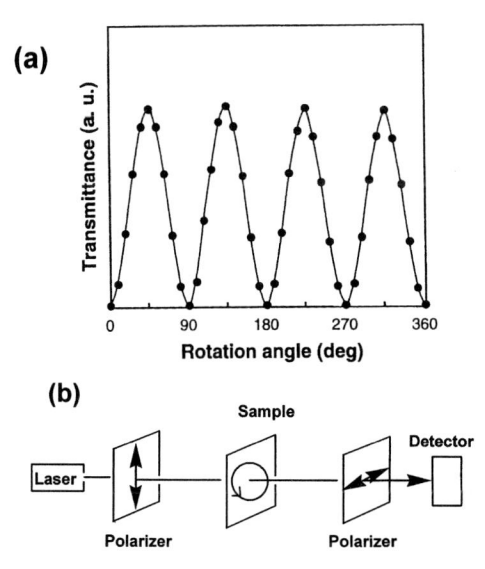

Fig. 5.8 Alignment of liquid-crystal copolymer MACB-CNB6 upon 30 min of exposure to polarized light at λ=366 nm (2.8 mW cm^{-2}) at 106 °C. (a) Transmittance of probe light (633 nm) through a 2 μm thick copolymer film placed between crossed polarizers as a function of the rotation angle. (b) Experimental set-up. Adapted from Wu et al. [45] with permission from Elsevier.

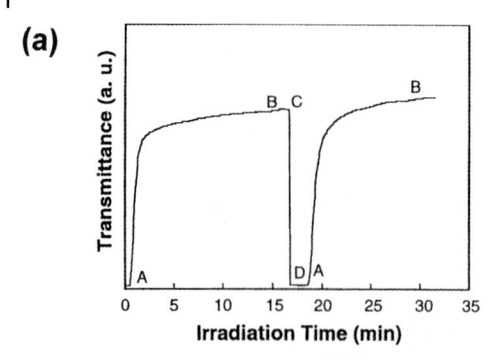

Fig. 5.9 (a) Transmittance response of copolymer MACB-CNB6 during irradiation with polarized light (A to B) and unpolarized light (C to D) at 106 °C. (b) One-year-old image stored in the liquid-crystal copolymer. The film was covered with a photo mask during irradiation with polarized light at $\lambda = 366$ nm (2.8 mW cm^{-2}) and 106 °C. Adapted from Wu et al. [45] with permission from Elsevier.

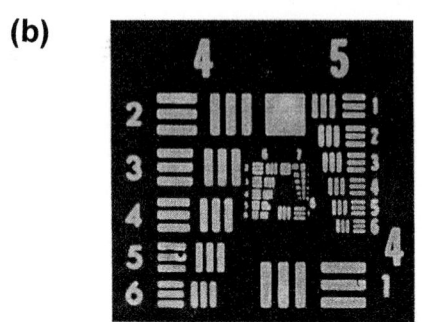

crystalline molecules per elementary isomerization process. The response time of the cells is determined by τ_{relax}, the relaxation time of the nematic phase. Values of τ_{relax} typically range from 50 to 300 ms [43] and so are several orders of magnitude longer than isomerization times, which are of the order of picoseconds. Figure 5.10 schematically depicts, for the case of azobenzene chromophores as the active entities at the surface, how irradiation with unpolarized light induces an alignment change from the homeotropic to the planar homogeneous state.

Notably, this kind of alignment change can also be accomplished by applying an electric field. On the other hand, alignment changes between planar homo-

Chart 5.9 Compounds forming nematic liquid-crystalline phases appropriate for photoalignment [43].

homeotropic   **planar**

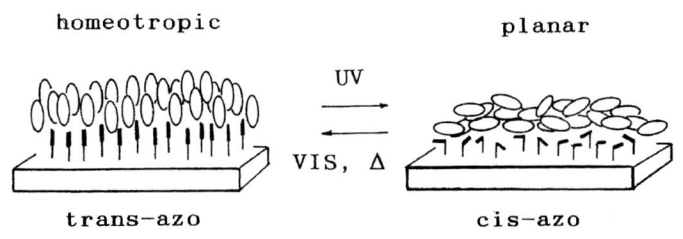

                              UV
                         ⟶
                         ⟵
                        VIS, Δ

trans-azo   **cis-azo**

Fig. 5.10 Light-induced surface-assisted alignment change in a liquid-crystal cell. Schematic depiction of the out-of-plane change from the homeotropic state to the planar homogeneous state upon exposure to unpolarized UV light. Adapted from Ichimura [43] with permission from Springer.

                    linearly polarized
                          light

homogeneous   homogeneous

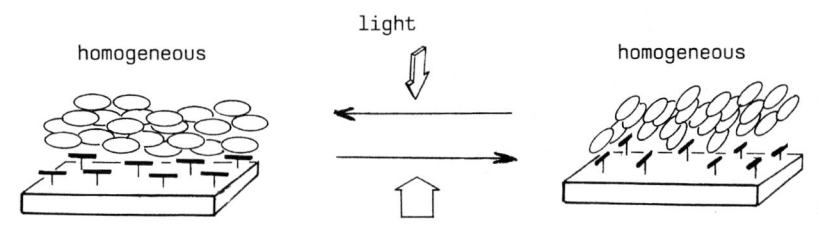

Fig. 5.11 Light-induced surface-assisted alignment change in a liquid-crystal cell. Schematic depiction of the in-plane change between homogeneous planar states under the influence of linearly polarized light. Adapted from Ichimura [43] with permission from Springer.

geneous states, not realizable with the aid of an electric field, can be achieved by employing linearly polarized light. An alignment change induced by an azimuthal in-plane reorientation of the photochromic groups is depicted schematically in Fig. 5.11.

It has been reported that cells fabricated with azobenzene-modified surfaces and operating on the basis of alternate irradiation with UV and visible light become inactive after about 2000 cycles, which is thought to be due to side reactions occurring with a quantum yield of about 10^{-4} [43].

5.6
Photomechanical effects

5.6.1
Bulk materials

The idea of transforming light into mechanical energy has fascinated many researchers. In the early studies, reviewed by Irie [11], contraction/expansion behavior in conjunction with isomerization of photochromic entities either admixed to or chemically incorporated into polymer films was found. However, the dimensional changes were only marginal, amounting to 1% or less, and on scrutiny, turned out in many cases to be due to the local increase in temperature arising from non-radiative transitions rather than to isomerization of the chromophores.

Large, real effects, on the other hand, were observed with hydrogels. A typical result is presented in Fig. 5.12, which shows how a polyacrylamide gel containing 1.9 mol% triphenylmethane leucocyanide swells upon irradiation with UV light at 25 °C [49]. The swelling is correlated to a 18-fold increase in the relative weight.

It can also be seen in Fig. 5.12 that in the dark the gel slowly attains the initial weight. More recently, rigid films (50×10×0.5 mm) of polyurethane–acrylate block copolymers containing nitrospiropyrans and nitro-bis-spiropyrans have been irradiated with 325 nm light at 20 °C in 5 min light/dark cycles [50]. The films expanded during irradiation and shrank in the dark with a response time of a few seconds in each case. The highest photomechanical responses were observed at a high acrylate content (72%), which rendered the system least elastic.

The possibility of converting light into mechanical energy has been impressively demonstrated with cross-linked liquid-crystalline polymeric systems containing azobenzene groups that were prepared by polymerizing previously aligned mixtures of acrylate 1-AC and diacrylate 2-AC (see Chart 5.10) [51].

Figure 5.13 shows how a film prepared from an 80/20 mol% mixture of 1-AC and 2-AC bends upwards towards the incident light ($\lambda=360$ nm). It becomes flat

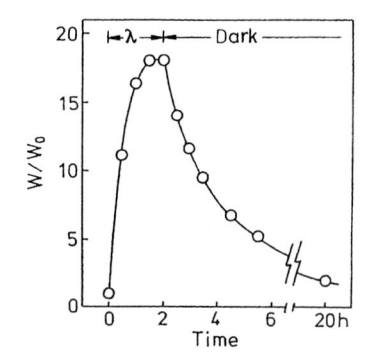

Fig. 5.12 Photomechanical effects. UV-light-stimulated dilatation of a polyacrylamide gel containing pendant triphenylmethane leucocyanide groups (1.9 mol%) at 25 °C. Adapted from Irie et al. [49] with permission from the American Chemical Society.

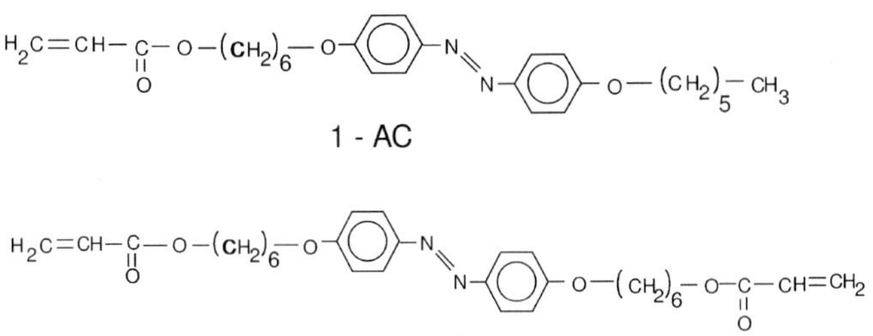

Chart 5.10 Monomers used to prepare cross-linked polymeric systems exhibiting photomechanical effects.

again upon irradiation at $\lambda=450$ nm. These processes are completed within 90 s. The anisotropic bending phenomenon caused by *trans-cis* isomerization may be explained in terms of a volume contraction. The latter is limited to a thin surface layer of the 10 μm thick film, in which the incident light is totally absorbed. Since the film mobility requires segment relaxation, the bending phenomenon can be observed with rigid films at $T>T_g$, in this case at $T=90\,°C$, or at room temperature with films swollen in a good solvent such as toluene.

The phenomenon of light-induced dimensional alterations in polymer films has been exploited for the generation of regular surface structures in azobenzene-containing polymers. The technique employed is based on the fact that azobenzene groups undergo reorientation due to repeated *trans-cis-trans* isomerization upon

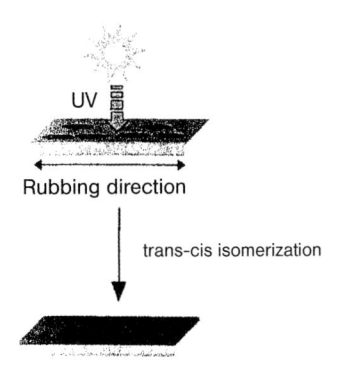

Fig. 5.13 Photomechanical effects. Schematic illustration of UV-light-induced bending of a cross-linked liquid-crystalline polymer film containing azobenzene groups. Light is absorbed at the upper surface layer of the film and causes anisotropic contraction. Adapted from Ikeda et al. [51] with permission from Wiley-VCH.

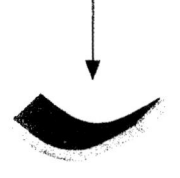

irradiation with polarized light (see Section 5.4) and that the target film is inhomogeneously irradiated. The reorientation results in a driving force that initiates mass transport from irradiated to unirradiated areas. The experimental set-up originally used to generate large surface gratings is shown in Fig. 5.14a [52, 53].

The gratings are optically inscribed onto the films with a single beam of an argon ion laser (488 nm, irradiation power between 1 and 100 mW), split by a mirror and reflected coincidently onto the film surface, which is fixed perpendicular to the mirror. The diffraction efficiency is monitored with the aid of a He-Ne laser beam (1 mW, $\lambda = 633$ nm). Changing the incident angle of the writing beam allows the intensity profile spacing on the sample, and thereby the grating spacing, to be changed. Under such conditions, irradiation of the polymer films for a few seconds at an intensity between 5 and 200 mW cm^{-2} produces reversible volume birefringence gratings with low diffraction efficiency. If

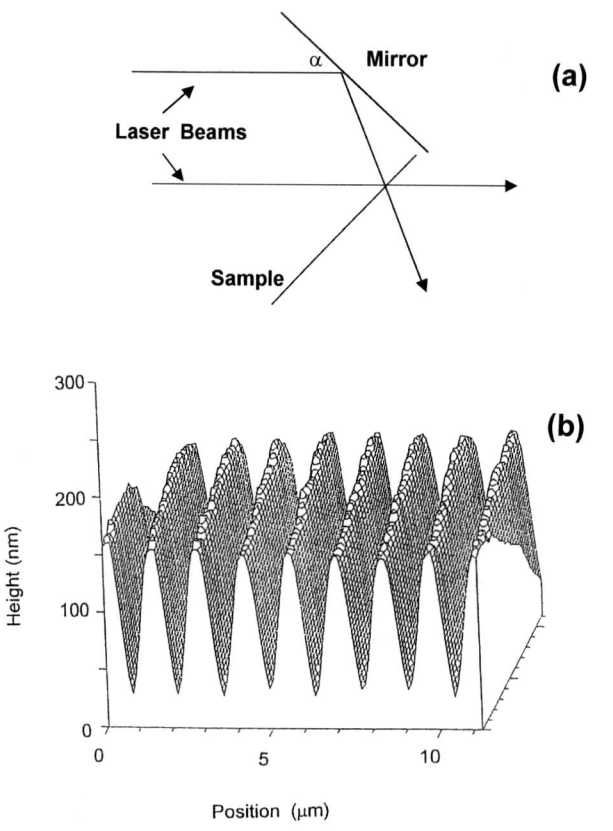

Fig. 5.14 Photomechanical effects. Generation of surface relief gratings in poly(4'-(2-acryloyloxy)ethylamino-4-nitroazobenzene) by light-induced mass transport. (a) Experimental set-up. (b) Sinusoidal surface relief profiles examined with the aid of an atomic force microscope. Adapted from Rochon et al. [53] with permission from the American Physical Society.

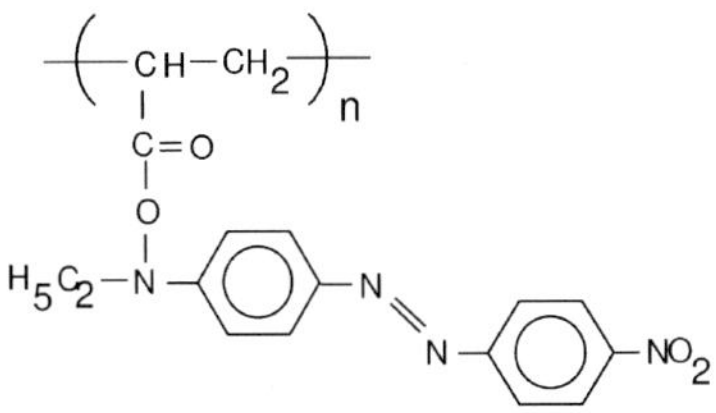

Chart 5.11 Chemical structure of poly(4'-(2-acryloyloxy)ethylamino-4-nitroazobenzene).

the film is exposed for a longer period (up to a few minutes), an irreversible process creates an overlapping and highly efficient surface grating. Thus, there is an initial rapid growth corresponding to the production of the reversible volume birefringence grating and a slower process which irreversibly creates surface gratings, observable by atomic force microscopy (AFM), with efficiencies of up to 50%. Figure 5.14 b shows a typical grating, generated in this case at the surface of a film of a polymer having the structure depicted in Chart 5.11.

Surface gratings have been generated in various azobenzene-modified polymers: epoxy polymers, polyacrylates, polyesters, conjugated polymers, poly(4-phenylazophenol), and cellulose [54–56].

5.6.2
Monolayers

Monolayers of a polypeptide consisting of two α-helical poly(L-glutamate)s linked by an azobenzene moiety (see Chart 5.12) become bent in the main

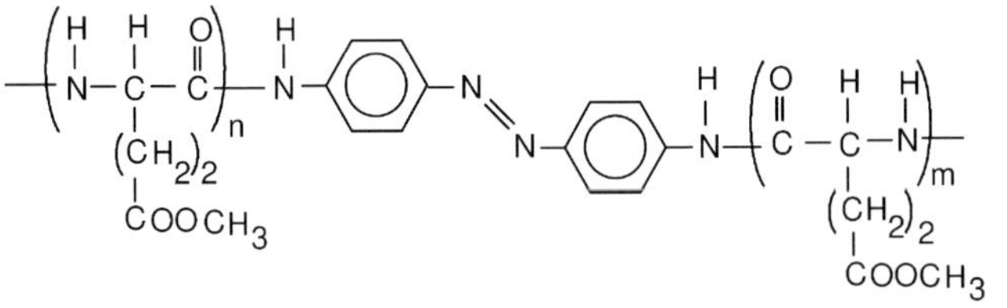

Chart 5.12 Chemical structure of a poly(L-glutamate) with in-chain azobenzene groups.

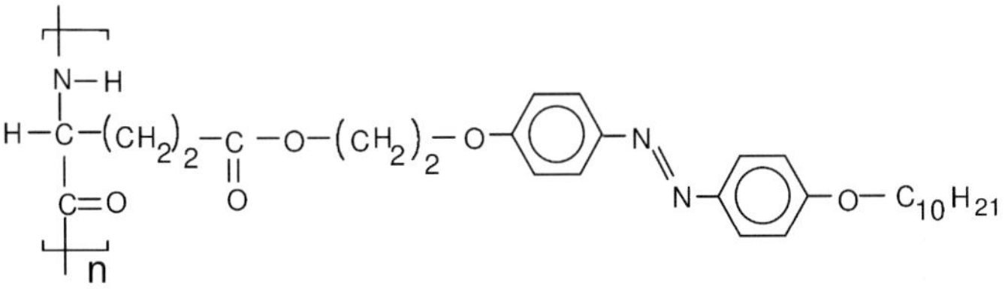

Chart 5.13 Chemical structure of a *hairy-rod-type* poly(glutamate) bearing pendant azobenzene groups.

chain to an angle of about 140° upon light-induced *trans-cis* isomerization. As a result, the area of the monolayer shrinks [57].

Photomechanical effects in monolayers have also been observed in other cases, for example with so-called *hairy-rod type poly(glutamate)s* (see Chart 5.13) [58].

5.7
Light-induced activation of second-order NLO properties

Apart from the aforementioned property alterations, photochromicity is frequently also connected with changes in nonlinear optical (NLO) properties. This is due to the fact that the two molecular species in a photochromic couple commonly exhibit different molecular NLO properties. Relevant studies have been performed with thin polymer films. For example, if spiropyran is transformed to merocyanine, the first hyperpolarizability β increases considerably. The second harmonic generation (SHG) increases by a factor of ten when a previously electric field-poled PMMA film doped with a spiropyran (see Chart 5.14) is irradiated at $\lambda = 355$ nm [59]. Subsequent irradiation at $\lambda = 514$ nm, at which merocyanine absorbs strongly, induces the reverse reaction, resulting in a drop of the SHG signal to almost zero. Figure 5.15 shows how the SHG signal changes in response to alternating irradiation with UV and visible light.

Clearly, the SHG signal decreases with increasing number of cycles, indicating that, in the absence of an external electric field, the chromophores become increasingly disorientated, i.e the NLO activity of the system is deactivated. Analogous behavior has been observed with a PMMA film doped with a furyl fulgide (see Chart 5.15). In this case, the ring-opening and -closure reactions need less free volume. Therefore, the matrix is less disturbed and the SHG signal decreases more slowly with increasing number of cycles.

Interestingly, the disorientation-induced distortion of the matrix can be avoided if the photoswitching is performed under an external electric field. This was demonstrated in the case of the polyimide of the structure shown in Chart 5.16 [60].

Here, the SHG signal decays under irradiation due to *trans-cis* isomerization and recovers almost completely in the dark after the light is switched off. The influence of the external electric field is thought to allow a compensation of the photo-induced distortion through photo-assisted poling.

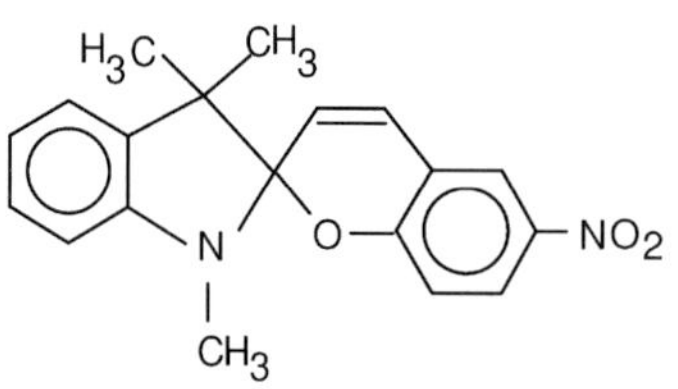

Chart 5.14 Chemical structure of 6-nitro-1′,3′,3′-trimethylspiro[2H-1-benzopyran-2,2′-indoline] [59].

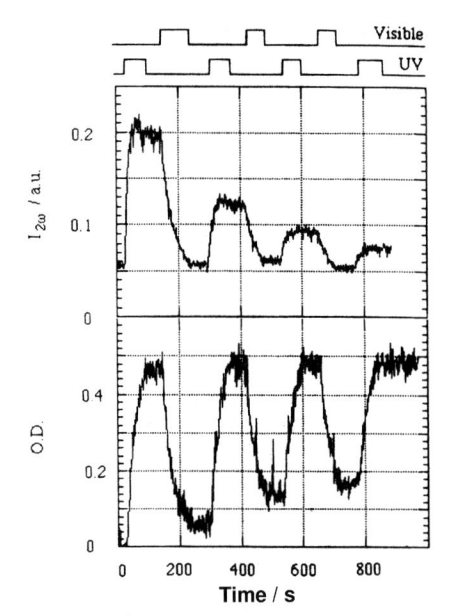

Fig. 5.15 Light-induced generation of second-order NLO properties in an electric field-poled PMMA film doped with 25 wt% of a spiropyran (see Chart 5.14). Alternating irradiation at $\lambda=355$ nm and $\lambda=514$ nm. Upper part: Second harmonic generation (SHG). Lower part: Optical absorption of the merocyanine isomer at $\lambda=532$ nm. Adapted from Atassi et al. [59] with permission from the American Chemical Society.

Chart 5.15 Chemical structure of furyl fulgide FF-1.

Chart 5.16 Chemical structure of a polyimide with pendant azobenzene groups.

5.8
Applications

5.8.1
Plastic photochromic eyewear

Besides classical inorganic glasses, there are certain optical plastics that are employed in the transparency and eyewear industry. For instance, thermoset resins based on allyl diglycol carbonate, poly(methyl methacrylate) derivatives, and bisphenol A polycarbonates have been used to produce commercial plastic non-photochromic and photochromic lenses. As far as has been disclosed by the manufacturers, indolinospironaphthoxazines, INSO, and pyridobenzoxazines,

Chart 5.17 Chemical structures of compounds that render plastic lenses photochromic.

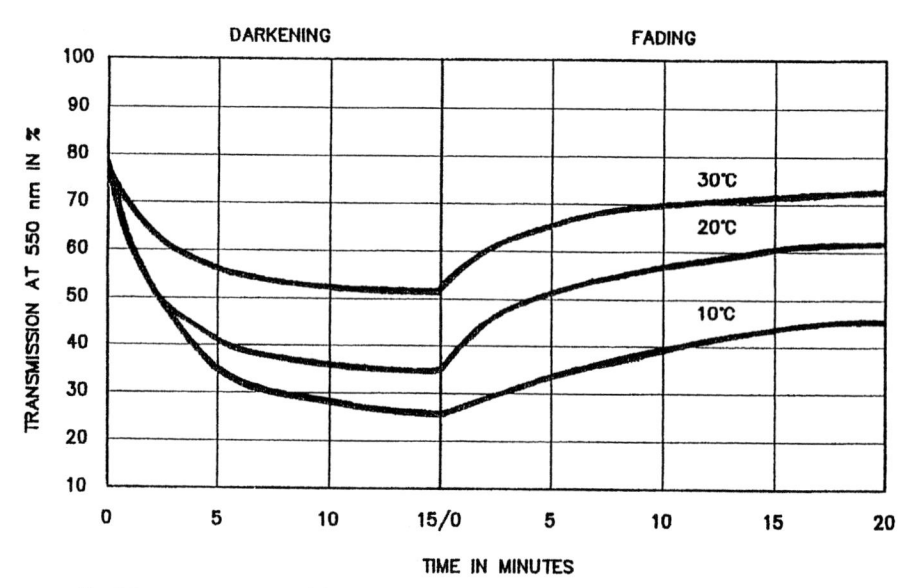

Fig. 5.16 UV activation and thermal bleach profiles at 10 °C, 20 °C, and 30 °C of a commercial photochromic lens based on indolinospironaphthoxazine. Adapted from Crano et al. [61] with permission from Springer.

QISO (see Chart 5.17), have received much attention among the compounds capable of rendering plastic lenses photochromic.

The photochromic compounds are incorporated at a concentration of 0.1–0.3%, either by admixing or by chemical bonding. In the latter case, modified compounds with appended polymerizable functionalities are employed. Photochromic lenses operate on the basis of UV activation and thermal bleaching as shown in Fig. 5.16.

As with most photochromic lenses, the performance of plastic photochromic lenses is temperature-dependent. In addition to variable light attenuation, photochromic lenses offer protection against UV light. Photochromic plastics coated onto classical glass lenses provide abrasion/scratch resistance and highly functional anti-reflectivity. For further details, the reader is referred to a review article [61].

5.8.2
Data storage

The availability of two states associated with the common photochromic process is a promising basis for erasable optical data storage systems, as outlined in a review article by Irie [62]. Besides sufficiently high quantum yields and rapid responses for both the forward and the reverse reaction, important requirements for device application include a high storage capacity, a long archival lifetime, and good intrinsic fatigue characteristics and cyclability, i.e. the number of times the interconversion can be made without significant performance loss. Obviously, a development of the recorded image should not be necessary.

Photochromic compound families that have been considered for employment in data storage systems include, for example, fulgides and diarylethenes. Compounds that have been examined, for instance, are the furyl fulgide FF-1 (see Chart 5.15) [63] and the diarylethene shown in Scheme 5.6. When dispersed in a polystyrene film, the latter system exhibited a strong fatigue resistance in a test using a low-power readout laser (633 nm, 20 nW). The initial optical density of 0.5 remained unchanged during more than 10^5 readout cycles [5, 64].

In this connection, the importance of fatigue resistance should be pointed out. If form A of a chromophoric couple A/B undergoes a side reaction with a quantum yield $\Phi_{side}=0.001$ and B converts to A without loss, 63% of the initial molecules of A will be decomposed after 1000 cycles. Thus, Φ_{side} should be less than 0.0001 if the system is expected to endure more than 10^4 cycles [65].

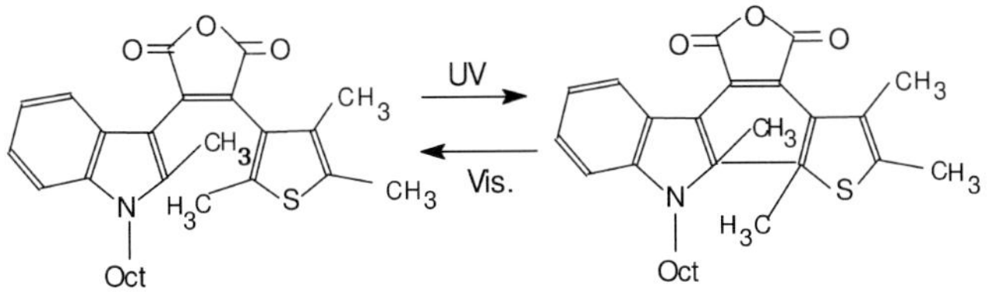

Scheme 5.6 Photoisomerization of 3-(1-octyl-2-methyl-3-indolyl)-4-(2,3,5-trimethyl-1-thienyl)maleic anhydride.

The search for materials appropriate for data storage has also been extended to liquid-crystalline copolymers containing photochromic moieties, and intensive studies have been focused on copolymers containing pendant azobenzene groups because of the possibility of generating anisotropy. Indeed, alignment alterations induced in such copolymers by exposure to linearly polarized light can be permanently frozen-in and stored. Since long durability is a prime requirement for information storage, materials with a high glass transition temperature (higher than 100 °C) seemed to be most appropriate [66]. However, in the case of a liquid-crystalline polyester (P6a12, see Chart 5.18) containing azobenzene side groups, holographically recorded gratings endured at room temperature over a period of several years, and up to 10^4 write-record-erase cycles could be accomplished [67, 68]. Notably, erasure is achieved by heating this polyester to approximately 80 °C. This temperature is much higher than the glass transition temperature of about 30 °C and corresponds to the clearing temperature, at which the liquid-crystalline domains form the mesophase melt.

Similarly, good long-term optical storage properties at room temperature have been reported for a liquid-crystalline copolymer composed of the moieties shown in Chart 5.19, with phase transitions at 48.7 °C (T_g), 83.2 °C (S_C), and 96.9 °C (S_A) [69].

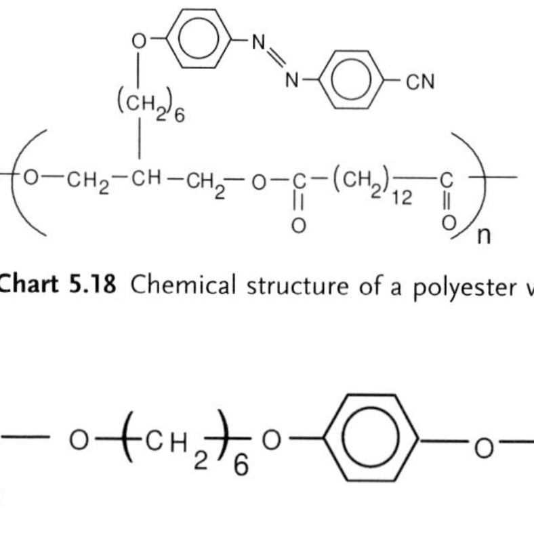

Chart 5.18 Chemical structure of a polyester with pendant azobenzene groups.

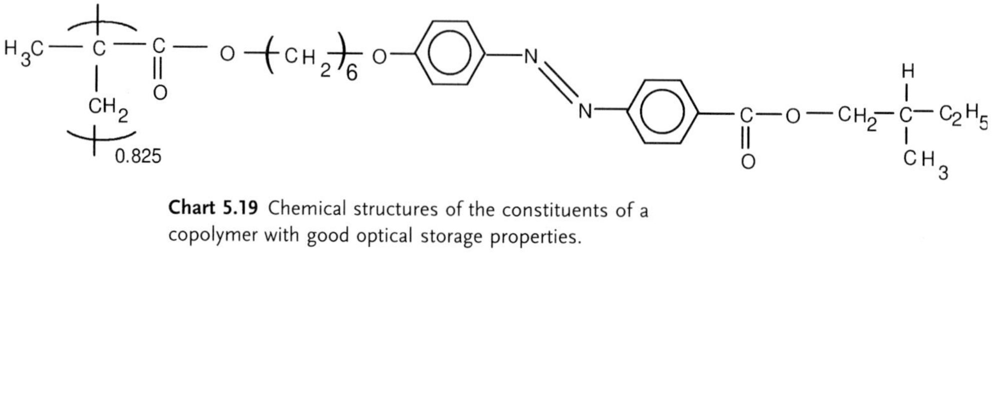

Chart 5.19 Chemical structures of the constituents of a copolymer with good optical storage properties.

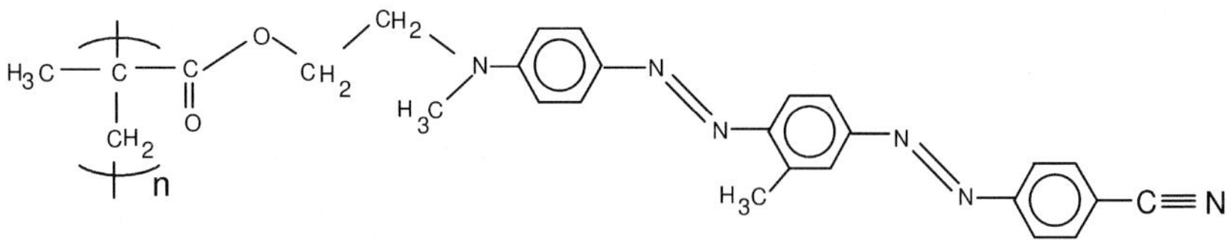

Chart 5.20 Chemical structure of a base unit of copolymers used for forgery-proof storage systems.

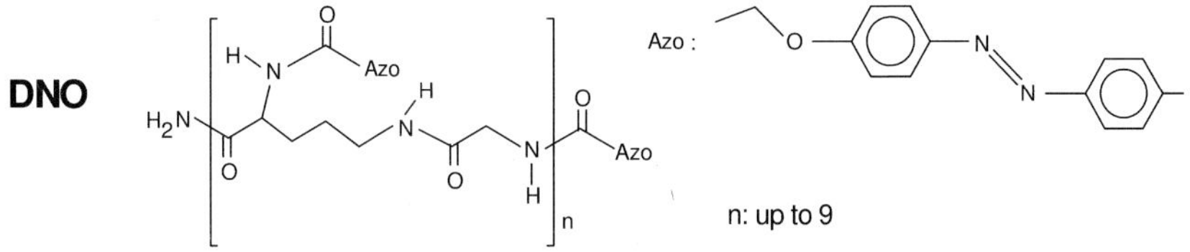

Chart 5.21 Chemical structure of oligopeptides with good optical storage properties.

Large induced birefringences [see Eq. (5-1)] up to $\Delta n = 0.36$ at 780 nm are obtained with liquid-crystalline copolymers containing the methyl methacrylate comonomer presented in Chart 5.20 [70, 71].

Since such copolymers possess, besides a high storage capacity, a high storage cyclability and, moreover, withstand temperatures up to 120 °C, they are utilized by Bayer Material Science for high-tech storage systems. The holography-related application potential of these materials includes forgery-proof storage systems, ID cards for access control to high security areas, etc. [72].

Regarding the heat resistance of potential storage materials, work on oligopeptides (see Chart 5.21) is also noteworthy. Holograms written in DNO films ($\lambda_{write} = 488$ nm, $\lambda_{read} = 633$ nm) remained stable at room temperature for up to one year and were not erased upon exposure to 80 °C for one month [73].

References

1 H. Bouas-Laurent, H. Dürr, *Organic Photochromism*, Pure Appl. Chem. 73 (2001) 639.
2 J. C. Crano, R. J. Guglielmetti (eds.), *Organic Photochromic and Thermochromic Compounds*, Vol. 1, Photochromic Families, Plenum Press, New York (1999).
3 G. H. Brown (ed.), *Photochromism, Techniques in Chemistry III*, Wiley-Interscience, New York (1971).
4 H. Dürr, H. Bouas-Laurent (eds.), *Photochromism: Molecules and Systems*, Elsevier, Amsterdam (1990).
5 M. Irie, Chem. Rev. 100 (2000) 1685.
6 Y. Yokoyama, Chem. Rev. 100 (2000) 1717.
7 G. Berkovic, V. Krongauz, V. Weiss, Chem. Rev. 100 (2000) 1741.
8 S. Kawata, Y. Kawata, Chem. Rev. 100 (2000) 1777.
9 N. Tamai, H. Miyasaka, Chem. Rev. 100 (2000) 1875.
10 C. B. McArdle (ed.), *Applied Photochromic Polymer Systems*, Blackie, Glasgow (1992).
11 M. Irie, Adv. Polym. Sci. 94 (1990) 27.

12 O. Nuyken, C. Scherer, A. Baindl, A. R. Brenner, U. Dahn, R. Gärtner, S. Kiser-Röhrich, R. Kollefrath, P. Matusche, B. Voit, Prog. Polym. Sci. 22 (1997) 93.
13 F. Ciardelli, O. Pieroni, *Photoswitchable Polypeptides*, in [21].
14 O. Pieroni, A. Fissi, G. Popova, Prog. Polym. Sci. 23 (1998) 81.
15 T. Kinoshita, Prog. Polym. Sci. 20 (1995) 527.
16 K. Ichimura, Chem. Rev. 100 (2000) 1847.
17 N. Hampp, Chem. Rev. 100 (2000) 1755.
18 J. A. Delaire, K. Nakatani, Chem. Rev. 100 (2000) 1817.
19 S. Xie, A. Natansohn, P. Rochon, Chem. Mater. 5 (1993) 403.
20 G. S. Kumar, G. Neckers, Chem. Rev. 89 (1989) 1915.
21 B. L. Feringa (ed.), *Molecular Switches*, Wiley-VCH, Weinheim (2001).
22 M. Irie, H. Tanaka, Macromolecules 16 (1983) 210.
23 M. Irie, W. Schnabel, *Light-Induced Conformational Changes in Macromolecules in Solution as Detected by Flash Photolysis in Conjunction with Light Scattering Measurements*, in B. Sedlacek (ed.), *Physical Optics of Dynamic Phenomena and Processes in Macromolecular Systems*, de Gruyter, Berlin (1985), p. 287.
24 M. Irie, M. Hosoda, Makromol. Chem. Rapid Commun. 6 (1985) 533.
25 M. S. Beattie, C. Jackson, G. D. Jaycox, Polymer 39 (1998) 2597.
26 M. Irie, W. Schnabel, Macromolecules 14 (1983) 1246.
27 J. Anzai, T. Osa, Tetrahedron 50 (1994) 4039.
28 O. Pieroni, F. Ciardelli, Trends in Polym. Sci. 3 (1995) 282.
29 I. Willner, Acc. Chem. Res. 30 (1997) 347.
30 X. Meng, A. Natansohn, P. Rochon, J. Polym. Sci., Polym. Phys. 34 (1996) 1461.
31 M. Eich, J. H. Wendorff, B. Reck, H. Ringsdorf, Makromol. Chem. Rapid Commun. 8 (1987) 59.
32 M. Eich, J. H. Wendorff, Makromol. Chem. Rapid Commun. 8 (1987) 467.
33 N. C. R. Holme, L. Nikolova, P. S. Ramanujam, S. Hvilsted, Appl. Phys. Lett. 70 (1997) 1518.
34 H. Ringsdorf, C. Urban, W. Knoll, M. Sawodny, Makromol. Chem. 193 (1992) 1235.
35 F. T. Niesel, J. Rubner, J. Springer, Makromol. Chem., Chem. Phys. 196 (1995) 4103.
36 Z. Seccat, P. Pretre, A. Knoesen, W. Volksen, V. Y. Lee, R. D. Miller, J. Wood, W. Knoll, J. Opt. Soc. Am. B. 15 (1998) 401.
37 Z. Seccat, J. Wood, W. Knoll, W. Volksen, R. D. Miller, A. Knoesen, J. Opt. Soc. Am. B 14 (1997) 829.
38 Z. Seccat, J. Wood, E. F. Aust, W. Knoll, W. Volksen, R. D. Miller, J. Opt. Soc. Am. B 13 (1996) 1713.
39 J. A. Delaire, K. Nakatani, Chem. Rev. 100 (2000) 1817.
40 F. H. Kreuzer, Ch. Bräuchle, A. Miller, A. Petri, *Cyclic Liquid-Crystalline Siloxanes as Optical Recording Materials*, in [48].
41 K. Ichimura, Y. Suzuki, T. Hosoki, K. Aoki, Langmuir 4 (1988) 1214.
42 T. Ikeda, S. Horiuchi, D. B. Karanjit, S. Krihara, S. Tazuke, Macromolecules 23 (1990) 36 and 42.
43 K. Ichimura, *Photoregulation of Liquid-Crystal Alignment by Photochromic Molecules and Polymeric Thin Films*, in [48].
44 (a) V. P. Shibaev, S. G. Kostromin, S. A. Ivanov, *Comb-Shaped Polymers with Mesogenic Side Groups as Electro- and Photooptical Active Media*, in [48], (b) V. P. Shibaev, A. Bobrovsky, N. Boiko, Prog. Polym. Sci. 28 (2003) 729.
45 Y. Wu, A. Kanazawa, T. Shiono, T. Ikeda, Q. Zhang, Polymer 40 (1999) 4787.
46 D. Creed, *Photochemistry and Photophysics of Liquid-Crystalline Polymers*, in V. Ramamurthy, K. S. Schanze (eds.), *Molecular and Supramolecular Organic and Inorganic Photochemistry*, Vol. 2, Marcel Dekker, New York (1998).
47 C. B. McArdle (ed.), *Side-Chain Liquid-Crystal Polymers*, Blackie, Glasgow (1989).
48 V. P. Shibaev (ed.), *Polymers as Electrooptical and Photooptical Active Media*, Springer, Berlin (1996).
49 M. Irie, D. Kungwatchakun, Macromolecules 19 (1986) 2476.
50 E. A. Gonzalez-de los Santos, J. Lozano-Gonzalez, A. F. Johnson, J. Appl. Polym. Sci. 71 (1999) 267.

51 T. Ikeda, M. Nakano, Y. Yu, O. Tsutsumi, A. Kanazawa, Adv. Mater. 15 (2003) 201.
52 D.Y. Kim, S.K. Tripathy, L. Li, J. Kumar, Appl. Phys. Lett. 66 (1995) 1166.
53 P. Rochon, E. Batalla, A. Natansohn, Appl. Phys. Lett. 66 (1995) 136.
54 T. Fukuda, K. Sumaru, T. Kimura, H. Matsuda, J. Photochem. Photobiol. A: Chem. 145 (2002) 35.
55 S. Yang, L. Li, A.L. Cholly, J. Kumar, S.K. Tripathy, J. Macromol. Sci., Pure Appl. Chem. A 38 (2001) 1345.
56 N.K. Viswanathan, S. Balasubramanian, J. Kumar, S.K. Tripathy, J. Macromol. Sci., Pure Appl. Chem. A 38 (2001) 1445.
57 M. Higuchi, N. Minoura, T. Kinoshita, Colloid Polym. Sci. 273 (1995) 1022.
58 H. Menzel, Macromol. Chem. Phys. 195 (1994) 3747.
59 Y. Atassi, J.A. Delaire, K. Nakatani, J. Phys. Chem. 99 (1995) 16320.
60 Z. Sekkat, P. Pretre, A. Knoesen, W. Volksen, V.Y. Lee, R.D. Miller, J. Wood, W. Knoll, J. Opt. Soc. Am. B 15 (1998) 401.
61 J.C. Crano, W.S. Kwak, C.N. Welch, *Spiroxazines and Their Use in Photochromic Lenses*, in [10].
62 M. Irie, *High-Density Optical Memory and Ultrafine Photofabrication*, Springer Series in Optical Sciences 84 (2002) 137.
63 J. Whittall, *Fulgides and Fulgimides – a Promising Class of Photochromes for Application*, in [10].
64 T. Tsujioka, F. Tatezono, T. Harada, K. Kuroki, M. Irie, Jpn. J. Appl. Phys. 33 (1994) 5788.
65 M. Irie, K. Uchida, Bull. Chem. Soc. Jpn. 71 (1998) 985.
66 R. Natansohn, P. Rochon, C. Barret, A. Hay, Chem. Mater. 7 (1995) 1612.
67 N.C.R. Holme, S. Hvilsted, P.S. Ramanujam, Appl. Optics 35 (1996) 4622.
68 N.C.R. Holme, S. Hvilsted, P.S. Ramanujam, Opt. Lett. 21 (1996) 1902.
69 Y. Tian, J. Xie, C. Wang, Y. Zhao, H. Fei, Polymer 40 (1999) 3835.
70 B.L. Lachut, S.A. Maier, H.A. Atwater, M.J.A. de Dood, A. Polman, R. Hagen, S. Kostromine, Adv. Mater. 16 (2004) 1746.
71 R.P. Bertram, N. Benter, D. Apitz, E. Soergel, K. Buse, R. Hagen, S.G. Kostromine, Phys. Rev. E 70 (2004) 041802-1.
72 *Forgery-Proof Information Storage, Genuine Security*, Bayer Scientific Magazine Research 16 (2004).
73 R.H. Berg, S. Hvilsted, P.S. Ramanujam, Nature 383 (1996) 506.

6
Technical developments related to photophysical processes in polymers

6.1
Electrophotography – Xerography

According to Schaffert's definition [1], electrophotography concerns the formation of images by the combined interaction of light and electricity, and xerography is a form of electrophotography that involves the development of electrostatic charge patterns created on the surfaces of photoconducting insulators. The term *xerography* originates from the Greek words *xeros* (dry) and *graphein* (to write), which together mean *dry writing*. The xerographic process invented by Carlson in 1938 [2] is the basis for copying documents with the aid of copying machines. The importance of xerography in our daily lives is unquestionable, in view of the ubiquitous employment of copying machines. At present, virtually all copiers use xerography. With the advent of semiconductor lasers and light-emitting diodes, xerography is also widely applied in desktop printing [3–8]. The principle of the xerographic process is outlined briefly in the following and depicted schematically in Fig. 6.1.

The essential part of a copying machine is the photoreceptor, which nowadays consists mostly of organic material. In order to make a copy of a document, the photoreceptor surface is first positively or negatively corona charged and subsequently exposed to the light reflected from the document. The resulting pattern of exposed and unexposed areas at the photoreceptor corresponds to areas where the corona charges were neutralized or remained unaltered, respectively. Electrostatically charged toner particles brought into contact with the exposed photoreceptor adhere exclusively to those areas that still carry charges. To complete the copying process, the toner particles are transferred to a sheet of paper, which is pressed onto the photoreceptor, and then fixed (fused) by a thermal (infrared) treatment.

Modern copying machines employ dual-layer photoreceptors (see Fig. 6.2). In this way, charge generation and charge transport are separated. The charge generation layer (CGL, 0.5–5.0 µm) is optimized for the spectral response and the quantum yield of charge carrier formation, and the charge transport layer (CTL, 15–30 µm) is optimized for the drift mobility of the charge carriers and for wear resistance.

Dual-layer systems have the advantages of high sensitivity, long process lifetime, and a reduction in the hysteresis of latent image formation. The transport layer requires the displacement of either electrons or holes. Since most trans-

Polymers and Light. Fundamentals and Technical Applications. W. Schnabel
Copyright © 2007 WILEY-VCH Verlag GmbH & Co. KGaA, Weinheim
ISBN: 978-3-527-31866-7

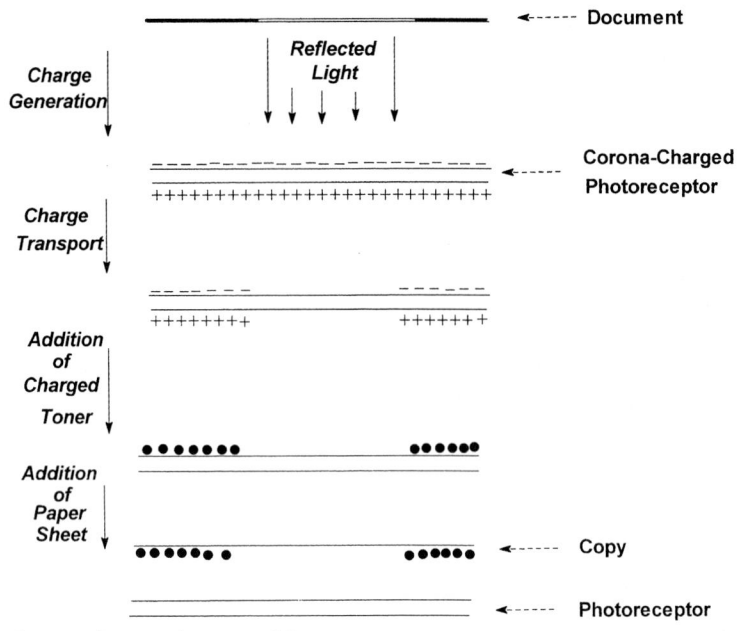

Fig. 6.1 Schematic depiction of the xerographic process for a positively corona-charged single-layer photoreceptor.

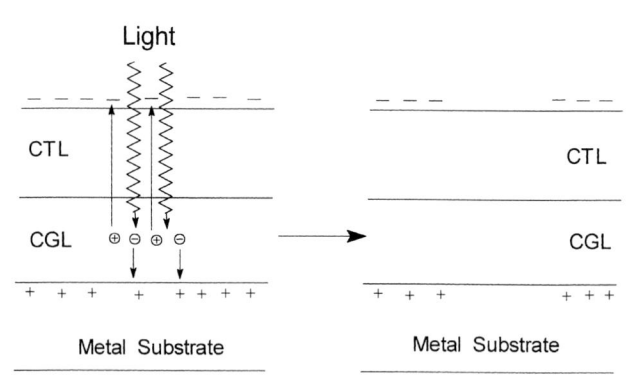

Fig. 6.2 Schematic depiction of the light-induced discharge process for a negatively corona-charged dual-layer photoreceptor. CGL and CTL denote the charge generation layer and the charge transport layer, respectively.

port layers are formulated to transport holes, dual-layer receptors are usually negatively charged.

Numerous compounds have been tested and applied commercially as charge-generation and charge-transport materials, as can best be seen from the book by Borsenberger and Weiss [4].

The first all-organic photoreceptor was a single-layer device consisting of a 1:1 molar mixture, an electron-donor polymer, poly(N-vinyl carbazole), and an electron acceptor, TNF (see Chart 2.1). A very effective dual-layer system, designated by the acronym TiO(F_4-Pc):TTA, contains a dispersion of tetrafluorotitanylphthalocyanine in poly(vinyl butyral) in the charge-generation layer, and a mixture of tris(p-tolylamine) and polycarbonate in the charge-transport layer. Highly sensitive charge-generation systems, appropriate for visible and also for near-infrared light, were obtained upon doping polymers with pigment particles of dyes. In this case, the CG layers consist of a light-sensitive crystalline phase dispersed in the polymeric matrix. Besides phthalocyanines, pigments employed comprise azo compounds, squaraines, and polycyclic aromatic compounds (the chemical structures of which are shown in Table 2.1). Improved sensitivities have sometimes been achieved with pigment mixtures. As a typical example, Fig. 6.3 presents results obtained with a dual-layer system [8, 9]. Here, the CG layer consisted of a dispersion of the triphenylamine triazo pigment AZO-3 (see Chart 6.1) in poly(vinyl butyral) in a 4:10 weight ratio, while the CT layer consisted of a mixture of bisphenol A polycarbonate and the triarylamine derivative MAPS (see Chart 6.1) in a 10:9 weight ratio.

Note that the value of the quantum yield of charge carrier formation is very high, about 0.45 at $F=3\times10^5$ V cm^{-1}, and remains practically constant over the investigated wavelength range from 470 to 790 nm. Interestingly, the quantum yield found for the single-layer system was about one order of magnitude lower. The very high quantum yield is interpreted in terms of exciton dissociation at the interface between the two layers and injection of practically all of the holes into the charge-transport layer.

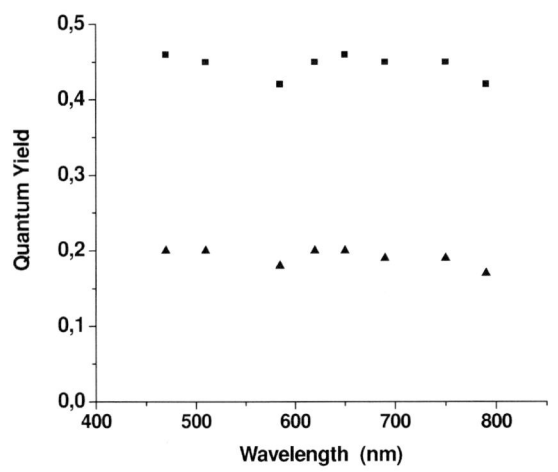

Fig. 6.3 Charge generation in a dual-layer photoreceptor system. The quantum yield of charge generation as a function of the wavelength of the incident light at $F=3\times10^5$ V cm^{-1} (■) and $F=0.8\times10^5$ V cm^{-1} (▲). See text for system characterization. Adapted from Williams [8] with permission from John Wiley & Sons, Inc.

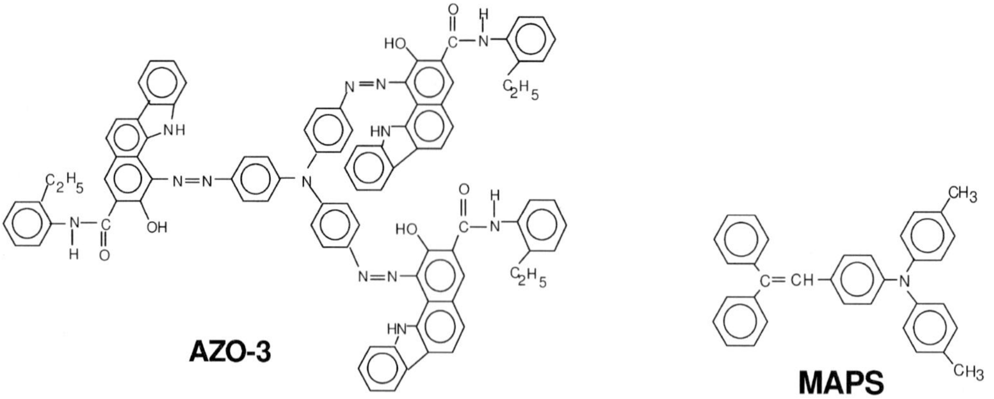

Chart 6.1 Chemical structures of the triphenylamine triazo pigment AZO-3 and the triarylamine derivative MAPS.

Regarding the charge-transport layers, materials for hole and electron transport have to be discriminated. A large number of hole-transport materials contain arylamine moieties. Moreover, polysilylenes are well-suited for hole transport. A key requirement for dual-layer systems is a high efficiency of charge injection from the generation layer into the transport layer. Moreover, it is important that the charge transport is not impeded by trapping and that the transit time is short compared to the time between exposure and development. For most applications, a hole mobility between 10^{-6} and 10^{-5} cm^2 V^{-1} s^{-1} is sufficient.

The requirements for electron-transport materials cannot be fulfilled easily. For instance, an appropriate compound should be weakly polar and have a low reduction potential, i.e. a high electron affinity. Actually, the electron affinity should be higher than that of molecular oxygen, which is always present. For this reason and because of some additional difficulties, electron-transport layers have not yet been used in commercial applications [4].

6.2
Polymeric light sources

One of the most fascinating developments in recent times concerns the generation of light with the aid of polymers. This development is characterized by two inventions, which are described in the following subsections: the polymeric light-emitting diode and the polymer laser.

6.2.1
Light-emitting diodes

6.2.1.1 General aspects

Polymeric light-emitting diodes operate on the basis of electroluminescence, i.e. luminescence generated by the application of high electric fields to thin polymer layers. Devices based on the electroluminescence of organic materials, commonly denoted as organic light-emitting diodes, OLEDs, are used, for example, for mini-displays in wrist watches and chip cards, for flexible screens, and for emitting wall paper. In contrast to liquid-crystal displays (LCDs), OLED displays can be seen from all viewing angles. OLED devices can be extremely thin, flexible, and of low weight. Moreover, production costs and energy consumption are low. Consequently, the potential for making large-area multicolor displays from easily processable polymers has initiated a large number of research pro-

Table 6.1 Poly(p-phenylene vinylene)s used in light-emitting diodes [11, 12, 20].

Polymer	Acronym	EL Maximum (nm)
	PPV	540
	PMPPV	560
	MEH-PPV	590
	PMCYH-PV	590
	PDFPV	600
	PPFPV	520

jects in the area of polymer light-emitting diodes, as has been documented by several reviews [10–23].

The phenomenon of polymer-based electroluminescence was first demonstrated in the case of poly(p-phenylene vinylene), PPV ($\pi-\pi^*$ energy gap: 2.5 eV) [24], and was later also observed with many PPV derivatives and other fully π-conjugated polymers. Typical representatives are shown in Tables 6.1 and 6.2. Table 6.1 relates to PPV and some of its derivatives, whereas Table 6.2 lists other classes of polymers that have been employed in LED work.

Table 6.2 Polymers employed in light-emitting diodes [10a].

Polymer class	Structure of typical polymer	Characteristics
Polythiophenes		p-Type (hole-transporting) polymers. Alkyl groups provide for solubility in organic solvents. Emission tunable from UV to IR through varying the substituent.
Poly-p-phenylenes	x = 3 – 20; x = 2 – 5; A = phenyl	p-Type polymers of rather high thermal stability, mostly used in the form of polymers containing oligo-p-phenylene sequences. Emit light in the blue wavelength range.
Polyfluorenes	R: typically hexyl, octyl, ethylhexyl	p-Type polymers of improved thermal and photostability (relative to PPV). Emit light primarily in the blue wavelength range.
Cyano polymers		Polymers, e.g. PPV derivatives, containing electron-withdrawing cyano groups. The latter provide for electron transport, thus complementing the hole-transport property.
Pyridine-containing polymers		Highly luminescent polymers soluble in organic solvents. High electron affinity affords improved electron transport. Quaternization of nitrogen allows manipulation of the emission wavelength.
Oxadiazole-containing polymers	OC_6H_{13}, $C_6H_{13}O$	Oxadiazole groups provide for efficient electron transport. Insertion of these groups into p-type polymers facilitates bipolar carrier transport.

(a)

| Metal Cathode |
| Polymer |
| ITO Anode |
| Glass Substrate |

(b)

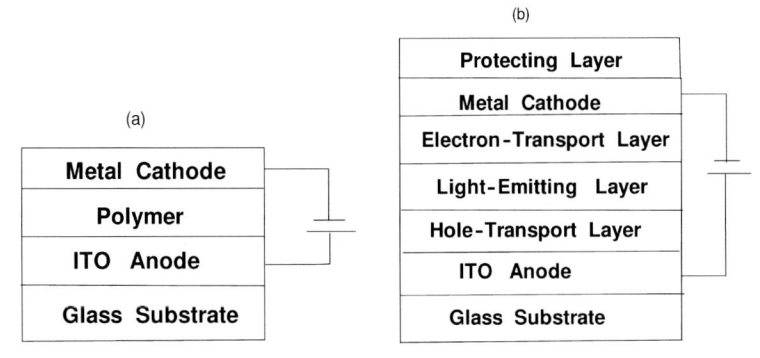

| Protecting Layer |
| Metal Cathode |
| Electron-Transport Layer |
| Light-Emitting Layer |
| Hole-Transport Layer |
| ITO Anode |
| Glass Substrate |

Fig. 6.4 (a) Structure of a single-layer polymer LED.
(b) Structure of a multilayer polymer LED.

In this connection, the reader is referred to a rather comprehensive review dealing with the various classes of polymers tested for LED application [10a] and to a list of appropriate commercially available materials [25].

As can be seen from Fig. 6.4a, an OLED consists, in the simplest case, of a polymer film placed between two electrodes, one of them being light-transparent such as indium tin oxide (ITO) and the other being a metal of low work function, e.g. barium, calcium or aluminum.

Holes and electrons are injected from the ITO electrode (anode) and the metal electrode (cathode), respectively. The energy level diagram under forward bias is shown in Fig. 6.5. More sophisticated OLEDs possess multilayer structures as shown in Fig. 6.4b.

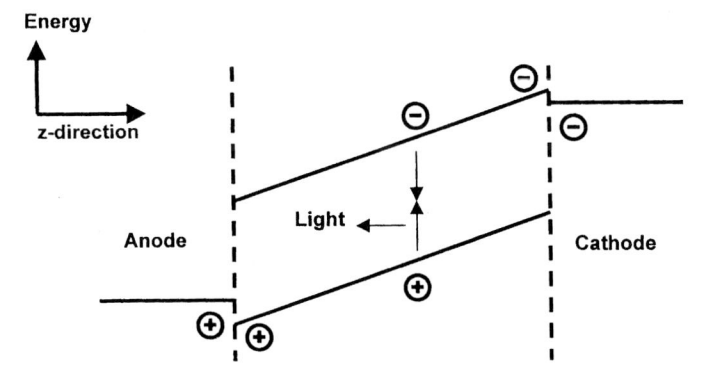

Fig. 6.5 Energy level diagram of a single-layer polymer LED under forward bias. The z-direction is parallel to the current direction and hence perpendicular to the layer. Adapted from Graupner [13] with permission from the Center for Photochemical Sciences, Bowling Green.

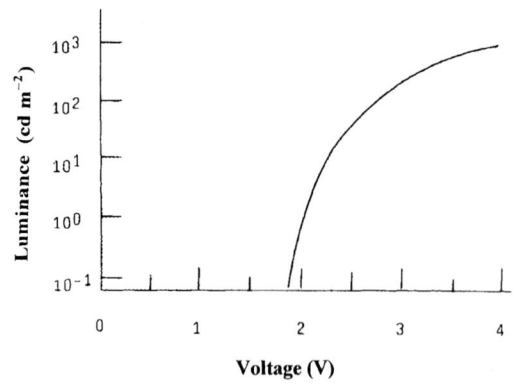

Fig. 6.6 Luminance–voltage characteristic for the polymer blend PCzDBT20/MEH-PPV (1/240). Adapted from Niu et al. [26] with permission from Wiley-VCH.

Chart 6.2 Polymers contained in the blend referred to in Fig. 6.6.

As can be seen from the typical luminance–voltage characteristic presented in Fig. 6.6, light generation requires a minimum voltage, the turn-on voltage, at which light emission commences.

The luminance increases drastically on further increasing the voltage immediately beyond the onset and later approaches saturation. The curve in Fig. 6.6 refers to a 240:1 blend of the polymers denoted as MEH-PPV and PCzDBT20 (see Chart 6.2) [26]. In this case, red light with a maximum at about 680 nm is emitted. Here, the turn-on voltage is quite low (<2 V) and the external quantum yield is rather high, $\Phi_{ext}=0.038$. Φ_{ext} represents the number of photons penetrating the device surface to the outside generated per injected electron. The availability of highly efficient OLEDs emitting light of the primary colors – red, green, and blue – is important for the realization of full color display applications.

6.2.1.2 Mechanism
The injection of charges from the electrodes into the bulk organic material is determined by various parameters. Since holes are injected into the highest occupied molecular orbital (HOMO) and electrons into the lowest unoccupied molecular orbital (LUMO), matching of energy levels is required. This is demon-

ITO anode/hole-transporting layer (HTL)/emitting layer (EML)/metal cathode
Chart 6.3 Structure of a two-layer OLED.

strated for a two-layer OLED of the structure shown in Chart 6.3 by the energy level diagram presented in Fig. 6.7 [12].

This diagram illustrates the equivalence of the valence band with the ionization potential (IP) and the HOMO as well as that of the conduction band with the electron affinity (EA) and the LUMO. Notably, electron and hole injection are controlled by the energy barrier between the contact and the organic material. In the absence of surface states and a depletion region due to impurity doping, the energy barriers are given by Eqs. (6-1) and (6-2).

$$\Delta E_h = IP - \Phi_{anode} \quad \text{(for holes)} \tag{6-1}$$

$$\Delta E_{el} = \Phi_{cathode} - EA \quad \text{(for electrons)} \tag{6-2}$$

Here, Φ_{anode} and $\Phi_{cathode}$ denote the work functions of the contact materials. Depending on the magnitude of ΔE, the current flow through an OLED can be either space-charge limited (SCL), i.e. transport-limited, or injection-limited. Prerequisites for SCL are that the injection barrier is rather low and that one of the contacts supplies more charge carries per unit time than can be transported through the organic material layer. Commonly, injection-limited conduction is described by Fowler-Nordheim (FN) tunneling into the transport band or by Richardson-Schottky (RS) thermionic emission [27, 28]. The FN model ignores image-charge effects and assumes tunneling of electrons from the contact through the barrier into a continuum of states. The RS model assumes that electrons capable of ejection from the contact have acquired sufficiently high thermal energies to cross the potential maximum resulting from the superposition of the external and the image-charge potentials. These models were developed for band-type materials. However, it turned out that they are inadequate for describ-

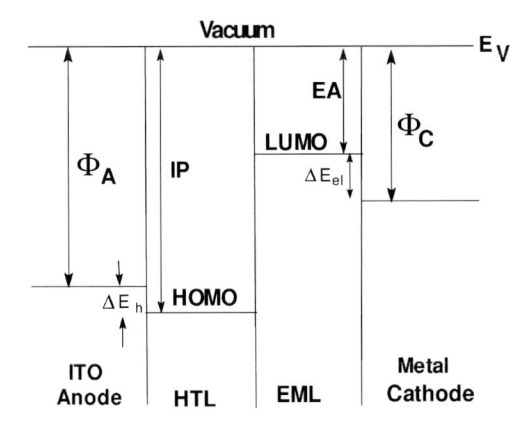

Fig. 6.7 Energy level diagram for a two-layer polymer LED, showing the ITO anode, the hole-transporting layer HTL, the emitting and electron-transporting layer EML, and the metal cathode. E_V denotes the vacuum potential.

ing the current–voltage dependence measured for disordered organic materials [29]. In organic materials, the charge carriers are not very mobile because they are localized and the transport involves localized discrete hopping steps within a distribution of energy states. For charge carrier injection of electrons from a metal contact into such *organic hopping systems*, a Monte Carlo simulation yielded excellent agreement with the experimentally observed dependence of the injection current on electric field strength and temperature [30, 31]. It is based on the concept of temperature and field-assisted injection from the Fermi level of an electrode into the manifold of hopping states. Under the influence of the applied electric field, the injected oppositely charged carriers migrate through the system towards the electrodes and a portion of them eventually combine to form excited electron-hole singlet states, so-called singlet *excitons*. The latter undergo radiative decay to only a small extent, that is to say, electroluminescence quantum yields, in terms of emitted photons per injected electron, are relatively low and amount to only a few per cent even in the best cases. Competing processes are operative, such as singlet-triplet crossing, singlet-exciton quenching, etc. Figure 6.8 shows typical photoluminescence and electroluminescence spectra recorded for PPV and two PPV derivatives.

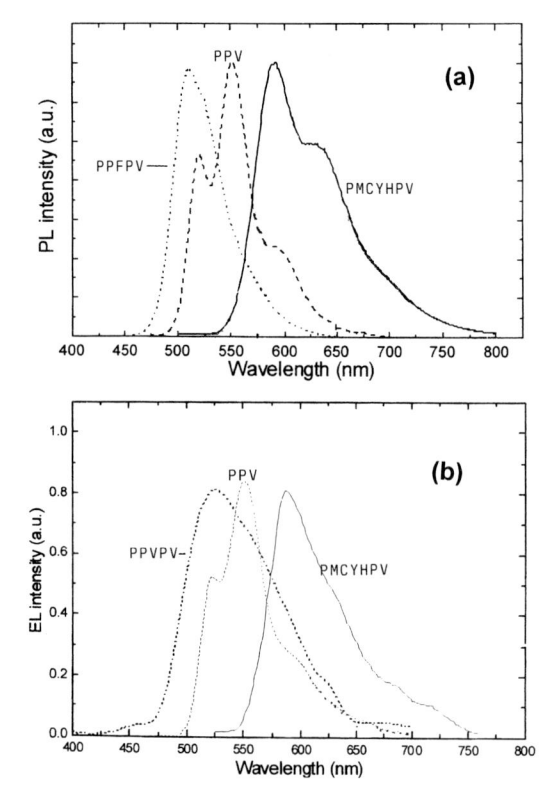

Fig. 6.8 Photoluminescence (a) and electroluminescence spectra (b) of PPV, PMCYH-PV, and PPFPV. Adapted from Shim et al. [11] with permission from Springer.

Obviously, in these cases, the maxima of both types of emission spectra are almost the same, indicating that the emission originates from the same species. In both cases, the peak position is red-shifted when strongly electron-donating groups are attached to the conjugated backbone of the polymer. Therefore, it is possible to tune the color of the electroluminescent emission by varying the chemical nature of the substituent. A blue color can be obtained by widening the $\pi-\pi^*$ gap through shortening the conjugation length and lowering the electron density in the conjugated backbone. In the case of PPFPV, the emission maximum lies in the greenish-blue region. Here, the strong electron-withdrawing influence of the perfluorobiphenyl group lowers the electron density in the

Table 6.3 Hole and electron transport materials employed in polymer LEDs [10a].

Chemical structure	Acronym
Hole transport materials:	
(structure)	TPD
(structure)	PPV
(structure)	PVK
(structure)	PMPS
Electron transport materials:	
(structure)	PBD
(structure)	Alq$_3$
(structure)	PMA-PBD

polymer chain and thus causes a shift of the maximum from 540 nm (PPV) to about 520 nm.

Notably, the major steps in the electroluminescence mechanism are injection, transport, and recombination of charge carriers. Good carrier transport and efficient recombination in the same material are antagonists, because the combination probability is low if the charge carriers swiftly migrate to the electrodes without interaction with their oppositely charged counterparts. A solution to this dilemma was found with devices consisting of several layers. In many cases, a layer allowing swift hole transport and blocking of the passage of electrons has been combined with a layer permitting only electron transport and serving as an emitting layer. Table 6.3 presents typical hole and electron transport materials [10a].

6.2.1.3 Polarized light from OLEDs

Provided that the macromolecules in a thin film employed as an emitting layer in a LED device are well oriented, the emitted light is largely polarized [31]. Regarding conjugated polymers, this phenomenon has attracted broad interest because low-cost techniques for chain alignment in such polymers are available. Polarized electroluminescence is useful for certain applications, for instance, for the background illumination of liquid-crystal displays (LCDs) [20, 32]. The first LED device emitting polarized light was realized with the stretch-oriented polythiophene PTOPT (see Chart 6.4) [33].

The methods commonly used for chain alignment in polymer films have been reviewed [34]. They comprise the Langmuir-Blodgett technique, rubbing of the film surface, mechanical stretching of the film, and orientation on pre-aligned substrates. As an example, electroluminescence spectra of the oriented substituted poly(p-phenylene) presented in Chart 6.5 are shown in Fig. 6.9a [35].

The device prepared by the Langmuir-Blodgett (LB) technique had the structure shown in Chart 6.6:

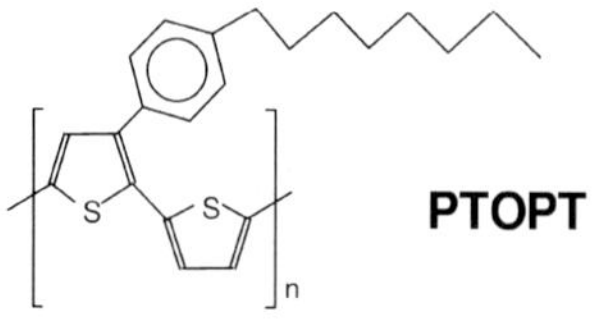

PTOPT

Chart 6.4 Chemical structure of poly[3-(4-octylphenyl)-2,2'-bithiophene], PTOPT.

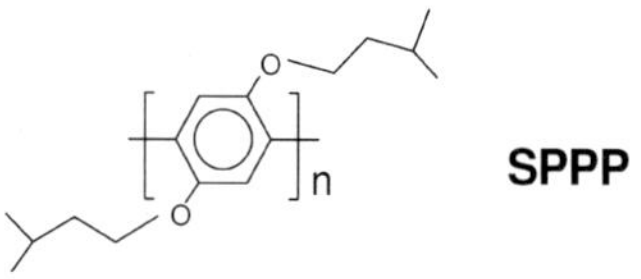

SPPP

Chart 6.5 Chemical structure of an oriented substituted poly(p-phenylene) [35].

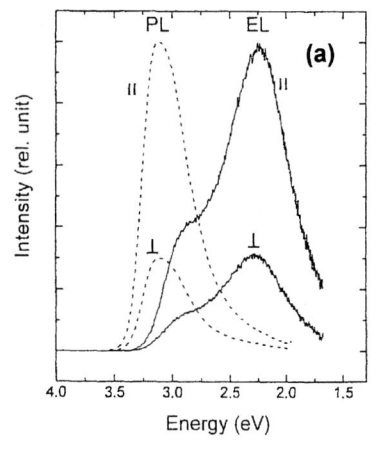

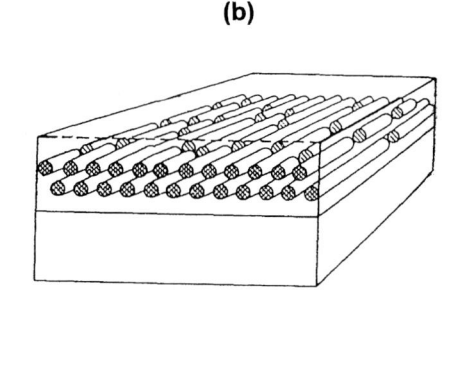

Fig. 6.9 (a) Electroluminescence spectra of the oriented substituted poly(p-phenylene) SPPP. The emission spectra were recorded with the polarization direction parallel and perpendicular to the dipping direction employed during preparation by the LB technique. (b) Schematic depiction of rigid rod-like macromolecules oriented parallel to the substrate plane. Adapted from Cimrova et al. [35] with permission from Wiley-VCH.

ITO anode/100 monolayers SPPP/Al cathode

Chart 6.6 Device used for recording the electroluminescence spectra depicted in Fig. 6.9a.

As demonstrated schematically in Fig. 6.9b, the rigid rod-like macromolecules are oriented parallel to the substrate plane and their backbones exhibit a preferential orientation along the dipping direction employed during LB processing.

From the emission spectra recorded with the polarization of the light parallel and perpendicular to the dipping direction, the polarization ratio can be estimated to be somewhat greater than three.

6.2.1.4 White-light OLEDs

In many cases, OLED devices have been developed that contain polymers as hole-transport media and low molar mass organic or inorganic compounds as emitting materials. This pertains, for instance, to certain white-light-emitting LEDs, two of them being exemplified here. The first case refers to a device containing CdSe nanoparticles in the emitting layer. These particles are embedded in a polymer, namely PPV. A device having the multilayer structure shown in Chart 6.7 produces almost white light under a forward bias of 3.5–5.0 V [36].

The second case refers to a device containing a platinum compound, such as FPt-1 or FPt-2, in the emitting layer (see Chart 6.8).

A device having the multilayer structure shown in Chart 6.9 emits white light with $\Phi_{ext}=1.9\%$ at a brightness of 100 cd m^{-2} (J = 2 mA cm^{-2}). The white light results from the simultaneous monomer (blue) and excimer (green to red) emission of the Pt compound [37].

ITO anode/PEI/(CdSe-PPV)/Al cathode

Chart 6.7 Device used to produce almost white light. PEI: poly(ethylene imine), $-(CH_2-CH_2-NH)_n-$.

FPt-1   FPt-2

Chart 6.8 Chemical structures of Pt-containing compounds used to produce white light.

ITO anode/PEDOT:PSS/(FPt2-CBP)/BCP/LiF/Al cathode

Chart 6.9 Device used to produce white light. PEDOT: poly(3,4-ethylenedioxythiophene), PSS: poly(styrene sulfonic acid), CBP: 4,4′-di(N-carbazolyl)-biphenyl (see Chart 6.10), BCP: bathocuproine (2,9-dimethyl-4,7-diphenyl-1-10-phenanthroline).

6.2.2
Lasers

6.2.2.1 General aspects

The term laser is an acronym (light amplification by stimulated emission of radiation) that denotes a technical device operating on the basis of the stimulated emission of light. A laser emits monochromatic, spatially coherent, and strongly polarized light. The essential parts of a laser device are an active material and a resonator, i.e. an optical feedback (see Fig. 6.10).

In classical laser systems, such as Ti:sapphire-based systems or semiconductor laser diodes, the active materials are inorganic compounds. In recent years, suitable organic active materials have been introduced [38–41]. These organic materials may be divided into two classes: host/guest systems consisting of a host material doped with organic dye molecules, and systems consisting of conjugated polymers. Typical dyes used in host/guest systems are rhodamines, coumarins, and pyrromethenes, and these are dissolved in polymeric hosts such as poly(methyl methacrylate) or methacrylate-containing copolymers. In some

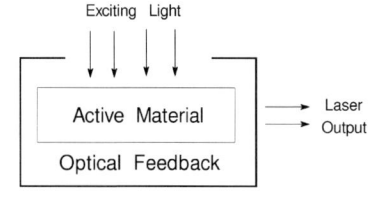

Fig. 6.10 Schematic illustration of an optically pumped laser device. Adapted from Kranzelbinder et al. [38] with permission from the Institute of Physics Publishing, Bristol, U.K.

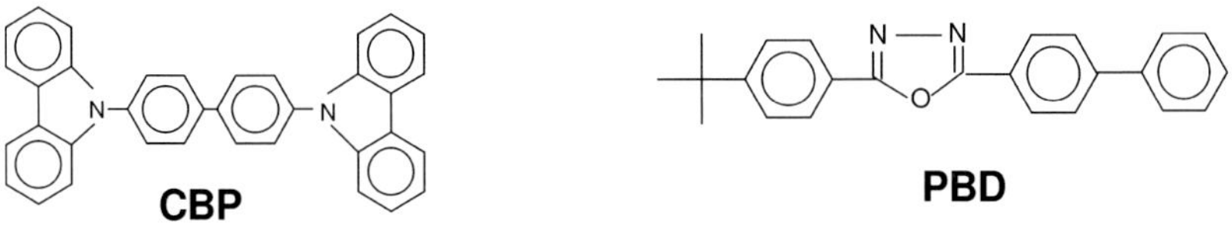

Chart 6.10 Chemical structures of 4,4'-di(N-carbazolyl)-biphenyl, CBP, and 2-(4-biphenyl)-5-(4-tert-butylphenyl)-1,3,4-oxadiazole, PBD.

Table 6.4 Conjugated polymers used as laser materials.

Polymer[a]	Chemical structure	Resonator	Excitation conditions	$I_{threshold}$ [b] ($\mu J\ cm^{-2}$)	Ref.
DOO-PPV	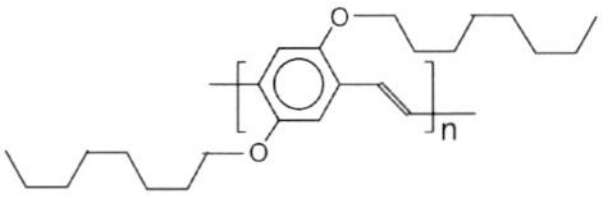	Microring	$\lambda = 532$ nm $\tau = 100$ ps	0.1	[43]
BEH-PPV	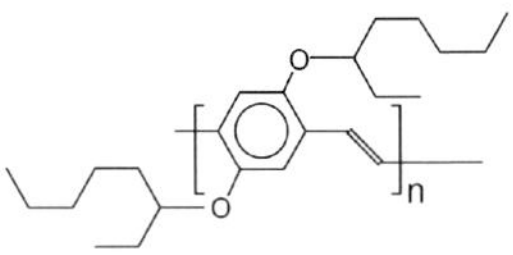	Microring	$\lambda = 555$ nm $\tau = 100$ fs	25	[44]
BuEH-PPV	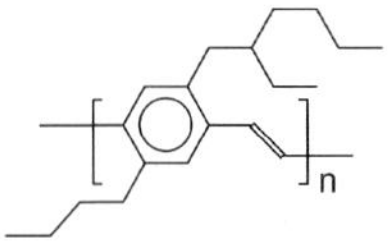	Microcavity	$\lambda = 435$ nm $\tau = 10$ ns	4.5	[45]
m-LPPP	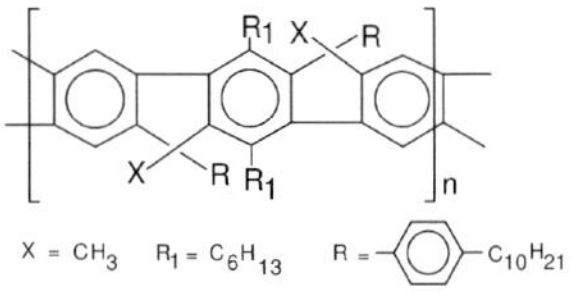	Flexible distributed feedback	$\lambda = 400$ nm $\tau = 150$ ps	3.7	[46]
PDOPT	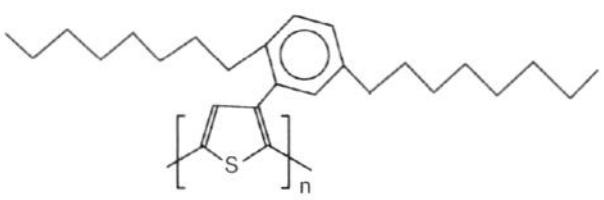	Microcavity	$\lambda = 530$ nm $\tau = 90$ fs	0.12	[47]

a) Acronyms used in this column: DOO-PPV: poly(2,5-dioctyloxy-p-phenylene vinylene); BEH-PPV: poly[2,5-di-(2'-ethylhexyloxy)-p-phenylene vinylene]; BuEH-PPV: poly[2-butyl-5-(2'-ethylhexyl)-p-phenylene vinylene]; m-LPPP: ladder-type poly(p-phenylene) bearing methyl groups; PDOPT: poly[3-(2,5-dioctylphenyl)thiophene].
b) Threshold pulse intensity for lasing.

cases, low molar mass materials have been employed as host materials, such as CBP or PBD (see Chart 6.10).

In systems of the type PBD/poly(p-phenylene vinylene) derivative, the host material, PBD, absorbs the pump light and transfers the excitation energy to the polymer, here the emitting guest [42]. Appropriate conjugated polymers cited in the literature are presented in Table 6.4.

It seems that m-LPPP, a ladder-type poly(p-phenylene), is one of the most promising materials for laser application. It is soluble in nonpolar organic solvents, thus enabling the facile preparation of thin layers on substrates that may possess structured uneven surfaces.

6.2.2.2 Lasing mechanism

At present, polymer lasers are operated by optical pumping, i.e. through the absorption of light by the active material. A four-level energy scheme similar to that used for organic laser dyes serves to explain the lasing mechanism in the case of conjugated polymers. As can be seen in Fig. 6.11, the absorption of a photon corresponds to a transition from the lowest vibronic level of the ground state S_0 to a higher-lying vibronic level of the singlet state S_1.

Rapid (non-radiative) internal conversion leads to the lowest vibronic excitation level of the S_1 manifold. Subsequent transition from this level to one of the vibronic excitation levels of the S_0 manifold is radiative and corresponds to either spontaneous or stimulated emission, SE. In terms of a simple model, stimulated emission is generated through the interaction of the excited molecules with other photons of equal energy. This process can only become important with respect to other competitive processes, such as spontaneous emission, when the concentration of excited states is very high, i.e. when the population of the upper state exceeds that of the lower state, a situation denoted by the term *population inversion*. In other words, the Boltzmann equilibrium of states must be disturbed. Notably, the lasing transition relates to energy levels that are not directly involved in the optical pumping process. The laser potential of an active material is characterized by Eq. (6-3).

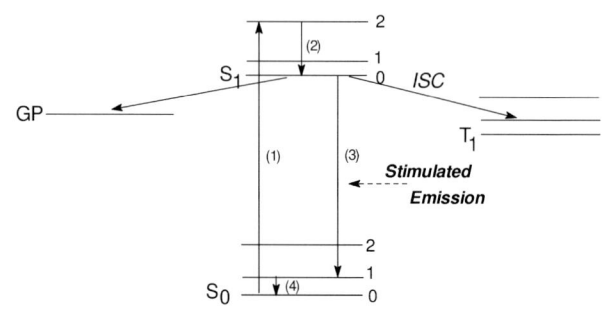

Fig. 6.11 Energy scheme illustrating stimulated emission in conjugated polymers.

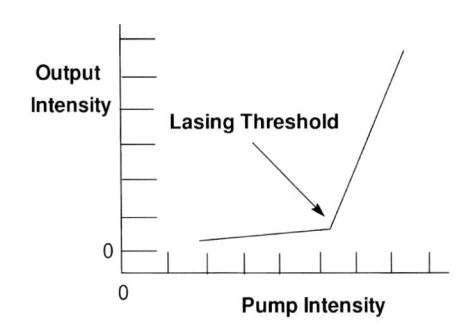

Fig. 6.12 Schematic depiction of the dependence of the intensity of the light emitted from a laser device on the intensity of the exciting light.

$$I_{out} = I_{in} \exp(\sigma N_{exc} L) \tag{6-3}$$

Here, I_{in} and I_{out} denote the intensities of the incoming and outgoing beam, respectively, σ is the cross-section for stimulated emission, N_{exc} is the concentration of excited S_1 states, and L is the path length of the light in the sample. The term $g_{net} = \sigma N_{exc}$ represents the net gain coefficient of the material.

As pointed out above, the transition from spontaneous to stimulated emission requires population inversion. In other words, SE becomes significant when N_{exc} exceeds a critical value, $N_{exc}(crit)$, which characterizes the lasing threshold. Experimentalists frequently denote the threshold in terms of the energy, or more exactly the intensity, $I_{threshold}$, of the excitation light pulse. Figure 6.12 shows a schematic depiction of the dependence of the laser output on the intensity of the excitation light pulse.

Typical $I_{threshold}$ values are given in Table 6.4. In films of conjugated polymers, $N_{exc}(crit)$ is about 10^{18} cm^{-3} if a resonator is not operative. Significantly, the employment of appropriate feedback structures lowers the threshold by several orders of magnitude.

6.2.2.3 Optical resonator structures

As has been pointed out above, a laser basically consists of an active material and a resonator. The latter enables the build-up of certain resonant modes and essentially determines the lasing characteristics. In most conventional devices, the optical feedback is provided by an external cavity with two end mirrors forming the resonator. With the advent of polymers as active materials, various new feedback structures were invented. Initially, a microcavity resonator device of the type shown schematically in Fig. 6.13a was employed [48].

This device consisted of a PPV layer placed between a highly reflective distributed Bragg reflector, DBR, and a vacuum-deposited silver layer functioning as the second mirror. The emission characteristics at different intensities of the pumping light are shown in Fig. 6.13b. At low intensity, the emission consisted of three different modes, whereas at high intensity it was concentrated into the mode of the highest gain. Moreover, the directionality of the emitted light was enhanced by increasing the intensity of the exciting light. Both effects were taken as evidence for

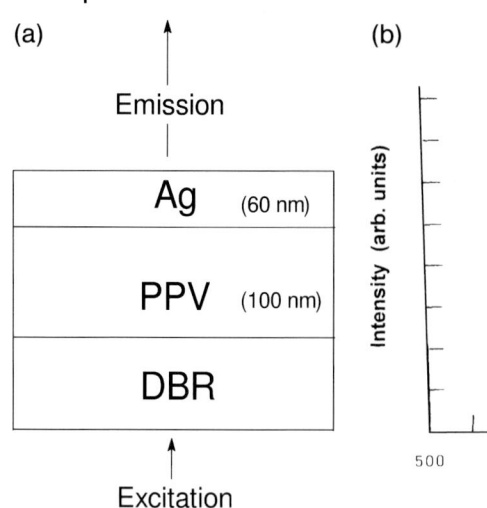

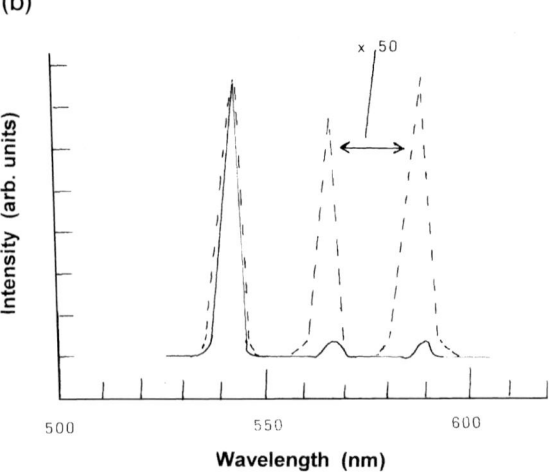

Fig. 6.13 The microcavity, a vertical cavity lasing device. (a) Schematic depiction of the device, consisting of a distributed Bragg reflector, a PPV layer, and a silver layer. (b) Spectra emitted at two different pump laser energies: $E_{exc}=0.05$ µJ/pulse (dashed line) and $E_{exc}=1.1$ µJ/pulse (solid line). Pulse duration: 200–300 ps. Adapted from Tessler et al. [48] with permission from McMillan Publishers Ltd.

the occurrence of lasing. During the ensuing development, resonators in the shape of microspheres, microrings, and flat microdisks were designed. As an example, Fig. 6.14a shows a schematic depiction of a cylindrical microring laser device with an outer diameter of $D=11$ µm and a lateral length of about 100 µm, consisting of a thin DOO-PPV film coated onto an optical fiber.

When the device was excited with 532 nm light pulses ($\tau=100$ ps) at an intensity below the lasing threshold (100 pJ/pulse), the spectrum shown in Fig. 6.14b, extending over about 100 nm, was emitted. Dramatic changes occurred when the intensity of the excitation light pulse exceeded the lasing threshold: the emission spectrum collapsed into several dominant microcavity modes [43].

Another device, the flexible distributed Bragg reflector laser with an active layer structure supporting second-order feedback, makes full use of the advantageous properties of polymers, namely flexibility, large-area fabrication, and low-cost processing [41, 42]. As can be seen in Fig. 6.15, the device consists of a one-dimensionally periodically structured flexible substrate coated with an m-LPPP layer, which acts as a planar wave guide. The substrate possesses a periodic height modulation with a period of $\Lambda=300$ nm.

The surface of the polymer layer exhibits a height modulation with the same period but a smaller amplitude (<10 nm). It should be pointed out that the polymer layer in the device considered here functions as a distributed Bragg reflector and the

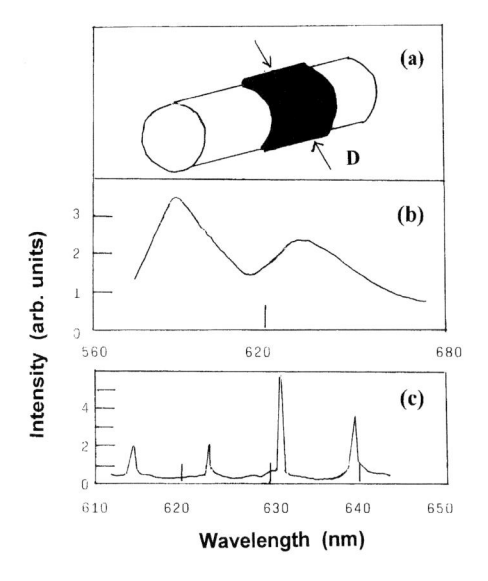

Fig. 6.14 Microring laser device (a) and spectra emitted at excitation light intensities below (b) and above (c) the threshold intensity. Active material: DOO-PPV coated onto an optical fiber. Adapted from Frolov et al. [43] with permission from the American Institute of Physics.

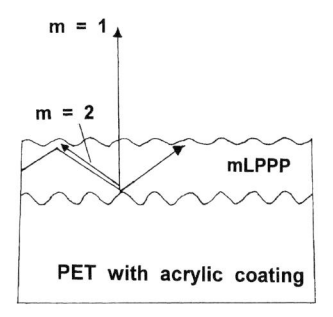

Fig. 6.15 Schematic illustration of a one-dimensionally patterned flexible distributed Bragg reflector laser device. Active layer: 400 nm m-LPPP. Substrate: 125 µm thick poly(ethylene terephthalate) film covered with acrylic coating. Adapted from Kallinger et al. [46] with permission from Wiley-VCH.

resonant modes for laser oscillation in this strongly frequency-selective feedback device correspond to the wavelength λ satisfying the Bragg condition [see Eq. (6-4)].

$$m\lambda = 2n\Lambda \qquad (6\text{-}4)$$

Here, m is the order of diffraction, n is the refractive index, and Λ is the grating period (height modulation period). Optical feedback is accomplished by way of the second-order diffraction mode ($m=2$), which is fed into the counter-propagating wave. The first-order light ($m=1$) is coupled out from the waveguide and propagates perpendicular to the film. Provided that the energy of the exciting light pulses (pulse duration: 150 fs, λ: 400 nm, spot size diameter: 200 µm) exceeds the threshold value, $E_{threshold}=1.5$ nJ, highly polarized laser light ($\lambda=488$ nm) is emitted perpendicular to the film plane. An improvement over this method of mode selection was achieved with the aid of two-dimensionally nano-patterned substrates [49]. The device depicted schematically in Fig. 6.16 emits a monomode beam perpendicular to its surface.

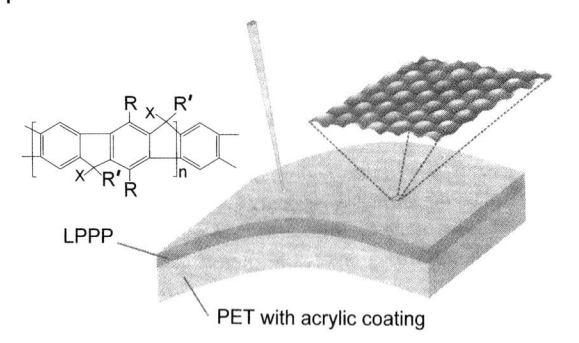

Fig. 6.16 Schematic illustration of a flexible polymer laser device consisting of an m-LPPP layer spin-coated onto a two-dimensionally structured flexible poly(ethylene terephthalate) substrate. The laser light is emitted perpendicular to the substrate. Adapted from Riechel et al. [49] with permission from the American Institute of Physics.

Compared to the one-dimensionally structured device, the lasing threshold is 30% lower and the divergence of the emission is drastically reduced. In accordance with the 2D laser operation, the emitted light is not polarized in this case.

6.2.2.4 Prospects for electrically pumped polymer lasers

At present, an electrically driven polymer laser has yet to be realized [39]. Nevertheless, low-cost polymer laser diodes could be an attractive alternative to the widely used inorganic laser diodes. In principle, an electrically pumped polymer laser could be realized with the aid of an appropriate feedback structure, provided that the excitation density, $N_{exc}(crit)$, i.e. the concentration of excitons, exceeded the lasing threshold (see Section 6.2.2.2). From research concerning optically pumped polymer lasers, it is known that $N_{exc}(crit)$ is about 10^{18} cm^{-3}. This value corresponds to a critical current density of 10^5 to 10^6 A cm^{-2} [50]. However, the highest current densities hitherto obtained are about 10^3 A cm^{-2}, i.e. several orders of magnitude below the required value. Therefore, besides the search for appropriate device structures and appropriate highly conducting materials, strategies aiming at an electrically pumped polymer laser are also concerned with achieving much higher exciton concentrations. An approach in this direction may lie in the application of sharp-edge shaped electrodes with the potential of generating locally very high electric fields, enabling the formation of locally very high charge carrier concentrations through field-induced emission.

6.3
Polymers in photovoltaic devices

Photovoltaic (PV) cells generate electric power when irradiated with sunlight or artificial light. Classical PV cells, based on inorganic semiconducting materials

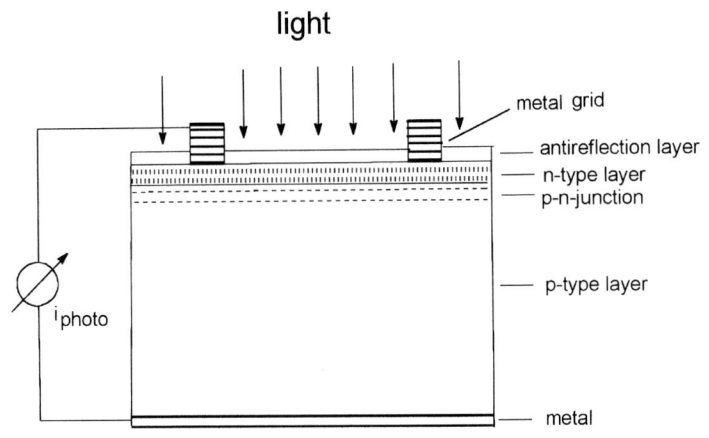

Fig. 6.17 Schematic depiction of a p-n homojunction crystalline silicon solar cell. Typical dimensions of commercial wafers: 10 cm×10 cm×0.3 mm. Adapted from Archer [67] with permission from the World Scientific Publishing Company.

such as silicon, GaAs, CdTe or $CuInSe_2$, consist of layers doped with small amounts of additives that provide n-type (electron) or p-type (hole) conductivity [51–59]. A "built-in" electric field exists across the junction between the two layers, which sweeps electrons from the n to the p side and holes from the p to the n side. Figure 6.17 shows the essential features of a (sandwich-structured) p-n homojunction silicon solar cell.

The absorption of photons having energies greater than the band gap energy promotes electrons from the valence to the conduction band, thus generating hole-electron pairs. The latter rapidly dissociate into free carriers that move independently of each other. As these approach the junction, they come under the influence of the internal electric field, which actually prevents recombination. At present, most of the industrially produced photovoltaic cells consist of monocrystalline or polycrystalline and to some extent of amorphous silicon (a-Si). Different types of junctions may be distinguished: homojunctions are p-n junctions formed by adjacent p- and n-doped regions in the same semiconductor of band gap U_g, whereas heterojunctions are formed between two chemically different semiconductors with different band gaps. Moreover, there are p-i-n junctions, which are formed by interposing an intrinsic undoped layer between p and n layers of the same semiconductor.

Certain organic materials also possess semiconductor properties and can be employed in PV cells, a fact that has recently been attracting growing interest since the advent of novel polymeric materials [22, 60–66]. Table 6.5 lists some typical polymers used in solar cells.

Criteria commonly used to characterize PV cells comprise J_{sc}, the short-circuit current density; V_{oc}, the open-circuit voltage; Φ_{cc}, the quantum efficiency for

Table 6.5 Chemical structures of semiconducting polymers used in organic solar cell devices [60–66].

Chemical structure	Acronym	Denotation
	MDMO-PPV	Poly[2-methoxy-5-(3′,7′-dimethyl-octyloxy)-1,4-phenylene vinylene]
	MEH-PPV	Poly[2-methoxy-5-(2′-ethyl-hexyl-oxy)-1,4-p henylene vinylene]
	MEH-CN-PPV	Poly[2-methoxy-5-(2′-ethyl-hexyl-oxy)-1,4-phenylene (1-cyano)vinylene]
	CN-PPV	Poly[2,5-di-n-hexyloxy-1,4-phenylene (1-cyano)vinylene]
	P3HT	Poly(3-hexylthiophene)
	POPT	Poly[3-(4′-octylphenyl)thiophene]
	PEOPT	Poly{3-[4′-(1″,4″,7″-trioxaoctyl)-phenyl]thiophene}
	PEDOT	Poly(3,4-ethylenedioxy thiophene)
	PDTI	Thiophene-isothianaphthene copolymer
	PTPTB	Benzothiadiazole-pyrrole copolymer

charge carrier generation, i.e. the number of electrons formed per absorbed photon; f_{fill}, the fill factor; and η_{mp}, the maximum power conversion efficiency. f_{fill} and η_{mp} are defined by Eqs. (6-5) and (6-6), respectively [67].

$$f_{fill} = i_{mp}V_{mp}/I_{sc}V_{oc} \tag{6-5}$$

$$\eta_{mp} = i_{mp}V_{mp}/D_r = f_{fill}(i_{sc}V_{oc}/D_r) \tag{6-6}$$

Here, i_{mp} and V_{mp} denote the current and the voltage at maximum power, and D_r (W cm^{-2}) is the incident solar irradiance.

Compared with inorganic PV cells, organic PV cells resemble the heterojunction type, apart from the fact that organic materials do not support the formation of a space-charge region at the junction. Figure 6.18 shows a schematic depiction of a cell simply formed by the superposition of two layers of semiconducting organic materials with different electron affinities and ionization potentials. One layer functions as the electron donor (p-type conductor) and the other as the electron acceptor (n-type conductor). In this case, the absorption of a photon is confined to a molecule or to a region of a polymer chain, where an excited state is created. This localized excited state is frequently termed an *exciton* (see Section 2.2.2). It refers to an electron-hole pair in semiconductor terminology. Charge separation at the interphase requires that the difference in energies of the hole states and the electron states exceeds the binding energy of the electron-hole pairs. This amounts to about 100 meV, and is much larger than the input energy required for charge separation in inorganic semiconductors. The efficiency of charge separation is critically determined by the exciton diffusion range, since after its generation the exciton must reach the junction in order to dissociate into two free charge carriers. Actually, the exciton diffusion range is at most a few nanometers and, therefore, a portion of the excitons generated in the bulk of the layer do not dissociate. In the course of efforts to overcome this flaw of flat-junction organic solar cells, new architectures consisting of phase-separated polymer blends were devised [68–70]. Figure 6.19 shows the structure of such a system and the charge transfer from an exciton at a donor/acceptor heterojunction. These blend systems consist of interpenetrating bicontinuous networks of donor and acceptor phases with domain sizes of 5–50 nm and provide donor/acceptor heterojunctions distributed throughout the layer thickness. In this case, the mean distance that the excitons have to travel to reach the interface is within the diffusion range and, therefore, efficiencies for

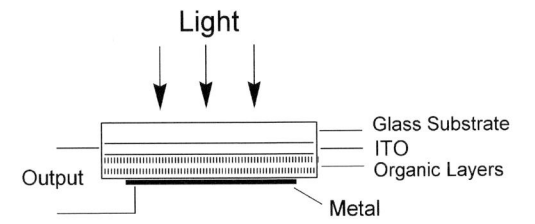

Fig. 6.18 Schematic depiction of a flat-heterojunction organic solar cell.

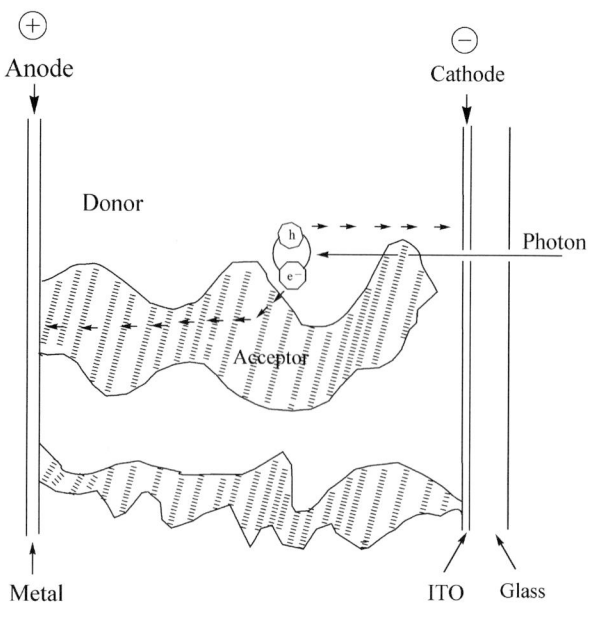

Fig. 6.19 Schematic diagram depicting charge transfer from an exciton at a donor/acceptor heterojunction in a composite of two conducting polymers.

the conversion of incident photons to electric current of over 50% have been achieved. Such systems can be formed, for example, from blends of donor and acceptor polymers such as MEH-PPV and CN-PPV [68, 69] or from composites of conducting polymers with buckminsterfullerenes such as MEH-PPV+C_{60} or poly(3-hexylthiophene) (P3HT)+C_{60} [70–74]. In the latter cases, the preparation of appropriate composites is facilitated by using fullerene derivatives with improved solubility such as PCBM, the structure of which is presented in Chart 6.11 [65, 75].

In typical experiments, thin (~100 nm) films of polymer blends were deposited by spin coating from a solution of the two polymers. Alternatively, two thin films of a hole-accepting and an electron-accepting polymer that had been deposited on ITO or metal substrates were laminated together in a controlled annealing pro-

Chart 6.11 Chemical structure of 1-(3-methoxycarbonyl)-propyl-1-phenyl-[6,6]C_{61}, PCBM.

Table 6.6 Performance characteristics of solar cells.

Material system	J_{sc} [a] (mA cm^{-2})	V_{oc} [b] (V)	f_{fill} [c]	η_{mp} [d] (%)	Φ_{cc} [e]	Ref.
P3HT/PCMB (1:0.8)	9.5	0.63	0.68	5.1		[70a]
P3HT/PCMB (1:1)	10.6	0.61	0.67	4.4		[70c]
MDMO-PPV/PCBM	5.25	0.82	0.61	2.5	0.50 (470 nm)	[70d]
POPT/MEH-CN-PPV	ca. 1	ca. 1	0.32	1.9	0.29	[76]
Amorphous silicon	19.4	0.89	0.74	12.7	~0.90	[61]
Monocrystalline silicon	42.4	0.71	0.83	24.7	>0.90	[61]

a) Short-circuit current density.
b) Open-circuit voltage.
c) Fill factor.
d) Maximum power conversion efficiency.
e) Quantum efficiency for charge carrier generation.

cess. In the latter case, a 20–30 nm deep interpenetration between the two layers was revealed by atomic force microscopy [76]. Performance characteristics of some of these organic PV cells and those of silicon cells are shown in Table 6.6.

Obviously, the performance of organic cells having bicontinuous network structures with quantum efficiencies of about 50% and power conversion efficiencies of about 5% remains far inferior to that of silicon cells, but is highly improved as compared to that of flat-junction organic cells, which have both quantum efficiencies and power conversion efficiencies of less than 0.1%.

In conclusion, for various reasons, certain organic materials and especially polymers are attractive for use in photovoltaics. There is the prospect of inexpensive production of large-area solar cells at ambient temperature, since high-throughput manufacture using simple procedures such as spin-casting or spray deposition and reel-to-reel handling is feasible. It is possible to produce very thin, flexible devices, which may be integrated into appliances or building materials. Moreover, it seems that new markets will become accessible with the aid of polymer-based photovoltaic elements. This concerns daily life consumer goods such as toys, chip cards, intelligent textiles, and electronic equipment with low energy consumption.

6.4 Polymer optical waveguides

6.4.1 General aspects

With the advent of semiconductor lasers, a new technique of information transmission based on optical fibers was developed [77]. Instead of propagating data electronically by the transport of electrons through coaxial copper cables, the new technique permits optical data transfer by laser light pulses guided through branching optical networks operated with the aid of optical fibers. Optical fibers consist of a highly transparent core and a surrounding cladding of refractive indices n_{core} and $n_{cladding}$, respectively. Provided that $n_{core} > n_{cladding}$, light entering the fiber at an angle $\theta < \theta_{max}$ is totally reflected at the cladding boundary and is thus transmitted through the fiber.

At present, copper conductors are still used in short-distance data communication. However, they can no longer cope with the high bandwidth demands of modern communication systems. Therefore, copper wiring systems are going to be replaced by high-bandwidth fiber-optic systems. The size and weight of optical fiber cables are significantly lower than those of coaxial copper wire cables, in which the single wires must be carefully isolated to prevent electromagnetic interference.

6.4.2
Optical fibers

6.4.2.1 Polymer versus silica fibers

Initially, the new fiber-optic technique was based solely on inorganic glass fibers, but in recent years polymeric optical fibers have also become attractive and appear to be in great demand for the transmission and the processing of optical communications compatible with the Internet [78–84]. As compared with silica fibers, polymer fibers have a larger caliber, are cheaper to prepare, and easier to process. However, because of their greater light attenuation and their lower frequency bandwidth for signal transmission, polymer fibers can only be employed in information networks over distances of several hundred meters. Typical properties of polymer and inorganic glass optical fibers are compared in Table 6.7.

Silica fibers are still unsurpassed as regards attenuation and bandwidth, but their diameter has to be kept rather small to provide for the required cable flexibility. Consequently, skillful hands and high precision tools are required to connect silica fibers in a time-consuming process. Polymer fibers have a much low-

Table 6.7 Typical properties of step-index optical fibers [85].

Property	PMMA [a]	Polycarbonate	Silica glass
Attenuation coefficient a (dB km^{-1}) [b]	125 at 650 nm	1000 at 650 nm	0.2 at 1300 nm
Transmission capacity C_{trans} (MHz km) [c]	<10	<10	10^2 to 10^3
Numerical aperture	0.3 to 0.5	0.4 to 0.6	0.10 to 0.25
Fiber diameter (mm)	0.25 to 10	0.25 to 10	9×10^{-3} to 1.25×10^{-1}
Maximum operating temperature (°C)	85	85	ca. 150

[a] Poly(methyl methacrylate).
[b] $a = (10/L) \log(P_0/P_L)$, P_0 and P_L: input and output power, L: fiber length.
[c] C_{trans}: product of bandwidth W_{band} and fiber length L, $C_{trans} = W_{band} \times L$, $W_{band} \approx 0.44\, L\, \Delta t^{-1}$, $\Delta t = (t_{out}^2 - t_{in}^2)^{1/2}$, t_{out} and t_{in}: width (FWHM) of output and input pulses.

er modulus than inorganic glass fibers and can, therefore, be of a much larger diameter without compromising their flexibility. Since their numerical aperture is larger, the acceptance angle, i.e. the light gathering capacity, is larger compared to that of glass fibers. Due to the large core diameter and the high numerical aperture, the installation of polymer optical fibers is facilitated and installation costs are much lower than for silica glass fiber networks. Hence, polymer optical fibers are suitable for short-distance data communication systems that require a large number of connections [85]. Generally, polymer optical fiber systems are applicable in local area networks (LANs), fiber-to-the-home systems, fiber-optic sensors, industrial environments, automotive applications, e.g. media-oriented system transport (MOST) devices, etc. Actually, data transmission rates increase in parallel with the number of devices connected to a system, and transmission rates of 400 Mbit s^{-1} or more are envisaged. With already existing and commercially available polymer optical fibers of a sufficiently large bandwidth, these requirements can be fulfilled. Another interesting field of application relates to lighting and illumination. In this context, end or point-source lighting and side- or line-lighting devices are to be discriminated. The former are used for motorway signaling and the latter for night illumination of buildings, to give typical examples [85].

The introduction of polymer optical fibers may have an impact on the development of next-generation light sources for optical communication. To date, the emission wavelength of semiconductor lasers has been adapted to the absorption characteristics of silica fibers. Since polymer optical fibers may be used in different wavelength regions, a change in an important boundary condition for light source engineering is anticipated.

6.4.2.2 Compositions of polymer optical fibers (POFs)

Polymer optical fibers have been prepared from various amorphous polymers, such as polycarbonate, poly(methyl methacrylate), polystyrene, and diglycol diallylcarbonate resin [79, 80]. In these cases, the light attenuation of the respective optical fibers is due to absorption by higher harmonics of C–H vibrations. Substitution of hydrogen by deuterium, fluorine or chlorine results in a shift of the absorption due to overtone vibrations to higher wavelengths and reduces the attenuation at key communication wavelengths, as is apparent from Table 6.8.

Table 6.8 Light attenuation (approximate values) caused by absorption due to overtone vibrations at key communication wavelengths, in units of dB km^{-1} [79].

λ (nm)	C–Cl	C–F	C–D	C–H
840	$<10^{-8}$	10^{-4}	10^{1}	10^{4}
1310	10^{-5}	10^{0}	10^{3}	10^{5}
1550	10^{-3}	10^{1}	10^{5}	10^{6}

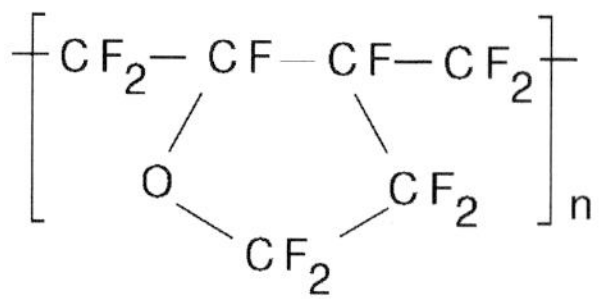

Chart 6.12 Chemical structure of a perfluorinated polymer used to make optical fibers.

Actually, commercial polymeric optical fibers made from a perfluorinated polymer (see Chart 6.12) exhibit an attenuation of 15 dB km^{-1} at $\lambda = 1300$ nm. Single-channel systems can be operated at a transmission rate of 2.5 Gbit s^{-1} over a distance of 550 m at $\lambda = 840$ or 1310 nm [79, 86]. Besides the intrinsic factors for optical propagation loss mentioned above, namely absorption and Rayleigh light scattering, there are extrinsic factors such as dust, interface asymmetry between core and cladding, variation in core diameter, etc., that may also affect the light transmission.

6.4.2.3 Step-index and graded-index polymer optical fibers

Table 6.7 presents the properties of large-core *step-index polymer optical fibers*, SI-POFs. They are characterized by a single refractive index, which extends over the entire core and changes abruptly at the core/cladding interface. SI-POFs possess a low bandwidth due to extensive pulse broadening. An increased bandwidth is achieved with *graded-index polymer optical fibers*, GI-POFs, which possess a refractive index profile over the core. Refractive index profiles can be obtained by special techniques, e.g. by polymerizing a mixture of two monomers differing in size and refractive index in rotating tubes or by photochemical partial bleaching of a dopant contained in a polymer [79].

6.4.3
Polymer planar waveguides

Planar, i.e. rectangular, waveguide components are applied in many photonic devices. They can be easily manufactured at low cost. Typical applications relate to computer backplanes combining electrical and optical cables [87], thermo-optical switches [88], optical splitters of multichannel high-density planar lightwave circuits [89], and polyimide-based electro-optical (EO) modulators [90].

6.4.4
Polymer claddings

Polymers also play a role in the case of specialized optical equipment, where the different parts are connected by silica fibers. This applies, for example, to instruments used for spectroscopic process analysis, i.e., for real-time control of chemical processes [91]. To prevent physical damage, the fibers are coated with poly(vinyl chloride) or acrylate-based polymers. Fibers coated with polyimide withstand temperatures up to 350 °C.

References

1 R. M. Schaffert, *Electrophotography*, 2nd Edition, Focal Press, London (1975).
2 C. F. Carlson, US Patent 2 297 691 (1942).
3 K. Y. Law, Chem. Rev. 93 (1993) 449.
4 P. M. Borsenberger, D. S. Weiss, *Organic Photoreceptors for Xerography*, Dekker, New York (1998).
5 P. M. Borsenberger, D. S. Weiss, *Organic Photoreceptors for Imaging Systems*, Dekker, New York (1993).
6 L. B. Schein, *Electrophotography and Development Physics*, 2nd Edition, Laplacian Press, Morgan Hills, CA, USA (1996).
7 L. B. Schein, *Electrophotography and Development Physics*, Springer, Berlin (1992).
8 E. M. Williams, *The Physics and Technology of Xerographic Processes*, Wiley, New York (1984).
9 M. Umeda, M. Hashimoto, J. Appl. Phys. 72 (1992) 117.
10 (a) L. Akcelrud, *Electroluminescent Polymers*, Prog. Polym. Sci. 28 (2003) 875; (b) K. Mullen, U. Scherf (eds.), *Organic Light-Emitting Devices: Synthesis, Properties and Applications*, Wiley, New York (2006).
11 H.-K. Shim, J.-I. Jin, *Light-Emitting Characteristics of Conjugated Polymers*, in K.-S. Lee (ed.), Polymers for Photonics Applications I, Springer, Berlin, Adv. Polym. Sci. 158 (2002) 193.
12 T. Bernius, M. Inbasekaran, J. O'Brien, W.-S. Wu, *Progress with Light-Emitting Polymers*, Adv. Mater. 12 (2000) 1737.
13 W. Graupner, *Science and Technology of Organic Light-Emitting Diodes*, The Spectrum 15 (2002) 20.
14 B. Ruhstaller, S. A. Carter, S. Barth, H. Riel, W. Riess, J. C. Scott, J. Appl. Phys. 89 (2001) 4575.
15 D. Y. Kim, H. N. Cho, C. Y. Kim, *Blue Light Emitting Polymers*, Prog. Polym. Sci. 25 (2000) 1089.
16 A. Greiner, *Design and Synthesis of Polymers for Light-Emitting Diodes*, Polym. Adv. Technol. 9 (1998) 371.
17 J. R. Sheats, Y. L. Chang, D. B. Roitman, A. Socking, *Chemical Aspects of Polymeric Electroluminescent Devices*, Acc. Chem. Res. 32 (1999) 193.
18 L. J. Rothberg, A. J. Lovinger, *Status and Prospects for Organic Electroluminescence*, J. Mater. Res. 11 (1996) 3174.
19 A. Kraft, A. Grimsdale, A. B. Holmes, *Electroluminescent Conjugated Polymers – Seeing Polymers in a New Light*, Angew. Chem. Int. Ed. 37 (1998) 402.
20 R. H. Friend, R. W. Gymer, A. B. Holmes, J. H. Burroughes, R. N. Marks, C. Taliani, D. D. C. Bradley, D. A. dos Santos, J. L. Bredas, M. Lögdlund, W. R. Salaneck, *Electroluminescence in Conjugated Polymers*, Nature 397 (1999) 121.
21 A. Bolognesi, C. Botta, D. Facchinetti, M. Jandke, K. Kreger, P. Strohriegl, A. Relini, R. Rolandi, S. Blumstengel, *Polarized Electroluminescence in Double-Layer Light-Emitting Diodes with Perpendicularly Oriented Polymers*, Adv. Mater. 13 (2001) 1072.
22 M. Schwoerer, H. C. Wolf, *Elektrolumineszenz und Photovoltaik*, Chapter 11 in M. Schwoerer, H. C. Wolf, Organische Molekulare Festkörper, Wiley-VCH, Weinheim (2005).
23 S. Miyata, H. S. Nalwa (eds.), *Organic Electroluminescent Materials and Devices*, Gordon & Breach, Amsterdam (1997).
24 J. H. Burroughes, D. D. C. Bradley, A. R. Brown, R. N. Marks, K. Mackay, R. H. Friend, P. L. Burns, A. B. Holmes, Nature 347 (1990) 539.
25 *OLED Cross Reference by Material Function*, H. W. Sands Corp. {http://www.hwsands.com/productslists/oled/cross_reference_material_function_oled.htm}.
26 Y.-H. Niu, J. Huang, Y. Cao, Adv. Mater. 15 (2003) 807.
27 J. Kalinowski, *Electronic Processes in Organic Electroluminescence*, in S. Miyata, H. S. Nalwa (eds.), Organic Electroluminescent Materials and Devices, Gordon & Breach, Amsterdam (1997), p. 1.
28 H. Bässler, Polym. Adv. Technol. 9 (1998) 402.
29 S. Barth, U. Wolf, H. Bässler, P. Müller, H. Riel, H. Vestweber, P. F. Seidler, W. Rieß, Phys. Rev. B 60 (1999) 8791.
30 (a) U. Wolf, V. I. Arkhipov, H. Bässler, Phys. Rev. B 59 (1999) 7507; (b) V. I. Ar-

khipov, U. Wolf, H. Bässler, Phys. Rev. B 59 (1999) 7514.

31 D. D. C. Bradley, R. H. Friend, H. Lindenberger, S. Roth, Polymer 27 (1986) 1709.

32 M. Grell, D. D. C. Bradley, M. Inbasekaran, E. R. Woo, Adv. Mater. 9 (1997) 798.

33 P. Dyreklev, M. Berggren, O. Inganäs, M. R. Andersson, O. Wennerström, T. Hjertberg, Adv. Mater. 7 (1995) 43.

34 M. Grell, D. D. C. Bradley, Adv. Mater. 11 (1999) 895.

35 V. Cimrova, M. Remmers, D. Neher, G. Wegner, Adv. Mater. 8 (1996) 146.

36 M. Gao, B. Richter, S. Kirstein, Adv. Mater. 9 (1997) 802.

37 B. W. D. Andrade, J. Brooks, V. Adamovich, M. E. Thompson, S. R. Forrest, Adv. Mater. 14 (2002) 1032.

38 G. Kranzelbinder, G. Leising, *Organic Solid-State Lasers*, Rep. Prog. Phys. 63 (2000) 729.

39 I. D. F. Samuel, G. A. Turnbull, *Polymer Lasers: Recent Advances*, Materials Today 7 (2004) 28.

40 U. Lemmer, A. Haugeneder, C. Kallinger, J. Feldmann, *Lasing in Conjugated Polymers*, in G. Hadziioannou, P. van Hutton (eds.), *Semiconducting Polymers: Chemistry, Physics and Engineering*, Wiley-VCH, Weinheim (2000), p. 309.

41 U. Lemmer, C. Kallinger, J. Feldmann, Phys. Blätter 56 (2000) 25.

42 Z. Bao, Y. M. Chen, R. B. Cai, L. Yu, Macromolecules 26 (1993) 5228.

43 S. V. Frolov, A. Fujii, D. Chinn, Z. V. Vardeny, K. Yoshino, R. V. Gregory, Appl. Phys. Lett. 72 (1998) 2811.

44 Y. Kawabe, Ch. Spielberg, A. Schülzgen, M. F. Nabor, B. Kippelen, E. A. Mash, P. Allemand, M. Kuwata-Gonokami, K. Takeda, N. Peyghambarian, Appl. Phys. Lett. 72 (1998) 141.

45 M. D. McGehee, R. Gupta, S. Veenstra, E. K. Miller, M. A. Diaz-Garcia, A. J. Heeger, Phys. Rev. B 58 (1998) 7035.

46 C. Kallinger, M. Hilmer, A. Haugeneder, M. Perner, W. Spirkl, U. Lemmer, J. Feldmann, U. Scherf, K. Müllen, A. Gombert, V. Wittwer, Adv. Mater. 10 (1998) 920.

47 T. Granlund, M. Theander, M. Berggren, M. Andersson, A. Ruzeckas, V. Sundstrom, G. Bjork, M. Granstrom, O. Inganas, Chem. Phys. Lett. 288 (1998) 879.

48 N. Tessler, G. J. Denton, R. H. Friend, Nature 382 (1996) 695.

49 S. Riechel, C. Kallinger, U. Lemmer, J. Feldmann, A. Gombert, V. Wittwer, U. Scherf, Appl. Phys. Lett. 77 (2000) 2310.

50 F. Hide, B. J. Schwartz, M. A. Diaz-Garcia, A. J. Heeger, Chem. Phys. Lett. 256 (1996) 424.

51 M. D. Archer, R. Hill (eds.), *Clean Electricity from Photovoltaics*, Imperial College Press, London (2001).

52 R. Messenger, G. Ventre, *Photovoltaic Systems Engineering*, CRC Press, Boca Raton, FL, USA (1999).

53 J. Perlin, *From Space to Earth: The Story of Solar Electricity*, Aatec Publications, Ann Arbor, MI, USA (1999).

54 R. H. Bube, *Photovoltaic Materials*, Imperial College Press, London (1998).

55 H.-J. Lewerenz, H. Jungblut, *Photovoltaik*, Springer, Berlin (1995).

56 M. A. Green, *Silicon Solar Cells: Advanced Principles and Practice*, Centre for Photovoltaic Devices and Systems, University of New South Wales, Sydney (1995).

57 S. R. Wenham, M. A. Green, M. E. Watt, *Applied Photovoltaics*, Centre for Photovoltaic Devices and Systems, University of New South Wales, Sydney (1995).

58 L. D. Partain (ed.), *Solar Cells and Their Applications*, Wiley-Interscience, New York (1995).

59 T. Markvart (ed.), *Solar Electricity*, Wiley, Chichester (1994).

60 (a) N. S. Sariftci *Plastic Photovoltaic Devices*, Materials Today 7 (2004) 36; (b) C. J. Brabec, V. Dyakonov, J. Parisi, N. S. Sariciftci (eds.), *Organic Photovoltaics, Concept and Realization*, Springer, Berlin (2003).

61 J. Nelson, (a) *Organic and Plastic Solar Cells*, Chapter IIe-2 in T. Markvart, L. Catañer (eds.), *Practical Handbook of Photovoltaics: Fundamentals and Applications*, Elsevier, Oxford (2003); (b) Materials Today 5 (2002) 20.

62 J. J. M. Halls, R. H. Friend, *Organic Photovoltaic Devices*, in Ref. [51] p. 377.

63 J.-F. Nierengarten, G. Hadziioannou, N. Armaroli, Materials Today 4 (2001) 16.

64 (a) C.J. Brabec, *Organic Photovoltaics: Technology and Markets*, Solar Energy Mater. Solar Cells 83 (2004) 273; (b) C.J. Brabec, N.S. Sariciftci, J. Keppler, Materials Today 3 (2000) 5.
65 A. Dhanabalan, J.K.J. van Duren, P.A. van Hal, J.L.J. van Dongen, R.A.J. Jannssen, Adv. Funct. Mater. 11 (2001) 255.
66 S.E. Shaheen, D. Vangeneugden, R. Kiebooms, D. Vanderzande, T. Fromherz, F. Padinger, C.J. Brabec, N.S. Sariciftci, Synth. Met. 121 (2001) 1583.
67 M.D. Archer, *The Past and Present*, in Ref. [51] p. 1.
68 J.J.M. Halls, C.A. Walsh, N.C. Greenham, E.A. Marseglia, R.H. Friend, S.C. Moratti, A.B. Holmes, *Efficient Photodiodes from Interpenetrating Networks*, Nature 376 (1995) 498.
69 G. Yu, J. Gao, J.C. Hummelen, F. Wudl, A.J. Heeger, Science 270 (1995) 1789.
70 (a) H. Hoppe, N.S. Sariciftci, *Morphology of Polymer/Fullerene Bulk Heterojunction Solar Cells*, J. Mater. Chem. 16 (2006) 45; (b) M. Al-Ibrahim, H.-K. Roth, U. Zhokhavets, G. Gobsch, S. Sensfuss, Solar Energy Mater. Solar Cells 85 (2005) 13; (c) G. Li, V. Shrotriya, J. Huang, Y. Yad, T. Moriarty, K. Emery, Y. Yang, Nature Mater. 4 (2005) 864; (d) S.E. Shaheen, C.J. Brabec, N.S. Sariciftci, F. Padinger, T. Fromherz, J.C. Hummelen, Appl. Phys. Lett. 78 (2001) 841.
71 I. Riedel, M. Pientka, V. Dyakonov, *Charge Carrier Photogeneration and Transport in Polymer-Fullerene Bulk-Heterojunction Solar Cells*, Chapter 15 in W. Brütting (ed.), *Physics of Organic Semiconductors*, Wiley-VCH, Weinheim (2005).
72 N. Armaroli, E. Barigeletti, P. Ceroni, J.-E. Eckert, J.-F. Nicoud, J.-F. Nierengarten, Chem. Commun. (2000) 599.
73 J.-E. Eckert, J.J. Nicoud, J.-F. Nierengarten, S.-G. Liu, L. Echegoyen, F. Barigelletti, N. Armaroli, L. Ouali, V. Krasnikov, G. Hadziioannou, J. Am. Chem. Soc. 122 (2000) 7467.
74 J.-F. Nierengarten, J.-E. Eckert, J.J. Nicoud, L. Ouali, V. Krasnikov, G. Hadziioannou, Chem. Commun. (1999) 617.
75 C.J. Brabec, V. Dyakonov, *Photoinduced Charge Transfer in Bulk Heterojunction Composites*, in Ref. [60b].
76 M. Granström, K. Petritsch, A.C. Arias, A. Lux, M.R. Andersson, R.H. Friend, Nature 395 (1998) 257.
77 H. Zanger, *Fiber Optics Communication and Other Applications*, McMillan, New York (1991).
78 H.S. Nalwa, *Polymer Optical Fibers*, American Scientific Publishers, Stevenson Ranch, CA, USA (2004).
79 W. Daum, J. Krauser, P.E. Zamzow, O. Ziemann, *POF – Polymer Optical Fibers for Data Communication*, Springer, Berlin (2002).
80 K. Horie, H. Ushiki, F.M. Winnik, *Molecular Photonics, Fundamentals and Practical Aspects*, Kodansha-Wiley-VCH, Weinheim (2000).
81 A. Weinert, *Plastic Optical Fibers, Principles, Components, Installation*, MCD Verlag, Erlangen (1999).
82 J. Hecht, *City of Light, The Story of Fiber Optics*, Oxford University Press, New York (1999).
83 T. Kaino, *Polymers for Light Wave and Integrated Optics*, L.A. Hornak (ed.), Dekker, New York (1992).
84 M. Kitazawa, *POF Data Book*, MCR Techno Research, Tokyo (1993).
85 M.A. de Graaf, *Transmissive and Emissive Polymer Waveguides for Communication and Illumination*, University Press Facilities, Eindhoven, The Netherlands (2002).
86 G.-D. Khoe, H. van den Boom, I.T. Monroy, *High Capacity Transmission Systems*, Chapter 6 in [78].
87 J. Moisel, J. Guttman, H.-P. Huber, O. Krumpholz, M. Rode, R. Bogenberger, K.-P. Kuhn, Opt. Eng. 39 (2000) 673.
88 N. Keil, H.H. Yao, C. Zawadski, W. Lösch, K. Satzke, W. Wischmann, J.V. Wirth, J. Schneider, J. Bauer, M. Bauer, Electron. Lett. 37 (2001) 89.
89 J.T. Kim, C.G. Choi, J. Micromech. Microeng. 15 (2005) 1140.
90 S. Ermer, *Applications of Polyimides to Photonic Devices*, in K. Horie, T. Yamashita (eds.), *Photosensitive Polyimides, Fundamentals and Applications*, Technomic, Lancaster, PA, USA (1995).
91 J. Andrews, P. Dallin, Spectroscopy Europe 15 (2003) 23.

Part II
Light-induced chemical processes in polymers

7
Photoreactions in synthetic polymers

7.1
Introductory remarks

According to the Grotthus-Draper law, chemical changes can only be produced in a system by absorbed radiation. It has been pointed out in Chapter 1 that light absorption involves electronic transitions. As regards organic molecules, such transitions occur with a high probability if some of the constituent atoms are arranged in special bonding positions. Such arrangements are termed *chromophoric groups* (Chapter 1, Table 1.1). They become resonant at certain light frequencies. Resonance gives rise to absorption bands in the absorption spectrum (Chapter 1, Figs. 1.4 and 1.5). The chemical activity of a chromophoric group may originate from two features: (a) The bonding strength between adjacent atoms is strongly reduced when an electron is promoted to a higher level. Therefore, a chemical bond can be cleaved if the atoms separate upon vibration. This type of monomolecular bond cleavage is a very rapid process (ca. 10^{-12} s) that cannot be prevented by any means after the absorption of a photon. (b) The electronic excitation leads to a relatively stable state. The lifetime of the excited state is so long (occasionally approaching the ms range) that, in the condensed phase, chromophoric groups have many encounters with the surrounding molecules, thus enabling bimolecular chemical interactions. Thereby, the original chemical bond is relinquished and a new bond is formed. This type of bond cleavage can be prevented by *energy quenching* (see Chapter 1), i.e. through energy transfer from the excited chromophore to an additive functioning as an energy acceptor. The bond scission processes mentioned above are energetically feasible since the photon energies associated with radiation of wavelengths ranging from 250 nm (4.96 eV) to 400 nm (3.1 eV) correspond to the bond dissociation energies of common covalent bonds, i.e. about 3.5 eV for C–H, C–C, and C–O bonds (in aliphatic compounds). Although these considerations apply to both small and large molecules, there are certain aspects pertaining to polymers that merit special attention and these are dealt with in this chapter. The subsequent sections are related overwhelmingly to phenomena associated with application aspects. Cross-linking and main-chain scission, for example, play key roles in lithographic applications, and photo-oxidation reactions are of prominent importance for the behavior of polymers in outdoor applications.

Polymers and Light. Fundamentals and Technical Applications. W. Schnabel
Copyright © 2007 WILEY-VCH Verlag GmbH & Co. KGaA, Weinheim
ISBN: 978-3-527-31866-7

It should be emphasized that a plethora of research papers and patents have been devoted to the field of photoreactions in synthetic polymers. However, only a few important results are highlighted in this chapter. For more detailed information, the reader is referred to relevant books and reviews [1–28].

7.1.1
Amplification effects

Photochemical reactions in polymers may result in amplification effects, as becomes obvious if we consider the example of the photochemical coupling of two molecules. In a system of linear chain macromolecules consisting of a large number of base units, the formation of a given small number of cross-links may lead to an enormous property change. This is so, because each cross-link connects two chains with many base units, which are all then affected. Consequently, the polymer may become insoluble in solvents if, on average, each macromolecule only contains one cross-link site. On the other hand, a property change is hardly detectable if the same number of cross-links is generated in a system consisting of small molecules because, in this case, each cross-link involves only two small molecules and leaves the other molecules unaffected.

7.1.2
Multiplicity of photoproducts

The deactivation of identical electronically excited chromophores can result in the cleavage of different chemical bonds. This common phenomenon is demonstrated for two polymers, polystyrene and poly(methyl methacrylate) in Schemes 7.1 and 7.2. Note that the bond cleavage probabilities are not equal, i.e., the quantum yields for the individual processes may differ by orders of magnitudes.

As indicated in Schemes 7.1 and 7.2, several different free radicals are generated upon exposure to light. These radicals undergo various reactions, e.g. hydrogen abstraction reactions, thereby generating new free radicals, and coupling reactions. In this way, a variety of products is eventually formed, as is demonstrated in Scheme 7.3 for the case of polystyrene.

Notably, this scheme does not cover all of the initially formed free radicals (see before, Scheme 7.1). Therefore, the number of photoproducts formed in the case of polystyrene exceeds that shown in Scheme 7.3.

Obviously, photochemical methods based on the direct absorption of light by the polymer can hardly be envisaged for chemical modifications of commercial polymers. Most practical applications, especially those devoted to photolithography, concern light-induced changes in the solubility of polymers as a consequence of intermolecular cross-linking or main-chain scission. In these cases, only reactions causing changes in the average molar mass are important because other photoreactions, and the resulting products, are ineffective with respect to the desired property change.

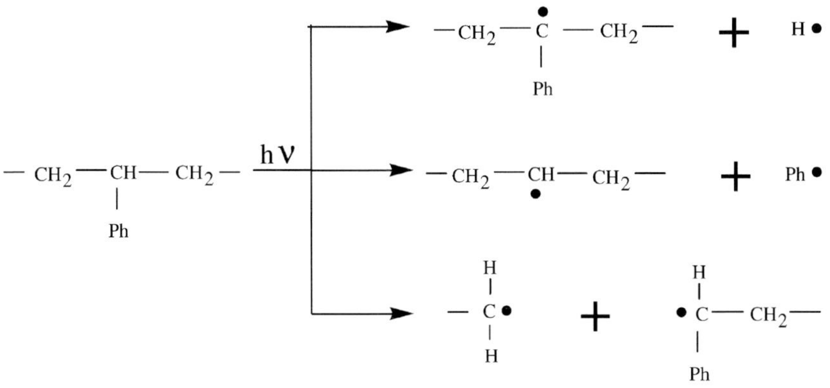

7.1 Introductory remarks

$$-CH_2-CH-CH_2- \xrightarrow{h\nu} \begin{cases} -CH_2-\overset{\bullet}{C}-CH_2- \;+\; H^\bullet \\ \quad\quad\quad | \\ \quad\quad\quad Ph \\[4pt] -CH_2-CH-CH_2- \;+\; Ph^\bullet \\ \quad\quad\quad\bullet \\[4pt] \;\;\;\;\;\;\;\;\; H \;\;\;\;\;\;\;\;\;\;\;\;\;\;\;\;\;\; H \\ \;\;\;\;\;\;\;\;\; | \;\;\;\;\;\;\;\;\;\;\;\;\;\;\;\;\;\;\; | \\ -\overset{|}{C}^\bullet \;\;+\;\; ^\bullet\overset{|}{C}-CH_2- \\ \;\;\;\;\;\;\;\;\; | \;\;\;\;\;\;\;\;\;\;\;\;\;\;\;\;\;\;\; | \\ \;\;\;\;\;\;\;\;\; H \;\;\;\;\;\;\;\;\;\;\;\;\;\;\;\;\;\; Ph \end{cases}$$

Scheme 7.1 Primary reactions in the photolysis of polystyrene [9].

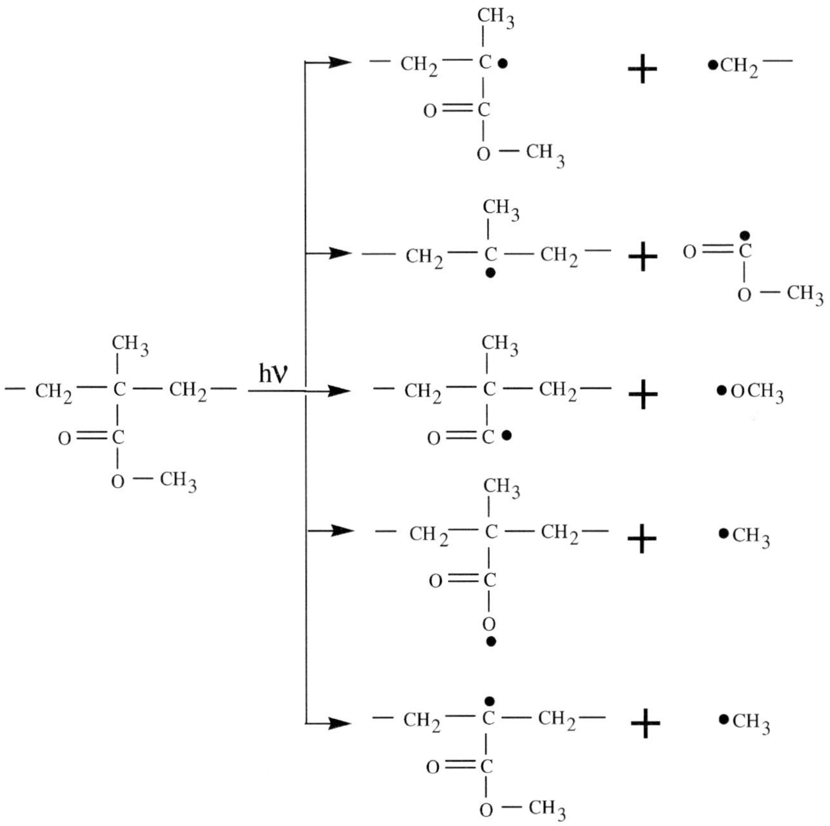

Scheme 7.2 Primary reactions in the photolysis of poly(methyl methacrylate) [14].

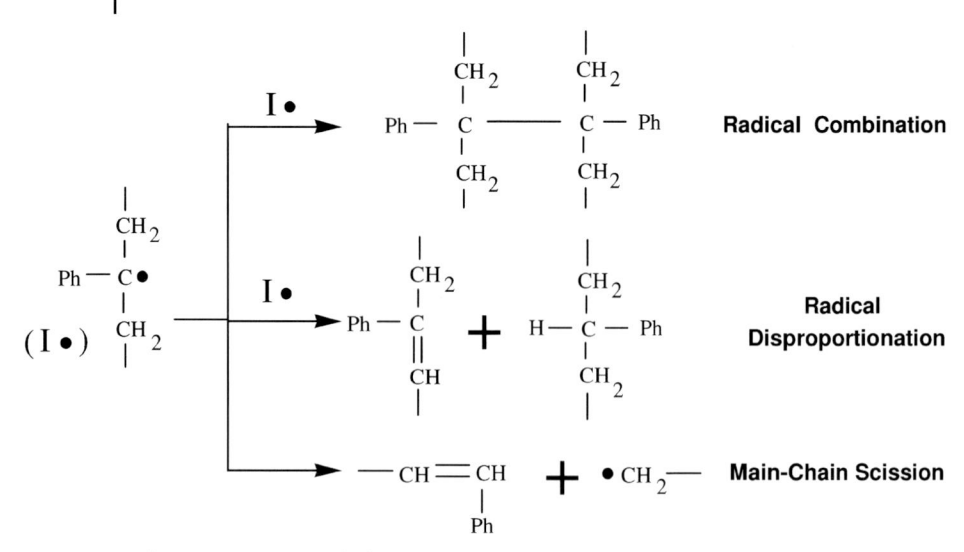

Scheme 7.3 Reactions of a benzyl-type macroradical formed in the photolysis of polystyrene [9].

7.1.3
Impurity chromophores

Commonly, commercial polymers contain impurities originating from the polymerization or from processing. These impurities, although mostly present in trace amounts only, play an undesired role, because they are capable of absorbing the near-UV portion (290–400 nm) of the solar radiation reaching the earth, and, therefore, jeopardize or curtail the stability of the polymers in outdoor applications, hastening degradation. According to the structures of their repeating units, some of the practically important linear polymers, such as polyethylene, polypropylene, and poly(vinyl chloride), should be transparent to light of $\lambda > 250$ nm. However, commercial polymer formulations contain impurity chromophores (see Table 7.1), which absorb UV light. Consequently, these formulations are subject to severe degradation in the absence of stabilizers.

Some of the chromophores shown in Table 7.1 are chemically incorporated into the polymers, such as carbonyl groups or carbon-carbon double bonds, whereas others are adventitiously dispersed, such as polynuclear aromatic compounds and metal salts. The latter are almost invariably present in many polymers. Oxygen-polymer charge-transfer complexes have been postulated as additional UV light-absorbing species. Apart from the latter, the impurity chromophores listed in Table 7.1 function as free radical generators, as illustrated in Scheme 7.4. Hydroperoxide groups, the most common and important of chromophores, yield highly reactive hydroxyl radicals. Carbonyl groups can give rise to the formation of various kinds of free radicals, as outlined in Section 7.1.4. Moreover, they may act as donors in energy-transfer processes, which also ap-

Table 7.1 Impurity chromophores commonly contained in commercial polyalkenes or poly(vinyl chloride)s.

Structure of chromophore	Denotation
−C(H)(OOH)−	Hydroperoxide group
>C=O	Carbonyl group
−C(H)=C(H)−C(=O)−	α,β-Unsaturated carbonyl group
−C(H)=C(H)−; −C(H)−C(H)=C(H)− with Cl; −CH$_2$−CH=CH−CH$_2$−Cl	Double bonds
−C(H)=C(H)−C(H)=C(H)−	Conjugated double bonds
(naphthalene, anthracene, rubrene structures)	Polynuclear aromatics (e.g. naphthalene, anthracene, rubrene)
Ti^{4+} Al^{3+} Fe^{3+}	Metal ions
$[RH\ldots O_2]_{CT}$	Charge-transfer complex

plies for polynuclear aromatic compounds. Metal salts produce free radicals by electron-transfer processes. In the case of poly(vinyl chloride), allyl-type chlorine atoms are split off.

Most of the radicals generated by photoreactions of impurity chromophores can abstract hydrogen atoms from the surrounding polymer. This applies especially to hydroxyl and chlorine radicals.

Dioxygen-polymer charge-transfer complexes are assumed to form hydroperoxide groups [Eq. (7-1)].

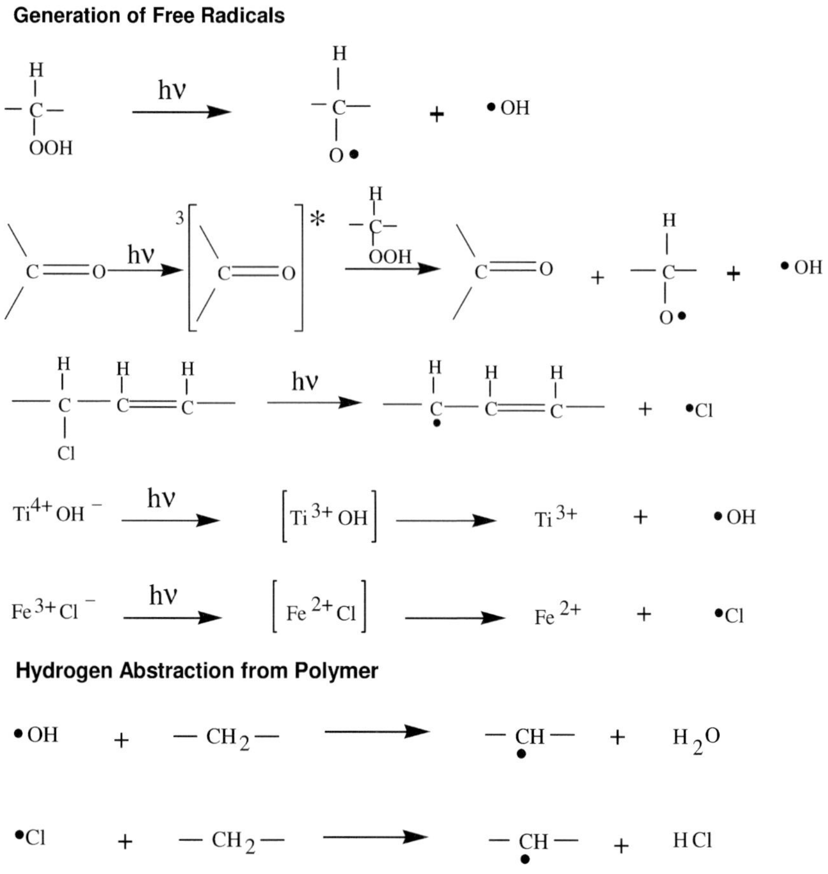

Generation of Free Radicals

$$-\overset{H}{\underset{OOH}{C}}- \xrightarrow{h\nu} -\overset{H}{\underset{O\bullet}{C}}- + \bullet OH$$

$$\overset{\diagdown}{\diagup}C=O \xrightarrow{h\nu} {}^3\left[\overset{\diagdown}{\diagup}C=O\right]^* \xrightarrow{-\overset{H}{\underset{OOH}{C}}-} \overset{\diagdown}{\diagup}C=O + -\overset{H}{\underset{O\bullet}{C}}- + \bullet OH$$

$$-\overset{H}{\underset{Cl}{C}}-\overset{H}{C}=\overset{H}{C}- \xrightarrow{h\nu} -\overset{H}{\underset{\bullet}{C}}-\overset{H}{C}=\overset{H}{C}- + \bullet Cl$$

$$Ti^{4+}OH^- \xrightarrow{h\nu} \left[Ti^{3+}OH\right] \longrightarrow Ti^{3+} + \bullet OH$$

$$Fe^{3+}Cl^- \xrightarrow{h\nu} \left[Fe^{2+}Cl\right] \longrightarrow Fe^{2+} + \bullet Cl$$

Hydrogen Abstraction from Polymer

$$\bullet OH + -CH_2- \longrightarrow -\underset{\bullet}{CH}- + H_2O$$

$$\bullet Cl + -CH_2- \longrightarrow -\underset{\bullet}{CH}- + HCl$$

Scheme 7.4 Generation of free radicals by photoreactions of impurity chromophores and ensuing hydrogen abstraction from the polymer.

$$[RH \cdots O_2]_{CT} \xrightarrow{h\nu} [RH^\oplus \cdots O_2^\ominus]_{CT} \rightarrow R^\bullet + {}^\bullet OOH \rightarrow ROOH \qquad (7\text{-}1)$$

7.1.4
Photoreactions of carbonyl groups

The detrimental environmental degradation of unstabilized commercial polymeric products consisting of polyethylene, polypropylene, poly(vinyl chloride), etc., is frequently due to very small amounts of ketonic carbonyl groups. Electronically excited ketone groups can undergo different processes, in particular the so-called Norrish type I and Norrish type II reactions, as illustrated in Scheme 7.5 for the case of a copolymer of ethylene and carbon monoxide:

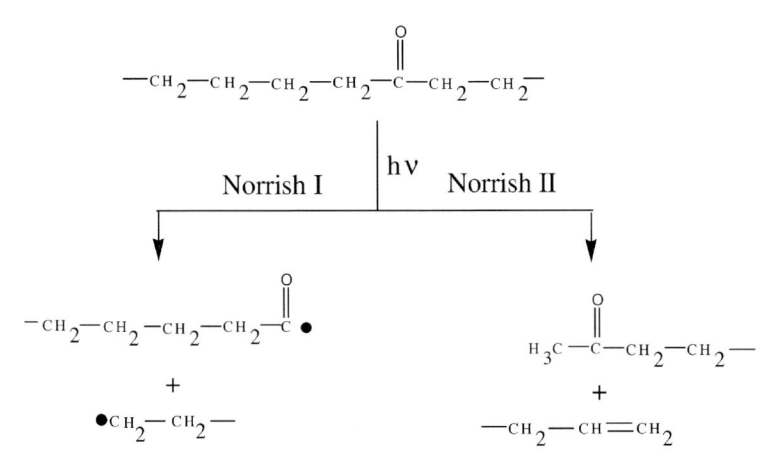

Scheme 7.5 Light-induced main-chain cleavage of polyethylene containing traces of carbonyl groups.

According to the Norrish type I reaction, a carbon-carbon bond in a position α to the carbonyl group is cleaved. The resulting ketyl radical is very likely to release carbon monoxide [Eq. (7-2)].

$$R-\overset{\overset{\displaystyle O}{\|}}{C}{}^{\bullet} \rightarrow R^{\bullet} + CO \qquad (7\text{-}2)$$

The Norrish type II process refers to a C–C bond cleavage initiated by the abstraction of a hydrogen in a γ-position with respect to the carbonyl group.

Note that Norrish-type reactions are not only of importance in relation to various polymers containing ketonic impurities, but they also play a dominant role in the photolysis of all polymers containing carbonyl groups as constituent moieties, such as polyacrylates, polymethacrylates, poly(vinyl acetate), polyesters, and polyamides.

7.2
Cross-linking

The formation of intermolecular cross-links, i.e. covalent bonds between different polymer chains, causes an increase in the average molar mass and eventually combines all of the macromolecules into a three-dimensional insoluble network. Cross-linking can be accomplished in various ways. Several methods rely on reactions of electronically excited pendant groups on the polymer chains, others on reactions of various kinds of reactive species in the ground state that are photogenerated in polymeric systems. Typical of the former reaction type are [2+2] cycloadditions that occur in the case of linear polymers bearing pendant C=C bonds; typical examples of the latter process are reactions of nitrenes generated in polymeric systems containing azide groups [17].

Photo-cross-linking of thick polymer films is a difficult task, because the penetration depth is limited to thin layers if the light is strongly absorbed. A high absorptivity, on the other hand, is required for effective photo-cross-linking. Therefore, only the photo-cross-linking of thin films (≤ 1 μm) is of practical importance. This process has found widespread application in photolithography (see Section 9.1). The following subsections are largely devoted to systems that have been employed for photolithographic applications, although some systems of as yet purely academic interest are also discussed.

7.2.1
Cross-linking by cycloaddition of C=C bonds

The reaction of an excited alkene molecule in its S_1 or T_1 state with an alkene molecule in its ground state produces a cyclobutane derivative [Eq. (7-3)].

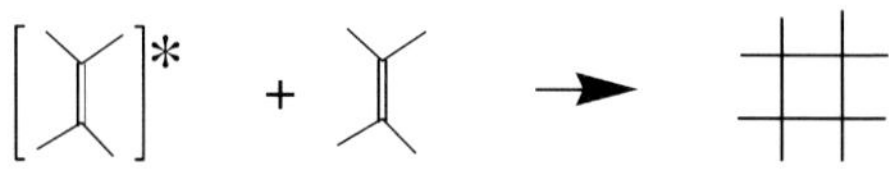

$$[\bowtie]^* + \bowtie \longrightarrow \#\!\!\# \tag{7-3}$$

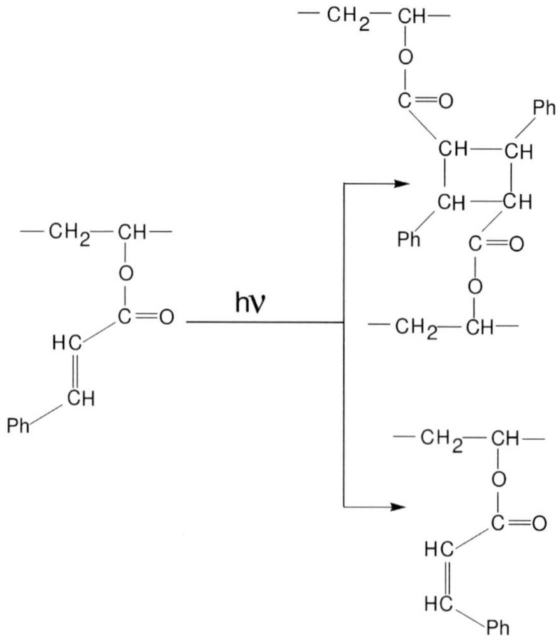

Scheme 7.6 Light-induced cross-linking and trans → cis isomerization of poly(vinyl cinnamate).

7.2 Cross-linking

In this reaction, which occurs in competition with isomerization, two π bonds are lost with the formation of two new σ bonds. Since two π electrons of each alkene molecule are involved, the reaction is called [$_\pi 2 +_\pi 2$], or simply [2+2] cycloaddition. As discovered by Minsk [29], linear polymers containing C=C bonds in pendant groups also undergo light-induced [2+2] cycloaddition reactions. This leads to the formation of intermolecular cross-links, as demonstrated here for the classical case of poly(vinyl cinnamate). Exposure of the polymer to UV light ($\lambda_{exp} = 365$ nm) results both in [2+2] cycloaddition and *trans*→*cis* isomerization (Scheme 7.6).

Besides cinnamate compounds, various other compounds containing C=C bonds also undergo light-induced cycloaddition reactions (see Chart 7.1).

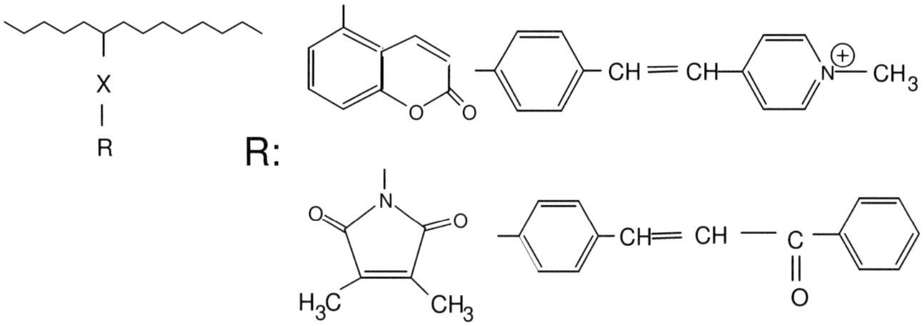

Chart 7.1 Structures of moieties suitable for the cross-linking of linear polymers through cycloaddition.

Scheme 7.7 shows, as a typical example, the photo-cross-linking of a co-polypeptide [30].

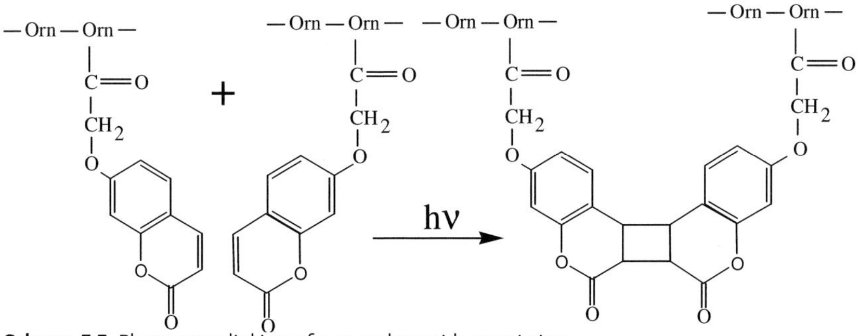

Scheme 7.7 Photo-cross-linking of a co-polypeptide consisting of L-ornithine and δ-7-coumaryloxyacetyl-L-ornithine residues [30].

7.2.2
Cross-linking by polymerization of reactive moieties in pendant groups

Photo-cross-linking of linear polymers can be achieved by light-induced polymerization of reactive moieties in pendant groups located on different macromolecules, a process analogous to the polymerization of low molar mass compounds, which is treated in Chapter 10. Provided that the pendant groups are capable of approaching to within the reaction distance, and their concentration is high enough, they undergo chain reactions, which can propagate by way of various mechanisms that are started with the aid of appropriate photoinitiators. From the technical point of view, free radical polymerizations of unsaturated carbon-carbon bonds are most important. In principle, cationic polymerizations involving the ring opening of epoxides and glycidyl ethers (see Chart 7.2) are also suitable.

Although, in contrast to free radical polymerizations, cationic polymerizations are unaffected by O_2, their importance is somewhat limited by the scarcity of appropriate macromolecules and suitable photoinitiators [3]. However, this does not apply to the photopolymerization of low molar mass epoxides (see Section 10.3). In this context, applications of photo-cross-linked epoxides in various fields such as stereolithography, volume holography, and surface coating are notable [16].

A typical example involving the polymerization of unsaturated pendant groups relates to the fixation of surface relief gratings that are optically inscribed with the aid of a 488 nm laser beam (see Section 5.6.1) onto a film of a copolymer bearing pendant azobenzene groups (chemical structure shown in Chart 7.3).

The generation of the relief gratings involves trans→cis isomerization of the pendant azobenzene groups, and the subsequent fixation is achieved by cross-linking with UV light at 80 °C, i.e., by polymerization of the acrylic groups with the aid of a photoinitiator (see Chart 7.4).

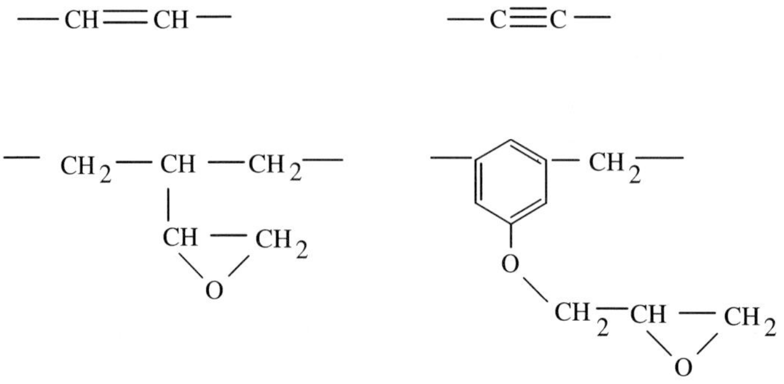

Chart 7.2 Structures of moieties suitable for cross-linking by photopolymerization.

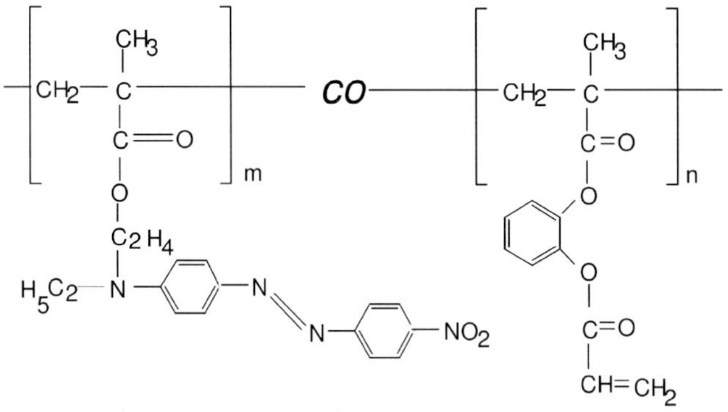

Chart 7.3 Co-monomers (1:1 molar ratio) contained in a polymer used to generate surface relief gratings.

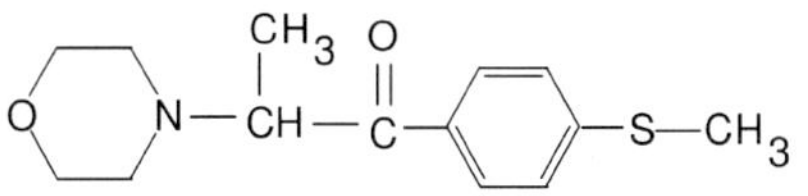

Chart 7.4 Chemical structure of 4'-(methylthio)-2-morpholino-propiophenone used as a photoinitiator in the cross-linking of the copolymer of Chart 7.3.

This process results in an improved thermal stability of the gratings [31]. Another example relates to the photo-cross-linking of a copolymer of the structure shown in Chart 7.5 [32].

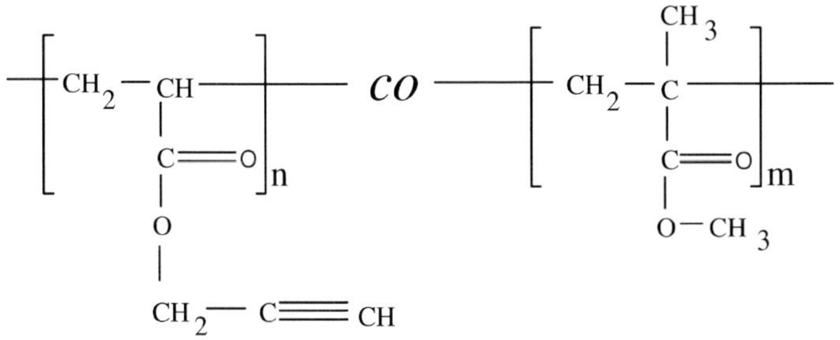

Chart 7.5 Chemical structure of a copolymer consisting of propargyl acrylate (34.5%, left) and methyl methacrylate (65.5%, right).

Here, the alkynyl side groups are polymerized to form a three-dimensional network when the copolymer is exposed to UV light (320–390 nm) in the presence of 5 mol% tungsten hexacarbonyl, $W(CO)_6$ (see also Subsection 10.2.2.4.1). The polymerization is presumed to be initiated by the formation of a η^2-alkyne tungsten pentacarbonyl complex, $\eta^2\text{-RC}\equiv\text{CR}'W(CO)_5$.

7.2.3
Cross-linking by photogenerated reactive species

This mode of photo-cross-linking has attracted attention for applications in resist technology, since it became apparent that the photodecomposition of organic azides in polymeric systems leads to insolubility. Azide groups can be chemically attached to polymer chains, as demonstrated here by two examples:

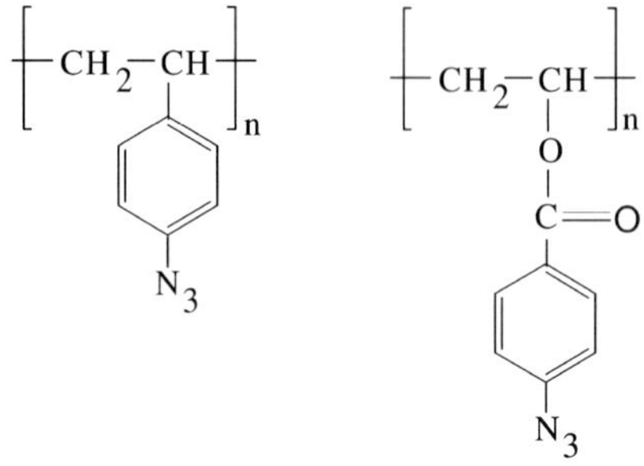

Chart 7.6 Base units of polymers bearing pendant azide groups.

Alternatively, bisazides, i.e. low molar mass compounds containing two azide groups, can be added to the polymer. Several commercially used bisazides are presented in Table 7.2. Many linear polymers can be photo-cross-linked with the aid of bisazides [17]. Of note in this context is poly(*cis*-isoprene), which contains some cyclized structures (Chart 7.7). It has been frequently applied as a resist material in photolithography applications.

A water-soluble bisazide (see Chart 7.8) is applicable for the photo-cross-linking of water-processable polymeric systems containing polyacrylamide or poly(vinyl pyrrolidone).

Table 7.2 Bisazides of practical importance for the photo-cross-linking of linear polymers [17].

Denotation	Chemical structure
2,6-Bis(4-azidobenzal)-4-methylcyclohexane	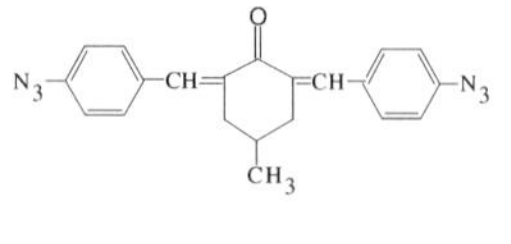 N_3—⟨ ⟩—CH=⟨cyclohexanone with CH_3⟩=CH—⟨ ⟩—N_3
4,4′-Diazidostilbene	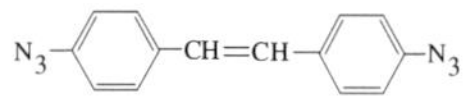 N_3—⟨ ⟩—CH=CH—⟨ ⟩—N_3
4,4′-Diazidobenzophenone	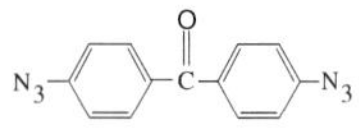 N_3—⟨ ⟩—C(=O)—⟨ ⟩—N_3
4,4′-Diazidobenzalacetone	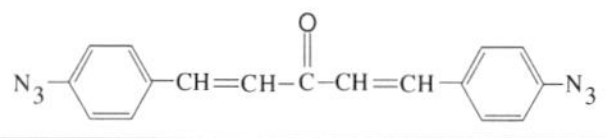 N_3—⟨ ⟩—CH=CH–C(=O)–CH=CH—⟨ ⟩—N_3

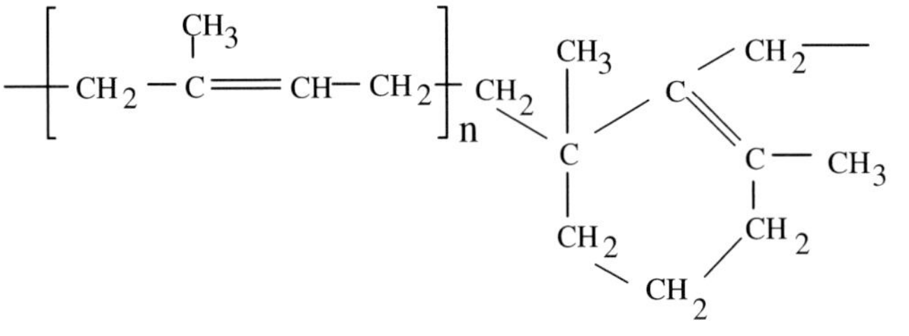

Chart 7.7 Cyclized structure in poly(cis-isoprene).

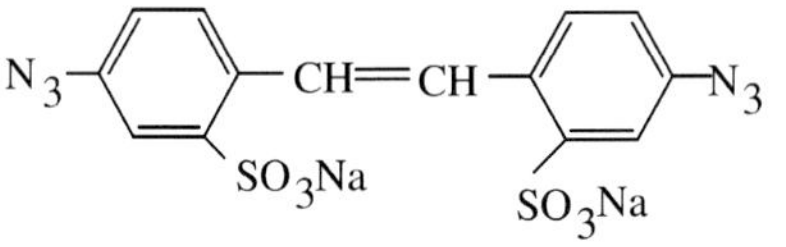

Chart 7.8 Chemical structure of a water-soluble bisazide.

When an azide group decomposes after absorption of a photon, an electrically neutral, very reactive intermediate called a nitrene is formed. Immediately after decomposition, the latter is in an electronically excited singlet state, which can decay to the ground state, the triplet nitrene [see Eqs. (7-4) and (7-5)].

$$RN_3 + h\nu \rightarrow {}^1(RN)^* + N_2 \tag{7-4}$$

$$^1(RN)^* \rightarrow {}^3(RN) \tag{7-5}$$

Both nitrene species are very reactive, since the nitrogen possesses only six valence electrons. Singlet nitrene can insert into C–H bonds of the polymer and, in the case of unsaturated polymers, can add to C=C bonds, both in single-step processes (Scheme 7.8).

As shown in Scheme 7.9, triplet nitrene can abstract a hydrogen atom from neighboring macromolecules, thus forming an amino radical and a carbon macroradical (reaction (a)). The two radicals have correlated spins and can,

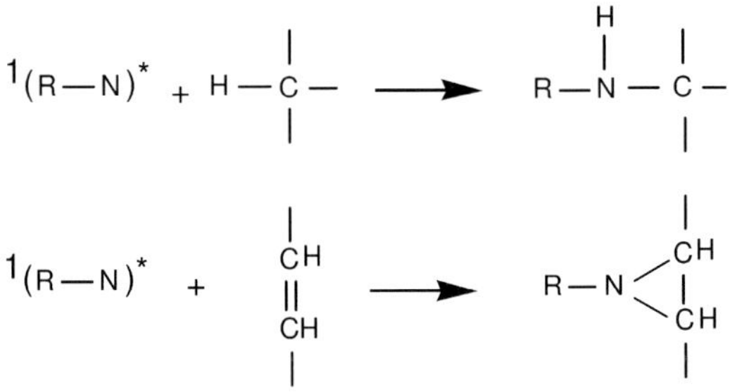

Scheme 7.8 Reactions of singlet nitrene with saturated and unsaturated polymers.

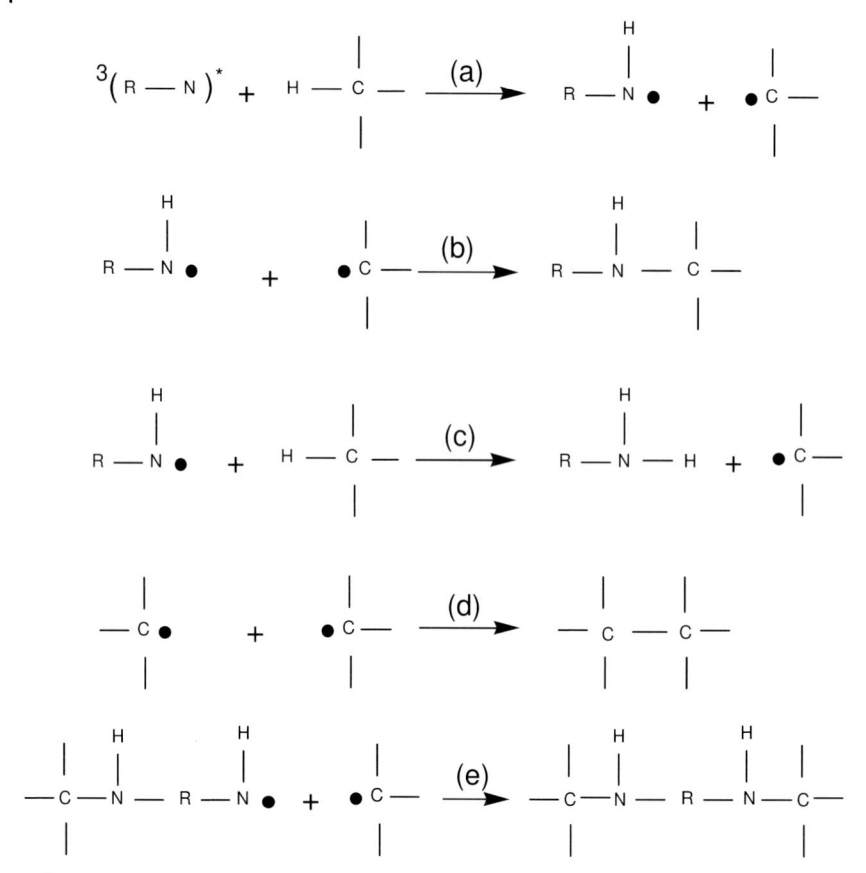

Scheme 7.9 Cross-linking of polymers through the reaction of triplet nitrene.

therefore, only couple after spin inversion (reaction (b)). The amino radical may also abstract a hydrogen atom from a different site to produce a primary amine (reaction (c)). Cross-links are formed by coupling reactions, namely by the combination of macroradicals (reaction (d)) and, if bisazides are employed, after the conversion of both azide groups according to reaction (e) [17].

Free radical mechanisms also serve to explain the photo-cross-linking of various polymers, such as that of polyethylene accomplished with the aid of light-absorbing additives such as benzophenone, quinone, benzoin, acetophenone, or their derivatives. When electronically excited by light absorption, these additives either directly abstract hydrogen from the polymer or decompose into free radicals capable of abstracting hydrogen, as shown in Schemes 7.10 and 7.11.

Macroradicals P$^{\bullet}$ can form cross-links by combination reactions according to Eq. (7-6).

$$P^{\bullet} + P^{\bullet} \rightarrow P-P \tag{7-6}$$

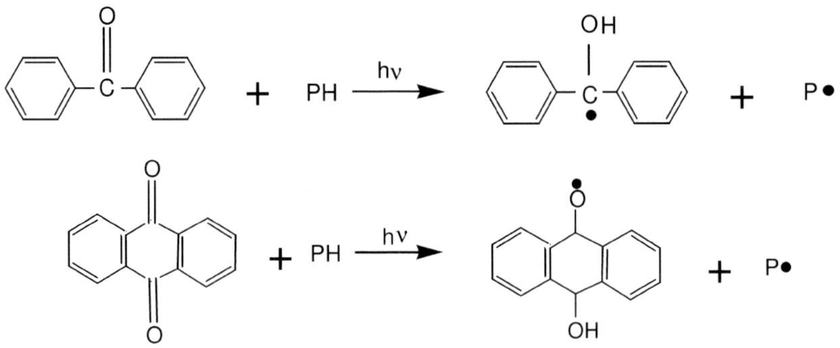

Scheme 7.10 Generation of macroradicals by the reaction of electronically excited benzophenone and anthraquinone with a polymer, PH.

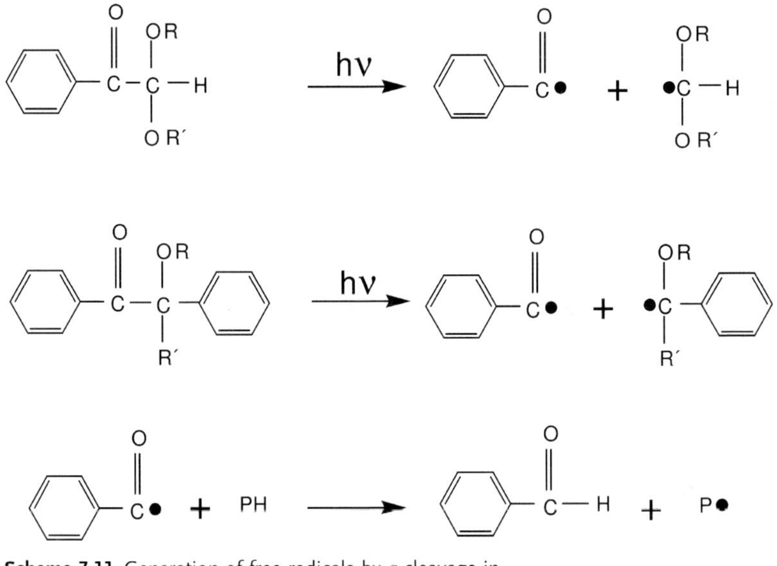

Scheme 7.11 Generation of free radicals by α-cleavage in electronically excited acetophenone and benzoin derivatives and subsequent formation of macroradicals P• by hydrogen abstraction from macromolecules, PH.

The occurrence of these reactions is restricted to the amorphous phase. Therefore, the photo-cross-linking process has to be performed at temperatures exceeding the crystalline melting point in the case of highly crystalline polymers such as polyethylene. The cross-linking efficiency can be strongly enhanced by the addition of small amounts of multifunctional compounds such as triallyl cyanurate, TAC (see Chart 7.9), or by the incorporation of special diene moieties into copolymers such as ethylene propylene diene copolymers (EPDM elastomers) [33].

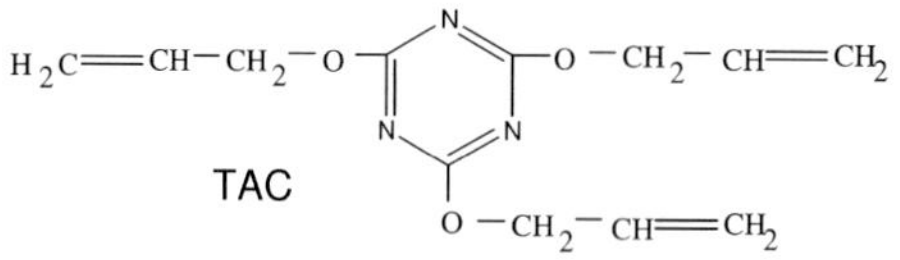

Chart 7.9 Chemical structure of triallyl cyanurate.

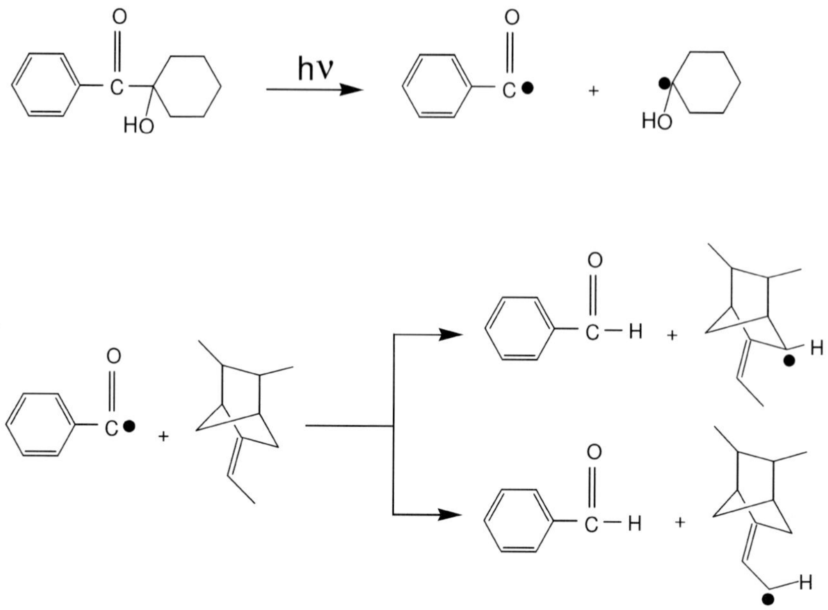

Scheme 7.12 Generation of pendant macroradicals acting as precursors for the cross-linking of an EPDM elastomer containing ethylidene norbornene moieties (other co-monomer moieties are not shown). Initiator: hydroxycyclohexyl phenyl ketone [33].

The reaction mechanism in this case is shown in Scheme 7.12. It is based on the fact that allyl-type hydrogens are readily abstracted by reactive radicals such as ketyl species. Side-chain macroradicals generated in this way combine to form intermolecular cross-links.

7.2.4
Cross-linking by cleavage of phenolic OH groups

Typical of this type of photo-cross-linking is the case of poly(4-hydroxystyrene) (see Chart 7.10) [34].

The deactivation of excited singlet phenolic groups proceeds by two main routes: cleavage of the O–H bonds and intersystem crossing to the triplet state, as shown in Scheme 7.13.

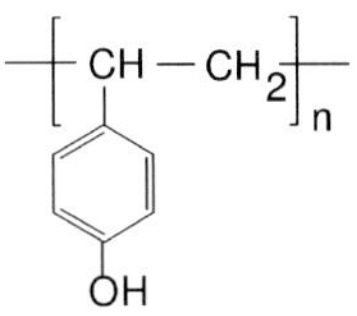

Chart 7.10 Chemical structure of poly(4-hydroxystyrene).

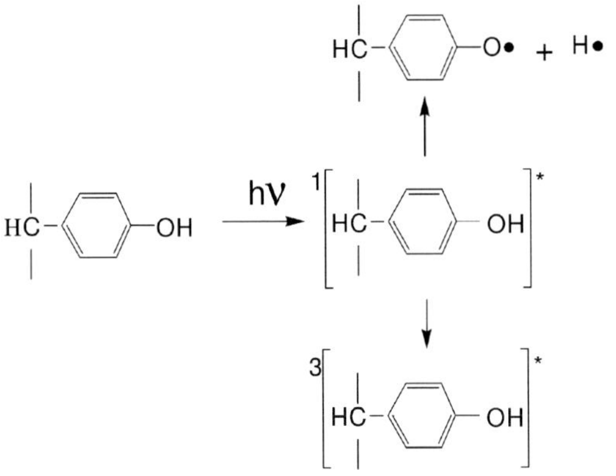

Scheme 7.13 Primary steps in the photolysis of poly(4-hydroxystyrene).

The phenoxyl radicals can couple to form cross-links (Scheme 7.14).

If dioxygen is present, additional phenoxyl radicals are formed by reaction according to Eq. (7-7), i.e. by the reaction of triplet excited phenolic groups with O_2.

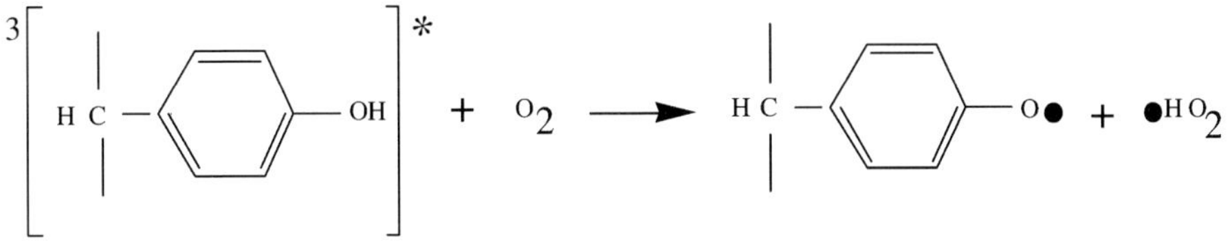

$$\text{[HC-C}_6\text{H}_4\text{-OH]}^* + O_2 \longrightarrow \text{HC-C}_6\text{H}_4\text{-O}^\bullet + {}^\bullet HO_2 \tag{7-7}$$

Therefore, the cross-linking quantum yield is significantly increased if the irradiation is performed in the presence of dioxygen.

7.3
Simultaneous cross-linking and main-chain cleavage of linear polymers

As has been pointed out in Section 7.1.2, polymers commonly undergo different kinds of bond ruptures simultaneously upon exposure to light, i.e. bond cleavage processes occur both in side chains and in the main chain of linear polymers. Bond rupture in side chains results in the formation of lateral macroradi-

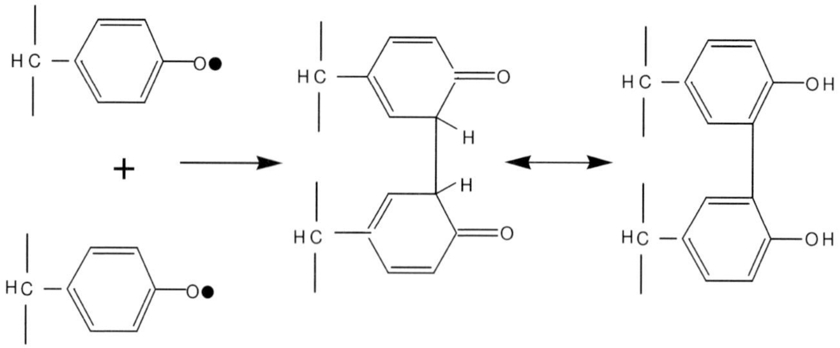

Scheme 7.14 Coupling of phenoxyl radicals.

cals, which can give rise to the release of low molar mass compounds and can also form inter- and intramolecular cross-links. Therefore, it is often the case that main-chain scission and cross-linking occur simultaneously. These processes cause changes in the molar mass distribution and in the average molar mass of the polymer, which has been treated theoretically [35–37]. The dependence of the weight-average molar mass M_w (g mol^{-1}) of linear polymers undergoing simultaneous main-chain cleavage and cross-linking on the absorbed dose D_{abs} (photons g^{-1}) is given by Eq. (7-8):

$$\frac{1}{M_{w,D}} = \frac{1}{M_{w,0}} + \left(\frac{\Phi(S)}{2} - 2\Phi(X)\right)\frac{D_{abs}}{N_A} \qquad (7\text{-}8)$$

where $\Phi(S)$ and $\Phi(X)$ denote the quantum yields for main-chain cleavage and cross-linking, respectively, and N_A is Avogadro's number. Equation (7-8) holds for the case that the initial molar mass distribution is of the most probable type and that main-chain ruptures and cross-links are randomly distributed along the polymer chains. Cross-linking predominates if $\Phi(S) < 4\Phi(X)$. In this case, the reciprocal average molar mass decreases, i.e. M_w increases with increasing absorbed dose. On the other hand, main-chain cleavage predominates if $\Phi(S) > 4\Phi(X)$. In this case, the reciprocal average molar mass increases, i.e. M_w decreases with increasing absorbed dose. In this context, it should be noted that predominant main-chain cleavage causes a deterioration of important mechanical properties that are related to the molar mass of the polymer. Several linear polymers are characterized with respect to the predominance of cross-linking or main-chain cleavage in Table 7.3.

Interestingly, polyacrylonitrile, poly(methyl acrylate), and polystyrene behave differently in the rigid state and in dilute solution. This may be explained in terms of lateral macroradicals being generated upon the release of side groups in a primary step. The combination of these radicals competes with decomposition through main-chain rupture. In dilute solution, where radical encounters are much less probable than in the rigid state, main-chain rupture predomi-

Table 7.3 Predominant effects upon UV irradiation of polymers in the absence of oxygen [27].

Polymer	Rigid state	Dilute solution
Poly(methyl methacrylate)	degradation	degradation
Poly(α-methyl styrene)	degradation	degradation
Poly(phenyl vinyl ketone)	degradation	degradation
Polyacrylonitrile	crosslinking	degradation
Poly(methyl acrylate)	crosslinking	degradation
Polystyrene	crosslinking	degradation

nates. This mechanism is illustrated for the case of polyacrylonitrile in Scheme 7.15.

When linear polymers undergo predominantly cross-linking, a three-dimensional insoluble network is formed. The absorbed dose at which the insoluble network begins to form is the *gel dose*, D_{gel}. It corresponds to an average of one cross-link per weight-average molecule [35], and a simple equation may be derived from Eq. (7-8) for the relationship between D_{gel} and $\Phi(X)$:

$$D_{gel} = \frac{N_A}{\Phi(X) M_{w,0}} \tag{7-9}$$

Equation (7-9) holds in the absence of main-chain scission, i.e. at $\Phi(S) = 0$. In this case, the reciprocal molar mass approaches infinity at the gel dose, i.e. $1/M_{w,D_{gel}} = 0$.

Quantum yields of photoproducts of selected polymers are presented in Table 7.4. It can be seen that both $\Phi(S)$ and $\Phi(X)$ are low (< 0.1). The quantum

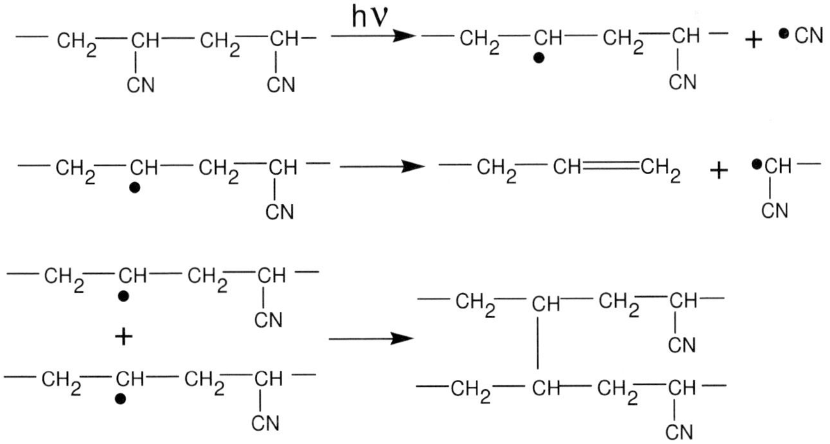

Scheme 7.15 Main-chain cleavage and cross-linking of polyacrylonitrile.

Table 7.4 Photoproduct quantum yields of polymers in the rigid state, determined at room temperature in vacuo [27].

Polymer	$\frac{\Phi(S)}{\Phi(X)}$	$\Phi(S) \times 10^2$	$\Phi(X) \times 10^2$	λ (nm)	Volatile products ($10^2\ \Phi$)
Poly-α-methylstyrene		0.1–0.6		253.7	α-methylstyrene H_2 (1.7×10^2)
Poly(methyl methacrylate)		1.2–3.9		253.7	CH_3OH (48); $HCOOCH_3$ (14); CO; H_2; CO_2
Poly(phenyl vinyl ketone)		6.0		313	
Poly(vinyl acetate)	1.4	6.6	4.7	253.7	CH_3COOH (1.0); CO_2 (0.65); CO (0.69); CH_4 (0.38)
Poly(ethylene terephthalate)	2.7	0.16	0.06	313	
Poly(methyl acrylate)	1.0	0.19	0.19	253.7	HCHO (2); CH_3OH (0.2); $HCOOCH_3$ (0.8)
Poly(p-methylstyrene)	0.52			253.7	H_2 (6); CH_4 (0.04)

yields of volatile products resulting from side-group degradation are also quite low for most polymers, apart from poly(methyl methacrylate).

7.4
Photodegradation of selected polymers

It is not intended to present a comprehensive treatise on the photoreactions in polymers in this book. Actually, many polymers exhibit analogous behavior. However, this certainly does not apply to poly(vinyl chloride) or polysilanes and, therefore, these two types of polymers are discussed to some extent in the following subsections.

7.4.1
Poly(vinyl chloride)

Poly(vinyl chloride), PVC, is one of the most widely used polymers. Commercial PVC products commonly contain plasticizers (up to 40%) such as phthalates or mellitates. If exposed to UV or solar radiation for prolonged periods, PVC products suffer from a deterioration of their mechanical and electrical properties and are eventually discolored [11, 19, 21]. Unsaturated moieties are believed to be the most important initiator species, with carbonyl groups as the next most important. The latter can undergo Norrish-type reactions (see Section 7.1.4). Moreover, excited carbonyl groups can transfer energy to unsaturated moieties or abstract hydrogens. In addition, hydroperoxide and peroxide groups formed during autoxidation of the polymer (see Section 7.5) can contribute to the initiation process [11].

7.4 Photodegradation of selected polymers

The discoloration is due to a dehydrochlorination process resulting in the formation of long conjugated polyene sequences in the polymer chain [Eq. (7-10)]. Polyenes can give rise to photo-cross-linking reactions.

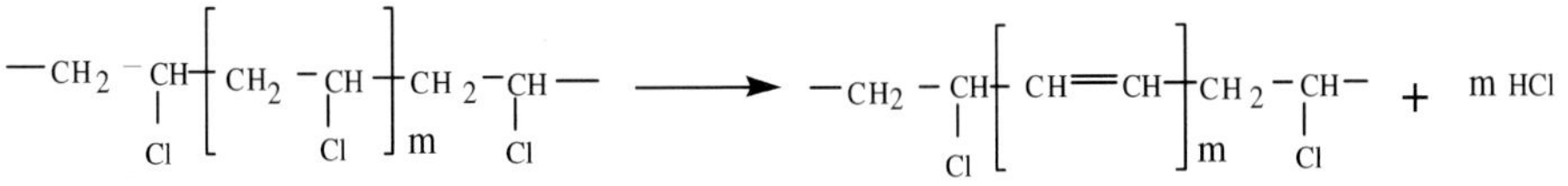

$$-CH_2-\underset{Cl}{CH}\left[CH_2-\underset{Cl}{CH}\right]_m CH_2-\underset{Cl}{CH}- \longrightarrow -CH_2-\underset{Cl}{CH}\left[CH=CH\right]_m CH_2-\underset{Cl}{CH}- + m\, HCl \quad (7\text{-}10)$$

It is generally accepted that the elimination of HCl occurs by way of a free radical chain reaction. As shown in the lower part of Scheme 7.16, chlorine atoms function as propagating species. Likely initiation mechanisms involving some of the impurity chromophores listed in Table 7.1 are presented in the upper part of Scheme 7.16.

The solar light-induced dehydrochlorination of PVC plasticized with phthalates has been reported to be sensitized by the plasticizer [38, 39]. In marked contrast, more recent work has revealed a weak protective effect of phthalates with respect to C–Cl bond cleavage and polyene formation. Phthalates are likely to quench electronically excited states of impurity chromophores [40].

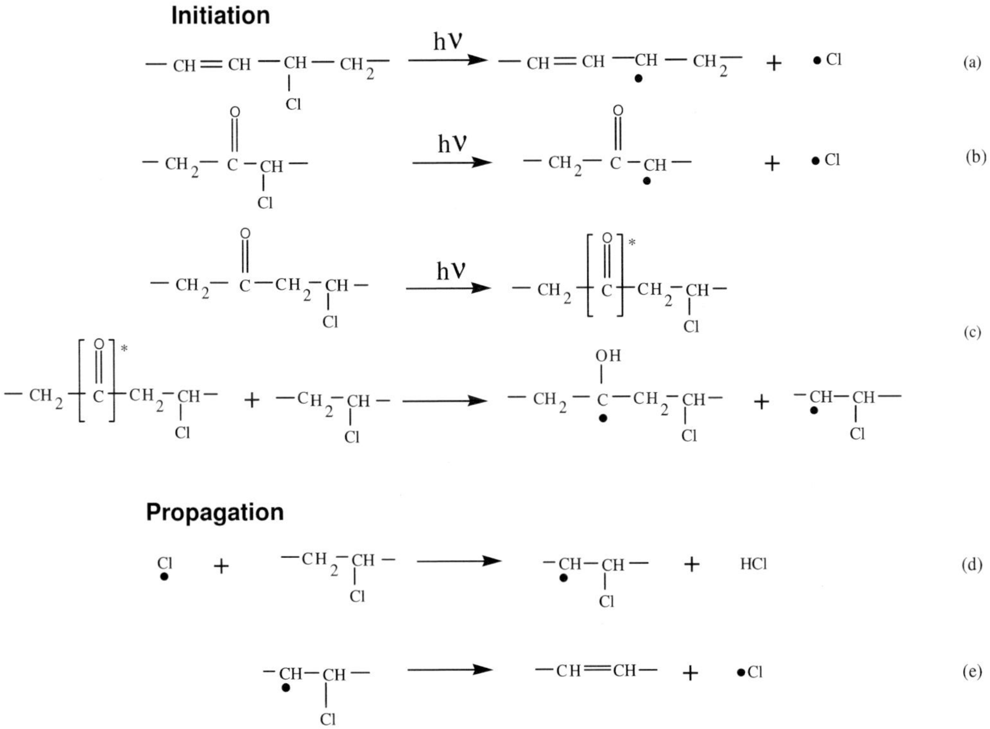

Scheme 7.16 Mechanism of the light-induced dehydrochlorination of poly(vinyl chloride).

7.4.2
Polysilanes

Polysilanes (alternative denotations: polysilylenes, poly-catena-silicons) of the general structure shown in Chart 7.11 exhibit an absorption band in a relatively long-wavelength region, i.e. between 300 and 400 nm, reflecting the σ-conjugation of electrons in the silicon chain.

In addition to their other interesting properties, polysilanes are photoconductive [41] (see Chapter 2) and, therefore, are attractive with regard to practical applications [42, 43]. However, to the detriment of their technical applicability, polysilanes show a pronounced trend to suffer photodegradation. Light absorption induces main-chain scission and extrusion of silylene, as depicted in Scheme 7.17.

The lifetime of the excited state giving rise to main-chain cleavage is shorter than 100 ps [44].

Chart 7.11 Chemical structure of a base unit of polysilane.

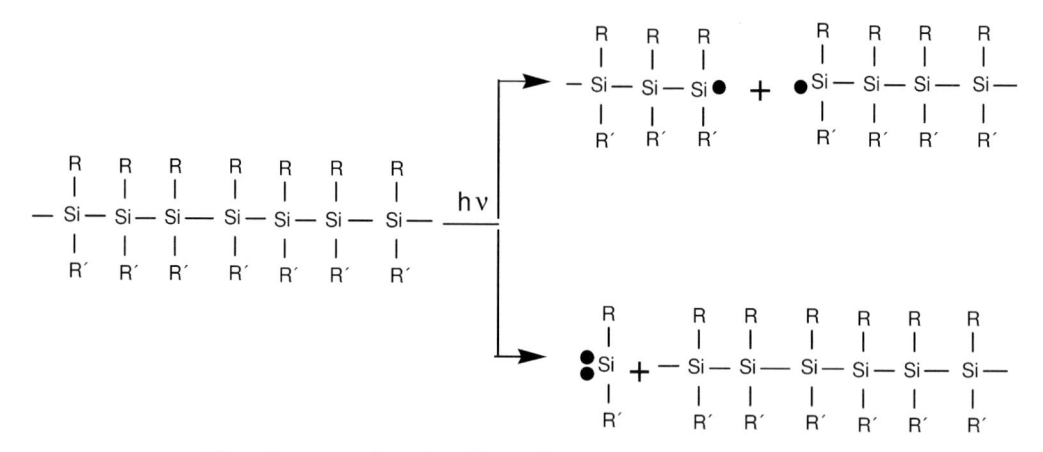

Scheme 7.17 Main-chain degradation of polysilanes.

7.5
Oxidation

Oxidation processes are initiated when polymers absorb visible or UV light in the presence of air [7, 12, 24-26]. In most cases, these processes occur as chain reactions, initiated by the light-induced generation of free radicals. Since some of the reaction products are chromophoric groups, capable of initiating new kinetic chains themselves, the oxidation becomes auto-accelerated during exposure. As a consequence of autoxidation, important mechanical properties of polymeric materials may suffer a sudden breakdown during continuous exposure to light. This is demonstrated in Fig. 7.1, which shows how the impact strength of an ABS polymer drops drastically after a certain exposure time [45].

The schematic representation in Fig. 7.2 shows how at first the oxygen uptake increases exponentially with increasing irradiation time, i.e. the reaction rate is accelerated. After prolonged irradiation, the autoacceleration is followed by an autoretardation stage due to a depletion in the O_2 concentration in the interior of the sample or to reaction products interfering with the propagation process.

The behavior depicted in Fig. 7.2 is observed with many polymers upon exposure to sunlight, including with commercial polyalkenes such as polyethylene and polypropylene. In the latter cases, impurity chromophores act as initiators of the autoxidation process (see Scheme 7.4 in Section 7.1.3). Important elementary reactions determining the autoxidation process are described in the following. Free radicals $R_X^\bullet$ formed during the initiation phase abstract hydrogen atoms from macromolecules PH, thus forming macroradicals $P^\bullet$ [Eq. (7-11)].

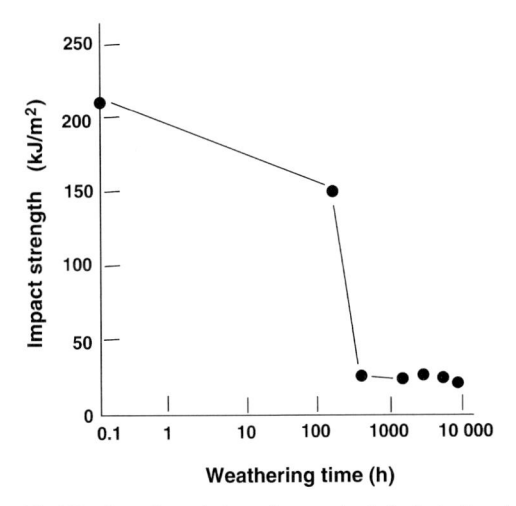

Fig. 7.1 Photodegradation of an acrylonitrile/butadiene/styrene (ABS) copolymer at 30 °C. Plot of the impact strength vs. the simulated natural exposure time (xenon-arc radiation, 0.55 W m^{-2} at 340 nm). Adapted from Davis et al. [45] with permission from Elsevier.

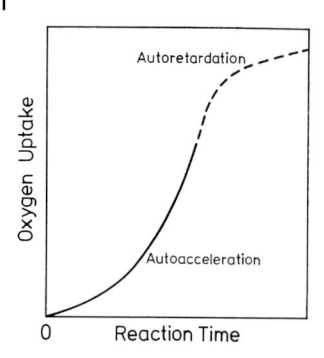

Fig. 7.2 Autoxidation of polymers. Schematic representation of the oxygen uptake as a function of time. Adapted from Schnabel [24] with permission from Carl Hanser.

$$R_X^\bullet + PH \rightarrow R_XH + P^\bullet \tag{7-11}$$

The ensuing chain reaction, which is propagated by the macroradicals, produces hydroperoxide groups (see Scheme 7.18).

$$P^\bullet + O_2 \rightarrow POO^\bullet$$
$$POO^\bullet + PH \rightarrow POOH + P^\bullet$$

Scheme 7.18 Propagation of the chain reaction in the autoxidation process.

Hydroperoxide groups can be photolytically cleaved, provided that the wavelength of the incident light is lower than about 300 nm [Eq. (7-12)].

$$POOH \xrightarrow{h\nu} PO^\bullet + {}^\bullet OH \tag{7-12}$$

The radicals generated in this way can initiate additional chain reactions (chain branching) by abstracting hydrogens from neighboring macromolecules, for instance by reaction according to Eq. (7-13):

$${}^\bullet OH + PH \rightarrow H_2O + P^\bullet \tag{7-13}$$

The kinetic chains are terminated by radical coupling reactions (see Scheme 7.19).

$$POO^\bullet + POO^\bullet \xrightarrow{(a)} Products$$
$$POO^\bullet + P^\bullet \xrightarrow{(b)} POOP$$
$$P^\bullet + P^\bullet \xrightarrow{(c)} P-P$$

Scheme 7.19 Termination reactions in the autoxidation process.

The combination of peroxyl radicals (reaction (a) in Scheme 7.19) is assumed to proceed via a tetroxide, P-O$_4$-P, a short-lived intermediate. Various reaction

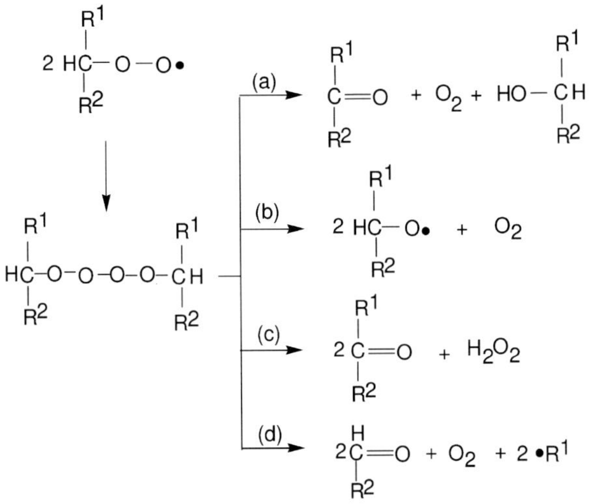

Scheme 7.20 Decay processes of secondary peroxyl radicals [46].

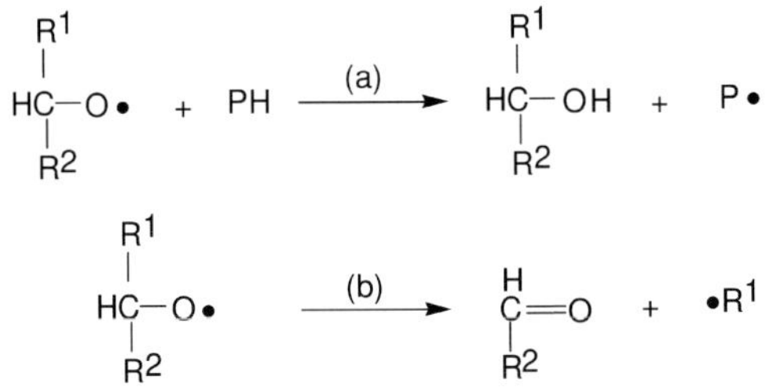

Scheme 7.21 Reactions of oxyl radicals.

paths that may be envisaged in the case of secondary peroxyl radicals are shown in Scheme 7.20 [46]. Reaction (a) in Scheme 7.20 refers to the so-called Russel mechanism. The extent to which each individual reaction occurs depends on the chemical nature of the polymer as well as on other parameters, particularly the temperature. The oxyl radicals formed by reaction (b) can abstract hydrogen in inter- and/or intramolecular reactions. Alternatively, they can decompose with the formation of carbonyl groups (see Scheme 7.21).

In conclusion, the salient features of the light-induced oxidation of polymers are the formation of hydroperoxide, peroxide, and carbonyl groups, the latter in the form of both aldehyde and keto groups. Moreover, certain reactions, such as reaction (d) in Scheme 7.20 and reaction (b) in Scheme 7.21, result in main-chain cleavage as far as the oxidation of linear macromolecules is concerned. Main-chain cleavage leads to a deterioration in certain important mechanical properties. Therefore, the photo-oxidation of polymers is deleterious and should be avoided in commercial polymers. Appropriate stabilization measures are discussed in Section 9.3.

7.6
Singlet oxygen reactions

The ground state of molecular oxygen ($^3O_2$) is a triplet state with two unpaired electrons. In addition to the reactions outlined in Section 7.5, $^3O_2$ can undergo energy-transfer reactions with many compounds, such as dyes and polynuclear aromatics, provided that the difference in the energy levels exceeds 94 kJ mol^{-1}. In these reactions, the first excited state of molecular oxygen, i.e. singlet oxygen ($^1O_2^*$), is formed, as is illustrated by the reaction of triplet excited carbonyl groups, present in a polymer, with $^3O_2$ according to Eq. (7-14).

$$^3(\text{C=O})^* + {}^3O_2 \longrightarrow {}^1(\text{C=O}) + {}^1O_2^* \qquad (7\text{-}14)$$

$^1O_2^*$ is unreactive towards saturated hydrocarbons, but reacts with unsaturated substances with a rate constant of 10^3 to 10^4 L mol^{-1} s^{-1} [47]. This reaction results in the insertion of hydroperoxide groups [Eq. (7-15)].

$$\text{alkene} \xrightarrow{{}^1O_2^*} \text{allyl hydroperoxides (OOH)} \qquad (7\text{-}15)$$

In conclusion, singlet oxygen plays a role in the photo-oxidative degradation of polymers containing olefinic unsaturations. Polymers that do not contain these groups, e.g. poly(vinyl chloride), poly(methyl methacrylate), polystyrene, etc., are unreactive [24].

7.7
Rearrangements

Certain organic molecules are modified by a rearrangement of some of their constituent groups upon light absorption. Typical processes that have gained importance in the polymer field are the photo-Fries rearrangement of aromatic esters, amides, and urethanes (see Scheme 7.22) and the o-nitrobenzyl ester rearrangement (see Scheme 7.23). In the latter case, nitronic acid forms as a long-lived intermediate. Its decay in polymeric matrices is non-exponential (kinetic matrix effect) up to temperatures exceeding the glass transition temperature range [49].

7.7 Rearrangements

Regarding linear polymers, rearrangements can involve the main chain, as in the case of a polycarbonate (see Scheme 7.24), or pendant groups, as in the case of poly(4-acetoxy styrene), which is converted into poly(3-acetyl-4-hydroxy sty-

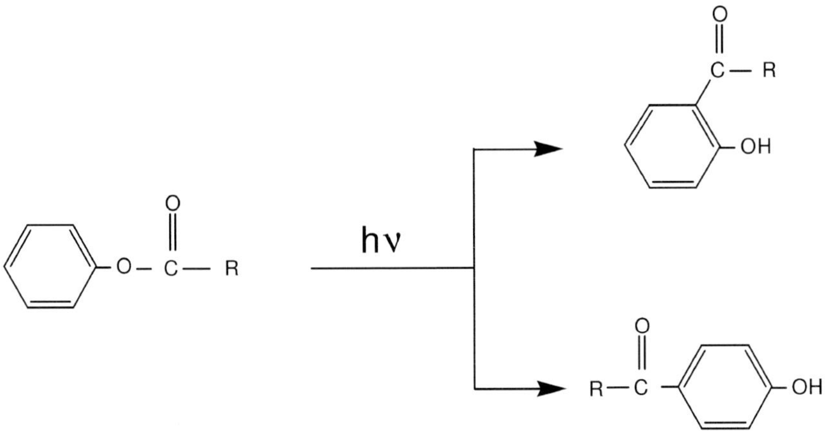

Scheme 7.22 Photo-Fries rearrangement of a carbonate.

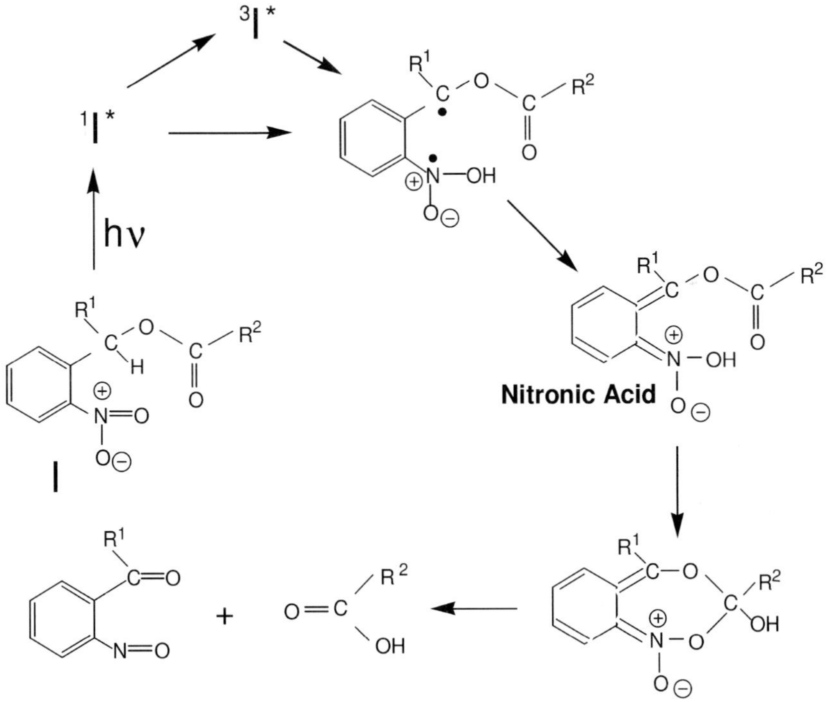

Scheme 7.23 Mechanism of the o-nitrobenzyl ester photo-rearrangement [48, 49].

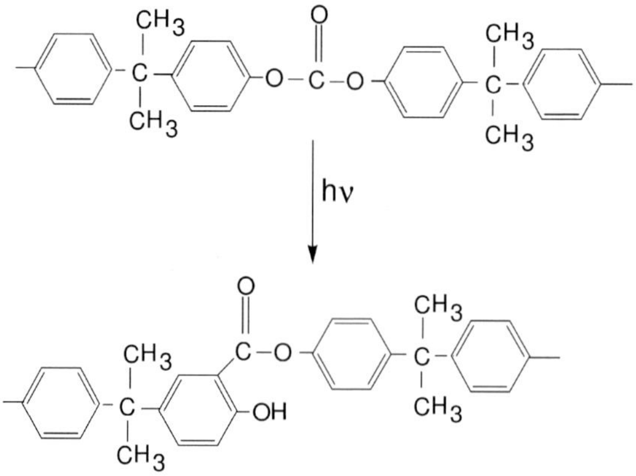

Scheme 7.24 Photo-rearrangement of a polycarbonate.

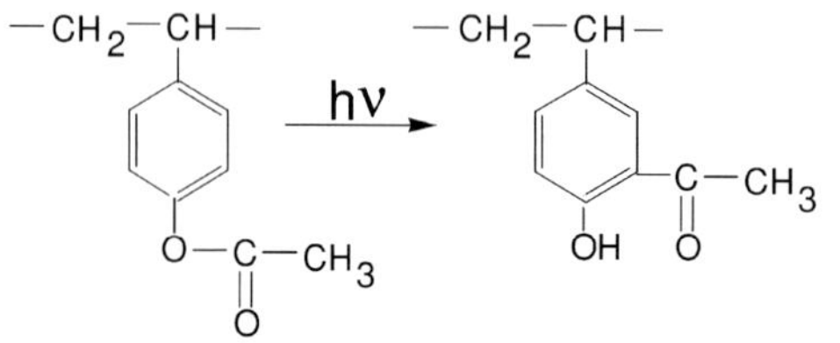

Scheme 7.25 Photo-rearrangement of poly(4-acetoxy styrene).

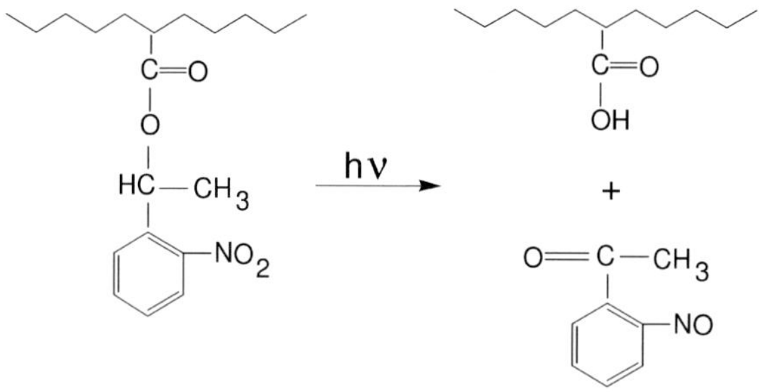

Scheme 7.26 Photo-rearrangement of polymers bearing o-nitrobenzyl pendant groups.

rene) (see Scheme 7.25), or with polymers bearing o-nitrobenzyl ester pendant groups (see Scheme 7.26).

Photo-rearrangements in polymers are important because they can lead to pronounced property changes. For example, polymers containing o-nitrobenzyl pendant groups become soluble in aqueous solution since benzyl ester groups are converted into carboxyl groups. Therefore, such polymers are applicable as positive-tone photoresists in lithographic processes [50, 51] (see Section 9.1).

References

1. (a) J.C. Salamone (ed.), *Polymeric Materials Encyclopedia*, CRC Press, Boca Raton, FL, USA (1996), (b) Abridgement of (a): J.C. Salamone (ed.), *Concise Polymeric Materials Encyclopedia*, CRC Press, Boca Raton, FL, USA (1999).
2. G. Scott, *Polymers and the Environment*, Royal Society of Chemistry, Cambridge (1999).
3. H.-J. Timpe, *Polymer Photochemistry and Photo-Crosslinking*, in R. Arshady (ed.), *Desk Reference of Functional Polymers: Synthesis and Applications*, American Chemical Society, Washington, D.C. (1997), p. 273.
4. S.I. Hong, S.Y. Joo, D.W. Kang, *Photosensitive Polymers*, in R. Arshady (ed.), *Desk Reference of Functional Polymers: Synthesis and Applications*, American Chemical Society, Washington, D.C. (1997), p. 293.
5. B. Rånby, B. Qu, W. Shi, *Photocrosslinking (Overview)*, in [1(a)], Vol. 7, p. 5155.
6. J. Paczkowski, *Photocrosslinkable Photopolymers (Effect of Cinnamate Group Structure)*, in [1(a)], Vol. 7, p. 5142.
7. J.F. Rabek, *Photodegradation of Polymers, Physical Characteristics and Applications*, Springer, Berlin (1996).
8. R.L. Clough, N.C. Billingham, K.T. Gillen (eds.), *Polymer Durability Stabilization, and Lifetime Prediction*, American Chemical Society, Washington, D.C., Advances in Chemistry Series 249 (1996).
9. W. Schnabel, I. Reetz, *Polystyrene and Derivatives, Photolysis*, in [1(a)], Vol. 9, p. 6786.
10. V.V. Krongauz, A.D. Trifunac, *Processes in Photoreactive Polymers*, Chapman & Hall, New York (1995).
11. A.L. Andrady, *Ultraviolet Radiation and Polymers*, in J.E. Mark, *Physical Properties of Polymers Handbook*, AIP Press, Woodbury, N.Y. (1995), Chapter 40.
12. G. Scott (ed.), *Atmospheric Oxidation and Antioxidants*, Elsevier, Amsterdam (1993).
13. N.S. Allen, M. Edge, *Fundamentals of Polymer Degradation and Stabilisation*, Elsevier Applied Science, London (1992).
14. Z. Osawa, *Photoinduced Degradation of Polymers*, in S.H. Hamid, M.B. Amin, A.G. Maadhah (eds.), *Handbook of Polymer Degradation*, Dekker, New York (1992).
15. H. Böttcher, J. Bendig, M.A. Fox, G. Hopf, H.-J. Timpe, *Technical Applications of Photochemistry*, Deutscher Verlag für Grundstoffindustrie, Leipzig (1991).
16. V. Strehmel, *Epoxies, Structures, Photoinduced Cross-Linking, Network Properties, and Applications*, in H.S. Nalwa (ed.), *Handbook of Photochemistry and Photobiology*, American Scientific Publishers, Stevenson Ranch, CA, USA (2003), Vol. 2, p. 2.
17. A. Reiser, *Photoreactive Polymers. The Science and Technology of Resists*, Wiley, New York (1989).
18. J. Guillet, *Polymer Photophysics and Photochemistry*, Cambridge University Press, Cambridge (1985).
19. C. Decker, *Photodegradation of PVC*, in E.D. Owen (ed.), *Degradation and Stabilization of PVC*, Elsevier Applied Science, London (1984), p. 81.
20. S. Tazuke, *Photocrosslinking of Polymers*, in N.S. Allen (ed.), *Developments in Polymer Photochemistry – 3*, Applied Science, London (1982), Chapter 2, p. 53.
21. E.D. Owen, *Photodegradation and Stabilization of PVC*, in N.S. Allen (ed.), *Developments in Polymer Photochemistry – 3*, Applied Science, London (1982), Chapter 5, p. 165.
22. Z. Ozawa, *Photodegradation and Stabilization of Polyurethanes*, in N.S. Allen (ed.), *Developments in Polymer Photochemistry – 3*, Applied Science, London (1982), Chapter 6, p. 209.
23. W. Schnabel, *Laser Flash Photolysis of Polymers*, in N.S. Allen (ed.), *Developments in Polymer Photochemistry – 3*, Applied Science, London (1982), Chapter 7, p. 237.
24. W. Schnabel, *Polymer Degradation, Principles and Practical Applications*, Hanser, München (1981), Chapter 4.
25. R. Arnaud, J. Lemaire, *Photocatalytic Oxidation of Polypropylenes and Polyundecanoamides*, in N.S. Allen (ed.), *Develop-*

ments in Polymer Photochemistry – 2, Applied Science, London (1981), Chapter 4, p. 135.
26 A. Garton, D. J. Carlsson, D. M. Wiles, *Photo-oxidation Mechanisms in Commercial Polyolefins*, in N. S. Allen (ed.), *Developments in Polymer Photochemistry – 1*, Applied Science, London (1980), Chapter 4, p. 93.
27 W. Schnabel, J. Kiwi, *Photodegradation*, in H. H. G. Jellinek (ed.), *Aspects of Degradation and Stabilization of Polymers*, Elsevier, Amsterdam (1979).
28 J. F. McKellar, N. S. Allen, *Photochemistry of Man-Made Polymers*, Applied Science, London (1979).
29 L. M. Minsk, J. G. Smith, W. P. Van Deusen, J. W. Wright, J. Appl. Polym. Sci. 11 (1959) 302.
30 K. Ohkawa, K. Shoumura, M. Yamada, A. Nishida, H. Shirai, H. Yamamoto, Macromol. Biosci. 1 (2001) 149.
31 H. Takase, A. Natansohn, P. Rochon, Polymer 44 (2003) 7345.
32 C. Badaru, Z. Y. Wang, Macromolecules 36 (2000) 6959.
33 B. Rånby, *Photoinitiated Modifications of Synthetic Polymers: Photocrosslinking and Surface Photografting*, in N. S. Allen, M. Edge, I. R. Bellobono, E. Selli (eds.), *Current Trends in Polymer Photochemistry*, Horwood, New York (1995), Chapter 2, p. 23.
34 K. Nakabayashi, R. Schwalm, W. Schnabel, Angew. Makromol. Chem. 195 (1992) 191.
35 A. Charlesby, *Atomic Radiation and Polymers*, Pergamon Press, Oxford (1960), Chapter 10.
36 O. Saito, *Statistical Theory of Crosslinking*, in M. Dole (ed.), *The Radiation Chemistry of Macromolecules*, Academic Press, New York (1972), Chapter 11.

37 C. L. Moad, D. J. Windzor, Prog. Polym. Sci. 23 (1998) 759.
38 I. S. Biggin, D. L. Gerrard, G. E. Williams, J. Vinyl Technol. 4 (1982) 150.
39 D. L. Gerrard, H. J. Bowley, K. P. J. Williams, I. S. Biggin, J. Vinyl Technol. 8 (1986) 43.
40 A. I. Balabanovich, S. Denizligil, W. Schnabel, J. Vinyl Add. Technol. 3 (1997) 42.
41 R. G. Kepler, J. M. Zeigler, L. A. Harrah, S. R. Kurtz, Phys. Rev. B 35 (1987) 2818.
42 R. D. Miller, J. Michl, Chem. Rev. 89 (1989) 1359.
43 R. D. Miller, *Radiation Sensitivity of Soluble Polysilane Derivatives*, in J. M. Zeigler, F. W. G. Fearon (eds.), *Silicon-Based Polymer Science. A Comprehensive Resource*, American Chemical Society, Washington, D.C. (1990), Advances in Chemistry Series 224, Chapter 24.
44 Y. Ohsako, C. M. Phillips, J. M. Zeigler, R. M. Hochstrasser, J. Phys. Chem. 93 (1989) 4408.
45 P. Davis, B. E. Tiganis, L. S. Burn, Polym. Degrad. Stab. 84 (2004) 233.
46 C. von Sonntag, *The Chemical Basis of Radiation Biology*, Taylor & Francis, London (1987), Chapter 4.
47 H. Bortolus, S. Dellonte, G. Beggiato, W. Corio, Eur. Polym. J. 13 (1977) 185.
48 K. H. Wong, H. Schupp, W. Schnabel, Macromolecules 22 (1989) 2176.
49 G. Feldmann, A. Winsauer, J. Pfleger, W. Schnabel, Macromolecules 27 (1994) 4393.
50 H. Barzynski, D. Sänger, Makromol. Chem. 93 (1981) 131.
51 E. Reichmanis, R. Gooden, C. W. Wilkins, H. Schonehorn, J. Polym. Sci. Polym. Chem. Ed. 21 (1983) 1075.

8
Photoreactions in biopolymers

8.1
Introductory remarks

Biopolymers play a key role in many light-triggered biological processes, such as in photomorphological processes in plants and in the photomovements of bacteria. Moreover, biopolymers participate in energy transduction processes related to the conversion of solar energy into chemical energy (photosynthesis) and to the conversion of chemical energy into light (bioluminescence). Apart from these beneficial effects, light can also have a harmful effect on polymers and cause chemical damage resulting in a deactivation of their biological activity. While the deleterious action is commonly restricted to UVB and UVC light (λ: 200–320 nm), i.e. to photons having energies high enough to cleave chemical bonds, the regulatory action relates to light of longer wavelengths, i.e. UVA (λ: 320–400 nm) and visible light. In the latter case, effective biopolymers contain chromophoric groups capable of absorbing light in the 400–800 nm wavelength region. This chapter, which deals with both modes of action of light, is organized according to the important biopolymer families of nucleic acids, proteins, lignins, and polysaccharides (see Chart 8.1). However, it should be kept in mind that, very often, members of these families exist in close proximity in biological objects, and are sometimes even linked by chemical bonds.

For relevant literature concerning the broad field of light-induced effects in biopolymers and biological objects, the reader is directed to several reviews and books [1–17].

The polymers presented in Chart 8.1 absorb UV light to quite different extents. Nucleic acids absorb more strongly than proteins. This can be seen in Fig. 8.1, which shows absorption spectra of aqueous solutions of DNA and bovine serum albumin, recorded at equal concentrations. In contrast to the rather strongly absorbing nucleotide residues in DNA, only a few of the amino acid residues in proteins absorb light measurably in the UV region. This pertains mainly to the *aromatic* amino acids phenylalanine, tyrosine, and tryptophan (see Chart 8.2).

Lignins, a major component of wood (15–30 wt%), are phenolic polymers based on three structural units, the content of which depends on the type of wood: *trans*-p-coumaryl alcohol (I), *trans*-coniferyl alcohol (II), and *trans*-sinapyl alcohol (III) (see Chart 8.3).

Polymers and Light. Fundamentals and Technical Applications. W. Schnabel
Copyright © 2007 WILEY-VCH Verlag GmbH & Co. KGaA, Weinheim
ISBN: 978-3-527-31866-7

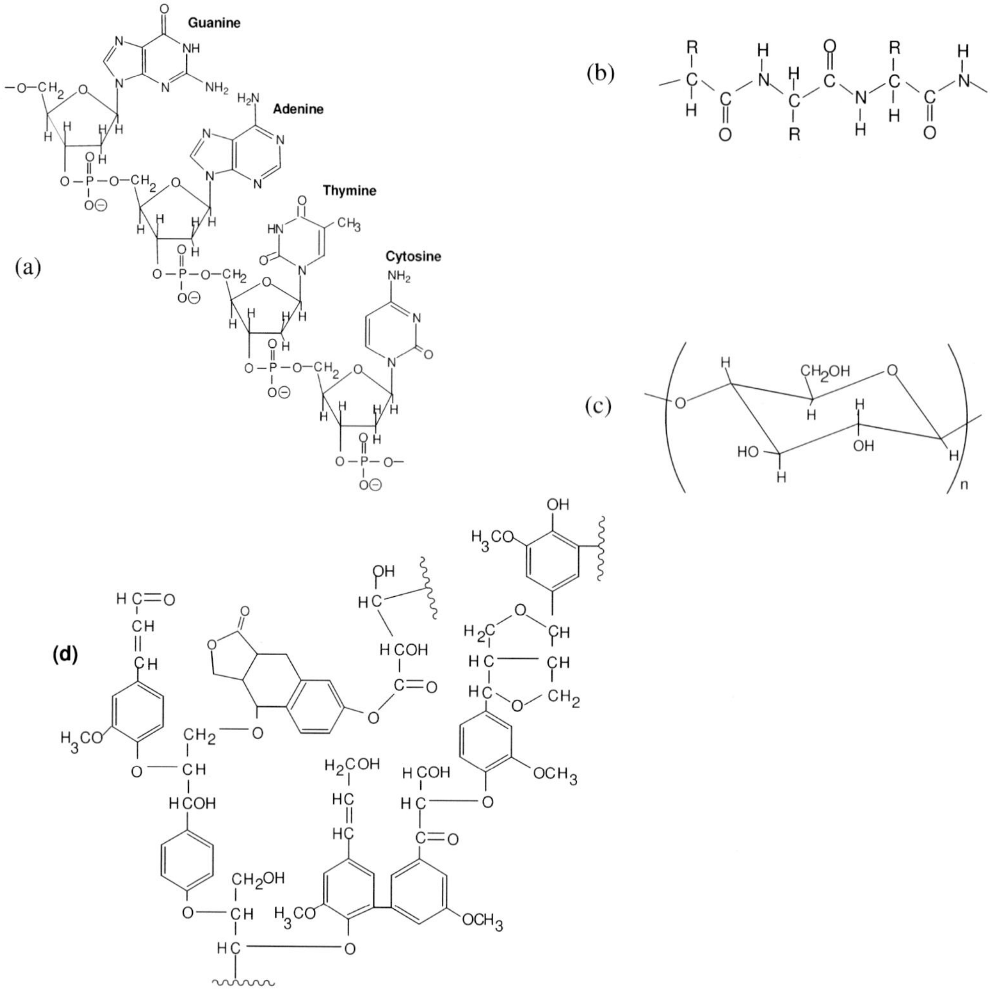

Chart 8.1 Biopolymer structures depicting (a) different nucleotides contained in human deoxyribonucleic acid, DNA, (b) part of a protein chain consisting of various amino acid residues with R being H (glycine), CH$_3$ (alanine), (CH$_2$)$_4$NH$_2$ (lysine), CH$_2$SH (cysteine), etc., (c) the base unit of the cellulose chain, representing the class of polysaccharides, and (d) part of a lignin with typical structural elements.

The optical absorption spectra of lignins extend into the visible wavelength region and exhibit peaks at about 205 and 280 nm and shoulders at 230 and 340 nm [18]. Polysaccharides such as cellulose and amylose essentially do not absorb light at $\lambda > 200$ nm. Very weak absorption bands observable in some cases in the region between 250 and 300 nm are due to intrinsic impurities such as acetal groups or carboxyl groups replacing hydroxyl groups [17, 19].

Special biopolymers containing covalently bound chromophoric groups absorb visible light and act as photoreceptors. They play a regulatory role in important

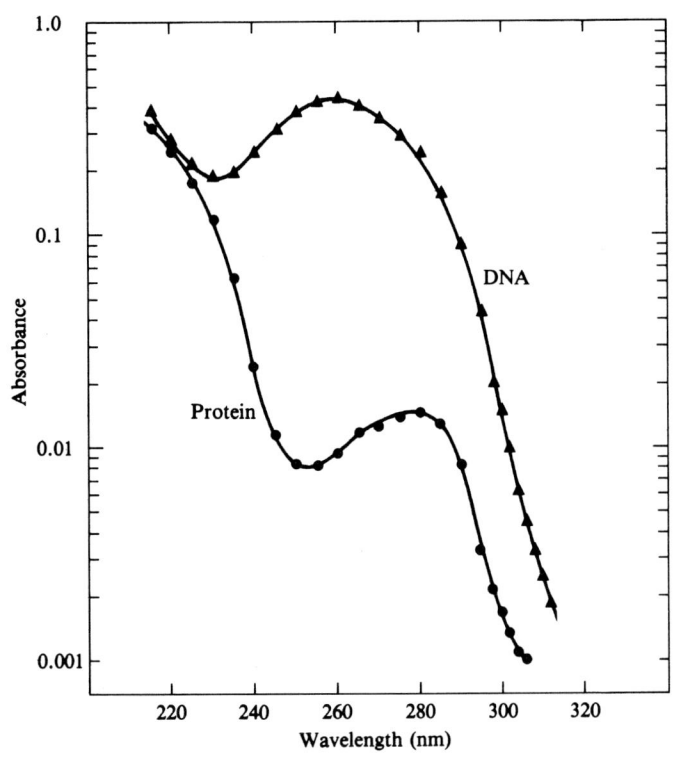

Fig. 8.1 Optical absorption spectra of aqueous solutions of a nucleic acid (calf thymus DNA) and a protein (bovine serum albumin), both recorded at a concentration of 1.97×10^{-2} g L^{-1}. Adapted from Harm [12] with permission from Cambridge University Press.

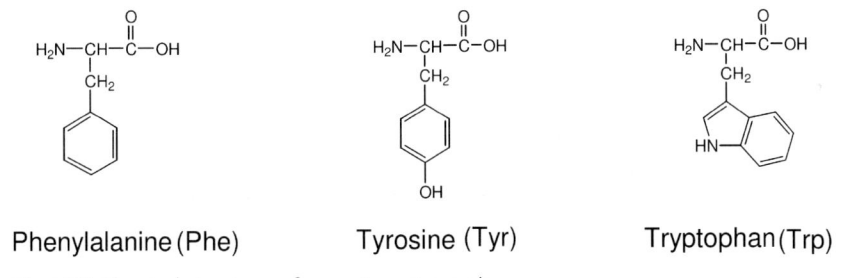

Phenylalanine (Phe)   Tyrosine (Tyr)   Tryptophan (Trp)

Chart 8.2 Chemical structures of aromatic amino acids.

biological processes. Typical photoreceptors are proteins belonging to the carotenoid (rhodopsin), phytochrome, and cryptochrome families. In this context, the chlorophyllic protein complexes are also of note. They function as light-harvesting antenna pigments and auxiliary cofactors in the photosynthetic process, and are,

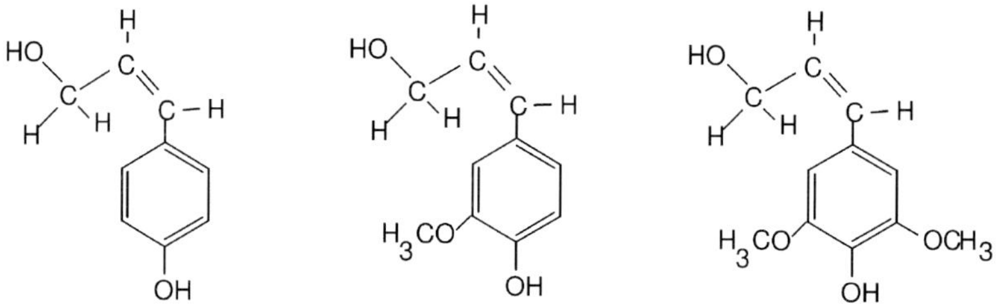

Chart 8.3 Substituted phenyl propanols that constitute the structural units of lignins.

Table 8.1 Photoactive chromophores (pigments) of photoreceptor proteins [9, 20–25].

Typical chromophore	Photoreceptor class	Typical functions
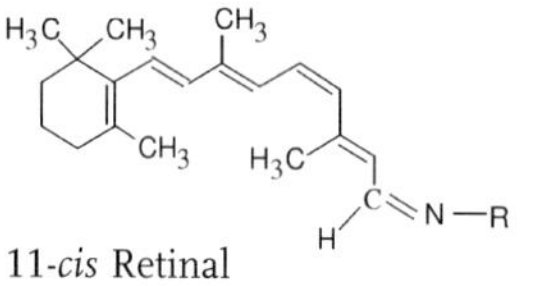 11-*cis* Retinal	Carotenoids	(a) Photoantennas in the photosynthetic system of plants. (b) Catalytic pigments in animal and bacterial rhodopsins.
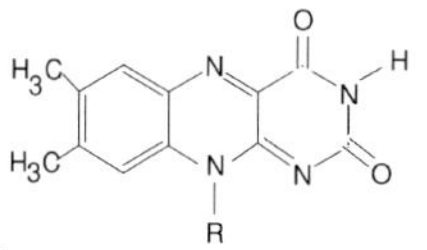 Flavin	Flavins	(a) Photoantennas in enzymes. (b) Cofactors for photolyase/blue-light photoreceptors.
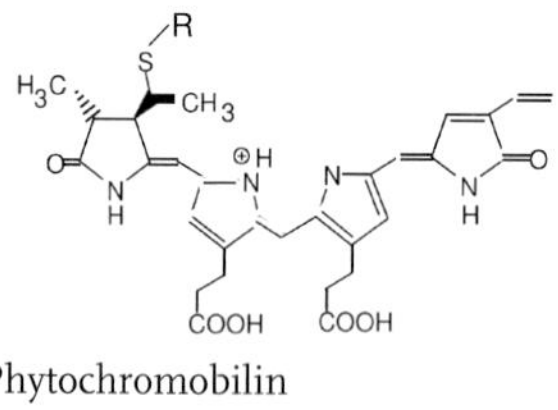 Phytochromobilin	Phytochromes	(a) Photoreceptors exerting morphogenic control in plants. (b) Accessory antennas in the light-harvesting complexes of photosynthetic systems.
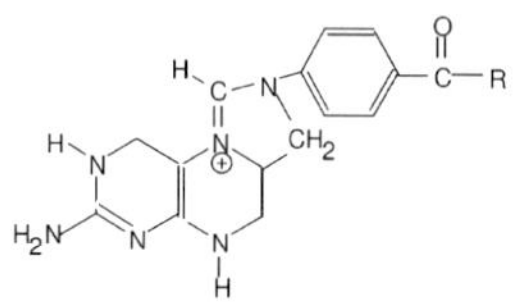 5,10-Methenyltetrahydrofolate (MTHF)	Pterins	Photoantennas in the majority of photolyase/cryptochrome blue-light photoreceptors.
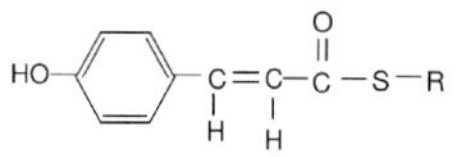 4-Hydroxycinnamate	Xanthopsins, Yellow Proteins	Sensory blue light receptors, water-soluble, controlling the life of bacteria in saline lakes.

Table 8.1 (continued)

Typical chromophore	Photoreceptor class	Typical functions
Stentorin	Naphthodianthrones, Blepharismins	Photosensors in ciliated protozoans exhibiting step-up photophobic and negative phototactic responses.
Chlorophyll a	Chlorophylls	Photoantennas in the light-harvesting complexes and electron donors in the reaction center of the photosynthetic system.

therefore, of profound biological importance. The chemical structures of typical chromophoric groups contained in these proteins are presented in Table 8.1.

In conclusion, proteins play a range of roles in relation to the exposure of biological objects to light of different wavelengths. UV light acts harmfully, since it causes chemical changes leading to the deactivation of specifically acting proteins such as enzymes. However, light-induced chemical changes might also trigger the synthesis of special proteins. As regards irradiation with visible light, it is most important that certain proteins serve as light-harvesting agents in photosynthesis and as photoreceptors and photosensors in photomorphogenic processes in plants. The various aspects are referred to briefly in the following sections.

8.2
Direct light effects

8.2.1
Photoreactions in deoxyribonucleic acids (DNA)

The energy-rich UV light portion of the terrestrial solar spectrum (λ: 280–400 nm) is harmful to most organisms and can even cause skin cancer in humans (basal and squamous cell carcinoma, melanoma). This is mainly due to light-induced chemical modifications in DNA bases, commonly termed UV-induced DNA lesions. The absorption of light converts the bases into their excited singlet or triplet states, from which chemical reactions can ensue. The resulting base modifications are accompanied by a change in the base-pairing properties, which, in turn, causes mutations [26–29]. There are a number of feasible photolesions based on the cleavage of chemical bonds with the concurrent generation of free radicals. Besides these, dimeric photoproducts may be formed in great abundance through a molecular rather than a free radical mechanism. Notably, pyrimidine bases are essentially involved in the generation of lesions of biological importance, although both purine and pyrimidine residues are rather strong absorbers in the far-UV region. Actually, the quantum yield of photodecomposition differs significantly. It amounts to about 10^{-4} for purines, i.e. one or two orders of magnitude lower than that for pyrimidines [12].

8.2.1.1 Dimeric photoproducts

The pyrimidine bases thymine (T) and cytosine (C) form dimers at sites with adjacent pyrimidine moieties, so-called dipyrimidine sites, in the DNA chain, which have been well characterized with respect to chemical structure and mutagenic potential. The dimerization presented in Scheme 8.1 is a $[2\pi+2\pi]$ cycloaddition (see Section 7.3) involving the two C(5)=C(6) double bonds, leading to cyclobutane structures denoted by the symbol T< >T, or generally Pyr< >Pyr.

The dimerization can, in principle, lead to three isomers: *cis-syn*, *trans-syn* I, and *trans-syn* II, but due to the constraints imposed by the DNA double strand, the *cis-syn* dimer shown in Scheme 8.1 is the major photoproduct [27].

Another type of dimeric lesions are pyrimidine–pyrimidone (Pyr[6-4]Pyr) dimers formed by a Paterno-Büchi-type reaction at dipyrimidine sites between the C(5)=C(6) double bond of the first pyrimidine and the C(4)=O carbonyl group of the second base. This kind of dimerization is demonstrated in Scheme 8.2 for the case of adjacent thymine moieties.

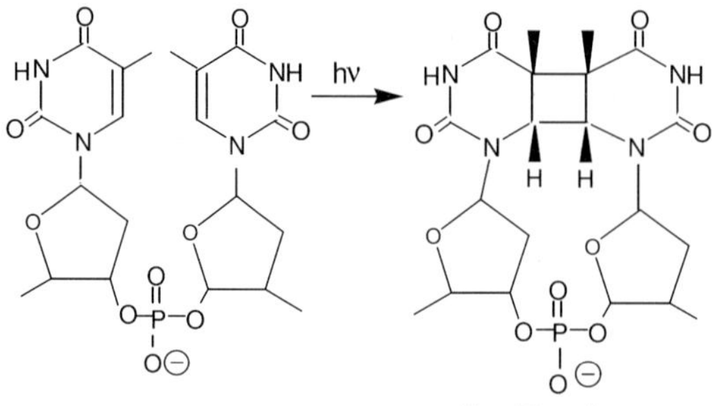

Scheme 8.1 Dimerization of adjacent thymine moieties in DNA by $[2\pi+2\pi]$ cycloaddition.

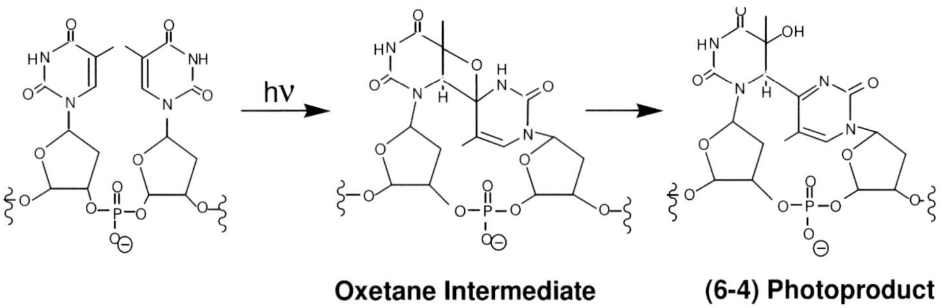

Scheme 8.2 Dimerization of adjacent thymine moieties in DNA by a Paterno-Büchi-type reaction.

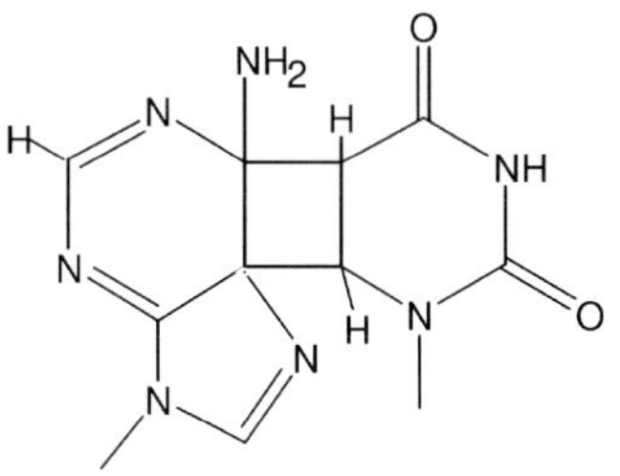

Chart 8.4 Structure of an adenine–thymine photodimer [29].

Analogous photoproducts may form between any types of adjacent pyrimidines, T-T, T-C, C-T, and C-C, except that the (6-4) photoproduct does not form at C-T sites. Adenine–thymine heterodimers (see Chart 8.4) have also been detected [29, 30].

The UV-induced generation of cyclobutane dimers is greatly dependent on double-helix conformational factors. In dormant spores of various bacillus species, for example, a group of small, acid-soluble proteins specifically bind to DNA, thereby enforcing a particular conformation that is unfavorable for the formation of harmful cyclobutane-type lesions. As a consequence, these dormant spores are much more resistant to UV radiation than the corresponding growing cells, in which DNA strands reassume conformations favorable for the formation of cyclobutane-type lesions [31].

Notably, photodimers of the cyclobutane type are cleaved by irradiation with far-UV light (240 nm) with a quantum yield of almost unity by way of the so-called [2+2] cycloreversion reaction. In living cells, dimer lesions can be repaired by the nucleotide excision repair pathway, which is based on the excision of a small piece of DNA around the lesion. Lesions not removed from the genome lead to cell death or mutagenesis.

8.2.1.2 Other DNA photoproducts

Additional photoproducts, commonly generated via free radical mechanisms, have been identified. These include single-strand breaks, cross-links between the strands of the same double helix and between different DNA strands and adjacent protein molecules, and the so-called photohydrates (see Chart 8.5).

8.2.2
Photoreactions in proteins

Gross changes in proteins due to UV irradiation include disturbance of the natural conformation, aggregation, and chain cleavage, all of which lead to denaturation. The structural proteins keratin (wool), collagen, elastin, and fibroin (silk) undergo losses in mechanical strength and elasticity (wool *tenders*) and sometimes color changes (*yellowing*). These changes are due to chemical alterations.

In order to assess possible photochemical events, one has to take into account that proteins are heterogeneously composed linear polymers (see Chart 8.1). The amino acid residues are connected by amide (peptide) bonds, $-CO-NH-$. Nature uses 20 amino acids to synthesize a great variety of proteins, which are characterized by amino acid sequence, size, and three-dimensional structure. Many proteins are intramolecularly cross-linked by disulfide links (R–S–S–R), i.e. they consist of several covalently connected chains. Alternatively, two or more protein chains can be linked by non-covalent forces. Proteins consisting of the 20 natural amino acids absorb light at $\lambda < 320$ nm. The low-wavelength portion of the terrestrial solar spectrum, extending to about 290 nm, is mainly absorbed by the aromatic amino acids (see Chart 8.2). Therefore, the sunlight-induced photochemistry of proteins essentially relates to these moieties. At $\lambda < 290$ nm, light is also absorbed by the other amino acid residues, which greatly increases the variety of possible bond ruptures. In view of these facts, it is clear that the photochemistry of proteins is extremely complex and, therefore, only certain aspects have been thoroughly investigated to date.

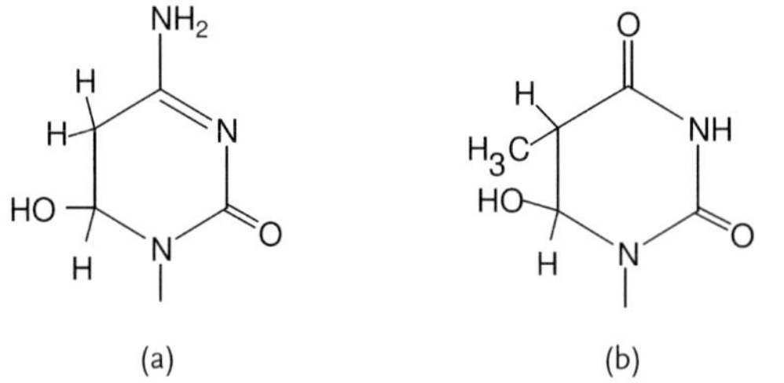

Chart 8.5 Photohydrates of cytosine (a) and of thymine (b) [30].

8.2.2.1 Chemical alterations by UV light

Tryptophan (Trp), tyrosine (Tyr), cystine (Cys), and phenylalanine (Phe) moieties play a determinant role regarding UV light-induced chemical alterations in many proteins. After the absorption of light by these moieties, in most cases mainly by Trp and Tyr, they undergo photoionization and participate in energy- and electron-transfer processes. This not only holds for structural proteins such as keratin and fibroin [11], but also for enzymes in aqueous media such as lysozyme, trypsin, papain, ribonuclease A, and insulin [7]. The photoionization of Trp and/or Tyr residues is the major initial photochemical event, which results in inactivation in the case of enzymes. A typical mechanism pertaining to Trp residues (see Scheme 8.3) commences with the absorption of a photon and the subsequent release of an electron. In aqueous media, the latter is rapidly solvated. By the release of a proton, the tryptophan cation radical $Trp^{•+}$ is converted to the tryptophan radical $Trp^{•}$.

In many proteins, such as a-lactalbumin, which consists of 123 amino acid moieties, the electron released from a Trp moiety is attached, by way of an intramolecular process, to a disulfide group of a cystine bridge in a position adjacent to the indole ring of the Trp moiety [32].

As shown in Scheme 8.4, the resulting disulfide anion radical dissociates into a thiolate ion R–S⁻ and a thiyl radical $R-S^{•}$. Proton transfer from the tryptophan cation radical to the thiolate ion leads to the tryptophan radical $Trp^{•}$ and the thiol RSH. The final stage of the process is governed by radical coupling, which may result in sulfenylation of the Trp moiety yielding Trp–S–R, or in intermolecular cross-linking, i.e. in the formation of enzyme dimers or trimers.

Disulfide bridges can also be ruptured by reaction with the triplet excited moieties $^{3}Trp^{*}$ or $^{3}Tyr^{*}$, the formation of which accompanies the electron release.

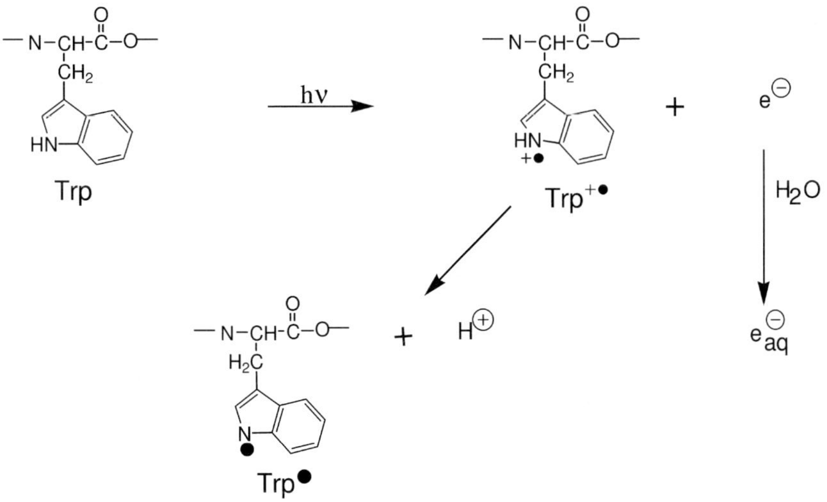

Scheme 8.3 Photolysis of proteins. Reactions involving tryptophan moieties [7].

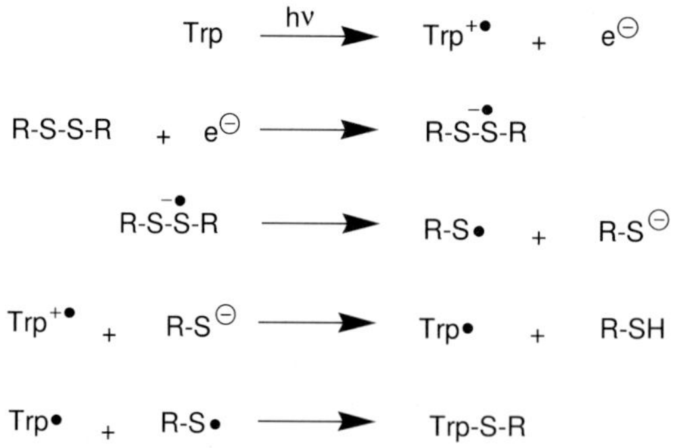

$$\text{Trp} \xrightarrow{h\nu} \text{Trp}^{+\bullet} + e^{\ominus}$$

$$\text{R-S-S-R} + e^{\ominus} \longrightarrow \text{R-S-S-R}^{-\bullet}$$

$$\text{R-S-S-R}^{-\bullet} \longrightarrow \text{R-S}^{\bullet} + \text{R-S}^{\ominus}$$

$$\text{Trp}^{+\bullet} + \text{R-S}^{\ominus} \longrightarrow \text{Trp}^{\bullet} + \text{R-SH}$$

$$\text{Trp}^{\bullet} + \text{R-S}^{\bullet} \longrightarrow \text{Trp-S-R}$$

Scheme 8.4 Rupture of cystine bridges by the attachment of electrons stemming from the photoionization of tryptophan [32, 33].

$$^3\text{Trp}^* + \text{R-S-S-R} \longrightarrow \text{Trp}^{+\bullet} + \text{R-S-S-R}^{-\bullet} \longrightarrow \text{R-S}^{\bullet} + {}^{\ominus}\text{S-R}$$

Scheme 8.5 Reaction of tryptophan triplets with cystine moieties.

In this process, the triplet species undergo an electron transfer with cystine moieties, thus forming the disulfide radical anion (see Scheme 8.5).

Intermediates occurring in these mechanisms have been identified by ESR measurements and by flash photolysis studies using optical absorption detection. For example, ESR measurements on wool keratins revealed the formation of sulfur-centered radicals of the structure $RCH_2S^{\bullet}$, which, in this case, are assumed to result from a reaction of electronically excited tyrosine moieties with cystine residues [11]. In many proteins, cross-links are formed. In the case of keratin and collagen, the cross-links are of the tryptophan-histidine and dityrosine types [11]. Cross-links formed by the combination of $R-S^{\bullet}$ or $R-S-S^{\bullet}$ radicals, both intermolecularly and intramolecularly, with incorrect sites are considered to be an important source of photoaggregation effects [8]. ESR measurements have also yielded evidence of C–H and C–N bond ruptures [8].

8.2.2.2 Formation of stress proteins

UV light induces the formation (expression) of so-called *stress proteins* in mammalian skin cells [34]. Stress proteins (shock proteins) are also generated by other stress factors, such as hyperthermia, and comprise a heterogeneous group of proteins with molar masses ranging from 10^4 to 1.1×10^5 g mol^{-1}. They function as molecular chaperones by transiently binding to unfolded proteins after synthesis as well as to denatured proteins in stressed cells, thus promoting their refolding and correct assembly. In this way, they protect proteins from misfolding and irreversible denaturation. The molecular mechanism of the formation of stress proteins has not yet been elucidated, although it is supposed that their formation is triggered by oxidative damage.

8.2.2.3 Effects of visible light – photoreceptor action

Photoreceptors, i.e. proteins containing chromophores absorbing visible light (see Table 8.1), play a key role in many light-triggered biological processes. For instance, in plants, they regulate and participate in energy transduction processes during the conversion of solar energy into chemical energy (photosynthesis), and trigger and support photomorphological processes. Moreover, photoreceptors are responsible for the photomovements of certain bacteria and regulate the circadian rhythm of higher animals. Circadian (circa = round about and dies = day) rhythms are oscillations in the biochemical, physiological, and behavioral functions of organisms with a periodicity of approximately 24 hours. Detailed information on this fascinating field is available from the cited literature [6, 9, 20, 22, 35–44]. Upon light absorption, the chromophores of photoreceptors undergo molecular transformations that result in the formation of signaling states in the protein. The regulatory action relates to UVA (λ: 320–400 nm) and visible light (λ: 400–800 nm). In most proteinaceous photoreceptor systems, such as cytochromes and phytochromes, the chromophores are covalently linked to the protein [35]. On the other hand, chlorophyll moieties are specifically associated with intrinsic proteins of the photosynthetic membrane, thus forming chlorophyll-protein (non-covalent) complexes.

Depending on their chemical nature, chromophores undergo different modes of light-induced molecular transformation. As can be seen in Table 8.2, the transformation modes include *trans-cis* isomerization, charge transfer, and energy transfer.

The chromophores act as photosensing-phototransducing devices, because they are not isolated but rather are embedded in and interacting with a molecular apoprotein framework. The latter senses the light-induced molecular modifications in the chromophores and, in turn, gives rise to the signaling state. The intimate interaction between chromophore and protein determines the physiological and spectroscopic properties of the photoreceptors. In recent years, photobiological research has been largely focused on photoreceptors and has revealed some very interesting results. This is illustrated here for the typical case of the family of phytochromes, which are present in plants and certain bacteria [20, 37–39]. Certain phytochromes exert morphogenic control functions in higher and lower plants, algae, and mosses, relating to, for example, blooming, the opening of hooks of shoots, or the germination of seeds. Other phytochromes function as accessory light-harvesting antennae in conjunction with the photosynthetic systems of certain algae. Plant phytochromes consist of polypeptide chains of about 1100 amino

Table 8.2 Transformation modes of chromophores in photoreceptors.

Transformation mode	Chromophores
trans-cis Isomerization	Retinals, 4-hydroxy-cinnamate, bilins
Charge transfer	Flavins, stentorins, blepharismins
Energy transfer	Pterins, flavins

Scheme 8.6 Mechanism of the $P_r \leftrightarrow P_{fr}$ photocycle for phytochromobilin. Adapted from [20].

acid moieties (molar mass $1.2–1.3\times10^5$ g mol^{-1}) and a single open-chain tetrapyrrole chromophore of the bilin family (see Table 8.1 and Scheme 8.6), which is covalently bound via an S-cysteine linkage to the apoprotein. The polypeptide chain is composed of two domains: the globular N (amino) terminal domain, bearing the chromophore, and the regulatory C (carboxyl) terminal domain [39]. The two domains are connected by a flexible protease-sensitive hinge region containing the Q (Quail) box. Active phytochrome entities are dimers, i.e. they consist of two polypeptide strands (see Fig. 8.2).

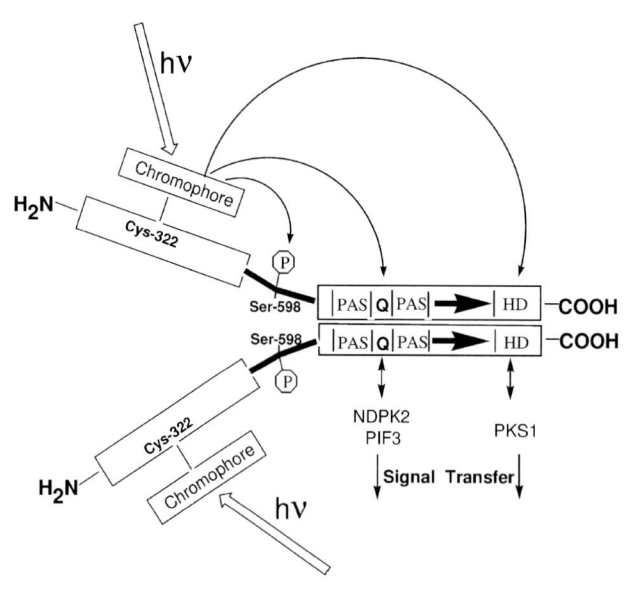

Fig. 8.2 Schematic illustration of the interdomain signal transmission in a dimeric oat phytochrome. Q: Quail box. PAS: Per-Arnt-Sim motif. Q and PAS constitute the regulatory core region. HD: Histidine kinase-related domain. PKS1: Phytochrome kinase substrate 1. NDPK2: Nucleoside diphosphate kinase 2. PIF3: Phytochrome interacting factor 3. Adapted from Bhoo et al. [39] with permission from Routledge/Taylor & Francis Group, LLC.

The photomorphogenic control functions are triggered by trans→cis and cis→trans double-bond isomerizations of the chromophore induced by red (r) and far-red (fr) light, respectively. The $P_r \rightarrow P_{fr}$ photocycle is illustrated in Scheme 8.6.

The P_r to P_{fr} isomerization induces a transformation from random to α-helical conformation in part of the N-terminal domain, and thus triggers a series of conformational changes in other structural peptide motifs, especially in the C-terminal domain (see Fig. 8.2). Here, certain regulatory sites become exposed and thus capable of interacting with signal transducer proteins such as PIF3 (phytochrome interacting factor 3), NDPK2 (nucleoside diphosphate kinase 2), etc. In this way, the enzymatic activity of these proteins is significantly increased. Moreover, the Q-box in the hinge region becomes uncovered, thus permitting the phosphorylation of the serine moiety in position 598 of the chain. The phosphorylation at Ser-598 exerts an accelerating effect on the association of PIF3 and NDPK2 and the phosphorylation of PKS1 (phytochrome kinase substrate 1). The latter is a protein that is complexed to the P_r state of the phytochrome and is released from the photoactivated P_{fr} state after phosphorylation to give downstream signals through a kinase cascade [39]. Recall that a kinase is an enzyme that catalyzes the phosphorylation of a substrate, here a protein. In conclusion, the light-induced isomerization of carbon-carbon double bonds in the chromophore causes a series of conformational changes within the two domains of the phytochrome. These changes trigger the association of signal transducer proteins with the phytochrome and allow phosphorylation and phosphate transfer at various sites. These are key steps initializing the downstream of processes that eventually result in transcriptional regulation.

8.2.2.4 Repair of lesions with the aid of DNA photolyases

The repair of dimer lesions, induced with the aid of light of relatively long wavelength that is not absorbed by the dimer sites (λ: 300–400 nm), is based on photoreceptor action, as dealt with in Section 8.2.2.3 above. It occurs if *DNA photolyases*, i.e. structure-specific (not sequence-specific) enzymes are present in the system during the irradiation [6]. Photolyases are proteins of 450-550 amino acids containing two non-covalently bound chromophore cofactors (see Chart 8.6).

One of the cofactors is always flavin adenine dinucleotide, FAD, and the second one is either methenyltetrahydrofolate, MTHF, or 8-hydroxy-7,8-didesmethyl-5-deazariboflavin, 8-HDF.

The repair of lesions by photolyases is the basis of the so-called *photoreactivation* of organisms. A striking example is the resurrection of UV-killed Escherichia coli by subsequent exposure to a millisecond light flash, which is demonstrated by the results shown in Fig. 8.3.

The reaction mechanism can be summarized as follows. In a dark reaction, the enzyme binds to DNA and flips out the pyrimidine dimer from the double helix into the active cavity. After the photochemical repair, the reaction partners are moved out of the cavity. As shown in Scheme 8.7, MTHF (or alternatively 8-HDF) is converted into an excited state, MTHF*, upon absorption of a photon.

Chart 8.6 Cofactors of photolyases.

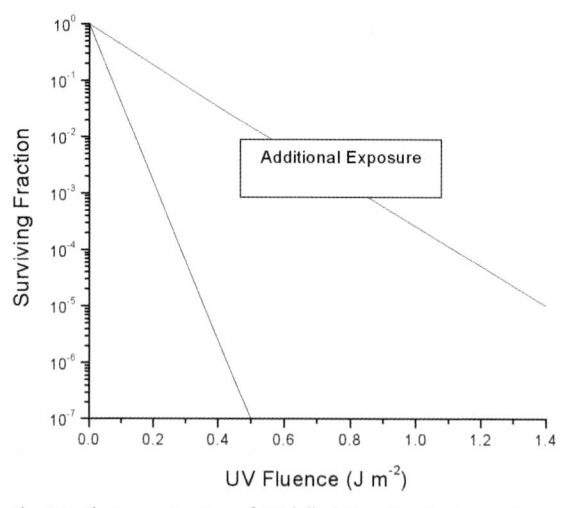

Fig. 8.3 Photoreactivation of UV-killed E. coli cells. Lower line: cells irradiated with UV light and plated on a growth medium. Upper line: UV-irradiated cells exposed to a 1 ms light flash before plating. Adapted from Sancar [6] with permission from the American Chemical Society.

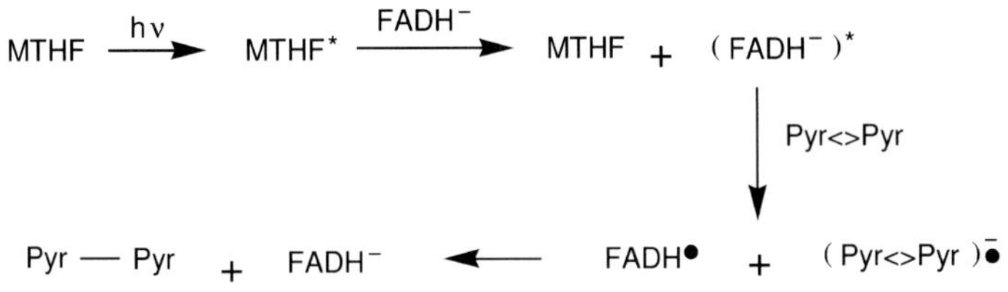

$$\text{MTHF} \xrightarrow{h\nu} \text{MTHF}^* \xrightarrow{\text{FADH}^-} \text{MTHF} + (\text{FADH}^-)^*$$

$$\downarrow \text{Pyr<>Pyr}$$

$$\text{Pyr}—\text{Pyr} + \text{FADH}^- \longleftarrow \text{FADH}^\bullet + (\text{Pyr<>Pyr})^{\bar{\bullet}}$$

Scheme 8.7 Reaction mechanism of the repair of pyrimidine dimer lesions in DNA with the aid of photolyases.

Excited reduced flavin, (FADH)*, formed by energy transfer from MTHF*, transfers an electron to Pyr < >.Pyr, the pyrimidine dimer. In a subsequent concerted reaction, the latter is split into two pyrimidines and an electron is transferred to the nascently formed FADH$^\bullet$.

8.2.3
Photoreactions in cellulose

It was pointed out in Section 8.1 that polysaccharides do not absorb light at $\lambda > 200$ nm. Therefore, photochemical alterations caused by light of longer wavelengths are due to the action of impurity chromophores. This also holds for cellulose, which is a major component of plants. Some plants, such as jute, flax, hemp, and cotton, contain up to 90% cellulose. *Neat* cellulose forms gaseous products (CO, CO_2, and H_2) upon exposure to UV light ($\lambda = 253.7$ nm). ESR studies have revealed the generation of H$^\bullet$ radicals and various carbon-centered free radicals. The degree of crystallinity of the cellulose fibrils is reduced [17]. If O_2 is present during the irradiation, carbonyl, carboxyl, and peroxide groups are formed, even at $\lambda > 340$ nm. Main-chain scission occurs and the brightness is reduced [45]. This is because irradiation at $\lambda < 360$ nm leads to homolysis of the previously formed hydroperoxide groups (see Scheme 8.8).

$$\text{RO} - \text{OH} \xrightarrow{h\nu} \text{RO}^\bullet + {}^\bullet\text{OH}$$

Scheme 8.8 Generation of hydroxyl radicals during the photolysis of hydroperoxide groups.

The OH radicals resulting from this process are very reactive, i.e. they abstract hydrogens from neighboring molecules and thus initiate further decomposition processes. For detailed information concerning the photochemistry of cellulose, the reader is referred to the relevant literature [17, 46].

8.2.4
Photoreactions in lignins and wood

Wood contains 15–30% lignin, an aromatic UV- and visible-light-absorbing polymer with a very complex structure (see Chart 8.1), and photochemical alterations of wood are essentially determined by reactions initiated by bond breakages in the

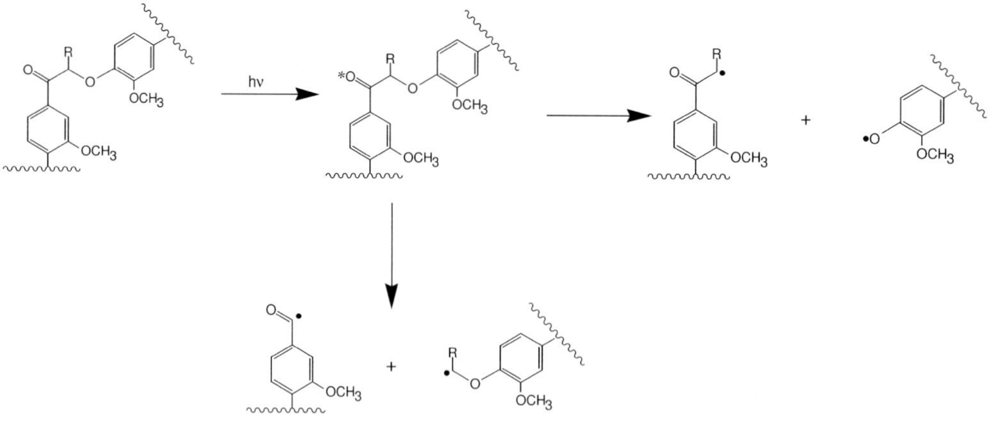

Scheme 8.9 Photoreactions of lignins.

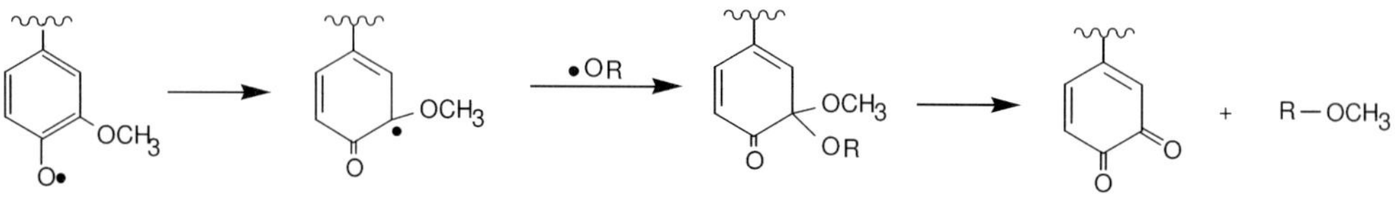

Scheme 8.10 Formation of quinoid structures in lignins.

lignin component. Due to a lack of systematic investigations, little is known about the complex mechanism of the photoreactions in lignins. Scheme 8.9 illustrates bond-breakage processes suggested in the literature [16, 47].

The formation of phenoxyl radicals has been revealed by ESR measurements. Phenoxyl radicals can be transformed into quinoid structures (see Scheme 8.10), which are thought to be responsible for the yellowing of the surfaces of wood products.

Because of the capability of lignins to absorb near-UV and visible light, even indoor yellowing and darkening of wood surfaces due to slow photooxidation processes is unavoidable. More detailed information concerning the photochemistry of lignins and wood is available in relevant review articles [16, 47].

8.3
Photosensitized reactions

Various applications are based on the indirect action of light on polymers contained in biological objects. Many biopolymers do not absorb visible light, and absorb UV light only to a limited extent. Therefore, sensitizers are used to accomplish light-induced chemical alterations. Sensitizers, which are in an electronically excited state after light absorption, either react directly with substrate polymers or decompose into fragments capable of reacting with the polymers.

Sensitizers can be employed for agricultural purposes as herbicides and insecticides, or for medical purposes as antibacterial and antiviral agents. Moreover, sensitizer-based methods serve as tools for the analysis of the interaction faces of polymer complexes and the sequence-selective photocleavage of double-stranded DNA. The ways in which photosensitized reactions are utilized are illustrated by the following typical examples. The first case relates to the photochemotherapy of cancer cells in superficial solid tumors [48]. The so-called *photodynamic therapy*, PDT, is based on the selective incorporation of a photosensitizer into tumor cells, followed by exposure to light (commonly at $\lambda = 600$ nm). Cytotoxic products, namely singlet oxygen, $^1O_2^*$, and superoxide radical anions, $^\bullet O_2^-$, are generated upon irradiation, and these are postulated to start a cascade of biochemical processes that inactivate neoplastic cells. The precise mechanism has not yet been elucidated [49]. However, it has been established that chemical alterations of the cytoskeleton trigger a sequence of reactions eventually causing cell apoptosis. The cytoskeleton consists of a complex array of highly dynamic protein structures that organize the cytoplasma of the cell. The basic proteinaceous constituents, having molar masses ranging from 4×10^4 to 7×10^4 g mol^{-1}, are microtubules and globular or linear microfilaments (actins and keratins, respectively). The cytoskeleton structure disorganizes and reorganizes continuously, depending on the shape and state of division of the cells, as well as on signals received from the environment. Assembly and disassembly of the cytoskeletal elements are severely disturbed or inhibited by light-induced damage. Chart 8.7 presents the chemical structures of several PDT sensitizers. Relevant research work has been reviewed [50].

The second example relates to photochemical cross-linking as a tool for studying metastable protein-nucleic acid and protein-protein assemblies [51–54]. Protein-protein and protein-nucleotide interactions are maintained by a multitude of weak, non-covalent interaction forces. From an analytical perspective, it is useful to stabilize such complexes by trapping the interaction partners by means of a cross-linking technique so as to generate covalent bonds between them. The process of protein assembly can be time-resolved in a snapshot manner, if the cross-linking period is significantly shorter than the lifetimes of intermediate stages reached during the complexing of two or more protein molecules, i.e. during dimerization or oligomerization, respectively. The method discussed here, denoted by the acronym PICUP (photo-induced cross-linking of unmodified proteins) in the case of the oligomerization of unmodified proteins, involves exposing the assemblies to a short, high-power laser pulse, thereby generating a number of cross-links that is sufficient to stabilize the interaction partners. The aim of the subsequent analysis is then to define binding sites by identifying the composition of the cross-linked domains of the partners. Mass spectrometry has been successfully applied for this purpose, and it appears that the desired information can be obtained more quickly and with greater sensitivity in this way than by NMR or X-ray crystallography [53]. The information obtained can be used as a basis for three-dimensional molecular modeling of protein oligonucleotide interfaces. Commonly, the cross-linking reaction is per-

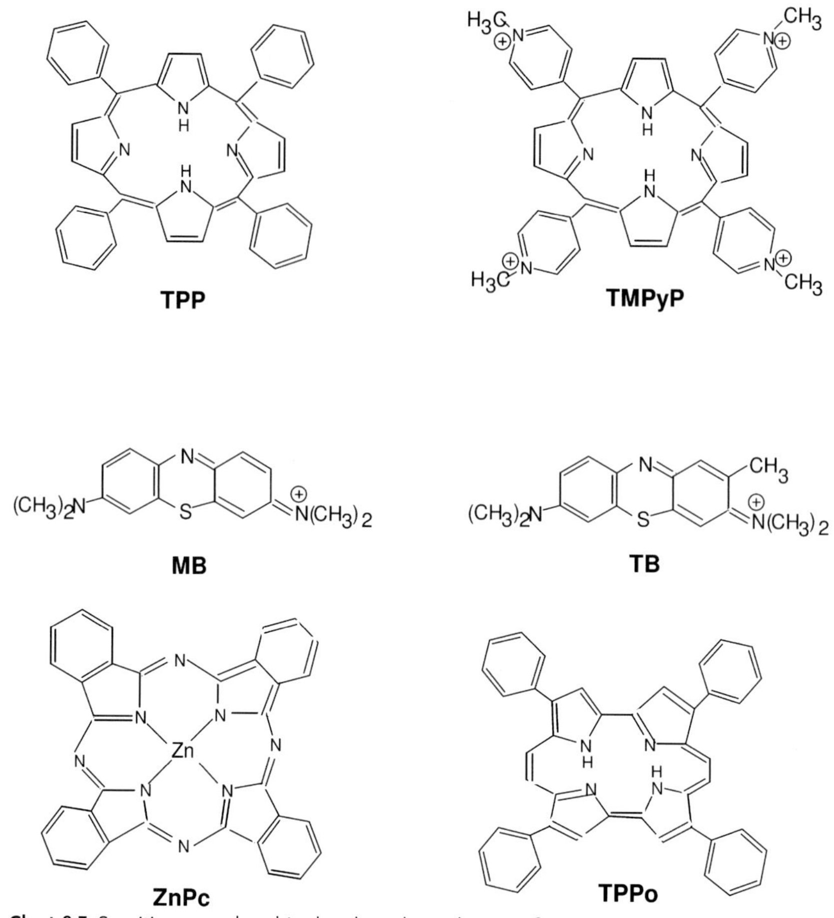

Chart 8.7 Sensitizers employed in the photochemotherapy of cancer cells. TPP: *meso*-tetraphenylporphine, TMPyP: *meso*-tetra(4-*N*-methylpyridyl)porphine, MB: methylene blue, TB: toluidine blue, ZnPc: zinc(II) phthalocyanine, TPPo: tetraphenylporphyrene.

formed with the aid of sensitizers that absorb light at wavelengths exceeding 300 nm, since photo-cross-linking by direct irradiation of the complexes with far-UV light suffers from serious disadvantages such as low cross-linking yield, strand breakage, and oxidation.

In studies of the dynamics of protein oligomerization in the context of investigations exploring amyloidoses, i.e. diseases including *Alzheimer's disease*, ruthenium(II) complexes are used [52, 55]. To this end, tris(2,2'-bipyridyl)dichlororuthenium(II), Ru(II)bpy$_3$Cl$_2$ (see Chart 8.8), and ammonium persulfate, (NH$_4$)$_2$S$_2$O$_8$, are homogeneously dispersed in an aqueous protein solution.

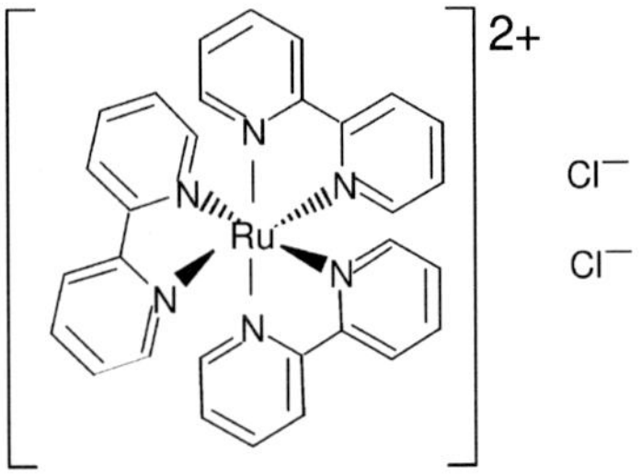

Chart 8.8 Structure of tris(2,2′-bipyridyl)dichloro ruthenium(II), Ru(II)bpy₃Cl₂.

$$Ru(II)bpy_3^{2+} \xrightarrow{h\nu} [Ru(II)bpy_3^{2+}]^*$$

$$[Ru(II)bpy_3^{2+}]^* + {}^-O_3S-O-O-SO_3^- \longrightarrow Ru(III)bpy_3^{3+} + SO_4^{2-} + {}^\bullet SO_4^-$$

Scheme 8.11 Photoreaction of Ru(II)bpy$_3^{2+}$ complexes with persulfate ions [53].

Table 8.3 Nucleobases bearing photosensitizer groups commonly used for nucleic acid/protein cross-linking studies [51, 53].

Structure of nucleobase	Denotation	λ_{max} (nm)	$\lambda_{operation}$ (nm)
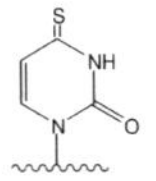	4-Thiouridine	330	> 300
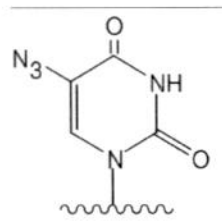 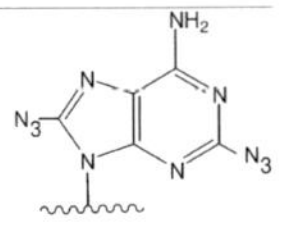	Azido-substituted nucleobases	280	> 300
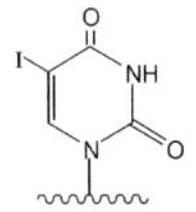 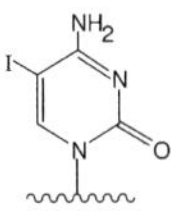	Iodouridine Iodocytidine	290/300	>300
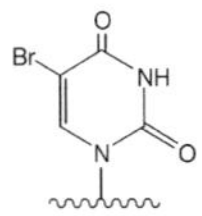	Bromouridine	275	> 300

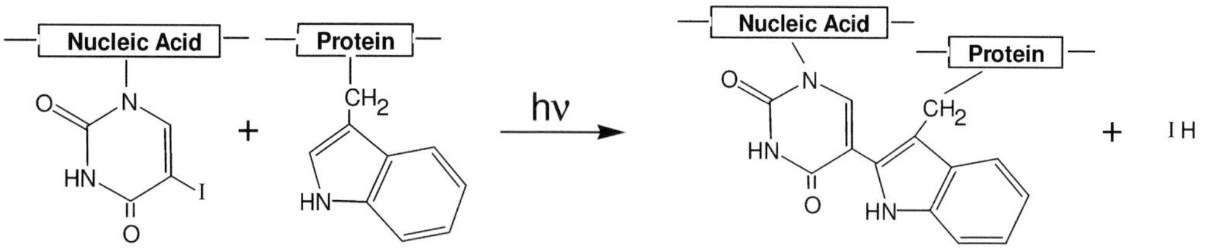

Scheme 8.12 Cross-linking of a nucleic acid with a protein by the reaction of a 5-iodouracil group with a tryptophan side group.

Upon photoexcitation, Ru(III) complexes and sulfate radicals are produced (see Scheme 8.11). Both resultant species, Ru(III)bpy$_3^{3+}$ and $^\bullet$SO$_4^-$, are potent one-electron oxidants and can generate protein radicals by hydrogen abstraction from protein molecules. The combination of the protein radicals leads to cross-links.

If nucleic acid/protein complexes are to be explored, photosensitive groups are synthesized and incorporated into the nucleic acids. Typical sensitizer-bearing nucleobases are shown in Table 8.3.

A typical cross-linking reaction is presented in Scheme 8.12.

A third example concerns the sequence-selective photocleavage of double-stranded DNA [14, 56–58]. The advantage of using photoreagents for this purpose is that they are inert in the dark and react only under irradiation with light of an appropriate wavelength that is not absorbed by neat DNA. Strand cleavage can be accomplished by attack of either sugar or nucleobase moieties. In the latter case, cleavage of DNA usually requires alkaline treatment after irradiation.

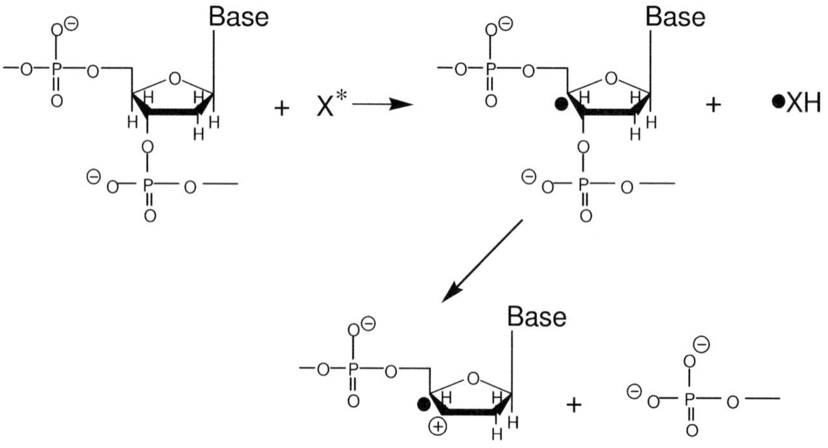

Scheme 8.13 Cleavage of a DNA strand following the abstraction of a hydrogen atom from a sugar moiety by an electronically excited photoreagent X*.

Scheme 8.14 Intra-chain hydrogen abstraction from the sugar moiety in poly(uridylic acid) involving an uracil radical formed by addition of an OH radical.

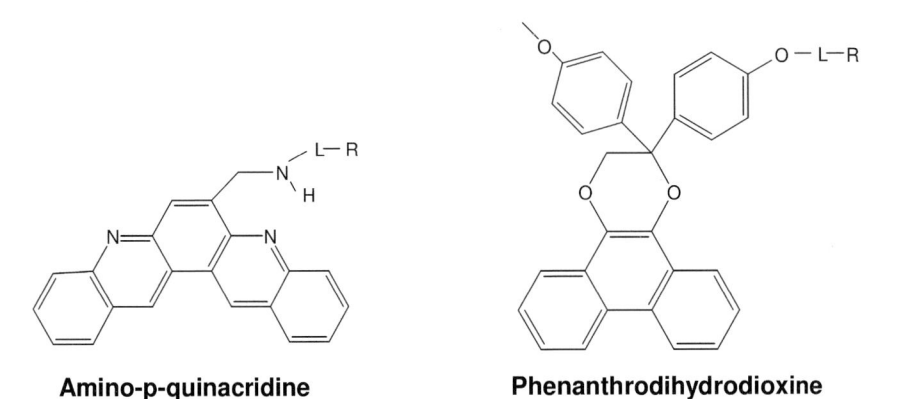

Anthraquinone

Naphthalene diimide

Amino-p-quinacridine

Phenanthrodihydrodioxine

Chart 8.9 Structures of typical photochemical nucleases used for sequence-specific cleavage of DNA strands. L: Linker. R: sequence-specific DNA-binding compound [56].

On the other hand, attack at a sugar moiety can lead to direct cleavage of the DNA strand. In this case, a common mechanism is based on hydrogen abstraction (see Scheme 8.13). The resulting sugar radicals can decompose by a variety of pathways to yield low molar mass products and DNA fragments.

Although mechanistic details, which are discussed in the relevant literature [14, 59, 60], cannot be dealt with here, the following aspect should at least be pointed out: an attack at the nucleobase might induce chemical alterations in the sugar moiety that eventually result in strand breakage. This applies, for example, to the intramolecular hydrogen abstraction suggested in the case of poly(uridylic acid) (see Scheme 8.14) [59].

The hydrogen abstraction process is, in principle, unselective, since abstractable hydrogens are present in all sugar moieties. Strand ruptures originating from attacks at the nucleobases are also intrinsically unselective. However, sequence selectivity can be accomplished if the photoreagent binds to one or a few sequences of the DNA strand. The focus of relevant research is on synthesizing conjugates composed of a photosensitizer group and a sequence-specific DNA-binding compound, also denoted as *photochemical nucleases* [56]. Appropriate photoactive groups (listed, e.g., in [14]) include complexes of transition metal ions such as Ru(II), Rh(III), and Co(II); polycyclic aromatic compounds such as anthraquinone and naphthalene diimide; porphyrins and related compounds (chlorins, sapphyrins); phthalocyanines, and fullerenes (see Chart 8.9).

References

1 W. M. Horspool, F. Lenci (eds.), *CRC Handbook of Organic Photochemistry and Photobiology*, 2nd Edition, Boca Raton, Florida (2004).

2 W. M. Horspool, P.-S. Song (eds.), *CRC Handbook of Organic Photochemistry and Photobiology*, 1st Edition, Boca Raton, Florida (1995).

3 H. Morrison (ed.), *Bioorganic Photochemistry*, Wiley, New York (1990).

4 A. R. Young, L. O. Bjorn, J. Moan, W. Nultsch (eds.), *Environmental UV Photobiology*, Plenum Press, New York (1993).

5 H. S. Nalwa (ed.), *Handbook of Photochemistry and Photobiology*, American Scientific Publ., Stevenson Ranch, California (2003).

6 A. Sancar, *Structure and Function of DNA Photolyase and Cryptochrome Blue-Light Photoreceptors*, Chem. Rev. 103 (2003) 2203.

7 L. I. Grossweiner, *Photochemistry of Proteins: A Review*, Curr. Eye Res. 3 (1984) 137.

8 K. M. Schaich, *Free Radical Initiation in Proteins and Amino Acids by Ionizing and Ultraviolet Radiation and Lipid Oxidation. – Part II: Ultraviolet Radiation and Photolysis*, CRC Crit. Rev. Food Sci. Nutr. 13 (1980) 131.

9 A. Sancar, *Cryptochrome: The Second Photoactive Pigment in the Eye and its Role in Circadian Photoreception*, Ann. Rev. Biochem. 69 (2000) 31.

10 N. L. Veksin, *Photonics of Biopolymers*, Springer, Berlin, Heidelberg (2002).

11 G. J. Smith, *New Trends in Photobiology (Invited Review). Photodegradation of Keratin and other Structural Proteins*, J. Photochem. Photobiol. B: Biol. 27 (1995) 187.

12 W. Harm, *Biological Effects of Ultraviolet Radiation*, Cambridge University Press, Cambridge (1980).

13 C. H. Nicholls, *Photodegradation and Photoyellowing of Wool*, in N. S. Allen (ed.), *Developments in Polymer Photochemistry – 1*, Appl. Science Publ., London (1980), Chapter 5, p. 125.

14 B. Armitage, *Photocleavage of Nucleic Acids*, Chem. Rev. 98 (1998) 1171.

15 J. Barber (ed.), *The Light Reactions*, Elsevier, Amsterdam (1987).

16 D. N. S. Hon, N. Shiraishi (eds.), *Wood and Cellulosic Chemistry*, 2nd Edition, Dekker, New York (2001).

17 P. J. Baugh, *Photodegradation and Photo-oxidation of Cellulose*, in N. S. Allen (ed.), *Developments in Polymer Photochemistry – 2*, Appl. Science Publ., London (1981), Chapter 5, p. 165.
18 A. Sakakibara, Y. Sano, *Chemistry of Lignin*, Chapter 4 in [16].
19 A. Bos, J. Appl. Polym. Sci. 16 (1972) 2567.
20 K. Schaffner, W. Gärtner, *Open-Chain Tetrapyrroles in Light Sensor Proteins: Phytochromes*, The Spectrum 12 (1999) 1.
21 G. E. O. Borgstahl, D. E. Williams, E. D. Getzoff, Biochemistry 34 (1995) 6278.
22 J. Hendriks, K. J. Hellingwerf, *Photoactive Yellow Protein, the Prototype Xanthopsin*, Chapter 123 in [1].
23 Y. Muto, T. Matsuoka, A. Kida, Y. Okano, Y. Kirino, FEBS Lett. 508 (2001) 423.
24 R. Dai, T. Yamazaki, I. Yamazaki, P. S. Song, Biochim. Biophys. Acta 1231 (1995) 58.
25 Y. Shichida, T. Yoshizawa, *Photochemical Aspects of Rhodopsin*, Chapter 125 in [1].
26 M. G. Friedel, *DNA Damage and Repair: Photochemistry*, Chapter 141 in [1].
27 S. Y. Wang (ed.), *Photochemistry and Photobiology of Nucleic Acids*, Academic Press, New York (1976).
28 F. Cadet, P. Vigny, *The Photochemistry of Nucleic Acids*, Vol. 1, Chapter 1 in [3].
29 D. L. Mitchell, D. Karentz, *The Induction and Repair of DNA Photodamage in the Environment*, p. 345 in [4].
30 D. L. Mitchell, *DNA Damage and Repair*, Chapter 140 in [1].
31 P. Setlow, Environ. Mol. Mutagen. 38 (2001) 97.
32 A. Vanhooren, B. Devreese, K. Vanhee, J. Van Beeumen, I. Hanssens, Biochem. 41 (2002) 11035.
33 D. V. Bent, E. Hayon, J. Am. Chem. Soc. 97 (1975) 2612.
34 F. Trautinger, *Stress Proteins in the Photobiology of Mammalian Cells*, Vol. 4, Chapter 5 in [5].
35 J. Breton, E. Naberdryk, *Pigment and Protein Organization in Reaction Center and Antenna Complexes*, Chapter 4 in [15].
36 H. Zuber, *The Structure of Light-Harvesting Pigment Protein Complexes*, Chapter 5 in [15].
37 K. Schaffner, S. E. Braslavski, S. E. Holzwarth, *Protein Environment, Photophysics and Photochemistry of Prosthetic Biliprotein Chromophores*, in H.-J. Schneider, H. Dürr (eds.), *Frontiers in Supramolecular Organic Chemistry and Photochemistry*, VCH, Weinheim (1991), p. 421.
38 S. E. Braslavski, W. Gärtner, K. Schaffner, *Phytochrome Photoconversion*, Plant Cell and Environment 6 (1997) 700.
39 S. H. Bhoo, P. S. Song, *Phytochrome: Molecular Properties*, Chapter 129 in [1].
40 G. Checcuci, A. Sgarbossa, F. Lenci, *Photomovements of Microorganisms: An Introduction*, Chapter 120 in [1].
41 S. C. Tu, *Bacterial Bioluminescence: Biochemistry*, Chapter 136 in [1].
42 V. Tozzini, V. Pellegrini, F. Beltram, *Green Fluorescent Proteins and Their Applications to Cell Biology and Bioelectronics*, Chapter 139 in [1].
43 N. K. Packham, J. Barber, *Structural and Functional Comparison of Anoxygenic and Oxygenic Organisms*, Chapter 1 in [15].
44 M. Salomon, *Higher Plant Phototropins, Photoreceptors not only for Phototropism*, in A. Batschauer (ed.), *Photoreceptors and Light Signalling*, Comprehensive Series in Photochemistry and Photobiology, Vol. 3, Royal Soc. Chem., Cambridge (2003), p. 272.
45 J. Malesic, J. Kolar, M. Strlic, D. Kocar, D. Fromageot, J. Lemaire, O. Hailland, Polym. Degrad. Stab. 89 (2005) 64.
46 D. N. S. Hon, *Weathering and Photochemistry of Wood*, Chapter 11 in [16].
47 B. George, E. Suttie, A. Merlin, X. Deglise, *Photodegradation and Photostabilisation of Wood – the State of the Art*, Polym. Degrad. Stab. 88 (2005) 268.
48 T. J. Dougherty, J. G. Levy, *Clinical Applications of Photodynamic Therapy*, Chapter 147 in [2].
49 B. W. Henderson, S. O. Gollnick, *Mechanistic Principles of Photodynamic Therapy*, Chapter 145 in [2].
50 A. Villanueva, R. Vidania, J. C. Stockert, M. Canete, A. Juarranz, *Photodynamic Effects on Cultured Tumor Cells. Cytoskeleton Alterations and Cell Death Mechanisms*, Vol. 4, Chapter 3 in [5].
51 K. Meisenheimer, T. Koch, Crit. Rev. Biochem. Mol. Biol. 32 (1997) 101.

52 G. Bitan, D. B. Teplow, Acc. Chem. Res. 37 (2004) 357.
53 H. Steen, O. N. Hensen, *Analysis of Protein-Nucleic Acid Interaction by Photochemical Crosslinking*, Mass Spectrom. Rev. (2002) 163.
54 B. Bartholomew, R. T. Tinker, G. A. Kassavetis, E. P. Geiduschek, Meth. Enzymol. 262 (1995) 476.
55 D. A. Fancy, I. Kodadek, Proc. Natl. Acad. Sci. USA 96 (1999) 6020.
56 A. S. Boutorine, P. B. Arimondo, *Sequence-Specific Cleavage of Double-Stranded DNA*, in M. A. Zenkova (ed.), *Artificial Nucleases*, Nucleic Acids and Molecular Biology, Vol. 13, Springer, Berlin (2004), p. 243.
57 T. Da Ros, G. Spalluto, A. S. Boutorine, R. V. Bensasson, M. Prato, *DNA-Photocleavage Agents*, Curr. Pharm. Design 7 (2001) 1781.
58 I. E. Kochevar, D. A. Dunn, *Photosensitized Reactions of DNA: Cleavage and Addition*, Vol. 1, Chapter 1, p. 299 in [3].
59 C. von Sonntag, *The Chemical Basis of Radiation Biology*, Taylor & Francis, London (1987), Chapter 9.
60 W. K. Pogozelski, D. T. Tullius, *Oxidative Strand Scission of Nucleic Acids: Routes Initiated by Hydrogen Abstraction from the Sugar Moiety*, Chem. Rev. 98 (1998) 1089.

9
Technical developments related to photochemical processes in polymers

9.1
Polymers in photolithography

9.1.1
Introductory remarks

In modern-day technical terminology, lithography denotes a technology used to pattern the surfaces of solid substrates. Lithography, as invented by Alois Senefelder in 1798, is a printing technique used by artists, who draw (*Greek: graphein*) directly onto a stone (*Greek: lithos*) surface with greasy ink, which adheres to the dry stone and attracts printing ink, while the background absorbs water and repels the printing ink. The patterning of surfaces with the aid of light is called photolithography. It serves to generate macrostructures in the millimeter range and is applied, for example, in the fabrication of printed circuit boards and printing plates. In its currently most important version, lithography, here denoted as *microlithography*, refers to the generation of microstructures on top of semiconductor (mostly silicon) wafers. Photomicrolithography has served as the essential tool in the information and electronic revolution. It is still unavoidable in the mass production of computer chips containing fine-line features, now in the sub-75 nm range, thus permitting an information density exceeding 10^9 integrated circuits (IC) per cm^2. This miniaturization technique is rendered possible by polymers, although they are not contained in the final products. Stimulated by the demand for further progress in the miniaturization of devices, outlined by the SIA International Roadmap [1], a large body of research and development still focuses on the improvement of the classical microlithographic techniques and the development of novel ones [2–4].

9.1.2
Lithographic processes

The lithographic process that is widely used to generate microstructures, especially in the context of the fabrication of microdevices, is shown schematically in Fig. 9.1. It is based on the interaction of electromagnetic or particle radiation with matter. Since direct irradiation of the substrate (e.g., silicon wafers) does

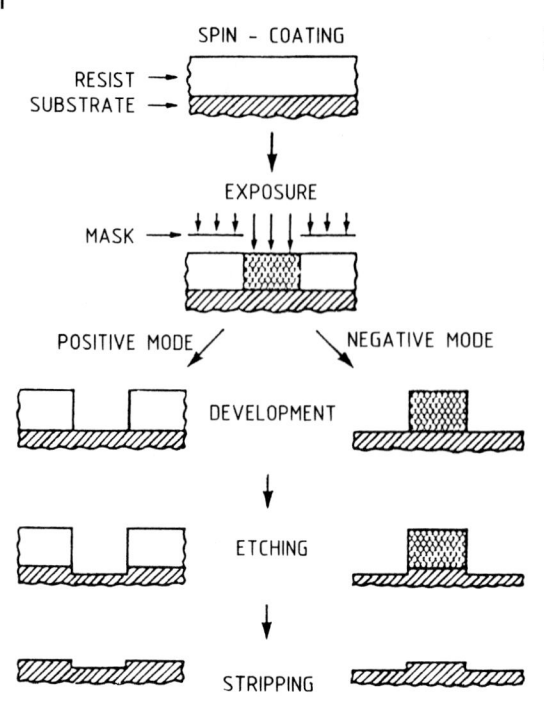

Fig. 9.1 Schematic illustration of the lithographic process.

not result in the generation of microstructures of the required quality, the technically utilized processes are performed with wafers coated with a thin layer of a radiation-sensitive material. The required fine-line structures are generated within this thin layer, essentially in two steps: irradiation through a stencil (here called the *mask*) and subsequent (commonly liquid) development. The radiation-sensitive material is called the *resist* (material) because it has to be resistant to etching agents, i.e. chemicals capable of reacting with the substrate. Etching is carried out after development, i.e. after the removal of either the irradiated or the unirradiated resist. All of these steps are illustrated in Fig. 9.1, which relates to photolithography. Most of the resists that have been employed to date are polymer-based, i.e. they consist wholly or partly of an amorphous polymer.

As regards the manufacture of microdevices photolithography is the key technology. On the other hand, charged particle beam lithography using electron or ion beams (e.g., H^+, He^{2+}, Ar^+) serves to fabricate photomasks. In this case, a computer-stored pattern is directly converted into the resist layer by addressing the writing particle beam.

In applying the process depicted in Fig. 9.1, the mask may either be placed directly onto the wafer (*contact printing*) or may be positioned a short distance in front of the wafer (*proximity printing*). In either case, the minimum feature size amounts to a couple of micrometers, and thus does not satisfy today's industrial demands. However, fine-line features down to the sub-micrometer range can be obtained with projection techniques, as described in the next subsection.

9.1.2.1 Projection optical lithography

Projection optical lithography has been the mainstream technology in the semiconductor industry for the last two decades [2]. Figure 9.2 shows a schematic depiction of an optical projection system consisting of a laser light source, a mask, a projection lens, and a resist-coated wafer. The projection of the pattern of the mask onto the resist layer provides a demagnification ratio of up to 4×.

Regarding a periodic fine structure assembly consisting of lines and spaces, the minimum line resolution of the pattern in terms of the minimum achievable feature size, LW_{min}, can be estimated with the aid of Eq. (9-1):

$$LW_{min} = \frac{k_1 \lambda}{NA} \tag{9-1}$$

Actually, LW_{min} is equal to p/2. Here, p denotes the pitch, i.e. the distance made up of a pair of lines and spaces. λ is the wavelength of the exposure light, and k_1 is a system factor that depends on various parameters such as resist response, pattern geometry in the mask, etc. NA is the numerical aperture given by Eq. (9-2).

$$NA = n \sin \Theta \tag{9-2}$$

Here, n is the refractive index and θ is the acceptance angle of the lens (see Fig. 9.2). According to Eq. (9-1), a decrease in LW_{min} can be accomplished by de-

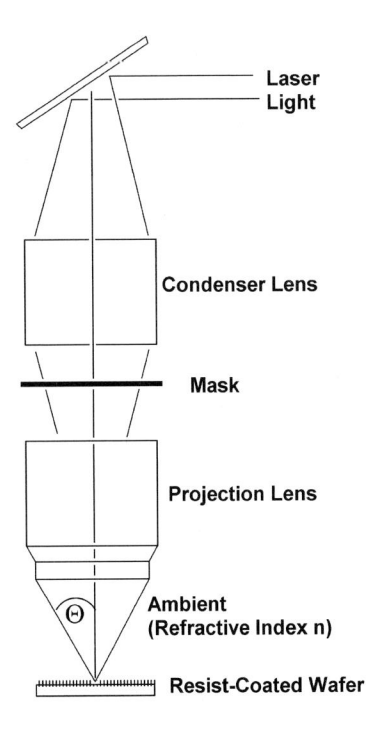

Fig. 9.2 Schematic illustration of an optical projection system.

creasing k_1 or λ or by increasing NA. In the past, all three approaches have been implemented in following industry's roadmap for the miniaturization of electronic devices [1]. For instance, a significant enhancement in resolution was achieved by using excimer lasers operating at short wavelengths: 248 nm (KrF), 193 nm (ArF), and 157 nm (F$_2$), as can be seen from Table 9.1. Sub-100 nm features can be generated with the aid of ArF and F$_2$ lasers, and sub-50 nm features with *extreme ultraviolet* (EUV) sources. The numerical aperture may be increased with the aid of lenses with increased acceptance angle. Most recently, a quite radical approach to enhanced resolution has been introduced, although not yet applied in manufacturing, namely *liquid immersion lithography* [5–7]. This new technology is based on an increase in the refractive index n by replacing the ambient gas (air, nitrogen) with a transparent liquid. Using water with $n=1.4366$ at $\lambda=193$ nm and $T=21.5\,°C$, the numerical aperture NA is increased by 44% at a given sin θ [2]. The revolutionary development in miniaturization becomes evident if one considers that the storage capacity of dynamic random access memory (DRAM) devices has been increased from less than 1 Megabit (1 Mb = 10^6 bit) to several Gigabit (1 Gb = 10^9 bit). This increase in storage capacity has been accomplished by lowering LW$_{min}$ from >1 μm to less than 0.07 μm.

A different approach, whereby the resolution may be improved by 50–100%, is based on the use of *phase-shifting transmission masks*. The latter contain opaque regions, as conventional masks do, but some of the apertures are covered with a transparent phase-shifting material, which reverses the phase of the light passing through them. The interaction of phase-shifted with non-phase-

Table 9.1 Correlation of radiation wavelength and minimum feature size in dynamic random access memory (DRAM) devices.

LW$_{min}$ (μm)	Light source	Wavelength (nm)
0.8	Hg discharge lamps	436 (g-line), 365 (i-line)
0.5	Hg discharge lamps	436, 365, 250
0.35	KrF excimer lasers	248
0.25	KrF excimer lasers	248
	ArF excimer lasers	193
0.18	ArF excimer lasers	193
0.090	F$_2$ excimer lasers	157
	ArF excimer lasers [a]	193
0.065	F$_2$ excimer lasers	157
	ArF excimer lasers [a]	193
0.045	EUV sources [b]	13.5 [c]

a) Using hard resolution enhancement technology (RET), including the immersion technique and phase-shift mask technology.
b) Laser- and discharge-produced plasmas [8], and compact electron-driven extreme ultraviolet (EUV) sources [9].
c) Si L-shell emission.

shifted light brings about destructive interference at the resist plane. This results in sharply defined contrast lines, because the resist is only sensitive to the intensity of the light and not to its sign [10].

9.1.2.2 Maskless lithography

The tools used for projection optical lithography, as described in the previous section, include very expensive parts, for instance the mask and the heavy (over 1000 kg) reduction lens. The projection of the image of the mask onto the silicon wafer requires such a heavy reduction lens. Moreover, the design and fabrication of the features of the mask are associated with high costs and long delays. The cost of the masks producing one chip can exceed $2 million. Innovations that have stemmed from these difficulties concern the development of maskless optical techniques. Actually, non-optical techniques such as electron-beam and ion-beam lithography have existed for many years. They are employed in photo-mask production, but are inappropriate for the large-scale production of chips. Novel techniques relating to optical projection are based on protocols differing from that described above in Section 9.1.2.1. *Zone-plate array lithography*, ZPAL, seems to play a prominent role among the novel techniques [3]. In ZPAL, an array of diffractive lenses focuses an array of spots onto the surface of a photoresist-coated substrate. This is accomplished by passing light from a continuous-wave laser through a spatial filter and a collimating lens to create a clean, uniform light beam. The latter is incident on a spatial light modulator, which replaces the mask. Under digital control, it splits the beam into individually controllable beamlets. Subsequently, the beamlets are passed through a telescope such that each is normally incident upon one zone plate in the array. By simple diffraction, the zone plate, consisting of circular concentric zones, focuses the light on a spot of the resist layer. The zones in the plate cause a phase shift of the transmitted light. The radii of the zones are chosen such that there is constructive interference at the focus. Lines and spaces with a density of 150 nm have been patterned with a ZPAL system operated at 400 nm. Sub-100 nm linewidths are expected to be realized with systems operating at lower wavelengths. At present, continuous-wave lasers emitting at $\lambda = 198$ nm are commercially available [3].

Imprinting lithography is another maskless technique capable of generating sub-100 nm patterns. It is essentially a nanomolding process, in which a transparent patterned template is pressed into a low-viscosity monomer layer dispensed onto the surface of a wafer. Thereby, the relief structure of the template is filled. After photopolymerization of the monomer with the aid of UV light (see Chapter 10), the template is separated, leaving a solid polymer replica of the template on the surface of the wafer. With the aid of subsequent etching processes, the pattern is fixed on the wafer's surface [4].

9.1.3
Resists

A resist material suitable for computer chip fabrication has to fulfil various requirements, the most important of which are the following. The material must be suited for spin casting from solution into a thin and uniform film that adheres to various substrates, such as metals, semiconductors, and insulators. It must possess high radiation sensitivity and high resolution capability. The aspect ratio of radiation-generated fine-line features (height-to-width ratio of lines) is desired to be high, but is limited by the risk of pattern collapse. Moreover, the resist material must withstand extremely harsh environments, for example, high temperature, strong acids, and plasmas.

On the aforementioned roadmap of progressive miniaturization, major advances in resolution have been achieved through the use of light of shorter wavelengths. New resist materials with low absorptivities (optical density less than 0.4) at these wavelengths had to be found, because near-uniform exposure throughout the resist layer needs to be maintained. For example, Novolak resists, which function well at 365 nm, are too opaque at 248 nm, and protected p-hydroxystyrene-based polymers that operate well at 248 nm are too opaque at 193 nm, at which acrylate- and cycloalkene-based polymers are used. At 157 nm, only transparent fluorocarbon-based polymers containing C–F bonds appear to operate satisfactorily.

Liquid development, which is commonly applied in lithographic processes, is based on the radiation-induced alteration of the solubility of the irradiated resist areas (see Fig. 9.1). Solubility is decreased by intermolecular cross-linking (negative mode) or increased by main-chain degradation of the polymer (positive mode). Moreover, radiation-induced chemical alterations of functional groups on the polymers can lead to a solubility change. Very importantly, radiation-induced conversion of additives controlling the solubility behavior of the polymer can also bring about the desired effect. For example, an additive that normally functions as a *dissolution inhibitor* may accelerate the dissolution after exposure to light. In the following subsections, typical resist systems are presented. Within the frame of this book, the aim is not to provide an exhaustive treatment of this subject. More information can be obtained from relevant review articles [1-25]. In this context, one should note that details of the compositions of resist systems and of the chemical nature of components are commonly withheld by the manufacturers.

9.1.3.1 Classical polymeric resists – positive and negative resist systems

The earliest photoresists used in integrated circuit manufacture consisted of polymers that were rendered insoluble by photo-cross-linking and thus operated in the negative tone mode. For instance, partially cyclized poly(*cis*-isoprene) containing a bisazide as additive served for a long time as the "workhorse" resist material in photolithography applications [15]. This system has already been described in Section 7.2.3. Subsequently, Novolak-based positively functioning sys-

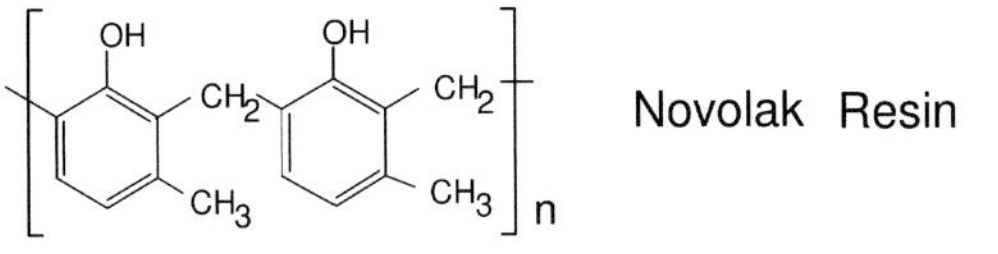

Chart 9.1 Chemical structure of Novolak resin.

tems (see Chart 9.1) were used as reliably performing "workhorse" resists for many years. Typical commercial formulations consist of a phenol-formaldehyde-type polycondensate containing a high proportion of cresol moieties and a dissolution inhibitor, e.g. a diazonaphthoquinone DNQ, commonly 2-DNQ. The polymer remains soluble since polycondensation is halted before the system becomes cross-linked. It dissolves very slowly in aqueous base. This dissolution process may be greatly enhanced upon irradiation.

As illustrated in Scheme 9.1, the photolysis of DNQ (quantum yield: 0.15–0.30) induces the release of nitrogen (N_2), which is followed by a Wolff rearrangement to give an indene ketene. In the presence of water, this reacts to form the corresponding 3-indene carboxylic acid. The latter accelerates the dissolution of the exposed areas of the coating on top of the wafer [13, 18].

Among the large family of classical resists, *polyimides* are renowned for their high temperature resistance (up to 500 °C) and their excellent electrical insulation properties. Therefore, polyimides are appropriate materials for mask fabrication, and can serve as passivation layers and interlayer dielectrics [20, 21, 25]. To this end, microstructures are generated from polyimide precursors, for instance, polyamic acid esters [26]. The ester groups contain reactive functions, e.g. carbon-carbon unsaturations. The unsaturated moieties can undergo cycloadditions or (in the presence of a photoinitiator) polymerization reactions upon exposure to UV light. In this way, the polyamic acid ester is cross-linked, thus acting in the negative tone mode. After removal of the unexposed material, imidization of the cross-linked polyamic acid ester by thermal treatment results in insoluble polyimide. The overall process is illustrated in Scheme 9.2.

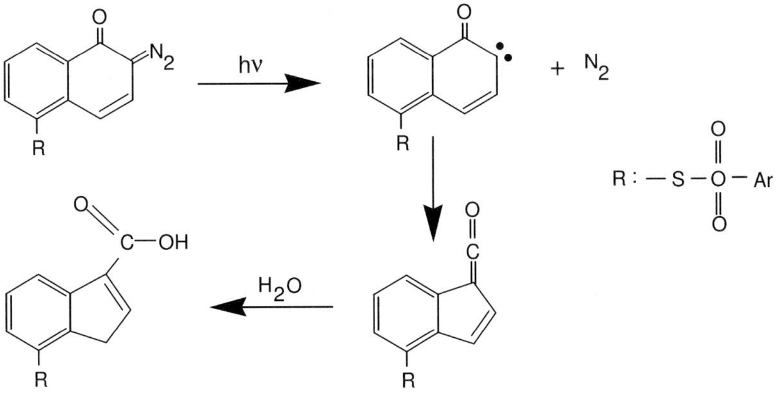

Scheme 9.1 Photolysis of 2-diazonaphthoquinone, 2-DNQ [13].

Scheme 9.2 Photo-cross-linking of polyamic acid esters and subsequent thermal imidization. R denotes a reactive group, e.g. –O–CH$_2$–CH=CH$_2$.

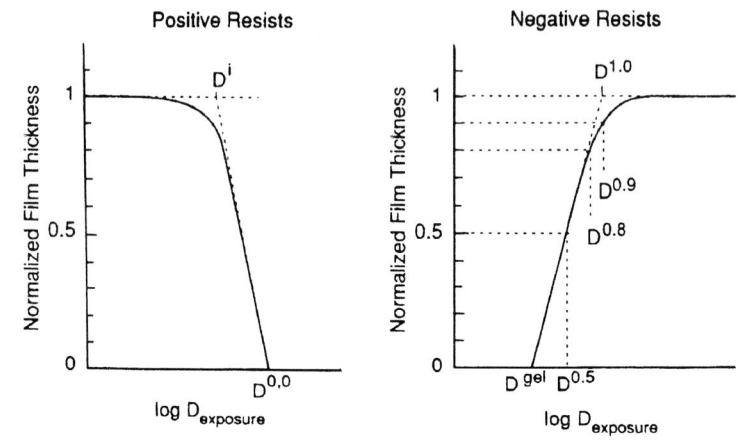

Fig. 9.3 Schematic representation of exposure characteristic curves for positive and negative resists. Adapted from Schlegel and Schnabel [27] with permission from Springer.

A host of resist systems that undergo changes in their solubility due to chemical alterations upon exposure to deep UV light (240–280 nm) has been described in the literature [11, 15, 16]. Tables 9.2 and 9.3 list some typical examples and commercially available resists, respectively. They also show sensitivity values of the resists.

By general convention, the sensitivity, S, is related to the thickness, d, of the resist layer measured after exposure and development, and is obtained from exposure characteristic curves, as are illustrated in Fig. 9.3. In the case of positively functioning resists, $S \equiv D_{exp}^{0.0}$ corresponds to the exposure light dose required to completely remove the irradiated polymer from the substrate, i.e. the dose at which the normalized thickness of the resist layer is equal to zero, $d_{irr}/d_0 = 0$. In the case of negatively acting resists, the sensitivity is reported as $S \equiv D_{exp}^{0.5}$ or sometimes as $S \equiv D_{exp}^{0.8}$ or $S \equiv D_{exp}^{0.9}$, corresponding to $d_{irr} = 0.5 \, d_0$, $d_{irr} = 0.8 \, d_0$, or $d_{irr} = 0.9 \, d_0$, respectively. D_{exp} is the product of light intensity and irradiation

Table 9.2 Sensitivities of deep UV positive-tone resists [15].

Polymer	S (mJ cm^{-2})[a]	λ (nm)[b]
Poly(methyl methacrylate)	3300	240
Poly(methylisopropenyl ketone)	700	280
Poly(perfluorobutyl methacrylate)	480	240
Poly(methyl methacrylate-co-glycidyl methacrylate)	250	250
Poly(methyl methacrylate-co-indenone)	20	240
Poly(butane sulfone)	5	185
Diazoquinone-containing Novolak resins	90	248

a) Sensitivity.
b) Wavelength of incident light.

time, and is commonly given in units of mJ cm^{-2}. A higher sensitivity, corresponding to a lower exposure dose, implies a faster production rate.

9.1.3.2 Chemical amplification resists

Past efforts to improve the fabrication of microdevices have been closely connected with attempts to increase the resist sensitivity, S. In the case of the resists described in Section 9.1.3.1, S is limited by the quantum yields, which are much low-

Table 9.3 Sensitivities S of some commercial deep UV resists ($\lambda \approx 250$ nm) [15].

Resist	S (mJ cm^{-2})	Tone
RD 2000N Poly(vinyl phenol) containing diazidodiphenyl sulfone [a]	20	Negative
Kodak KTFR Cyclized polyisoprene rubber containing azide	20	Negative
AZ-1350J Novolak resin containing diazonaphthoquinone	90	Positive

a) 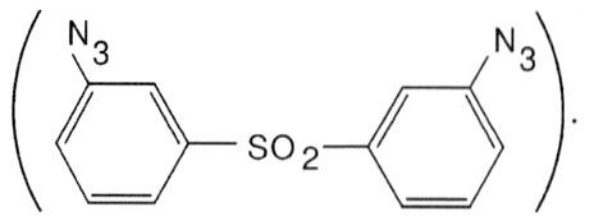 (structure: two azide-substituted phenyl groups linked by SO$_2$)

er than unity, typically 0.2–0.3. Quantum yields can rarely be increased. In the best case, S would be improved by a factor of three to five, if the quantum yield could be increased to unity, the maximum value. Therefore, the introduction into lithography in the early 1980s of processes based on the concept of chemical amplification represented a truly significant advance [28]. Chemical amplification means that a single photon initiates a cascade of chemical reactions. This applies, for instance, to the photogeneration of a Brønsted (protonic) acid capable of catalyzing the deprotection of functional groups attached to the backbone of linear polymers such as PBOCSt or PTBVB (see Chart 9.2).

The protonic acid is formed upon irradiation with UV light (e.g., at $\lambda = 248$ nm) when the polymers contain a small amount of an appropriate acid generator such as an iodonium or sulfonium salt (see Scheme 9.3).

Upon baking the exposed resist system at elevated temperatures (>100 °C), the photogenerated acid catalyzes the cleavage of C–O bonds, as illustrated in Scheme 9.4. The deprotected polymer host is soluble in aqueous base developers. Typical turnover rates for one acid molecule are in the range of 800–1200 cleavages. Resists thus amplified may attain a photosensitivity of 1–5 mJ cm^{-2} [14], thus significantly surpassing the sensitivity of non-amplified commercial resists (see Table 9.3).

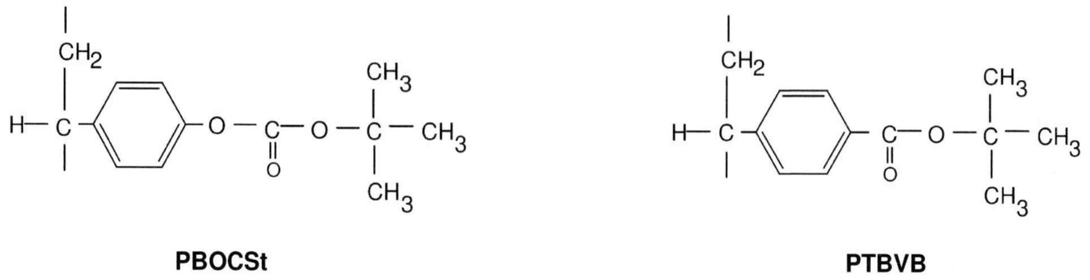

PBOCSt

PTBVB

Chart 9.2 Chemical structures of poly(t-butoxycarbonyl oxystyrene), PBOCSt, and poly(t-butyl-p-vinyl benzoate), PTBVB.

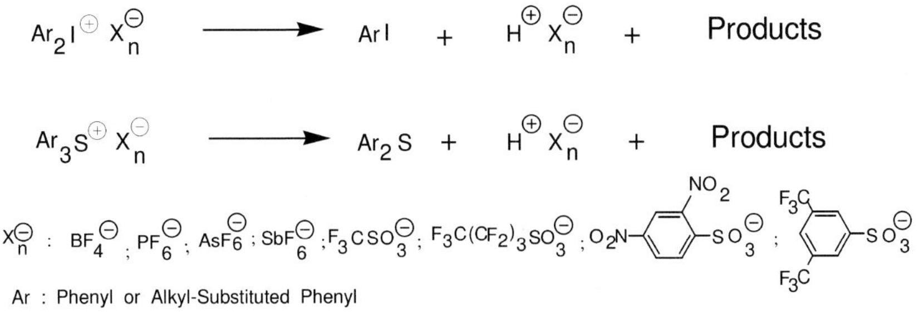

$$Ar_2I^{\oplus} X_n^{\ominus} \longrightarrow ArI + H^{\oplus} X_n^{\ominus} + \text{Products}$$

$$Ar_3S^{\oplus} X_n^{\ominus} \longrightarrow Ar_2S + H^{\oplus} X_n^{\ominus} + \text{Products}$$

$X_n^{\ominus}$: $BF_4^{\ominus}$; $PF_6^{\ominus}$; $AsF_6^{\ominus}$; $SbF_6^{\ominus}$; $F_3CSO_3^{\ominus}$; $F_3C(CF_2)_3SO_3^{\ominus}$; $O_2N\text{-}C_6H_3(NO_2)\text{-}SO_3^{\ominus}$; $(F_3C)_2C_6H_3\text{-}SO_3^{\ominus}$

Ar : Phenyl or Alkyl-Substituted Phenyl

Scheme 9.3 Proton generation by photolysis of diphenyliodonium and triphenylsulfonium salts. For a detailed mechanism, see Section 10.3.

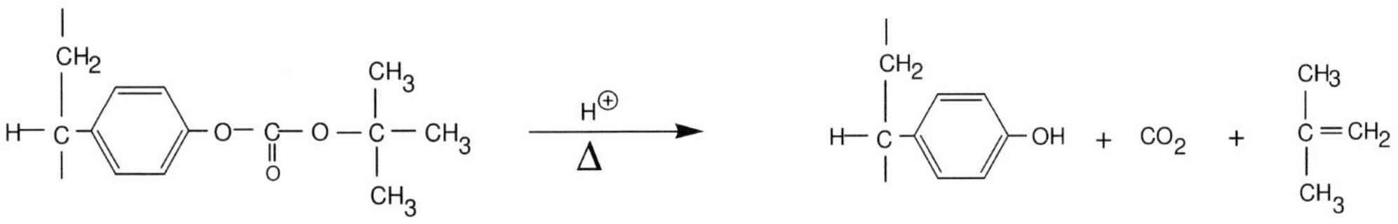

Scheme 9.4 Acidolysis of PBOCSt, a protected poly(p-hydroxystyrene).

Resist systems based on PBOCSt turned out to be very sensitive towards airborne impurities. These difficulties were overcome by employing another chemically amplified resist, a random copolymer consisting of p-hydroxystyrene and t-butyl acrylate (see Chart 9.3).

This system, denoted as *Environmentally Stable Chemical Amplification Positive Photoresist*, ESCAP, has become the standard 248 nm resist in device manufacture by leading chip makers. It is capable of printing features with a density of 125 nm [29].

Photogenerated acids can also catalyze various other reactions, e.g. the cross-linking of polymers containing epoxide groups (see Chart 9.4), or Claisen and pinacol rearrangements in polymers as shown in Scheme 9.5. Resist systems operating on the basis of these reactions have been proposed [12, 13].

Besides the onium salts considered above, various other organic compounds are capable of acting as acid generators [27]. Typical examples are presented in Table 9.4.

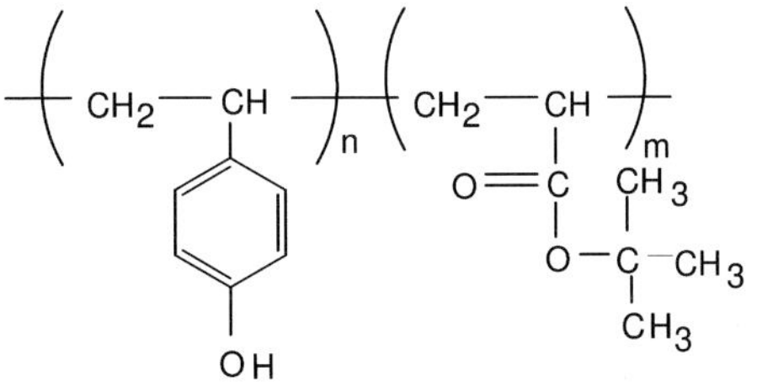

Chart 9.3 Structure of a random copolymer forming the host polymer of ESCAP [29].

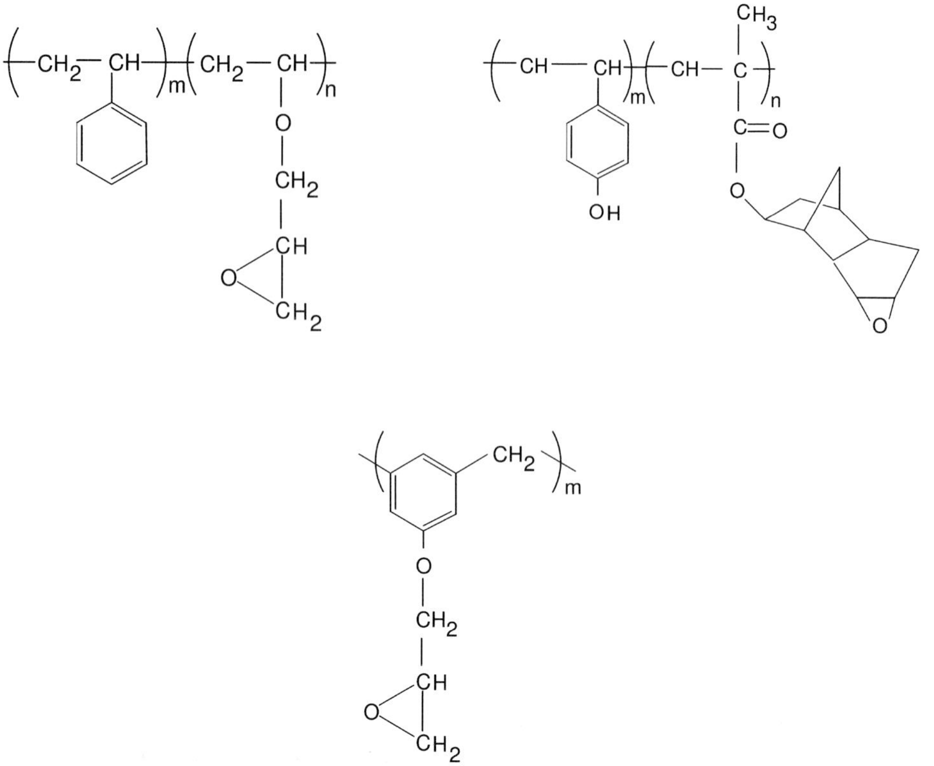

Chart 9.4 Structures of polymers containing epoxide groups capable of undergoing photoacid-catalyzed cross-linking [12].

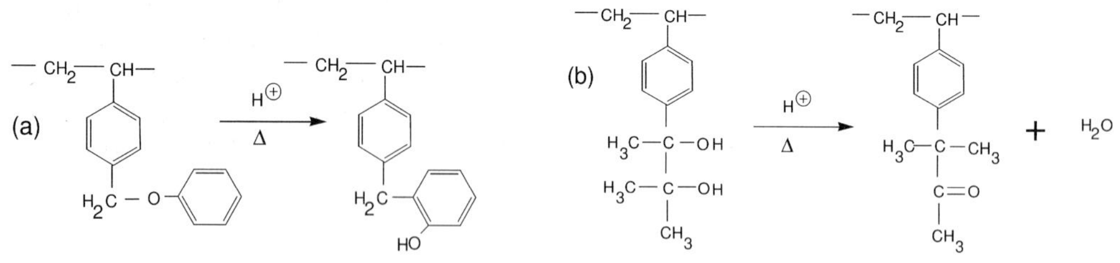

Scheme 9.5 Acid-induced Claisen (a) and pinacol (b) rearrangements [12].

9.1.3.3 Resists for ArF (193 nm) lithography

ArF lithography, employing ArF lasers emitting 193 nm light, has been developed with the aim of generating sub-100 nm features. Since the industrially widely used 248 nm resists containing aromatic (e.g., hydroxystyrene) moieties are too opaque at 193 nm, novel polymers of much lower absorptivity at this wavelength are needed. These polymers are required to withstand dry etching

Table 9.4 Organic photoacid generators.

Acid generators	Acid
o-Nitrobenzyl sulfonates	HO–SO$_2$–R
Imino sulfonates	HO–SO$_2$–R^3
2-Aryl-4,6-bis(trichloromethyl)triazines	HCl
o,o'-Dibromophenols	HBr

Table 9.5 Chemical amplification resists applicable in 193 nm lithography

Resist system	Chemical structure of typical base units	References
Random copolymers of norbornene methyl-cyclopentyl ester and norbornene hexafluoro-isopropanol		[29] [30]
Random co- and terpolymers containing norbornene derivatives and maleic anhydride		[31]
Alternating copolymers of vinyl ether and maleic anhydride		[32]
Random co- and terpolymers containing acrylate or methacrylate moieties with pendant alicyclic groups		[32, 33]

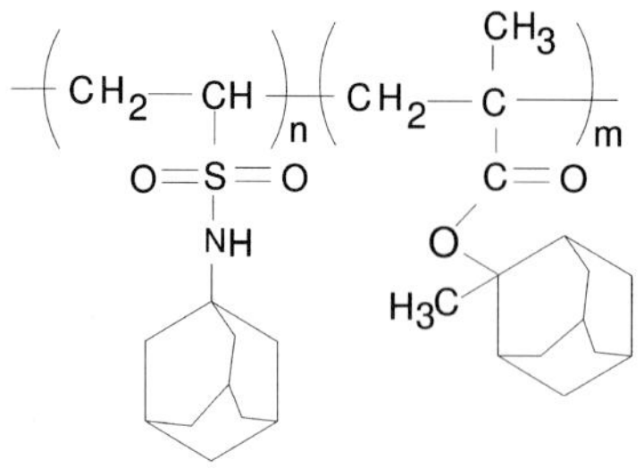

Chart 9.5 Structure of poly[N-(1-adamantyl)vinylsulfonamide-co-(2-methyl)adamantyl methacrylate], a random copolymer that absorbs light only weakly at $\lambda = 193$ nm [33].

agents and to be base-soluble when chemical amplification based on the deprotection of carboxylic or phenolic groups is the imaging mechanism of choice.

Table 9.5 presents families of random copolymers with cycloaliphatic structures in the main chain or in side groups that are appropriate for lithographic applications. Cycloaliphatic moieties, such as adamantyl groups, offer etch durability, while carboxylic acid groups, which become available through amplified deprotection processes, impart base solubility.

The components of the copolymers are cycloaliphatic monomers (norbornene), and vinyl ether, maleic anhydride, acrylate, and methacrylate. In addition, vinyl sulfonamides have been used as co-monomers in the synthesis of random copolymers capable of functioning as acid-amplified resists. An example is presented in Chart 9.5. A high sensitivity, $S = 2$ mJ cm^{-2}, was measured for a copolymer (Chart 9.5) with $n = 0.4$ and $m = 0.6$ (resist thickness $d = 220$ nm, developer: aqueous tetramethylammonium hydroxide solution). Triphenylsulfonium perfluoro-1-butane sulfonate served as acid generator [33].

Notably, the liquid immersion technique (see Section 9.1.2.1) in conjunction with high refractive index fluids can be applied to generate 32 nm structures (see Fig. 9.4a) [7b].

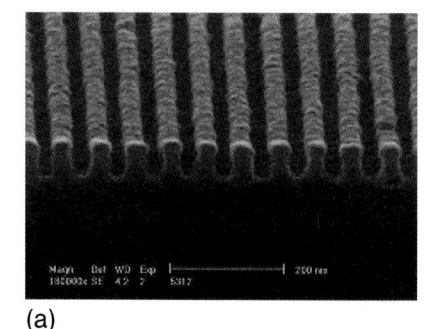

Fig. 9.4 (a) 32 nm line and space structures (X-SEM graphs) generated by means of 193 nm immersion lithography, and (b) 60 nm structures generated by means of 157 nm lithography. Adapted from Mulkens et al. [7b] and Hohle et al. [39], respectively, with permission from the author (a) and from Carl Hanser (b).

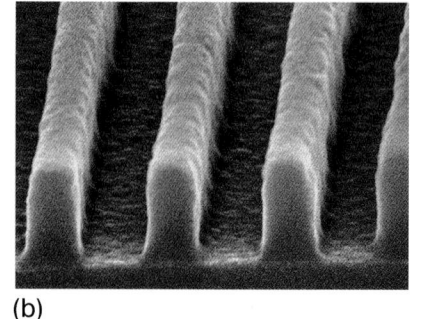

9.1.3.4 Resists for F$_2$ (157 nm) lithography

Photoresists employed at 248 nm and 193 nm are too opaque at 157 nm, the wavelength of light emitted by F$_2$ lasers. However, sufficiently transparent fluorocarbon-based polymers containing non-absorbing C–F bonds operate satisfactorily at 157 nm [30, 34]. Therefore, new fluoropolymers, also functioning as acid-amplified resists, were synthesized. Chart 9.6 shows the structures of copolymers containing 4-(2-hydroxy hexafluoro isopropyl) styrene units.

At $\lambda = 157$ nm, the fluorine-containing homopolymers and copolymers presented in Chart 9.7 and in Scheme 9.6 have absorption coefficients ranging from 3.0 to 4.0 µm^{-1} [35–40]. At a resist thickness lower than 100 nm, they turned out to be capable of imaging 40 nm lines with a 100 nm pitch [35].

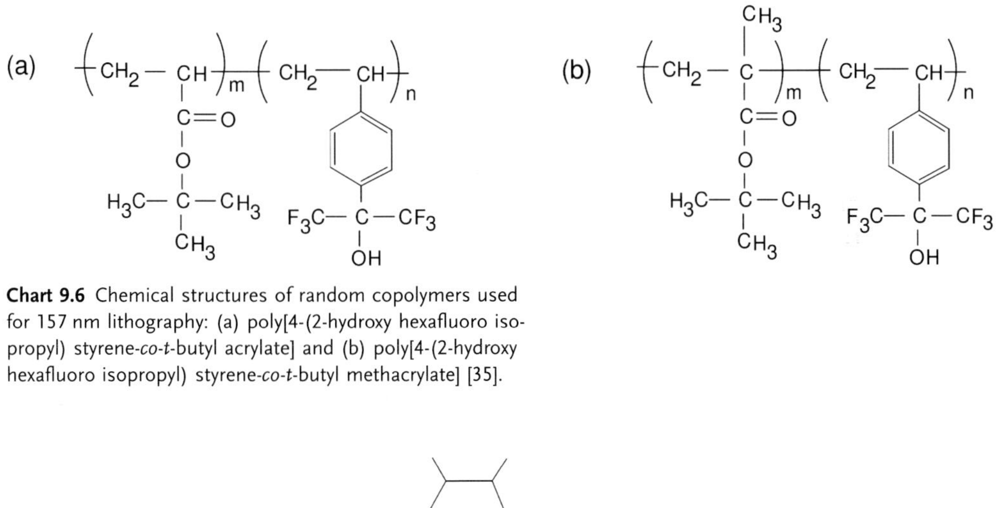

Chart 9.6 Chemical structures of random copolymers used for 157 nm lithography: (a) poly[4-(2-hydroxy hexafluoro isopropyl) styrene-*co*-*t*-butyl acrylate] and (b) poly[4-(2-hydroxy hexafluoro isopropyl) styrene-*co*-*t*-butyl methacrylate] [35].

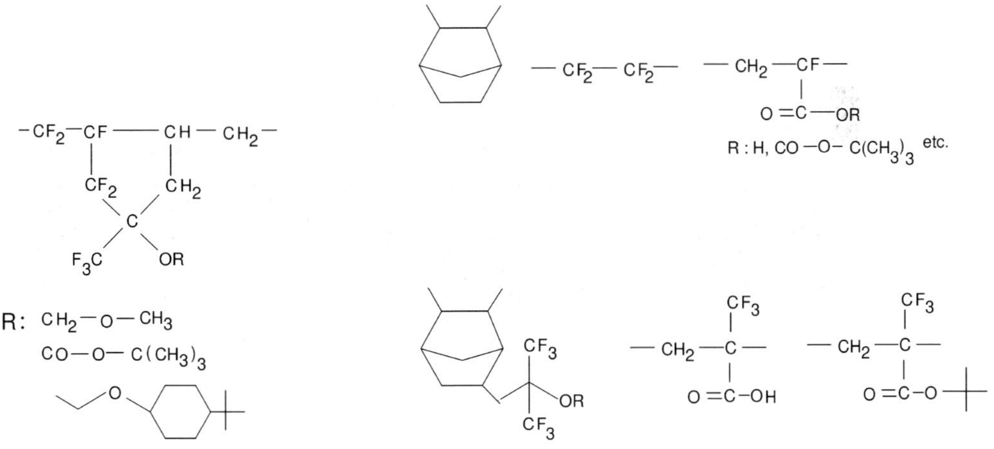

Chart 9.7 Chemical structures of monomer moieties of homopolymers and random copolymers capable of acting as 157 nm resists [36–38].

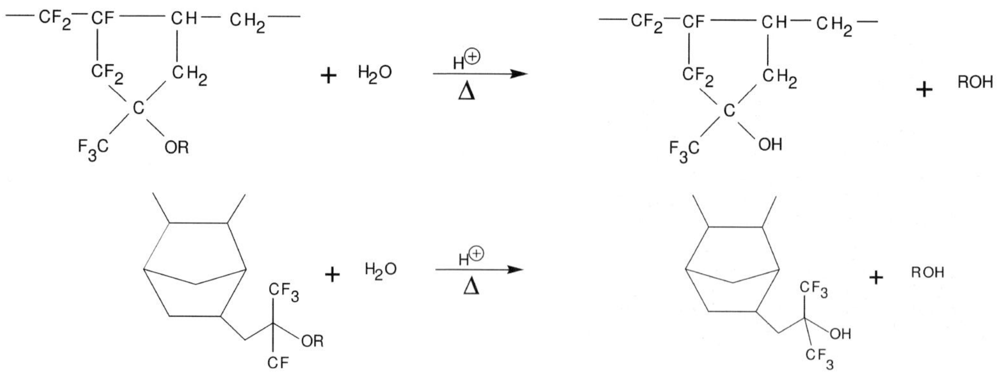

Scheme 9.6 Acidolysis of polymers appropriate for 157 nm lithography [39, 40].

Here, the excellent performance of these polymers is demonstrated by the 60 nm structures shown in Fig. 9.4b.

Absorption coefficients of about 0.5 µm^{-1} allow imaging of 200 nm thick films. At present, however, there are problems concerning pattern development. Moreover, difficulties regarding lenses and masks have to be resolved. As yet, CaF$_2$ is the only feasible lens material, since fused quartz is not transparent at 157 nm. However, CaF$_2$ is crystalline and, therefore, intrinsically birefringent. Consequently, lenses have to be made from elements with different crystal orientations.

9.1.4
The importance of photolithography for macro-, micro-, and nanofabrication

Photolithography is industrially employed also for the generation of macrostructures of dimensions up to several millimeters. Typical examples in this context include the fabrication of printed circuit boards, picture tubes, and printing plates. For details, the reader is referred to the literature [21]. Actually, printing plates are mostly made from photopolymer systems functioning on the basis of photopolymerization of appropriate monomers. This aspect is dealt with in Section 11.5. Currently, photolithography continues to play a dominant role in the semiconductor industry with regard to the production of microdevices. However, with miniaturization being extended to nanofabrication, methods using extreme ultraviolet (EUV) radiation ($\lambda = 13$ nm) and soft X-rays (synchrotron radiation) might become important in the future. In addition to the fact that photolithography involves high capital and operational costs, it is not applicable to nonplanar substrates. To overcome this disadvantage, alternative methods have been developed. At present, *soft lithography* seems to be a promising new technique for micro- and nanofabrication. The soft lithographic process consists of two parts: the fabrication of elastomeric elements (masters), i.e., stamps or molds, and the use of these masters to pattern features in geometries defined

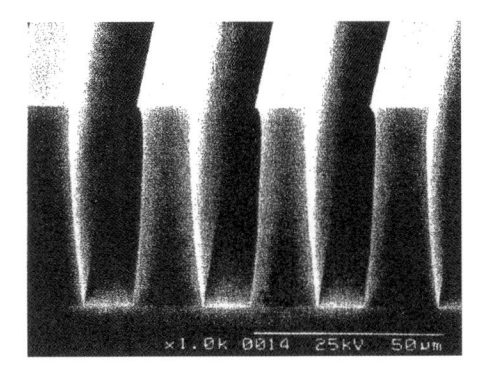

Fig. 9.5 High aspect ratio microstructures (height: 50 µm, spacing: 15 µm). Resist system: Novolak/DNQ (see Section 9.1.3.1). Adapted from Maciossek et al. [44] with permission from Leuze.

by the masters' relief structure. The formation of a master includes a photolithographic step, i.e., the relief structure is generated by shining light through a printed mask onto the surface of a photoresist film. After development, the latter is subsequently impressed in an elastomer [41–43].

Photomicrolithography also plays a major role in the field of *micromachining*, whereby photofabrication provides a tool for making inexpensive, high aspect ratio microstructures having dimensions of several micrometers. For example, height-to-width ratios as high as 18:1, at a resist thickness of up to several hundred µm, and minimum feature sizes down to 3 µm, can be realized with a negative-tone resist containing epoxide groups (see Chart 9.4). Cross-linking of the irradiated resist is achieved through a photoacid-amplified mechanism [44, 45]. In this case, irradiations can be performed at 365 nm. Metallization of the polymer patterns (with steep edges, more than 88°) by galvanization or other means, and subsequent removal of the polymer, results in metal structures, which opens up a plethora of applications. Additional resist systems tested in relation to this technique include the positive-tone system Novolak/DNQ (see Section 9.1.3.1 and Fig. 9.5) and negative-tone polyimides (see Section 9.1.3.1).

Notably, the patterning of thick layers, commonly consisting of multiple coats of spun-cast polymer, necessitates a high transparency of the resist system. Therefore, care has to be taken that the maximum exposure depth exceeds the thickness of the layer. In special cases, the initiator/sensitizer is photobleached, thus causing the penetration depth of the incident light to increase during exposure.

9.2
Laser ablation of polymers

9.2.1
General aspects

9.2.1.1 Introductory remarks

Material can be ejected when a laser beam or, more generally speaking, a high intensity light beam is directed onto a polymer sheet. On the basis of this phenomenon, commonly called *laser ablation*, mechanical machining such as cutting and drilling of polymeric materials is possible. Moreover, microstructures can be generated with laser beams of small diameter. Since its discovery, there have been attempts to utilize laser ablation as a photolithographic tool [46, 47]. However, because of several disadvantages, such as contamination of the surrounding surfaces with debris, carbonization, and insufficient sensitivity, it has not become a serious competitor to conventional photolithographic techniques, at least as far as the use of readily available polymers is concerned. At present, there is growing interest in exploiting laser ablation for various practical applications, such as laser desorption mass spectrometry or laser plasma thrusters for the propulsion of small satellites. Moreover, basic research is still focused on the mechanism of laser ablation. The increasing importance of laser ablation has been recognized by two renowned scientific journals, which have published special issues devoted to various aspects of this interesting field [48, 49]. Most published laser ablation work concerns the irradiation of polymers with femto- or nanosecond pulses provided by excimer lasers operating at wavelengths of 157, 193, 248, 308, and 351 nm. In more recent work, diode-pumped solid-state Nd:YAG lasers, generating 10 ns light pulses at the harmonic wavelengths 532, 355, and 266 nm (pulse energy: several mJ) have also been applied, especially for the *micromachining* of plastics [50].

9.2.1.2 Phenomenological aspects

The ablation is quantified by means of the ablation rate, i.e. the ablated depth per pulse. Generally, the ablation rate is insignificant at fluences below a *threshold fluence*. Above this threshold, the ablation rate increases dramatically. This is demonstrated in Fig. 9.6 [51] for a commercial polyimide. It can also be seen in Fig. 9.6 that the threshold fluence decreases with shortening wavelength.

A sharp rise in the etch rate at the threshold is found only at the lowest laser wavelength (193 nm). At higher wavelengths, the curves bend smoothly upwards in an exponential fashion, indicating that there is also ablation below the threshold fluence point obtained by extrapolating the linear portion of the curve to zero ablation rate. This was corroborated by a study on poly(methyl methacrylate) concerning the so-called *incubation effect* [52]. The latter refers to the phenomenon of the polymer surface being etched less deeply by the initially applied pulses than by subsequent pulses of the same fluence. Actually, material

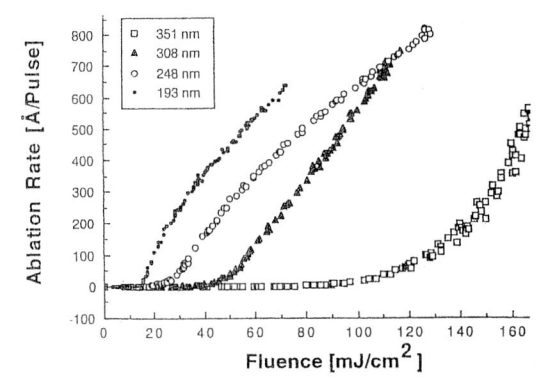

Fig. 9.6 Laser ablation of polyimide (Kapton™) at different wavelengths (given in the graph). The ablation rate obtained by single-shot experiments as a function of the fluence. The changes in the film thickness were measured with the aid of a quartz crystal microbalance. Adapted from Küper et al. [51] with permission from Springer.

is even ejected during the incubation period. However, it cannot be released because of insufficient formation of gaseous products. The latter are needed to build up a pressure sufficient for the ejection of large fragments. Therefore, the initially etched pit is refilled. Evidence for the ejection of fragments was obtained with the aid of acoustic signals, detected in the 2–85 MHz range [53]. In Fig. 9.7, it can be seen that the longitudinal 20 MHz signal increases drastically in the fluence range around the threshold deduced from ablation depth and temperature measurements.

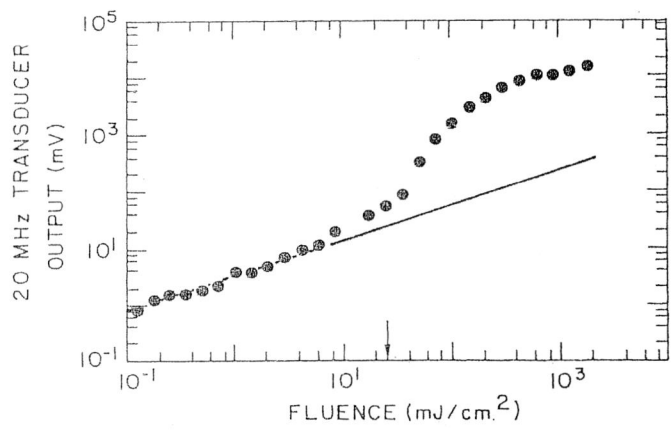

Fig. 9.7 Laser ablation of polyimide (Kapton™) at $\lambda = 193$ nm. The longitudinal acoustic signal (20 MHz) received by a piezoelectric transducer as a function of the fluence. The arrow indicates the threshold fluence obtained by recording the signal voltage produced at a pyroelectrical crystal (LiTaO$_3$). Adapted from Gorodetsky et al. [53] with permission from the American Institute of Physics.

The signal increase is interpreted as arising from the transfer of momentum of the ablated particles to the remaining substrate. The particles acquire a kinetic energy of $E_{kin} = mv^2/2$ (v: particle velocity, of the order of 10^5 cm s^{-1}, m: particle mass). The force exerted by the ablated particles on the sample surface gives rise to a pulse of acoustic energy, which propagates through the sample. The signal detected below the threshold is thought to be of thermoelastic and, to some extent, of photoelastic origin.

9.2.1.3 Molecular mechanism

Both photochemical and photothermal reactions contribute to the release of volatile fragments, a process that leads to the breakage of a certain number of chemical bonds in the polymer within a short period. A versatile model that addresses the fact that ablation always requires the application of a large number of laser pulses, and that rationalizes the dependence of the ablation rate on fluence, wavelength, pulse length, and irradiation spot size, has been proposed by Schmid et al. [54]. Accordingly, the absorption of laser light leads to the electronic excitation of chromophoric groups in the polymer. The subsequent deactivation processes involve both direct bond breakage in the excited state and relaxation, i.e. internal conversion to a highly excited vibrational state of the electronic ground state. In the latter case, the interaction with surrounding molecules can lead to thermal activation, resulting in further bond breakage. The chemical alterations that accompany these reactions lead to modified chromophores with absorption cross-sections differing from those of the original ones. If the number of broken bonds exceeds a threshold value, a thin layer of the polymer is ablated, and the ablated material forms a *plume* that expands three-dimensionally and continues to absorb laser radiation. The ablation plume consists of gaseous organic products and particulate fragments, and, in the case of biological tissues, also of water vapor and water droplets. The expansion of the plume into the surrounding air is coupled with the generation of acoustic transients that, for high volumetric energy densities, evolve into shock waves [55]. In principle, simultaneous multi-photon absorption may also be involved in laser ablation of neat polymers, although it seems to be important only at the large pulse fluences attained with sub-ps pulses.

9.2.2
Dopant-enhanced ablation

Ablation can be significantly enhanced by the presence of dopants, i.e. by additives that strongly absorb laser light. Dopant-enhanced ablation is important in cases in which the laser light is only weakly absorbed by the polymer matrix. Typical examples of such systems are poly(methyl methacrylate) containing acridine or tinuvin-328 (λ_{exc} = 308 or 351 nm) [56, 57] and nitrocellulose doped with stilbene-420, coumarin-120, or rhodamine 6G (λ_{exc} = 337 nm) [58]. In these cases, different mechanisms can become operative [57]. Degradation of the polymer matrix can be caused by thermal energy transferred from the dopant to the polymer. In other words, most of the electronically excited dopant molecules deactivate through vibronic relaxation (internal conversion) to vibronically excited ground states, from which energy is transferred to surrounding macromolecules. Alternatively, the additive may be excited to higher electronic levels by multi-photon absorption and subsequently decompose into various fragments, which leads to explosive decomposition of the polymer matrix.

9.2.3
Polymers designed for laser ablation

Novel photopolymers have been developed to overcome certain disadvantages, such as debris contamination and insufficient sensitivity, encountered in the application of laser ablation in lithographic techniques. Of note in this context are novel linear polymers containing photochemically active chromophores in the main chain [59]. In relation to the 308 nm laser light generated by XeCl excimer lasers, polymers containing triazene or cinnamylidene malonic acid groups were found to be much more appropriate than a commercial polyimide (see Chart 9.8):

The TC and CM polymers decompose exothermically at well-defined positions. Thereby, gaseous products are formed, which carry away the larger fragments. In the case of the triazene polymer (see Scheme 9.7), the fragmentation pattern has been analyzed with the aid of time-of-flight mass spectrometry.

A comparison of characteristic ablation parameters (see Table 9.6) reveals that the polymer containing triazene groups possesses a lower threshold fluence and a higher etch rate than the other two polymers and is, therefore, most appropriate for technical processes based on laser ablation of polymers.

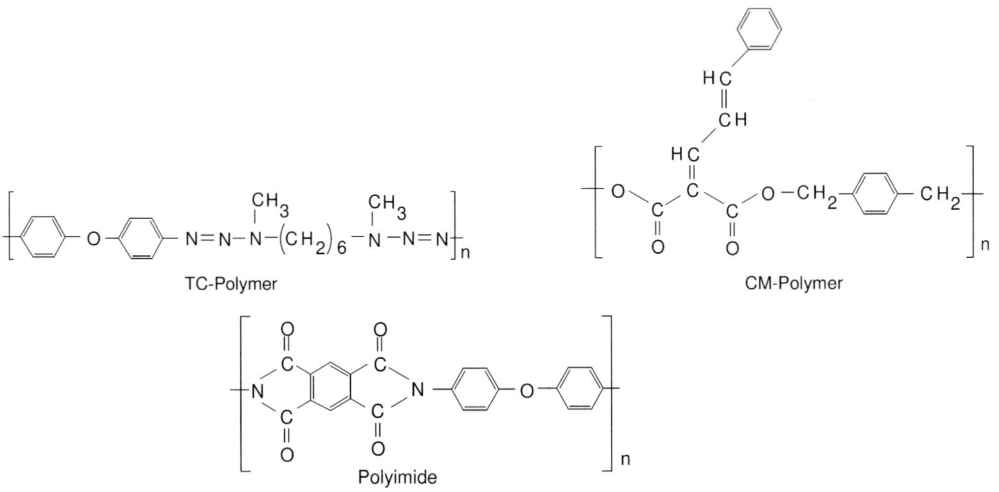

Chart 9.8 Chemical structures of polymers appropriate for laser ablation at $\lambda = 308$ nm.

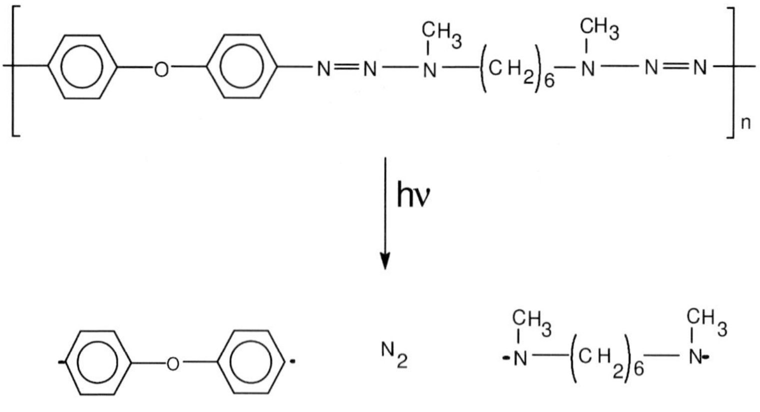

Scheme 9.7 Laser decomposition of the TC polymer [59].

Table 9.6 Ablation parameters of polymers [59].

	TC Polymer	CM Polymer	Polyimide [a]
a_{linear} (cm^{-1}) [b]	100 000	102 000	95 000
$F_{threshold}$ (mJ cm^{-2}) [c]	27	63	60
D (nm/pulse) [d]	267	90	61

a) 125 μm Kapton™.
b) Linear absorption coefficient.
c) Threshold fluence.
d) Etch rate at $F = 100$ mJ cm^{-2}.

9.2.4
Film deposition and synthesis of organic compounds by laser ablation

Thin films with special chemical and physical properties can be deposited on a substrate upon irradiating a target material located in a closed system in the neighborhood of the substrate with a laser beam [60, 61]. A schematic depiction of such a set-up with a target–substrate distance of 20 mm is shown in Fig. 9.8. Besides silicon wafers, appropriate substrate materials include ZnSe, KBr, and quartz.

In deposition studies with polyacrylonitrile, it was found that the composition of the deposited films could be controlled by varying the laser wavelength and the fluence per pulse. Films containing varying amounts of cyano side groups have been generated in this way [63, 64]. Moreover, poly(tetrafluoroethylene) and poly(methyl methacrylate) have been used as target materials for the deposition of thin films [65, 66]. Films possessing an Si-C network structure have been obtained by laser ablation of poly(dimethylsilane) or hexaphenyldisilane (see Chart 9.9). With blends of these two compounds, films of increased hardness were obtained [67, 68].

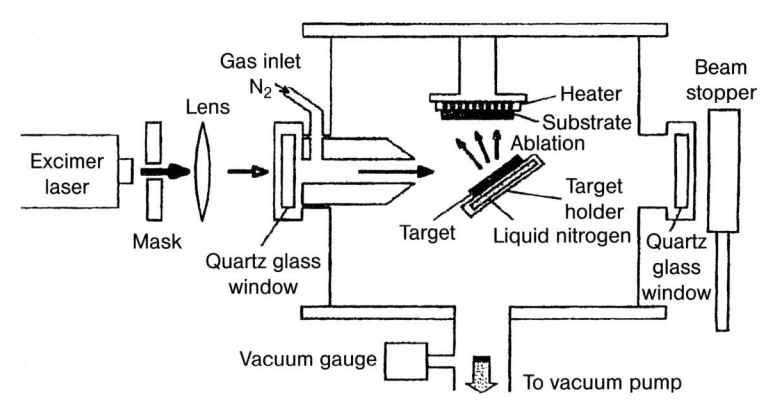

Fig. 9.8 Schematic representation of a set-up used for film deposition with the aid of laser ablation. Adapted from Nishio et al. [62] with permission from the Editorial Office of J. Photopol. Sci. Technol.

Chart 9.9 Chemical structures of poly(dimethylsilane), left, and hexaphenyldisilane, right.

3,4,9,10-Perylenetetracarboxylic dianhydride, PTCDA, has been used as a target material for the generation of films consisting essentially of polyperinaphthalene (see Chart 9.10) [62, 69]. Films annealed at 350 °C immediately after deposition possessed an electrical conductivity of 10^{-3} S cm^{-1}.

Proteins such as collagen (see Chart 9.11), keratin, and fibroin have also been successfully employed as target polymers in the generation of films [70]. The primary structure of the target protein is retained in the deposited film, as was inferred from IR spectroscopic analysis. Interestingly, relevant research led to the application of lasers for medical purposes. Nowadays, excimer laser beams are frequently employed by ophthalmologists for the purpose of *keratectomy*, i.e.

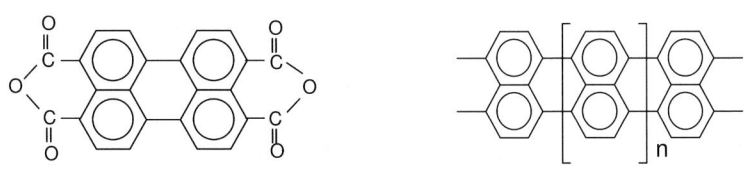

Chart 9.10 Chemical structures of 3,4,9,10-perylenetetracarboxylic dianhydride, left, and polyperinaphthalene, right.

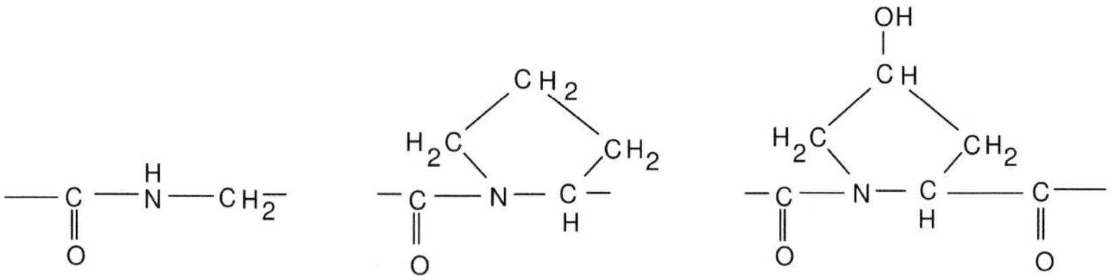

Chart 9.11 Chemical structures of base units contained in collagen.

cornea reprofiling and sculpting. As a matter of fact, a large portion of the cornea consists of a collagen hydrogel.

9.2.5
Laser desorption mass spectrometry and matrix-assisted laser desorption/ionization (MALDI)

Laser beam ablation in conjunction with mass spectrometry has contributed greatly to the progress in polymer analysis made in recent years [71]. *Laser desorption mass spectrometry* (LDMS) refers to the irradiation of a polymer surface with a high-power laser beam and the subsequent mass analysis of the ablated species. For this purpose, the ablated species are ionized by irradiation with another laser beam or with an electron beam. Typical LDMS work pertains to the characterization of polyamide-6,6 [72] and perfluorinated polyethers [73], and to the detection of additives in polymers [73, 74]. A particular kind of LDMS, called *matrix-assisted laser desorption/ionization (MALDI)*, has recently become quite important [75–77]. The development of the analysis of proteins by means of MALDI has been recognized by the award of the Nobel prize for chemistry to K. Tanaka in 2002. MALDI is characterized by specific sample preparation techniques and low fluences in order to create the analyte ions. Fundamentally, the analyte is embedded within a solid matrix in a molecularly dispersed state by placing a droplet of a solution containing analyte and matrix compound on a substrate and subsequently vaporizing the solvent. Alternatively, a layered target may be formed by casting solutions of both analyte and matrix on a substrate. This target is then placed in the source of a mass spectrometer, and the ablation of both matrix and analyte molecules is induced by irradiation with a laser beam (usually at $\lambda = 337$ nm, at which the matrix absorbs the laser light). The ablated neutral analyte molecules are cationized in the gas phase by reaction with protons (e.g., analytes bearing amine functions) or metal cations (e.g., oxygen-containing analytes react with Na^+; unsaturated hydrocarbons react with Ag^+). The resulting ions are extracted into the mass spectrometer for mass analysis. Most of the matrix materials used in polymer MALDI are aromatic organic acids that can readily supply protons, such as 2,5-dihydroxybenzoic acid, α-cyano-hydroxycinnamic acid, ferulic acid, indole acrylic acid, or *trans*-retinoic acid. If metal cationization is required, the source of the appropriate metal must be

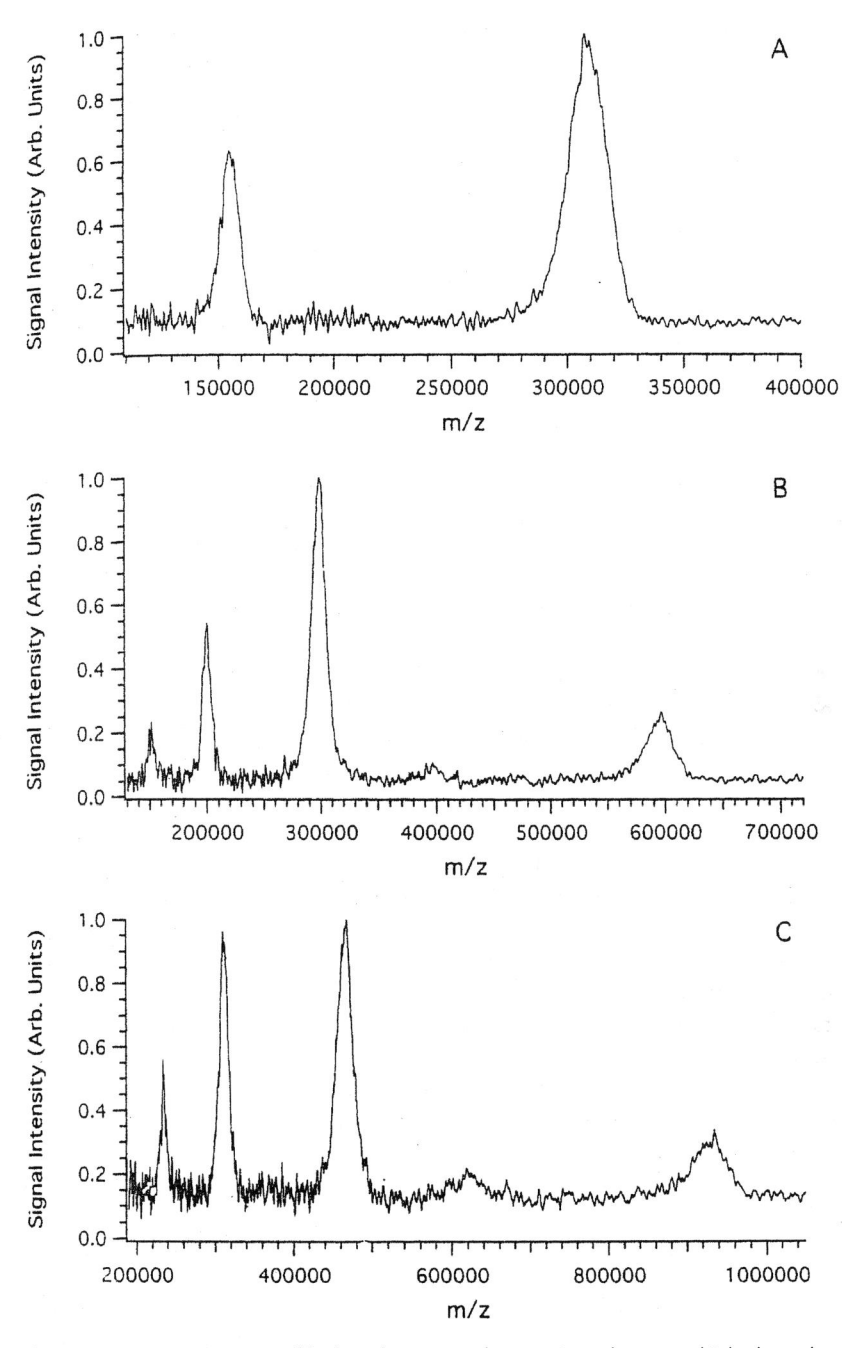

Fig. 9.9 MALDI mass spectra of high molar mass polystyrene samples with nominal molar masses of 3.1×10^5 (A), 6.0×10^5 (B), and 9.3×10^5 (C). The peaks at lower mass-to-charge ratios relate to multiply-charged ions. Adapted from Schriemer et al. [78] with permission from the American Chemical Society.

contained within the matrix. The mechanisms of ionization in MALDI are not yet well understood. In many cases, cations are likely to form rather stable complexes with ablated analyte molecules in the gas phase.

An outstanding quality of polymer MALDI is that it offers the possibility of measuring molar masses. Very accurate values can be obtained for oligomers with molar masses up to several thousand g mol^{-1}, but the determination of much higher molar masses is difficult. Nevertheless, the successful analysis of a polystyrene sample of molar mass 1.5×10^6 g mol^{-1} has been claimed [78]. Typical MALDI mass spectra of high molar mass polystyrene samples are shown in Fig. 9.9.

For more detailed information concerning this interesting field, the reader is referred to relevant literature reviews [79–82].

9.2.6
Generation of periodic nanostructures in polymer surfaces

The possibility of generating periodic sub-100 nm line structures in polyimide by direct laser ablation was demonstrated as long ago as 1992 [83]. Structures with a period of 167 nm and line widths varying from 30 to 100 nm were produced by 248 nm laser irradiation by means of an interferometric technique. The polyimide film was exposed to 500–800 laser shots at a pulse fluence ranging from 34 to 58 mJ cm^{-2}. Work of this kind is important because of possible applications in the fabrication of optical microdevices such as high-speed photonic switches or gratings for coupling light into waveguides. Actually, *grating couplers* can be easily produced by laser ablation at any position of the waveguide, which implies good prospects for employment in the industrial fabrication of waveguides. This aspect has been outlined in work concerning the generation of periodic nanostructures in PDA-C$_4$UC$_4$, a polydiacetylene (for the chemical structure, see Table 3.5), by UV laser pulses (248 nm, 130 fs) [84, 85].

9.2.7
Laser plasma thrusters

A potential application of polymer laser ablation concerns the propulsion of small satellites (1–10 kg) used in space science [86]. Laser plasma thrusters, LPTs, operating with small, powerful diode lasers emitting in the near-infrared wavelength range (930–980 nm) have been proposed. Polymers intended to serve as fuel for a thruster are required to possess a large momentum coupling coefficient, C_m, defined by Eq. (9-3):

$$c_m = \frac{m\Delta v}{W} \quad (9\text{-}3)$$

Here, $m\Delta v$ is the target momentum of the laser-ejected material, and W is the energy absorbed by the polymer per laser pulse. The triazene polymer (TC polymer), dealt with in Section 9.2.3, doped with carbon, seems to be a promising fuel candidate for application in LPTs for microsatellites. This was concluded on the basis of a high absorption coefficient at 930 nm, a large C_m value, a low threshold fluence, and a high ablation rate [59].

9.3
Stabilization of commercial polymers

9.3.1
Introductory remarks

No polymer is capable of withstanding prolonged exposure to solar radiation. Therefore, commercial polymers are stabilized with small amounts of additives denoted as *light stabilizers*. Research and development concerning light stabilizers dates back to the time when polymers first became constructive materials and industrial companies started to fabricate a plethora of plastic items. Actually, the development of efficient light stabilizers has been a critical factor in relation to the growth of the plastics industry. Mechanistic aspects regarding the photodegradation of polymers are outlined in Chapter 7, where it is shown that the absorption of a photon by a chromophoric group generates an electronically excited state and that the latter can undergo various deactivation modes. Commonly, chemical deactivation results in the formation of free radicals, which are reactive and attack intact molecules. Extremely important in this context are reactions involving molecular oxygen. The aims of the strategies that are currently employed to stabilize commercial polymers are to interfere with the absorption of light, with the deactivation of excited states, and with the reactions of free radicals. Therefore, stabilizers may be divided into three classes: *UV absorbers*, *energy quenchers*, and *radical scavengers*. It should be noted, however, that a stabilizer molecule may protect a polymer by more than one mechanism. Radical scavengers are commonly denoted as *chain terminators*, *chain breakers*, or *antioxidants*.

Screening is the most obvious and historically most familiar method of protection. Surface painting, which serves as a means of protection for many materials, is not applicable for most plastics because of incompatibility problems. However, intrinsic screening is widely applied. It is based on the addition of effective light absorbers, denoted as *pigments*, i.e. hyperfinely dispersed compounds with extinction coefficients that significantly exceed those of the polymers. Most prominent in this context is carbon black. Other pigments and fillers of industrial importance include ZnO, MgO, $CaCO_3$, $BaSO_4$, and Fe_2O_3. Light stabilizers for commercial polymers are required to be physically compatible with the polymers. They should not readily be transformed into reactive species. Moreover, they should not alter the mechanical or other physical properties of the polymer before, during, or after exposure to light. For instance, they should be resistant to discoloration. The different classes of light stabilizers are

9.3.2
UV absorbers

9.3.2.1 Phenolic and non-phenolic UV absorbers

UV absorbers (UVAs) are colorless compounds having high absorption coefficients in the UV part of the terrestrial solar spectrum. They transform the absorbed radiation energy into harmless thermal energy by way of photophysical processes involving the ground state and the excited state of the molecule. Typical UVAs are listed in Tables 9.7 and 9.8.

Effective UVAs are required to have absorption maxima lying between 300 and 380 nm, preferably between 330 and 350 nm, and an inherent photostability. Various UVAs, including derivatives of benzotriazoles, 1,3,5-triazines, and oxanilides, fulfil these requirements and are, therefore, widely applied in coatings [87].

Table 9.7 Typical phenolic UV absorbers capable of forming an intramolecular hydrogen bond [87, 107].

Denotation	Chemical structure
o-Hydroxybenzophenones	R_1: H, alkyl; R_2: H, alkyl, phenyl; R_3: H, alkyl; R_4: H, butyl. R: CH_3, C_8H_{17}, $C_{12}H_{25}$
2-(2-Hydroxyphenyl)benzotriazoles	R_1: H, CH_3, C_4H_9 etc. R_2: CH_3, C_4H_9 etc.
2-(2-Hydroxyphenyl)-1,3,5-triazines	R: C_6H_{13}; C_8H_{17}; $CH_2CH(OH)CH_2OC_4H_9$ etc.
Phenyl salicylates	

Table 9.8 Typical non-phenolic UV absorbers [87].

Denotation	Chemical structure
Cyanoacrylates	RO—⌬—CH=C(CN)(C(=O)—O—CH₃); (Ph)₂C=C(CN)(C(=O)—O—R)
Oxanilides	structure with $R_1: C_2H_5$, $R_2: C_{12}H_{25}$

9.3.2.2 Mechanistic aspects

Efficient phenolic UVAs are characterized by a planar structure and a capacity to form intramolecular hydrogen bonds, i.e. O···H···O or O···H···N bridges, which allow intramolecular proton tunneling in the excited state. The process, referred to in the literature as *excited-state intramolecular proton transfer* (ESIPT), is illustrated in Scheme 9.8.

The formation of the tautomeric form S_1' by proton tunneling proceeds with a rate constant of about 10^{11} s^{-1}. The subsequent processes, namely dissipation of energy by internal conversion (IC) to the ground state S_0' of the tautomeric form and regeneration of the original ground state S_0 by reverse proton transfer (RPT) are complete within 40 ps. Mechanisms based on intramolecular H-tunneling have been proposed for benzotriazoles and 1,3,5-triazines, as well as for (non-phenolic) oxanilides (see Scheme 9.9).

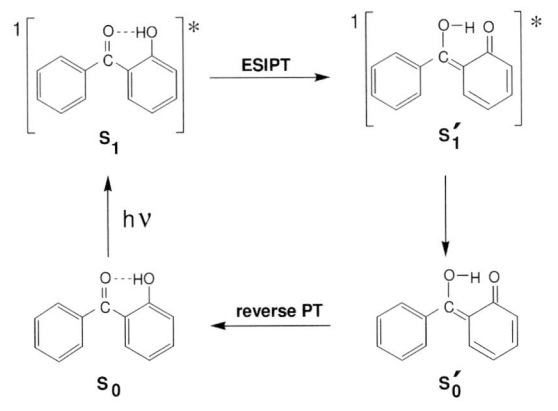

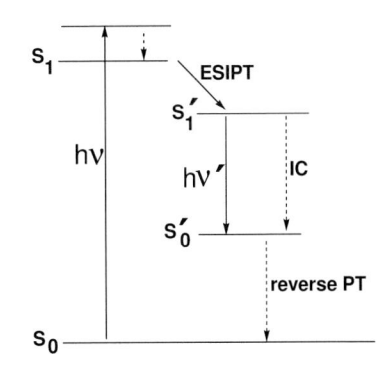

Scheme 9.8 Excited-state intramolecular proton transfer (ESIPT) in the case of 2-hydroxybenzophenone.

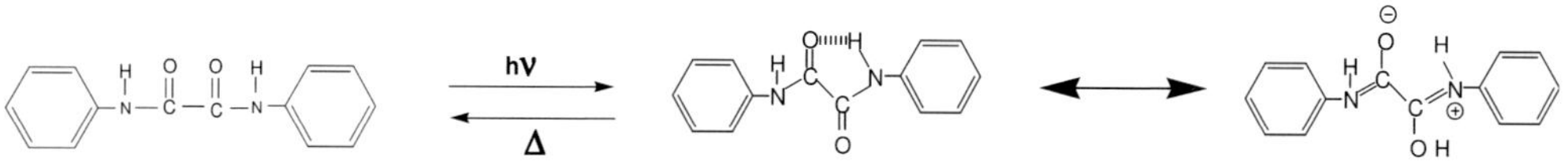

Scheme 9.9 Excited-state intramolecular proton transfer (ESIPT) in the case of oxanilides.

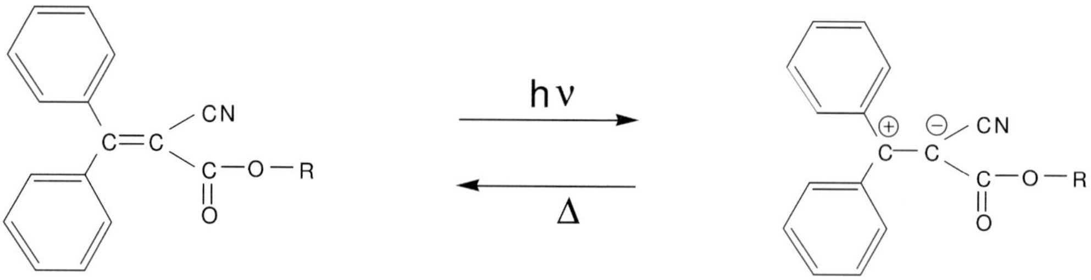

Scheme 9.10 Light-induced intramolecular charge separation in the excited state in cyanoacrylates.

A mechanism involving intramolecular charge separation after photoexcitation serves to explain the UVA properties of (non-phenolic) cyanoacrylates (see Scheme 9.10).

9.3.3
Energy quenchers

Energy quenchers accept energy from excited chromophores tethered to polymers and thus prevent harmful chemical transformations. Commonly, the generally undesired chemical deactivation of the excited chromophore through bond rupture (e.g., via Norrish type I and II processes) or rearrangements (e.g., via the photo-Fries rearrangement) and energy transfer to the quencher are competing processes (see Scheme 9.11):

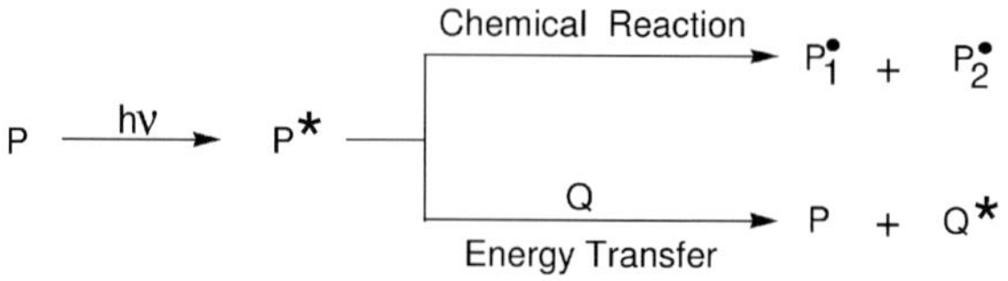

Scheme 9.11 Schematic illustration of the action of energy quenchers.

Therefore, the photodegradation of polymers cannot be completely suppressed by energy quenchers. Energy transfer from P* to Q is possible, if the energy level of the excited state of the chromophore is higher than that of the quencher. Excited quencher molecules are deactivated to the ground state by emission of light or dissipation of thermal energy (see Scheme 9.12).

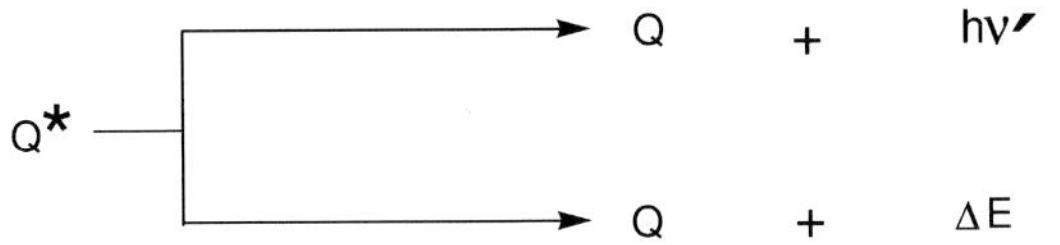

9.3 Stabilization of commercial polymers

Scheme 9.12 Schematic illustration of the deactivation of excited quencher molecules.

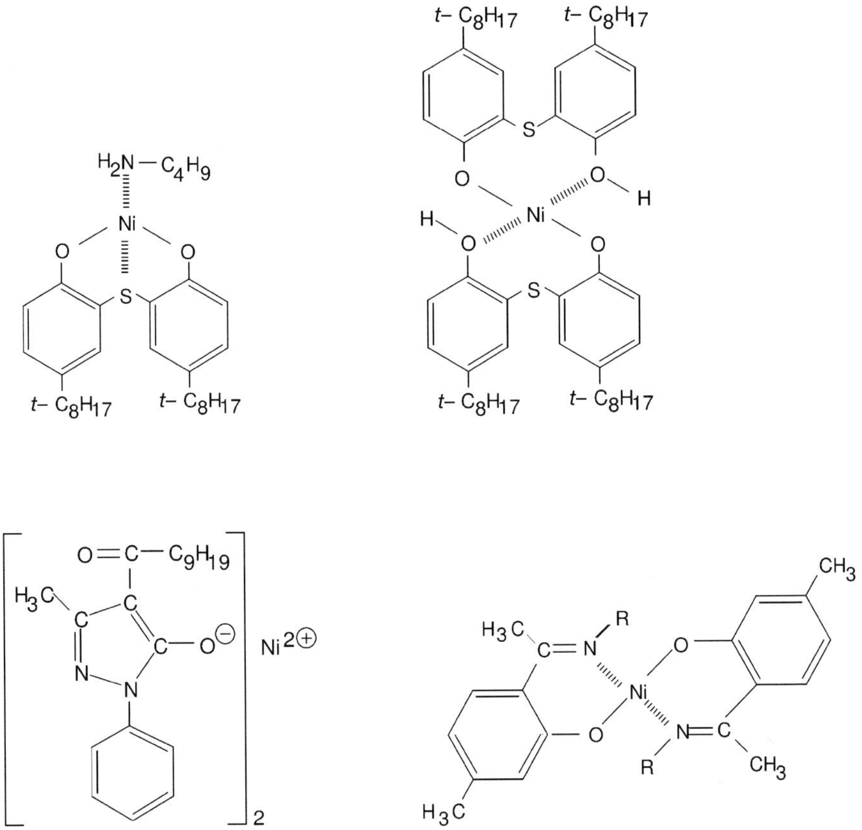

Chart 9.12 Chemical structures of typical nickel chelates used as quenchers in polyalkenes [93].

The importance of quenchers derives mainly from their ability to interact with excited carbonyl groups, which are present in many thermoplastics, especially in polyalkenes. Commercially available energy quenchers include complexes and chelates of transition metals, such as those shown in Chart 9.12.

It may be the case that energy quenchers also act as UVAs, i.e. that they also protect the polymer by light absorption.

9.3.4
Chain terminators (radical scavengers)

Chain terminators interrupt the propagation of the oxidative chain reaction [reactions (a) and (b) in Scheme 9.13; see also Scheme 7.18] and thus prevent deterioration of the mechanical properties of polymers.

The chain propagation would be totally prevented if all macroradicals P• generated during the initiation stage were scavenged according to reaction (c). However, reaction (a) proceeds at a relatively large rate, even at ambient temperature and low O_2 pressure. Therefore, in practically relevant situations, the concentration of P• will be much lower than that of POO• [99]. Consequently, an effective chain terminator is required to react rapidly with POO• (reaction (d)) and the products of this reaction must be inert towards the polymer. Hindered amines based on the 2,2′,6,6′-tetramethylpiperidine (TMP) structure (see Chart 9.13) satisfactorily fulfil these requirements, especially in the case of polyalkenes. In the literature, they are referred to as *hindered amine stabilizers* (HASs) or frequently also as *hindered amine light stabilizers* (HALSs). The stabilizing power of a typical HAS is demonstrated by the results shown in Fig. 9.10.

Hindered amine stabilizers are transparent to visible and terrestrial UV light (300–400 nm). In polymeric matrices, they are oxidized in a sacrificial reaction by way of a not yet fully understood mechanism to stable nitroxyl (aminoxyl) radicals >N–O•. A mechanism based on the reaction of HASs with alkyl hydroperoxides and alkyl peroxyl radicals is presented in Scheme 9.14 [87].

A mechanism involving charge-transfer complexes formed by HAS, polymer, O_2, and ROO•, i.e. [HAS···O_2], [polymer···O_2], [HAS···ROO•], has been proposed [111]. It is considered to contribute in the early stages of the hindered amine stabilization mechanism [87]. The oxidation of TMP derivatives as illustrated in Scheme 9.15 commences when the polymer is processed. It continues later, when the polymer is exposed to light.

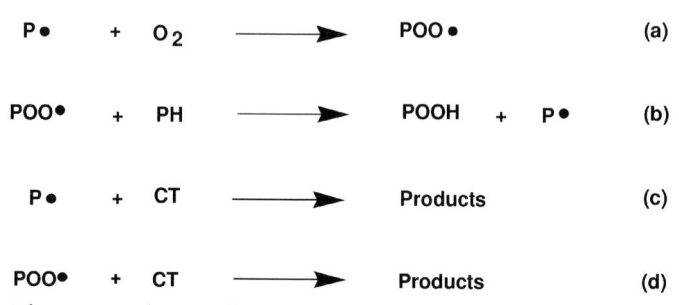

P•	+	O_2	⟶	POO•		(a)
POO•	+	PH	⟶	POOH + P•		(b)
P•	+	CT	⟶	Products		(c)
POO•	+	CT	⟶	Products		(d)

Scheme 9.13 Schematic illustration of elementary reactions occurring in a polymeric matrix containing O_2 and a radical scavenger (chain terminator, CT).

Chart 9.13 Chemical structures of typical commercial hindered amine stabilizers [109].

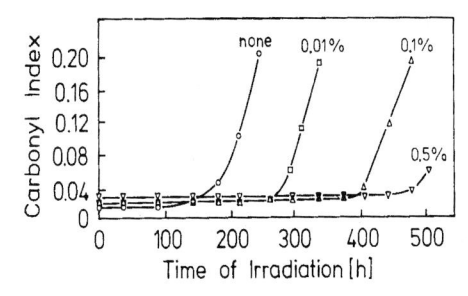

Fig. 9.10 Photooxidation of a commercial polypropylene in the absence and presence of a typical HAS (for chemical structure, see Chart 9.13, uppermost). Adapted from Schnabel [110] with permission from Carl Hanser.

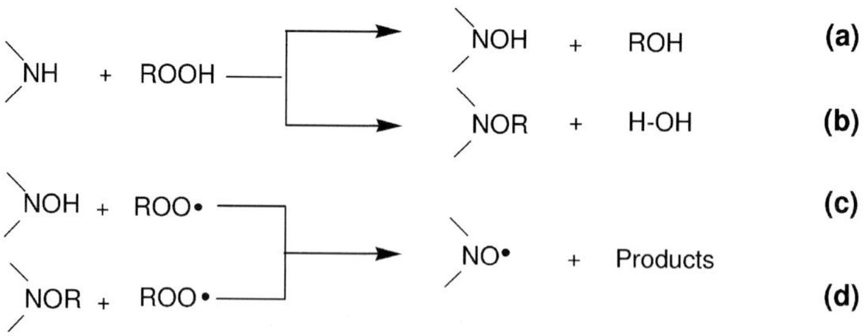

Scheme 9.14 Schematic illustration of the oxidation of hindered amine stabilizers by alkyl hydroperoxides and alkyl peroxyl radicals [87].

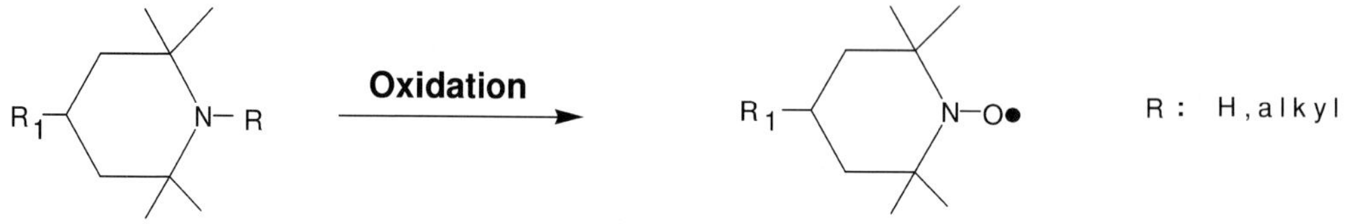

Scheme 9.15 Oxidation of a 2,2′,6,6′-tetramethylpiperidine (TMP) derivative to the corresponding nitroxyl radical, i.e. the piperidinoxyl radical TMPO.

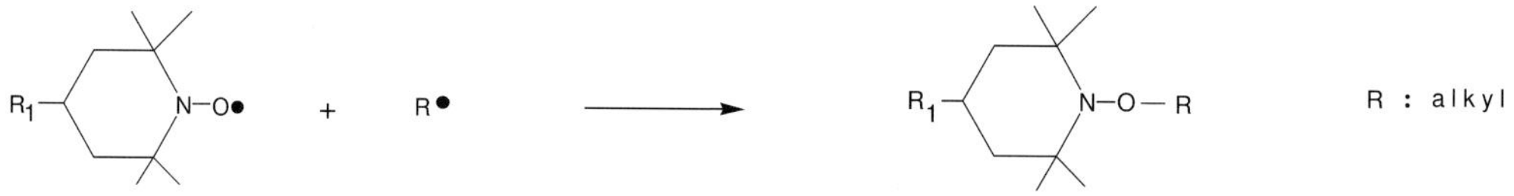

Scheme 9.16 Formation of amino ethers by the reaction of TMPO with alkyl radicals.

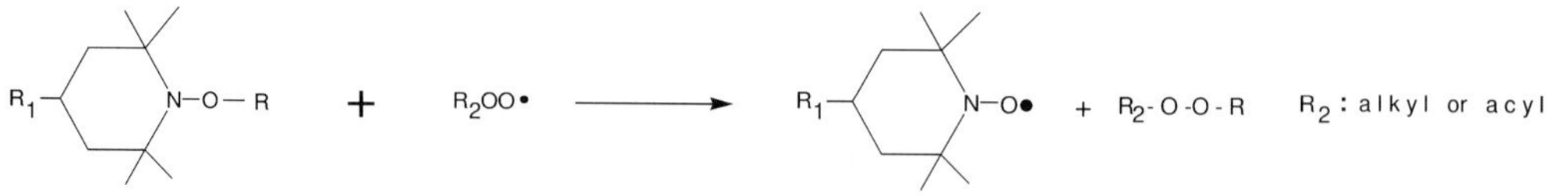

Scheme 9.17 Regeneration of nitroxyl radicals by the reaction of amino ethers with alkyl peroxyl or acyl peroxyl radicals.

The reaction of TMPO with alkyl radicals yields amino ethers, as illustrated in Scheme 9.16.

Amino ethers are capable of reacting with peroxyl radicals, thereby regenerating nitroxyl radicals. This is considered to be the reason for the high stabilizer efficiency of many hindered amines (see Scheme 9.17).

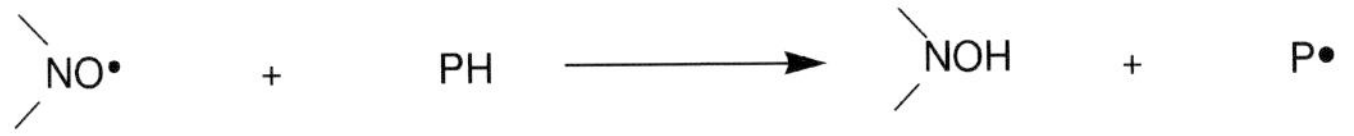

$$\text{\textbackslash}NO^\bullet + PH \longrightarrow \text{\textbackslash}NOH + P^\bullet$$

Scheme 9.18 Reaction of nitroxyl radicals with polymers.

Besides the beneficial role that nitroxyl radicals play in the stabilization of polyalkenes, hydrogen abstraction according to Scheme 9.18 may have an adverse effect [87].

The macroradicals $P^\bullet$ generated in this process can initiate oxidative chain reactions and thus reduce the stabilizing power of hindered amines.

9.3.5
Hydroperoxide decomposers

Besides hindered amines (see Section 9.3.4), there are compounds that are capable of functioning as long-term hydroperoxide decomposers. These include alkyl and aryl phosphites, and organosulfur compounds such as dialkyl dithiocarbamates, dithiophosphates, and dithioalkyl propionates (see Chart 9.14).

These compounds are commonly used to stabilize thermoplastic polymers during processing in the melt at temperatures up to 300 °C. Their contribution to the long-term stabilization of polymers at ambient temperatures is small but not negligible. Phosphite stabilizers destroy hydroperoxides stoichiometrically in a sacrificial process, as shown in Scheme 9.19.

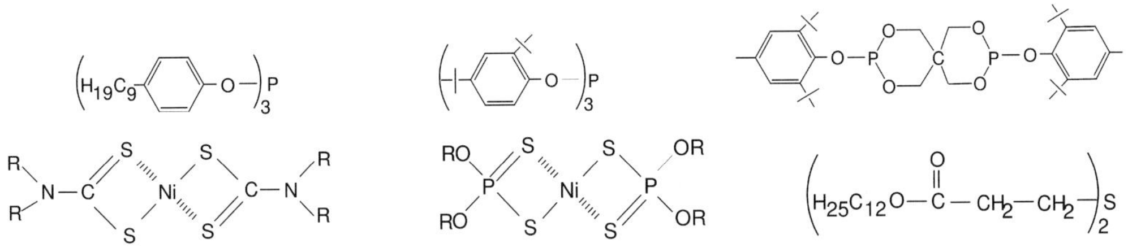

Chart 9.14 Chemical structures of hydroperoxide decomposers [93, 94].

$$(RO)_3P + ROOH \longrightarrow (RO)_3P=O + ROH$$

Scheme 9.19 Reaction of phosphites with hydroperoxides.

$$\left(R_1-O-\overset{O}{\overset{\|}{C}}-CH_2-CH_2\right)_{\!2}\!S + ROOH \longrightarrow \left(R_1-O-\overset{O}{\overset{\|}{C}}-CH_2-CH_2\right)_{\!2}\!S=O + ROH$$

$$\left(R_1-O-\overset{O}{\overset{\|}{C}}-CH_2-CH_2\right)_{\!2}\!S=O + ROOH \longrightarrow \left(R_1-O-\overset{O}{\overset{\|}{C}}-CH_2-CH_2\right)_{\!2}\!S\overset{O}{\underset{O}{\diagdown\!\!\!\diagup}} + ROH$$

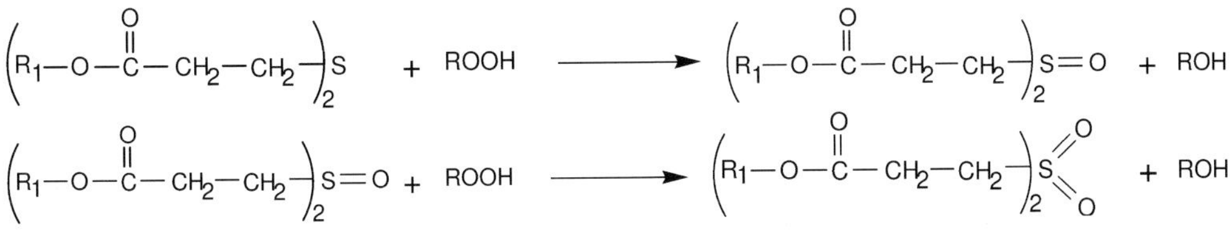

Scheme 9.20 Reaction of dialkyl dithiopropionates with hydroperoxides.

Metal dialkyl dithiocarbamates are oxidized to sulfur acids, which act as ionic catalysts for the non-radical decomposition of hydroperoxides. When the metal is nickel or another transition metal, they also function as UVAs. Dialkyl dithiopropionates are oxidized by hydroperoxides, as shown in Scheme 9.20.

9.3.6
Stabilizer packages and synergism

Frequently, different classes of light stabilizers are combined to optimize stabilizing efficiency [112]. For example, UVAs and HALSs, used in combination, often provide better photostability than either class alone. Light stabilizers are also used in combination with additives that protect the polymers against thermal degradation during processing, such as hindered phenols and phosphates [113]. Consequently, various bifunctional and trifunctional photostabilizers have been synthesized and some have been selected for use in commercial applications (see Chart 9.15).

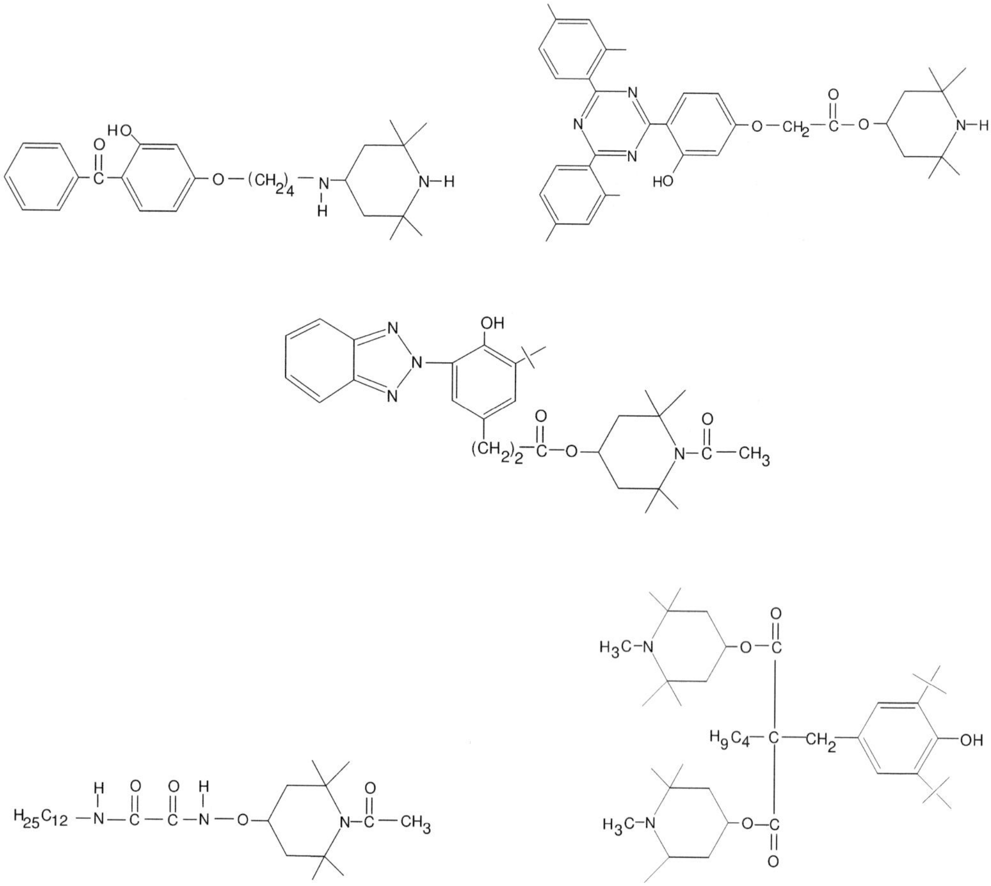

Chart 9.15 Chemical structures of typical bifunctional stabilizers [87].

In the context of multifunctionality, carbon black, a polycrystalline material, merits special mention. The surface layer of carbon black particles may contain quinones, phenols, carboxy phenols, lactones, etc. Therefore, apart from being a powerful UV absorber and a quencher of excited states (such as those of carbonyl groups), carbon black acts as a scavenger of free radicals in chain-breaking reactions and as a hydroperoxide decomposer [114, 115]. In polyethylene, carbon black forms a complex with macroradicals [115].

9.3.7
Sacrificial consumption and depletion of stabilizers

All polymer systems eventually undergo a loss in durability during long-term outdoor application. However, the presence of stabilizers at concentrations between 0.25 and 3.0% provides for longevity. The ultimate outdoor lifetime of polymer articles, such as coatings, is determined by the sacrificial consumption and/or depletion of the stabilizers. During outdoor application, the concentration of the active form of the stabilizer is continually reduced and eventually reaches a level below the critical protection value determining the ultimate lifetime of polymer coatings. The term *sacrificial consumption* refers to the chemical alterations that stabilizer molecules undergo in protecting the polymer matrix. Stabilizer molecules are also consumed by direct or sensitized photolysis (e.g., by the attack by free radicals), photooxidation, reactions with atmospheric pollutants, etc., processes that are covered by the term *depletion*. Stabilizer depletion can also be caused by physical loss, i.e. by migration of the stabilizer molecules. This relates, for example, to coatings, in which stabilizer molecules may migrate from the clearcoat to the basecoat or plastic substrate. These problems may be alleviated by the use of physically persistent stabilizers. High molar mass stabilizers ($M > 500$ g mol^{-1}), including oligomers with appropriate molecular structures ($M = 3500–5000$ g mol^{-1}), are sufficiently physically persistent and do not evaporate at the elevated temperatures of curing [87].

In the case of UV absorbers forming intramolecular hydrogen bonds, the loss of stabilizer efficiency may be due to the interruption of intramolecular hydrogen bonds and the formation of intermolecular hydrogen bonds with H-acceptors (carbonyl groups) generated by photooxidation of the polymer matrix. Thus, the

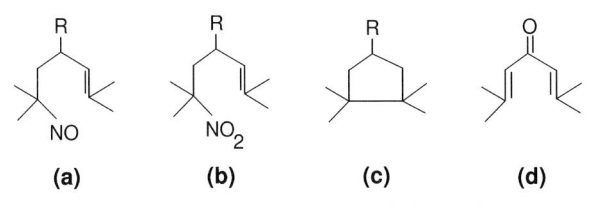

Chart 9.16 Nitroso (a) and nitro compounds (b), and nitrogen-free compounds ((c) and (d)) formed during the photolysis of hindered amines [117].

ESIPT mechanism (see Section 9.3.2.2) can no longer be repeated. Regarding hindered amine stabilizers, depletion is caused by the reaction of acyl radicals, stemming from Norrish reactions, with nitroxyl radicals. Nitroso and nitro compounds (see Chart 9.16) are formed when nitroxyl radicals are photolyzed [117].

References

1 Semiconductor Industry Association, *The International Roadmap for Semiconductors*, International SEMATECH, Austin TX (2003). See also: P. M. Zeitzoff, Solid State Technol. (2004), January, p. 30, and P. Gargini, R. Doering, Solid State Technol. (2004), January, p. 72.

2 M. Rothschild, *Projection Optical Lithography*, materials today 8 (2005) 18.

3 R. Menon, A. Patel, D. Gil, H. I. Smith, *Maskless Lithography*, materials today 8 (2005) 26.

4 (a) D. J. Resnick, S. V. Sreenivasan, C. G. Willson, *Step & Flash Imprint Lithography*, materials today 8 (2005) 34. (b) S. V. Sreenivasan, C. G. Willson, N. E. Schumaker, D. J. Resnick, Proc. SPIE 4688 (2002) 903.

5 M. Switkes, M. Rothschild, J. Vac. Sci. Technol. B 19 (2001) 2353.

6 B. J. Lin, Proc. SPIE 4688 (2002) 11.

7 (a) S. Owa, H. Nagasaka, J. Microlith. Microfabr. Microsyst. 3 (2004) 97. (b) J. Mulkens, P. Graupner, Proc. 2nd Symposium on Immersion Lithography, Bruges, Belgium (2005).

8 U. Stamm, J. Phys. D: Appl. Phys. 37 (2004) 3244.

9 A. Egbert, A. Ostendorf, B. N. Chichkov, T. Missala, M. C. Schürman, K. Gäbel, G. Schriever, U. Stamm, Lambda Highlights 61 (2002) 6.

10 M. D. Levenson, Microlithogr. World 6 (1992) 6.

11 H. Ito, *Functional Polymers for Microlithography: Nonamplified Imaging Systems*, Chapter 2.3 in R. Arshady (ed.), *Desk Reference of Functional Polymers. Syntheses and Applications*, American Chemical Society, Washington, D.C. (1997).

12 H. Ito, *Functional Polymers for Microlithography: Amplified Imaging Systems*, Chapter 2.4 in R. Arshady (ed.), *Desk Reference of Functional Polymers. Syntheses and Applications*, American Chemical Society, Washington, D.C. (1997).

13 N. P. Hacker, *Photoresists and Their Development*, in V. V. Krongauz, A. D. Trifunac (eds.), *Processes in Photoreactive Polymers*, Chapman & Hall, New York (1995), p. 368.

14 M. Madou, *Fundamentals of Microfabrication*, CRC Press, Boca Raton, FL, USA (1997).

15 A. Reiser, *Photoreactive Polymers. The Science and Technology of Resists*, Wiley, New York (1989).

16 C. G. Willson, *Organic Resist Materials – Theory and Chemistry*, in L. F. Thompson, C. G. Willson, M. J. Bowden (eds.), *Introduction to Microlithography*, American Chemical Society, Washington, D.C., 1st ed. (1983), p. 87; 2nd ed. (1994), p. 139.

17 W. S. DeForest, *Photoresist Materials and Processes*, McGraw-Hill, New York (1975).

18 L. E. Bogan, *Novolak Resins for Microlithography*, in J. C. Salamone (ed.), *Concise Polymeric Materials Encyclopedia*, CRC Press, Boca Raton, FL, USA (1999).

19 J. N. Helbert (ed.), *Handbook of VLSI Microlithography, Principles, Tools, Technology and Application*, 2nd Edition, Noyes Publ., Park Ridge, NJ, USA (2001).

20 J. J. Lai, *Polymer Resists for Integrated Circuit (IC) Fabrication*, in J. J. Lai (ed.), *Polymers for Electronic Applications*, CRC Press, Boca Raton, FL, USA (1989).

21 J. H. Bendig, H.-J. Timpe, *Photostructuring*, in H. Böttcher (ed.), *Technical Applications of Photochemistry*, Deutscher Verlag für Grundstoffindustrie, Leipzig (1991), p. 172.

22 P. Rai-Choudhury, *Handbook of Microlithography, Micromachining and Microfabrication, Vol. 1: Microlithography, Vol. 2: Micromachining and Microfabrication,*

SPIE Optical Engineering Press, Bellingham, Washington (1997).
23 H.J. Levinson, W.H. Arnold, *Optical Lithography*, Chapter 1 in [22].
24 H. Ito, E. Reichmanis, A. Nalamasu, T. Ueno (eds.), *Micro- and Nanopatterning of Polymers*, ACS Symposium Series 706, American Chemical Society, Washington, D.C. (1998).
25 K. Horie, T. Yamashita (eds.), *Photosensitive Polyimides, Fundamentals and Applications*, Technomic, Lancaster, PA, USA (1995).
26 H. Ahne, R. Rubner, *Applications of Polyimides in Electronics*, Chapter 2 in [25].
27 L. Schlegel, W. Schnabel, *Polymers in X-ray, Electron-Beam and Ion-Beam Lithography*, in J.P. Fouassier, J.F. Rabek (eds.), *Radiation Curing in Polymer Science and Technology*, Vol. 1, Elsevier Applied Science, London (1993).
28 H. Ito, G. Willson, Polym. Eng. Sci. 23 (1983) 1012.
29 H. Ito, *Chemical Amplification Resists: Inception, Implementation in Device Manufacture, and New Developments*, J. Polym. Sci.: Part A, Polym. Chem. 41 (2003) 3863.
30 W. Li, P.R. Varanasi, M.C. Lawson, R.W. Kwong, K.J. Chen, H. Ito, H. Truong, R.D. Allen, M. Yamamoto, E. Kobayashi, M. Slezak, Proc. SPIE 5039 (2003) 61.
31 J.-C. Jung, M.-H. Jung, G. Lee, H.-H. Baik, Polymer 42 (2001) 161.
32 H.W. Kim, S.J. Choi, S.G. Woo, J.T. Moon, J. Photopolym. Sci. Technol. 13 (2000) 419.
33 T. Fukuhara, Y. Shibasaki, S. Ando, S. Kishimura, M. Endo, M. Sasago, M. Ueda, Macromolecules 38 (2005) 3041.
34 A.K. Bates, M. Rothschild, T.M. Bloomstein, T.H. Fedynyshyn, R.R. Kunz, V. Liberman, M. Switkes, *Review of Technology for 157 nm Lithography*, IBM J. Res. & Dev. 45 (2001) 605.
35 T.H. Fedynyshyn, R.R. Kunz, R.F. Sinta, M. Sworin, W.A. Mowers, R.B. Goodman, S.P. Doran, Proc. SPIE 4345 (2001) 296.
36 S. Kodama, T. Kaneko, Y. Takebe, S. Okada, Y. Kawaguchi, N. Shida, S. Ishikawa, M. Toriumi, T. Itani, Proc. SPIE 4690 (2002) 76.
37 T. Yamazaki, T. Furukawa, T. Itani, T. Ishikawa, M. Koh, T. Araki, M. Toriumi, T. Kodani, H. Aoyama, T. Yamashita, Proc. SPIE 5039 (2003) 103.
38 H. Ito, G.M. Wallraff, P. Brock, N. Fender, H. Truong, G. Brayta, D.C. Miller, M.H. Sherwood, R.D. Allen, Proc. SPIE 4345 (2001) 273.
39 C. Hohle, M. Sebald, T. Zell, Metalloberfläche 57 (2003) 45.
40 F. Houlihan, R. Sakamuri, A. Romano, D. Rentkiewicz, R.R. Dammel, W. Conley, D. Miller, M. Sebald, N. Stepanenko, K. Markert, U. Mierau, I. Vermeir, C. Hohle, T. Itani, M. Shigematsu, E. Kawaguchi, Proc. SPIE 5376 (2004) 134.
41 Y. Xia, G.M. Whitesides, *Soft Lithography*, Angew. Chem. Int. Ed. 37 (1998) 550.
42 J.A. Rogers, R. Nuzzo, *Recent Progress in Lithography*, materials today 8 (2005) 50.
43 B. Gates, *Nanofabrication with Molds and Stamps*, materials today 8 (2005) 44.
44 A. Maciossek, G. Grützner, F. Reuther, Galvanotechnik 91 (2000) 1988.
45 M. Madou, *Surface Micromachining*, Chapter 5 in [14].
46 R. Srinivasan, V. Mayne-Banton, Appl. Phys. Lett. 41 (1982) 576.
47 Y. Kawamura, K. Toyoda, S. Namba, Appl. Phys. Lett. 40 (1982) 374.
48 H. Masuhara, N.S. Allen (eds.), *Photoablation of Materials*, J. Photochem. Photobiol. A: Chem., Special Issue 145 (2001) 3.
49 S. Georgiou, F. Hillenkamp (eds.), *Laser Ablation of Molecular Substrates*, Chem. Rev., Special Issue, 103, (2003) 2.
50 A. Ostendorf, K. Körber, T. Nether, T. Temme, Lambda Physics Highlights 57 (2000) 1.
51 S. Küper, J. Branon, K. Brannon, Appl. Phys. A 56 (1993) 43.
52 R. Srinivasan, B.J. Braren, K.G. Casey, J. Appl. Phys. 68 (1990) 1842.
53 G. Gorodetsky, T.G. Kazyaka, R.L. Melcher, R. Srinivasan, Appl. Phys. Lett. 46 (1985) 828.
54 H. Schmid, J. Ihlemann, B. Wolff-Rottke, K. Luther, J. Troe, Appl. Phys. A 83 (1998) 5458.

55 A. Vogel, V. Venugopalan, *Mechanisms of Pulsed Laser Ablation of Biological Tissues*, Chem. Rev. 103 (2003) 577.
56 R. Srinivasan, B.J. Braren, R.W. Dreyfus, L. Hadel, D.E. Seeger, J. Opt. Soc. Am. B3 (1986) 785.
57 R. Srinivasan, B.J. Braren, Appl. Phys. A 45 (1988) 289.
57 C.E. Kosmidis, C.D. Skordoulis, Appl. Phys. A 56 (1993) 64.
58 T. Lippert, C. David, J.T. Dickinson, M. Hauser, U. Kogelschatz, S.C. Langford, O. Nuyken, C. Phipps, J. Robert, A. Wokaun, J. Photochem. Photobiol. A: Chem. 145 (2001) 145.
59 J. Hobley, S. Nishio, K. Hatanaka, H. Fukumura, *Laser Transfer of Organic Molecules*, Chapter 5 of Vol. 2, in H.S. Nalwa (ed.), *Handbook of Photochemistry and Photobiology*, American Scientific Publishers, Stevenson Ranch, CA, USA (2003).
60 D.B. Chrisey, A. Piqué, R.A. McGill, J.S. Horwitz, B.R. Ringeisen, D.M. Bubb, P.K. Wu, *Laser Deposition of Polymer and Biomaterial Films*, Chem. Rev. 103 (2003) 533.
62 (a) S. Nishio, Y. Tsujine, A. Matsuzaki, H. Sato, T. Yamabe, J. Photopol. Sci. Technol. 13 (2000) 163. (b) S. Nishio, K. Tamura, Y. Tsujine, T. Fukao, M. Nakano, A. Matsuzaki, H. Sato, T. Yamabe, J. Photochem. Photobiol. A: Chem. 145 (2001) 165.
63 S. Nishio, T. Chiba, A. Matsuzaki, H. Sato, Appl. Surf. Sci. 106 (1996) 132.
64 S. Nishio, T. Chiba, A. Matsuzaki, H. Sato, J. Appl. Phys. 79 (1996) 7198.
65 G.B. Blanchet, S.I. Shah, Appl. Phys. Lett. 62 (1993) 1026.
66 S.G. Hansen, T.E. Robitaille, Appl. Phys. Lett. 52 (1988) 81.
67 M. Suzuki, Y. Nakata, H. Nagai, K. Goto, O. Nishimura, T. Oku, Mat. Sci. Eng. A 246 (1998) 36.
68 M. Suzuki, M. Yamaguchi, L. Ramonat, X. Zeng, J. Photochem. Photobiol. A: Chem. 145 (2001) 223.
69 M. Yudasaka, Y. Tasaka, M. Tanaka, H. Kamo, Y. Ohki, S. Usamai, S. Yoshimura, Appl. Phys. Lett. 64 (1994) 3237.
70 Y. Tsuboi, N. Kimoto, M. Kabeshita, A. Itaya, J. Photochem. Photobiol. A: Chem. 145 (2001) 209.
71 S.D. Hanton, Chem. Rev. 101 (2001) 527.
72 A.C. Cefalas, N. Vassilopoulos, E. Sarantopoulou, K. Kollis, C. Skordoulis, Appl. Phys. A: Mater. Sci. Process. 70 (2000) 21.
73 M.S. de Vries, H.E. Hunziker, Polym. Prepr. 37 (1996) 316.
74 Q. Zhan, R. Zenobi, S.J. Wright, P.R.R. Langridge-Smith, Macromolecules 29 (1996) 7865.
75 K. Tanaka, H. Waki, Y. Ido, S. Akita, Y. Yoshida, T. Yoshida, Rapid Commun. Mass. Spectrom. 2 (1988) 151.
76 M. Karas, F. Hillenkamp, Anal. Chem. 60 (1988) 2299.
77 U. Bahr, A. Deppe, M. Karas, F. Hillenkamp, U. Giessman, Anal. Chem. 64 (1992) 2866.
78 D.C. Schriemer, L. Li, Anal. Chem. 68 (1996) 2721.
79 H.J. Raeder, W. Schrepp, Acta Polym. 49 (1998) 272.
80 M.W.F. Nielsen, Mass Spectrom. Rev. 18 (1999) 309.
81 R. Zenobi, R. Knochenmuss, Mass Spectrom. Rev. 17 (1998) 337.
82 H. Pasch, W. Schrepp, *MALDI-TOF Mass Spectrometry of Synthetic Polymers*, Springer, Heidelberg (2003).
83 H.M. Phillips, D.L. Callahan, R. Sauerbrey, G. Szabo, Z. Bor, Appl. Phys. A 54 (1992) 158.
84 G.I. Stegeman, R. Zanoni, C.T. Seaton, Mater. Res. Soc. Symp. Proc. 109 (1988) 53.
85 J.-H. Klein-Wiele, M.A. Mader, I. Bauer, S. Soria, P. Simon, G. Marowsky, Synth. Metals 127 (2002) 53.
86 C. Phipps, J. Luke, *Diode Laser-Driven Microthrusters: A New Departure for Micropropulsion*, AIAA Journal 40/1, January (2002).
87 J. Pospišil, S. Nêpurek, *Photostabilization of Coatings. Mechanisms and Performance*, Prog. Polym. Sci. 25 (2000) 1261.
88 J. Pospišil, S. Nêpurek, *Highlights in the Inherent Chemical Activity of Polymer Stabilizers*, in H.S. Hamid (ed.), *Handbook of Polymer Degradation*, Dekker, New York (2000), p. 191.

89 J. Pospíšil, *Aromatic and Heteroaromatic Amines in Polymer Degradation*, Adv. Polym. Sci. 124 (1995) 87.

90 J. Pospíšil, *Chain-Breaking Antioxidants in Polymer Stabilization*, in G. Scott (ed.), *Developments in Polymer Stabilization*, Applied Science Publishers, London (1979), p. 1.

91 J. Pospíšil, P. P. Klemchuk (eds.), *Oxidation Inhibition in Organic Materials*, Vols. I and II, CRC Press, Boca Raton, FL, USA (1990).

92 H. Zweifel, *Plastics Additives Handbook*, 5th Edition, Hanser, München (2001).

93 H. Zweifel, *Stabilization of Polymeric Materials*, Springer, Berlin (1998).

94 G. Scott, *Polymers and the Environment*, Royal Society of Chemistry, Cambridge (1999).

95 G. Scott (ed.), *Atmospheric Oxidation and Antioxidants*, Elsevier, London (1993).

96 G. Scott, *Mechanisms of Polymer Degradation and Stabilisation*, Elsevier Applied Science, London (1990).

97 A. Valet, *Stabilization of Paints*, Vincentz, Hannover (2000).

98 V. Ya. Shlyapintokh, *Photochemical Conversion and Stabilization of Polymers*, Hanser, München (1984).

99 Yu. A. Shlyapnikov, S. G. Kiryushkin, A. P. Marin, *Antioxidative Stabilization of Polymers*, Taylor & Francis, London (1996).

100 W. W. Y. Lau, P. J. Qing, *Polymeric Stabilizers and Antioxidants*, Chapter 4 in R. Arshady, *Desk Reference of Functional Polymers. Syntheses and Applications*, American Chemical Society, Washington, D.C. (1997).

101 N. Grassie, G. Scott, *Polymer Degradation and Stabilisation*, Cambridge University Press, Cambridge (1985).

102 R. Gächter, H. Müller (eds.), *Plastics Additives*, 3rd Edition, Hanser, München (1990).

103 J. F. Rabek, *Photostabilization of Polymers*, Elsevier Applied Science, London (1990).

104 B. Rånby, J. F. Rabek, *Photodegradation, Photooxidation and Photostabilization of Polymers*, Wiley, London (1975).

105 S. Al-Malaika, A. Golovoy, C. A. Wilkie (eds.), *Specialty Polymer Additives*, Blackwell, Oxford (2001).

106 F. Gugumus, *The Many-sided Effects of Stabilizer Mass on UV Stability of Polyolefins*, Chapter 9 in [105].

107 R. E. Lee, C. Neri, V. Malatesta, R. M. Riva, M. Angaroni, *A New Family of Benzotriazoles: How to Modulate Properties within the Same Technology*, Chapter 7 in [105].

108 C. Decker, *Photostabilization of UV-Cured Coatings and Thermosets*, Chapter 8 in [105].

109 J. Sedlář, *Hindered Amines as Photostabilizers*, Chapter 1 of Vol. II in [91].

110 W. Schnabel, *Polymer Degradation, Principles and Practical Applications*, Hanser, München (1981).

111 F. Gugumus, Polym. Degrad. Stab. 40 (1993) 167.

112 S. Yachigo, *Synergistic Stabilization of Polymers*, in S. H. Hamid, M. B. Amin, A. G. Maadhah (eds.), *Handbook of Polymer Degradation*, Dekker, New York (1992), p. 305.

113 J. P. Galbo, *Light Stabilizers (Overview)*, in J. C. Salamone (ed.), *Concise Polymeric Materials Encyclopedia*, CRC Press, Boca Raton, FL, USA (1999), p. 749.

114 N. S. Allen, J. M. Pena, M. Edge, C. M. Liauw, Polym. Degrad. Stab. 67 (2000) 563.

115 J. M. Pena, N. S. Allen, M. Edge, C. M. Liauw, I. Roberts, B. Valange, Polym. Degrad. Stab. 70 (2000) 437.

116 E. C. D. Nunes, A. C. Babetto, J. A. M. Agnelli, Polim. Cienc. Tecnol., April/June (1997) 66.

117 D. M. Wiles, J. P. T. Jensen, D. J. Carlson, Pure Appl. Chem. 55 (1983) 165.

Part III
Light-induced synthesis of polymers

10
Photopolymerization

10.1
Introduction

While the previous chapters have demonstrated how light can affect the physical behavior of polymers and chemically modify or degrade them, this chapter shows how light can be used as a tool to make polymers. In other words, various kinds of polymers can be synthesized by light-induced chemical processes, a technique commonly denoted by the term *photopolymerization*. In accordance with the widely accepted terminology, *polymerization* denotes a chain reaction (chain polymerization), and consequently *photopolymerization* refers to the synthesis of polymers by chain reactions that are initiated upon the absorption of light by a polymerizable system. Notably, light serves only as an initiating tool. It does not interfere with the propagation and termination stages of the chain process. Both radical and ionic chain polymerizations can be photoinitiated, provided that appropriate initiators and monomers are employed. It is common practice to add small amounts of *photoinitiators* to formulations to be polymerized. Photoinitiators are compounds that are thermally stable and capable of absorbing light with relatively high absorption coefficients in the UV and/or visible wavelength ranges. Industrially employed photopolymerization processes overwhelmingly rely on the use of easily available UV light sources emitting in the 300–400 nm wavelength range. Actually, many highly efficient UV photoinitiators, which are stable in the dark, are commercially available. The handling of UV-sensitive systems is easy and does not require special precautions such as safety light conditions, which are mandatory for the application of systems sensitive to visible light. In many cases, photoinitiation can replace other initiation techniques, including thermochemical or electrochemical initiation. Photoinitiation parallels initiation by high-energy radiation, such as γ-radiation or electron beam radiation. Initiation by high-energy radiation proceeds in the absence of initiators, but is less specific than photoinitiation, since high-energy radiation simultaneously generates various kinds of free radicals of differing reactivity as well as free ions.

Both free radical and ionic polymerizations are restricted to certain types of monomers. Many olefinic and acrylic monomers are readily polymerizable by a free radical mechanism, whereas other compounds such as oxiranes (epoxides)

Polymers and Light. Fundamentals and Technical Applications. W. Schnabel
Copyright © 2007 WILEY-VCH Verlag GmbH & Co. KGaA, Weinheim
ISBN: 978-3-527-31866-7

and vinyl ethers are solely polymerizable by a cationic mechanism. Photopolymerizations can be readily performed at ambient or at an even much lower temperature. Moreover, solvent-free formulations can be used. Therefore, there are important technical applications, for instance in the field of curing of coatings and printing inks. Technical aspects are described in Chapter 11.

According to the large number of publications and patents concerned with photopolymerization that continue to appear, this field is still expanding. This remarkable development has been documented in various books and reviews [1–40].

10.2
Photoinitiation of free radical polymerizations

10.2.1
General remarks

The synthesis of macromolecules by the free radical chain polymerization of low molar mass compounds, denoted as monomers, commences with the generation of free radicals, which is conveniently performed through photoreactions of initiator molecules. The subsequent processes, i.e. propagation, including chain transfer, and termination, are thermal (dark) reactions, which are not affected by light. The simplified overall mechanism is described in Scheme 10.1.

Two types of compounds are employed as photoinitiators of free radical polymerizations, which differ in their mode of action of generating reactive free radicals. Type I initiators undergo a very rapid bond cleavage after absorption of a photon. On the other hand, type II initiators form relatively long-lived excited triplet states capable of undergoing hydrogen-abstraction or electron-transfer reactions with co-initiator molecules that are deliberately added to the monomer-containing system.

10.2.2
Generation of reactive free radicals

10.2.2.1 Unimolecular fragmentation of type I photoinitiators
Typical type I photoinitiators are listed in Table 10.1. Most of them contain aromatic carbonyl groups, which act as chromophores. Since the dissociation energy of the C–C bond adjacent to the benzoyl group is lower than the excitation energy of the excited state, these compounds undergo rapid bond cleavage resulting in the formation of a pair of radicals, one of them being a benzoyl-type radical (see Scheme 10.2).

Phosphinoyl radicals are much more reactive towards olefinic compounds than carbon-centered radicals. For example, the rate constants for the addition of diphenylphosphinoyl radicals (see Scheme 10.2) to vinyl monomers are of the order 10^6 to 10^7 M^{-1} s^{-1}, i.e. one or two orders of magnitude larger than those

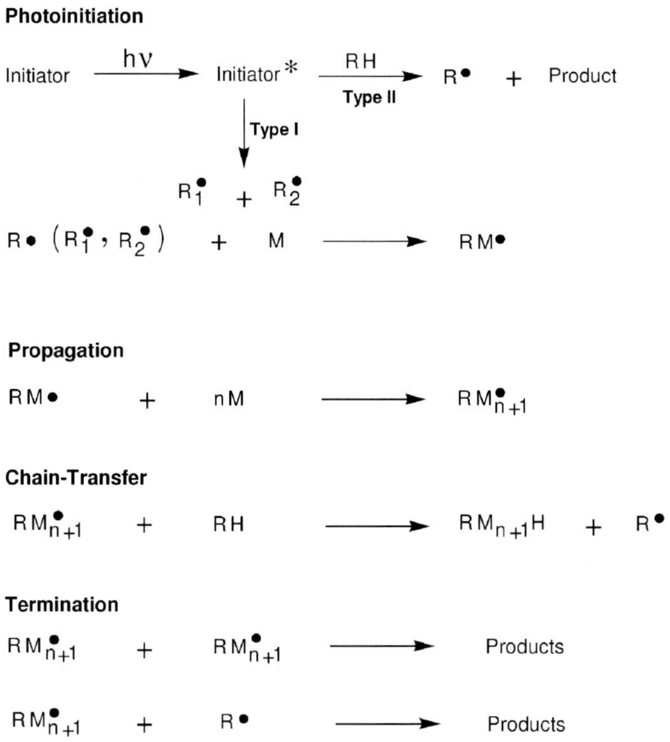

Photoinitiation

Initiator $\xrightarrow{h\nu}$ Initiator* $\xrightarrow{RH}$ R• + Product
$\qquad\qquad\qquad\qquad\quad$ Type II

$\qquad\qquad\qquad$ ↓ Type I

$\qquad\qquad\qquad$ R$_1^{\bullet}$ + R$_2^{\bullet}$

R• (R$_1^{\bullet}$, R$_2^{\bullet}$) + M $\longrightarrow$ RM•

Propagation

RM• + nM $\longrightarrow$ RM$_{n+1}^{\bullet}$

Chain-Transfer

RM$_{n+1}^{\bullet}$ + RH $\longrightarrow$ RM$_{n+1}$H + R•

Termination

RM$_{n+1}^{\bullet}$ + RM$_{n+1}^{\bullet}$ $\longrightarrow$ Products

RM$_{n+1}^{\bullet}$ + R• $\longrightarrow$ Products

Scheme 10.1 Reaction scheme illustrating the photoinitiated free radical polymerization of monomer M, commonly a compound with a C=C bond.

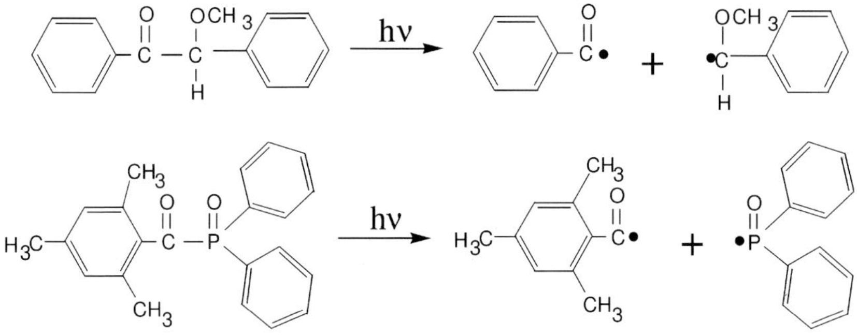

Scheme 10.2 Photofragmentations by α-cleavage of benzoin methyl ether and 2,4,6-trimethylbenzoyl diphenylphosphine oxide.

Table 10.1 Chemical structures of typical type I free radical photoinitiators.

Class	Chemical structure	
Benzoin and benzoin ethers	[structures of benzoin and benzoin ether]	R: methyl, ethyl, ethyl, isopropyl, n-butyl, isobutyl
Benzil ketals	[structure]	R: methyl
Acetophenones	[two structures]	
Hydroxyalkylphenones	[structure]	
Phenylglyoxylates	[structure with –O–CH$_3$]	
S-Phenyl thiobenzoates	[structure]	
O-Acyl-α-oximo ketones	[structure with C=N–O]	
Morpholino-acetophenones	[two structures]	
Acylphosphine oxides	[two structures with OCH$_3$]	
Acylphosphonates	[structure with OCH$_3$]	
Halogenated compounds	[three structures with Cl, Br]	

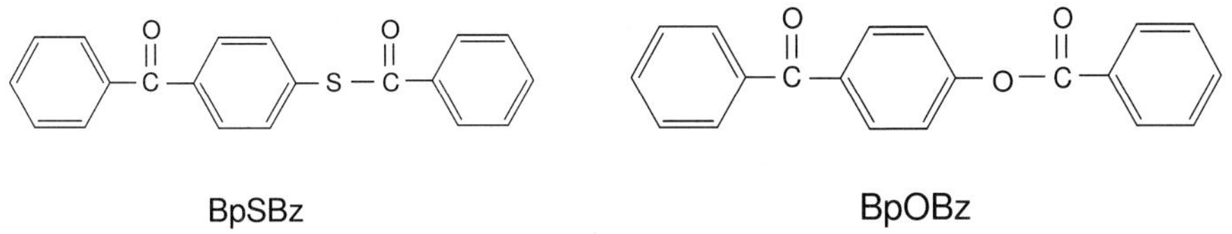

Chart 10.1 Chemical structures of BpSBz and BpOBz.

for benzoyl or other carbon-centered radicals formed by the photolysis of benzoin or the other compounds listed in Table 10.1 [41, 42].

In spite of the large number of available photoinitiators [4], the search for new initiators is ongoing. For example, S-(4-benzoyl)phenylthiobenzoate, BpSBz, has been found to be a type I photoinitiator. Upon exposure to light it is cleaved into free radicals (quantum yield: 0.45), which initiate the polymerization of methyl methacrylate. In contrast, BpOBz (see Chart 10.1) is not cleaved. It forms a long-lived triplet state rather than free radicals [43].

10.2.2.2 Bimolecular reactions of type II photoinitiators

Typical type II initiators containing carbonyl chromophores are listed in Table 10.2. Upon photon absorption, they form long-lived triplet states, which do not undergo α-cleavage reactions because the triplet energy is lower than the bond dissociation energy. The triplet species can, however, react with suitable co-initiators (see Table 10.3). For example, benzophenone and other diaryl ketones abstract hydrogen atoms from other compounds such as isopropanol, provided that the triplet energy exceeds the bond dissociation energy of the C–H bond to be broken.

Type II initiators containing carbonyl groups can also undergo electron-transfer reactions, which lead to hydrogen abstraction after an intermediate *exciplex* (excited complex) has been formed between the diaryl ketone radical anion and the amine radical cation, as illustrated in Scheme 10.3.

10.2.2.3 Macromolecular photoinitiators

Both type I and type II initiator moieties (see Chart 10.2) can be chemically incorporated into macromolecules as pendant groups through the copolymerization of conventional monomers and monomers containing the initiator moieties. In the curing of surface coatings, the use of macromolecular photoinitiators provides for a good compatibility of the initiator in the formulation. Moreover, the migration of the initiator to the surface of the material is prevented, which results in low-odor and non-toxic coatings.

In this context, linear polysilanes are also worthy of note. As reported in Section 7.4.2 (see Scheme 7.17), light absorption induces the formation of silyl radicals by main-chain scission, in addition to the extrusion of silylene. Free radical

10 Photopolymerization

Table 10.2 Chemical structures of typical type II photoinitiators.

Class	Chemical structure
Benzophenone derivatives	
Thioxanthone derivatives	
1,2-Diketones (benzils and camphorquinone)	
α-Keto coumarins	
Anthraquinones	
Terephthalophenones	
Water-soluble aromatic ketones	

Table 10.3 Chemical structures of amines functioning as co-initiators for type II free radical photoinitiators.

Denotation	Chemical structure
Methyl diethanolamine	$H_3C-N(CH_2-CH_2OH)_2$
Triethanolamine	$HO-CH_2-CH_2-N(CH_2-CH_2OH)_2$
Ethyl 4-(dimethylamino)benzoate	$(H_3C)_2N-C_6H_4-C(O)-O-CH_2-CH_3$
n-Butoxyethyl 4-(dimethylamino)benzoate	$(H_3C)_2N-C_6H_4-C(O)-O-CH_2-CH_2-O-(CH_2)_3CH_3$

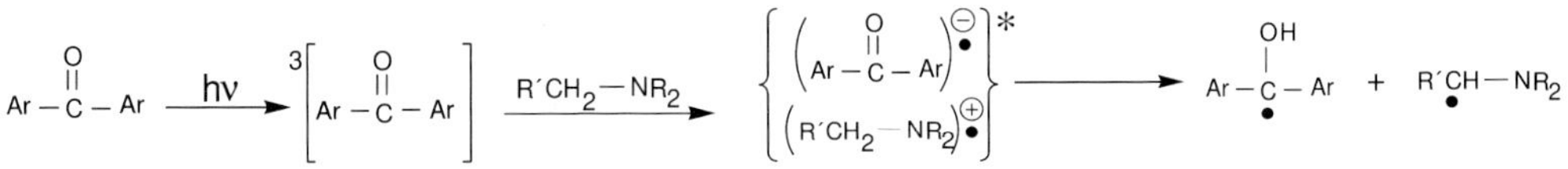

Scheme 10.3 Generation of reactive free radicals with the aid of type II initiators, exemplified by the reaction of a triplet-excited diaryl ketone with a tertiary amine.

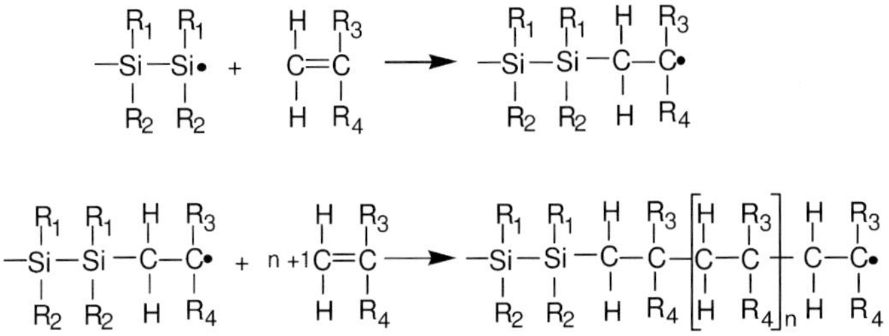

Scheme 10.4 Initiation of the polymerization of unsaturated compounds by reaction with photogenerated macrosilyl radicals.

chain polymerization is initiated if polysilanes are photolyzed in the presence of unsaturated monomers such as methyl methacrylate and styrene (see Scheme 10.4) [44]. As in the case of benzoin, the quantum yield for initiation, Φ_i, is of the order of 0.1. Φ_i represents the number of kinetic chains initiated per photon absorbed by the initiator. The rate constants for the addition of silyl radicals to unsaturated compounds are quite large (8×10^7 and 2×10^8 M^{-1} s^{-1} for methyl methacrylate and styrene, respectively) [45].

10.2.2.4 Photoinitiators for visible light

At present, visible-light-sensitive polymerizable systems are used for special applications in conjunction with visible-light-emitting lasers of low cost and excellent performance. Typical such applications are maskless photoimaging processes, such as *laser direct imaging, LDI,* and *computer-assisted design, CAD, systems* which are used for the imaging of printed circuit boards. Additional visible light applications include the production of holograms and color printing [2–4]. In the literature, a large number of photoinitiator systems appropriate for visible light exposure have been proposed. Of importance for practical applications are some organometallic initiators, various dye/co-initiator systems, and some *a*-diketones, which are dealt with in the following sections.

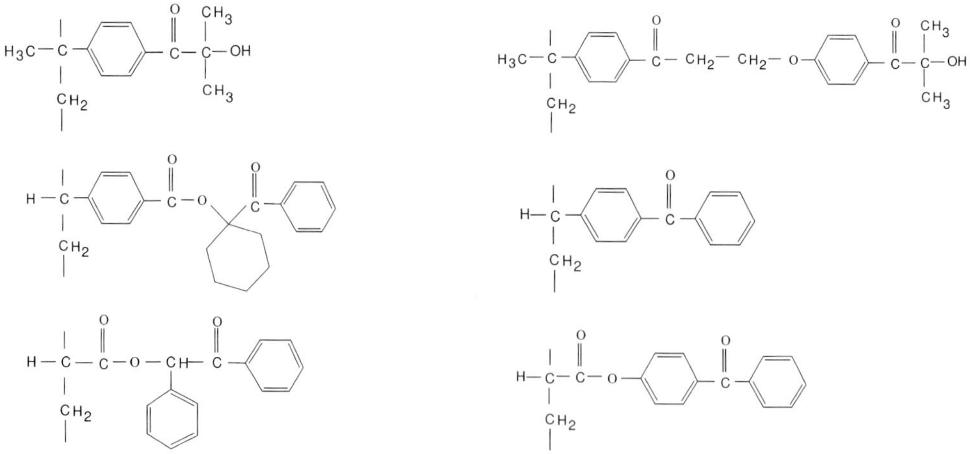

Chart 10.2 Chemical structures of photosensitive moieties contained in typical macromolecular photoinitiators.

10.2.2.4.1 Metal-based initiators

There is a large group of metal-based compounds capable of initiating the free radical photopolymerization of unsaturated compounds (see Table 10.4) [23, 24].

By virtue of their absorption characteristics, many of the compounds listed in Table 10.4 can be employed in conjunction with visible light sources. As the research in organometallic chemistry gained momentum, the potential advantages of organometallic complexes as photoinitiators were also explored, and two such compounds, a ferrocenium salt and a titanocene, were commercialized (see Chart 10.3).

Table 10.4 Typical metal-based photoinitiators [9, 24, 46].

Class	Example [a]
Transition metal ions	Fe^{2+}, V^{2+}, V^{3+}, V^{4+}, UO_2^{2+}
Transition metal inorganic complexes	L_2VOCl, L_3Mn, $L_3Fe(SCN)_3$, L_3Ru^{2+}
Transition metal organometallic complexes, including ferrocenium salts and titanocene derivatives	In conjunction with a co-initiator such as CCl_4: $Mn_2(CO)_{10}$, $Fe(CO)_5$, $Cr(CO)_6$, $W(CO)_6$, $Mo(CO)_6$, $Mo(CO)_5Py$, $CpMn(CO)_3$
	In conjunction with hydroperoxides: (η^6-arene)(η^5-cyclopentadienyl)iron(II) hexafluorophosphate [b]
	bis(η^5-cyclopentadienyl)-bis[2,6-difluoro-3-(1H-pyrr-1-yl)phenyl]titanium [b]
Non-transition metal complexes	$Al(C_2H_5)_3$

[a] L: ligand such as acetylacetonyl (acac), Cp: cyclopentadienyl.
[b] see Chart 10.3.

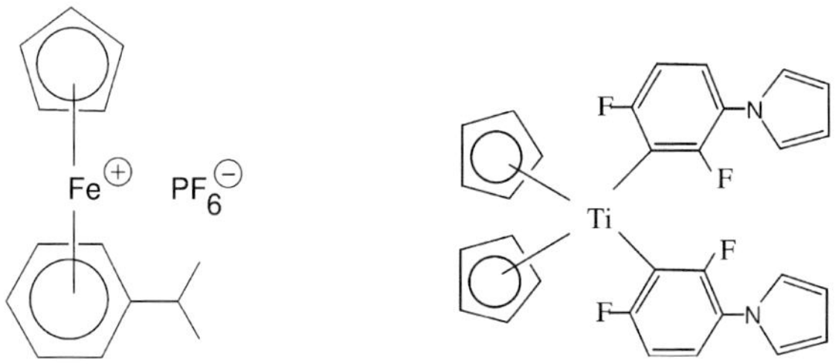

Chart 10.3 Chemical structures of (η^6-arene)(η^5-cyclopentadienyl)iron(II) hexafluorophosphate (left) and bis(η^5-cyclopentadienyl)-bis[2,6-difluoro-3-(1H-pyrr-1-yl)phenyl]titanium (right).

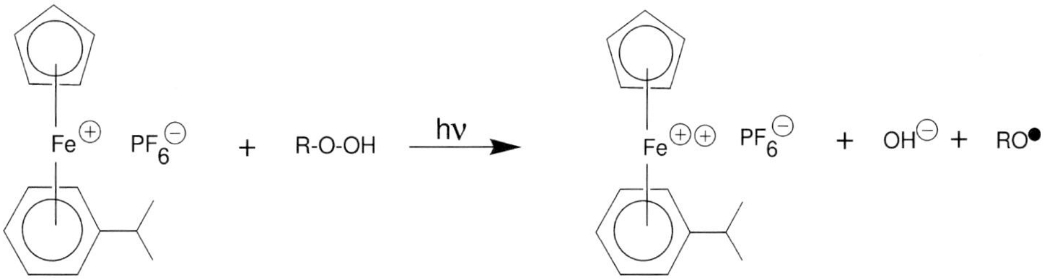

Scheme 10.5 Generation of free radicals upon irradiation of a ferrocenium salt in the presence of an alkyl hydroperoxide.

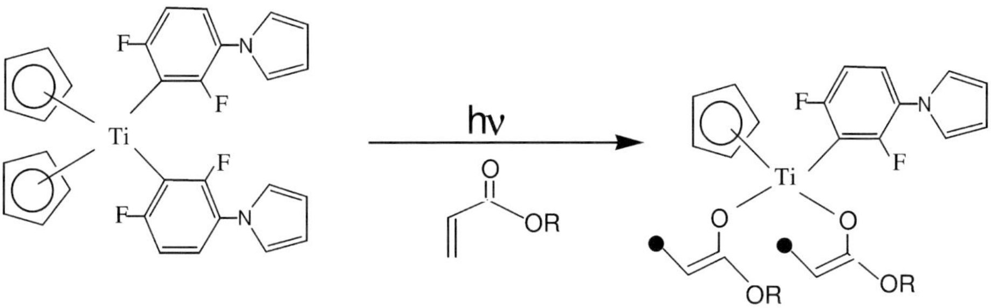

Scheme 10.6 Photoinitiation of the free radical polymerization of an alkyl acrylate with the aid of a fluorinated titanocene [47].

When the ferrocenium salts are applied in conjunction with alkyl hydroperoxides such as cumyl hydroperoxide, they yield, on exposure to light, reactive free radicals, as shown in Scheme 10.5.

The fluorinated titanocene presented in Chart 10.3 is a very effective photoinitiator that functions without a co-initiator when irradiated with visible light. It is thermally stable (decomposition at 230 °C) and absorbs light up to 560 nm with maxima at 405 and 480 nm [2]. According to mechanistic studies, the complex undergoes an isomerization upon absorption of a photon. In the presence of an unsaturated monomer, the resulting coordinatively unsaturated isomer undergoes a ligand-exchange reaction to yield a biradical capable of initiating the polymerization of further monomer molecules (see Scheme 10.6) [47].

10.2.2.4.2 Dye/co-initiator systems

Dye molecules in an electronically excited state are capable of undergoing electron-transfer reactions with appropriate compounds, denoted as *co-initiators* [2, 8, 12, 15]. The free radical ions, formed by electron transfer, or the free radicals, formed by the decomposition of the radical ions, can initiate the polymerization of monomers. In principle, the excited dye molecule can be reduced or oxidized, i.e. it can accept an electron from the co-initiator, CI, or it can transfer an electron to the CI [see Eqs. (10-1) and (10-2)].

$$D^* + CI \rightarrow (D \cdots CI)^* \rightarrow D^{(-)\bullet} + CI^{(+)\bullet} \qquad (10\text{-}1)$$

$$D^* + CI \rightarrow (D \cdots CI)^* \rightarrow D^{(+)\bullet} + CI^{(-)\bullet} \qquad (10\text{-}2)$$

The electron transfer is thermodynamically allowed if the free energy, ΔG, calculated by the Rehm-Weller equation [Eq. (10-3)] is negative.

$$\Delta G = F[E^{ox}_{1/2} - E^{red}_{1/2}] - E_S - \Delta E_c \qquad (10\text{-}3)$$

Table 10.5 Chemical structures of typical photoreducible dyes.

Family	Denotation	Chemical structure	λ_{max} (nm) [a]
Acridines	Acriflavin		460
Xanthenes	Rose Bengal		565
Thiazenes	Methylene blue		645
Cyanines	Cyanine dye		490–700 depending on n

a) Maximum of absorption band.

Table 10.6 Chemical structures of typical co-initiators employed in dye-sensitized free radical polymerization [2].

Family	Chemical structure	Denotation
Amines	N(CH$_2$–CH$_2$–OH)$_3$ Ph–NH–CH$_2$–C(O)–OH	Triethanolamine, N-phenylglycine
Phosphines and arsines	Ph$_3$P, Ph$_3$As	Triphenylphosphine, triphenylarsine
Borates	[Ph$_3$B–C$_4$H$_9$]$^-$ X$^+$	Triphenylbutylborate
Organotin compounds	H$_3$C–Sn(CH$_3$)$_2$–CH$_2$–Ph	Benzyltrimethylstannane
Heterocyclic compounds	oxazole, thiazole rings	Oxazole, thiazole

Here, F is the Faraday constant, $E^{ox}_{1/2}$ and $E^{red}_{1/2}$ are the oxidation and reduction potentials of the donor and acceptor, respectively, E_S is the singlet-state energy of the dye, and ΔE_c is the coulombic stabilization energy. Typical dyes and co-initiators are presented in Tables 10.5 and 10.6, respectively.

For practical applications, initiator systems functioning on the basis of dye reduction are most important. Scheme 10.7 illustrates how free radicals are formed with the aid of a co-initiator of the tertiary amine type. In this case, the amino radical cation, formed by electron transfer, loses a proton to give an α-aminoalkyl radical, which initiates the polymerization.

10.2.2.4.3 Quinones and 1,2-diketones

In conjunction with hydrogen donors such as dimethylaniline and triethylamine, benzils and various quinones, such as anthraquinone, 9,10-phenanthrene quinone, and camphor quinone (see Chart 10.4), can be used as visible-light-sensitive photoinitiators [8]. Some of these compounds are used to cure dental restorative systems (see Section 11.3). Another application concerns the curing of waterborne pigmented latex paints, which do not contain *volatile organic compounds (VOCs)* [48].

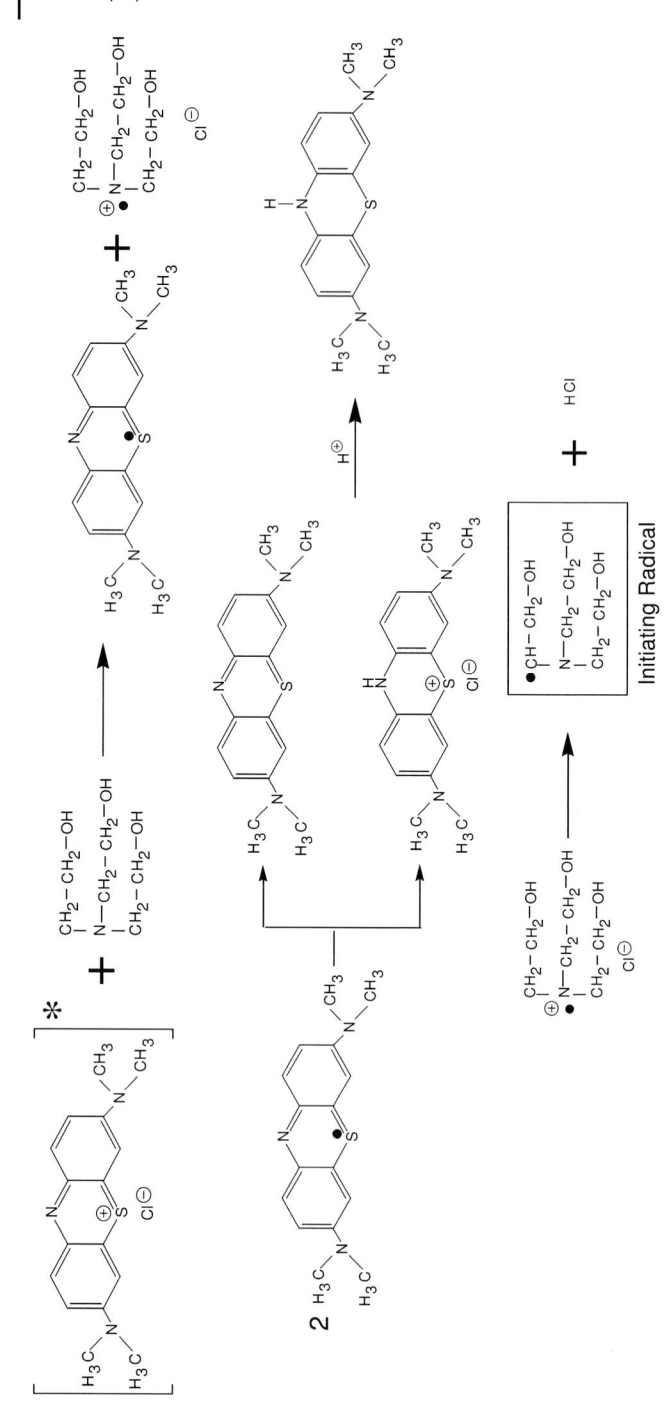

Scheme 10.7 Generation of free radicals by the photoreduction of methylene blue with triethanolamine [2].

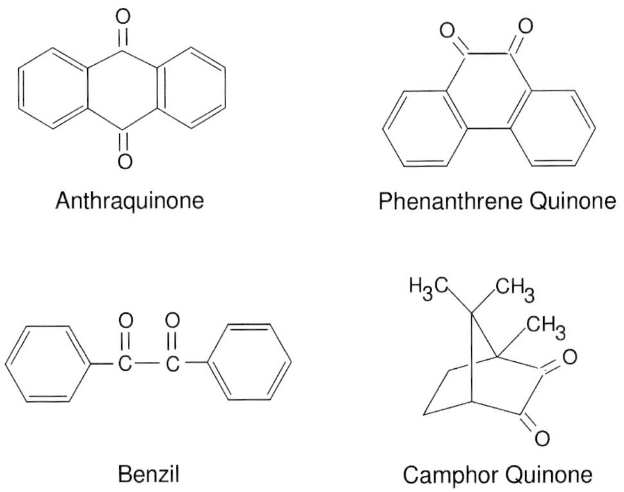

Anthraquinone

Phenanthrene Quinone

Benzil

Camphor Quinone

Chart 10.4 Chemical structures of quinones and 1,2-diketones.

10.2.2.5 Inorganic photoinitiators

Inorganic materials such as titanium dioxide, TiO_2, and cadmium sulfide, CdS, can initiate the polymerization of unsaturated compounds upon exposure to light [49–51]. For the photoinitiation of the polymerization of methyl methacrylate by nanosized titanium dioxide [49, 50], the mechanism presented in Scheme 10.8 has been proposed. Accordingly, electrons released upon absorption of light by the TiO_2 particles are trapped at the hydrated surface of the particles by $Ti^{4(+)}OH$ groups. $Ti^{3(+)}OH$ formed in this way can react with molecular oxygen to form $^{\bullet}O_2^{(-)}$. The latter combines with $H^{(+)}$ to yield $HOO^{\bullet}$. When two $HOO^{\bullet}$ radicals combine, H_2O_2 is formed, which can react with $^{\bullet}O_2^{(-)}$. This reaction yields the polymerization initiator, i.e. very reactive $^{\bullet}OH$ radicals. Actually, this is a photocatalytic mechanism, since the inorganic particles are not consumed during the process.

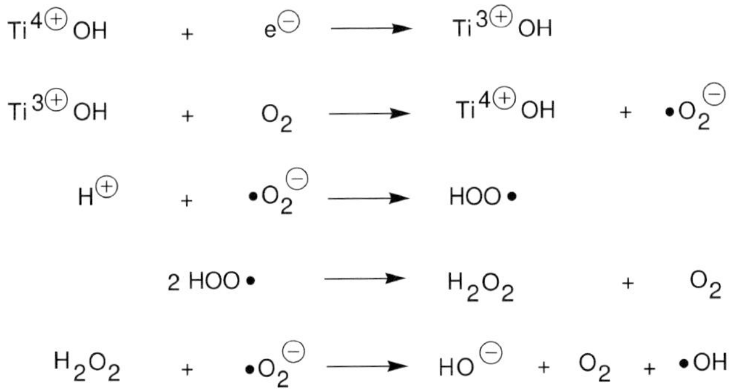

$Ti^{4\oplus}OH \quad + \quad e^{\ominus} \quad \longrightarrow \quad Ti^{3\oplus}OH$

$Ti^{3\oplus}OH \quad + \quad O_2 \quad \longrightarrow \quad Ti^{4\oplus}OH \quad + \quad ^{\bullet}O_2^{\ominus}$

$H^{\oplus} \quad + \quad ^{\bullet}O_2^{\ominus} \quad \longrightarrow \quad HOO^{\bullet}$

$2\ HOO^{\bullet} \quad \longrightarrow \quad H_2O_2 \quad + \quad O_2$

$H_2O_2 \quad + \quad ^{\bullet}O_2^{\ominus} \quad \longrightarrow \quad HO^{\ominus} \quad + \quad O_2 \quad + \quad ^{\bullet}OH$

Scheme 10.8 Generation of reactive free radicals during the absorption of light by titanium dioxide.

10.3
Photoinitiation of ionic polymerizations

10.3.1
Cationic polymerization

10.3.1.1 General remarks

The virtues of photoinitiated cationic polymerization are rapid polymerization without oxygen inhibition, minimal sensitivity to water, and the ability to polymerize vinyl ethers, oxiranes (epoxides), and other heterocyclic monomers (see Table 10.7) that do not polymerize by a free radical mechanism.

In analogy to free radical polymerizations (see Scheme 10.1), cationic polymerizations proceed as chain reactions involving initiation and propagation. However, in many cases, there is no termination by neutralization, and the growing chains are only terminated by nucleophilic impurities contained in the

Table 10.7 Chemical structures of monomers polymerizable by a cationic mechanism [2, 7].

10.3 Photoinitiation of ionic polymerizations

Table 10.8 Chemical structures of typical cationic photoinitiators [2, 27, 52, 53].

Class	Chemical structure[a]
Diazonium salts	Ph–N≡N⊕ X⊖
Diaryl iodonium salts	Ph–I⊕–Ph X⊖ ; R₁–C₆H₄–I⊕–C₆H₄–R₂ X⊖, R₁, R₂: H, Alkyl etc.
Triaryl sulfonium salts	Ph₃S⊕ X⊖ ; Ph₂S⊕–C₆H₄–S–Ph X⊖ ; Ph₂S⊕–C₆H₄–S–C₆H₄–S⊕Ph₂ X⊖
5-Arylthianthrenium salts	(thianthrenium with Ar) X⊖
Dialkylphenacyl sulfonium salts	Ar–C(=O)–CH₂–S⊕(R₁)(R₂) X⊖ ; R₁ and R₂: alkyl (C₄H₉, C₁₂H₂₅, C₁₈H₃₇ etc.)
N-Alkoxy pyridinium and isoquinolinium salts	pyridinium–N⊕–O–R X⊖ ; isoquinolinium–N⊕–O–R X⊖, R: alkyl
Phosphonium salts	Ph₃P⊕–CH₂–(pyrenyl) X⊖ ; Ph₃P⊕–CH₂–C(=O)–Ph X⊖
Ferrocenium salts	[Cp–Fe–(C₆H₄–iPr)]⊕ X⊖
Phenacyl anilinium salts	Ar–C(=O)–CH₂–N⊕(CH₃)(CH₃)(C₆H₅) X⊖
Triaryl cyclopropenium salts	(triaryl cyclopropenium, R substituents) X⊖
Sulfonyloxy ketones	Ph–C(=O)–CH(CH₃)–SO₂–C₆H₄–CH₃
Silyl benzyl ethers	Cl–C₆H₃(NO₂)–CH₂–O–SiPh₃

a) $X^{(-)}$ denotes a non-nucleophilic anion such as $BF_4^{(-)}$, $PF_6^{(-)}$, $AsF_6^{(-)}$, $SbF_6^{(-)}$, $CF_3SO_3^{(-)}$, $CF_3(CF_2)_3SO_3^{(-)}$, $(C_6F_5)_4B^{(-)}$, $(C_6F_5)_4Ga^{(-)}$.

10.3.1.2 Generation of reactive cations

Reactive cations can be generated via three different routes: (a) by direct photolysis of the initiator, (b) by sensitized photolysis of the initiator, and (c) by free radical mediation. These routes are described below.

10.3.1.2.1 Direct photolysis of the initiator

Crivello's pioneering work on onium salt-type photoinitiators (sulfonium and iodonium salts) gave great impetus to investigations of cationic polymerizations [5, 6]. A common feature of mechanisms proposed in relation to onium salt-type initiators of the general structure $\{(A–B)^{(+)}X^{(-)}\}$ is the generation of Brønsted acids (superacids) of the structure $H^{(+)}X^{(-)}$ based on non-nucleophilic anions $X^{(-)}$. These superacids play a prominent role in the initiation process. However, radical cations $A^{(+)\bullet}$, formed by light-induced bond cleavage, may also react with the polymerizable monomers. According to the general mechanism

$$(A\text{-}B)^{\oplus} + h\nu \longrightarrow A^{\oplus\bullet} + B^{\bullet}$$

$$A^{\oplus\bullet} + RH \longrightarrow AH^{\oplus} + R^{\bullet}$$

$$AH^{\oplus} \longrightarrow A + H^{\oplus}$$

Scheme 10.9 Photolysis of an onium ion $(A–B)^{(+)}$.

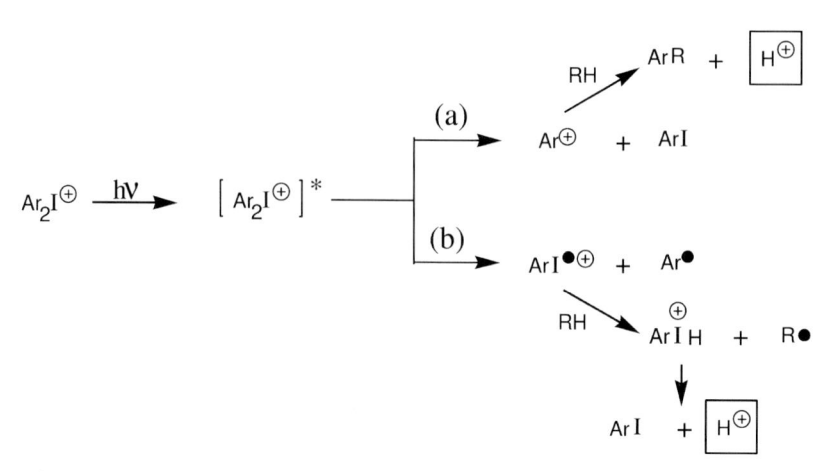

$$Ar_2I^{\oplus} \xrightarrow{h\nu} [Ar_2I^{\oplus}]^{*}$$

(a) $\longrightarrow Ar^{\oplus} + ArI \xrightarrow{RH} ArR + H^{\oplus}$

(b) $\longrightarrow ArI^{\bullet\oplus} + Ar^{\bullet} \xrightarrow{RH} ArIH^{\oplus} + R^{\bullet} \longrightarrow ArI + H^{\oplus}$

Scheme 10.10 Photolysis of a diaryl iodonium ion involving both heterolytic (a) and homolytic (b) Ar–I bond rupture.

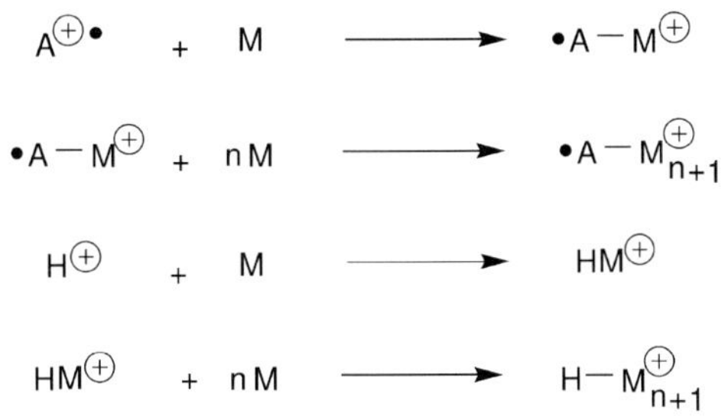

10.3 Photoinitiation of ionic polymerizations

$$A^{(+)\bullet} + M \longrightarrow {}^\bullet A - M^{(+)}$$

$$^\bullet A - M^{(+)} + nM \longrightarrow {}^\bullet A - M^{(+)}_{n+1}$$

$$H^{(+)} + M \longrightarrow HM^{(+)}$$

$$HM^{(+)} + nM \longrightarrow H - M^{(+)}_{n+1}$$

Scheme 10.11 Reactions of a radical cation $A^{(+)\bullet}$ and a proton $H^{(+)}$ with a polymerizable monomer.

of the photolysis, shown in Scheme 10.9, the radical cation $A^{(+)\bullet}$ may abstract a hydrogen from surrounding molecules RH. The resulting cation $AH^{(+)}$ then releases a proton.

The detailed mechanism of the photolysis of a diaryl iodonium ion presented in Scheme 10.10 may serve here as a typical example, since the scope of this book does not permit the discussion of mechanistic details concerning the photolysis of all of the initiators compiled in Table 10.8. Details concerning the photolysis of initiators and mechanisms of the initiation of cationic polymerizations are available in review articles [2, 27].

Both the initially formed radical cation and the proton are potential initiating species for the reaction with a polymerizable monomer M (see Scheme 10.11).

10.3.1.2.2 Sensitized photolysis of the initiator

If onium salts do not or only weakly absorb light at $\lambda > 300$ nm, then photosensitizers, PS, that absorb strongly at long wavelengths may be employed in conjunction with the onium salts. In most cases, energy transfer from PS* to $(A-B)^{(+)}$ can be excluded. However, PS* can be oxidized by the onium ion, i.e. radical cations PS$^{+\bullet}$ can be formed by electron transfer from the electronically excited photosensitizer PS* to the onium ion (see Scheme 10.12), provided that the free energy, ΔG, of this reaction has a sufficiently high negative value.

Regarding the cationic polymerization of an appropriate monomer, three initiation routes are feasible (see Scheme 10.13): (a) $PS^{(+)\bullet}$ reacts directly with M, (b) $PS^{(+)\bullet}$ abstracts a hydrogen from a surrounding molecule RH to form the proton-releasing $PSH^{(+)}$ ion, (c) $PS^{(+)\bullet}$ combines with radical $B^\bullet$ thus forming the cation $B-PS^{(+)}$. Protons released from $PSH^{(+)}$ ions, as well as $B-PS^{(+)}$ ions, are likely to add to M.

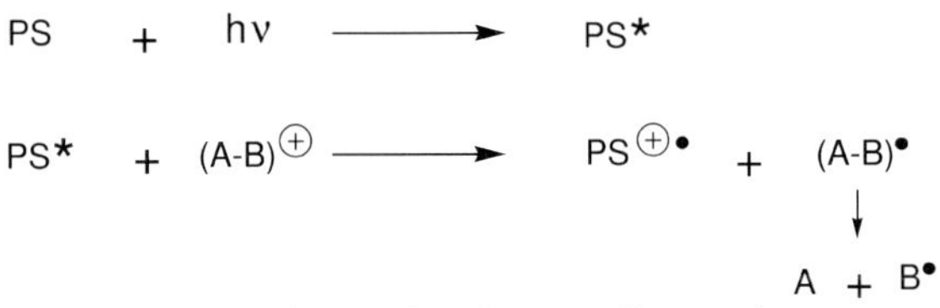

$$PS + h\nu \longrightarrow PS^*$$

$$PS^* + (A-B)^{(+)} \longrightarrow PS^{(+)\bullet} + (A-B)^\bullet$$
$$\downarrow$$
$$A + B^\bullet$$

Scheme 10.12 Oxidation of an electronically excited sensitizer PS* by an onium ion $(A-B)^{(+)}$.

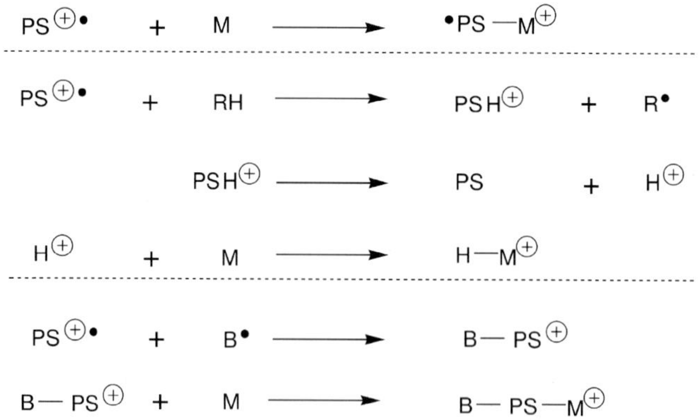

Scheme 10.13 Possible initiation routes in the cationic polymerization of monomer M. Initiating system: onium salt/sensitizer.

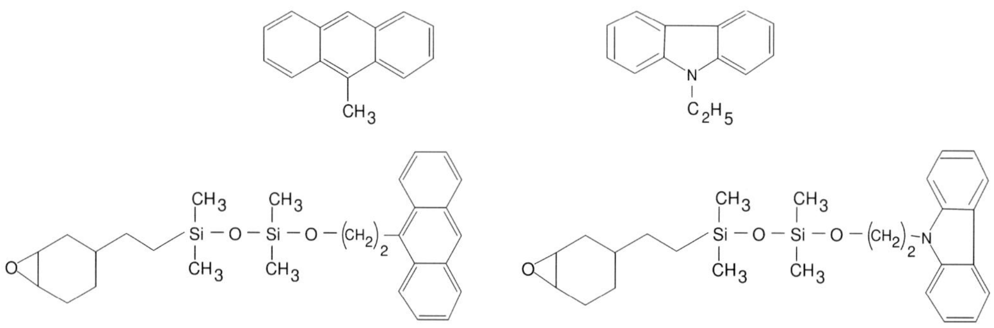

Chart 10.5 Typical electron-transfer photosensitizers that may be applied in conjunction with onium salts [54].

Derivatives of anthracene and carbazole are typical electron-transfer photosensitizers. Of practical interest are derivatives containing cationically polymerizable epoxide groups (see Chart 10.5) [54]. During the ring-opening photopolymerization of epoxides, these sensitizers are covalently incorporated into the polymeric network and cannot be removed by extraction. Therefore, the potential risk of toxic effects of the sensitizers is strongly diminished.

10.3.1.2.3 Free-radical-mediated generation of cations

10.3.1.2.3.1 Oxidation of radicals

A large number of carbon-centered free radicals, which are formed by photolysis or thermolysis of commercially available free radical initiators, can be oxidized by onium ions $(A-B)^{(+)}$ by reaction according to Eq. (10-4):

$$-\overset{|}{\underset{|}{C}}{}^{\bullet} + (A-B)^{\oplus} \longrightarrow -\overset{|}{\underset{|}{C}}{}^{\oplus} + (A-B)^{\bullet} \qquad (10\text{-}4)$$

Carbocations generated in this way can add directly to appropriate monomers (e.g., tetrahydrofuran, cyclohexene oxide, n-butyl vinyl ether) or can form Brønsted acids by abstracting hydrogen from surrounding molecules. This method, which is commonly referred to as *free-radical-promoted cationic polymerization*, is quite versatile, because the user may rely on a large variety of radical sources. Some of them are compiled in Table 10.9.

A sufficiently high negative value of the free energy, ΔG, is required for the occurrence of reaction according to Eq. (10-4). ΔG, in units of kJ mol^{-1}, can be estimated with the aid of Eq. (10-5), the modified Rehm–Weller equation.

$$\Delta G = f_c(E^{ox}_{1/2} - E^{red}_{1/2}) \qquad (10\text{-}5)$$

Here, $E^{ox}_{1/2}$ and $E^{red}_{1/2}$ denote the half-wave potentials, in units of V, of oxidation and reduction of the carbon-centered radical and of the onium ion $(A-B)^{(+)}$, re-

Table 10.9 Free radicals that may be employed in free-radical-promoted cationic polymerizations.

Photoinitiator	Electron-donating free radical	Generation of radical
Benzoin		Direct
Phenylazotriphenylmethane		Direct
Polysilanes		Direct
Benzophenone/RH		Indirect: $^3(Ph_2C=O)^* + RH \longrightarrow Ph_2\overset{\bullet}{C}\text{-}OH + R\bullet$
Acylphosphine oxides/CH_2=CHR		Indirect: $R\bullet + CH_2=CHR \longrightarrow R-CH_2-\overset{\bullet}{C}HR'$ $R: Ph-\overset{O}{\overset{\|}{C}}\bullet$ and $\bullet\overset{O}{\overset{\|}{P}}\overset{R_1}{\underset{R_2}{\diagdown}}$

Table 10.10 The importance of the reduction potential with regard to the reaction of onium ions with 2-hydroxypropyl radicals [26].

Species	$E^{ox}_{1/2}(V)$	$E^{red}_{1/2}(V)$	$E^{ox}_{1/2} - E^{red}_{1/2}(V)$
$H_3C-\overset{\bullet}{\underset{OH}{C}}-CH_3$	−1.2		
triphenylsulfonium cation		−1.1	−0.1
N-ethoxy-2-methylpyridinium cation (N−O−C$_2$H$_5$, CH$_3$)		−0.7	−0.5
N-ethoxyisoquinolinium cation (N−O−C$_2$H$_5$)		−0.5	−0.7
diphenyliodonium cation		−0.2	−1.0

spectively. The conversion factor, f_c, is equal to 97 kJ mol^{-1} V^{-1}. On the basis of the reduction potentials listed in Table 10.10, it becomes evident why 2-hydroxypropyl radicals are oxidized much more efficiently by N-ethoxypyridinium and diphenyliodonium ions than by triphenylsulfonium ions.

10.3.1.2.3.2 Addition-fragmentation reactions

The addition of a free radical to the carbon-carbon double bond of an allylic group that forms part of an onium ion can induce disintegration of the onium salt, thus giving rise to the release of an inert compound and a reactive radical cation. Allylic compounds employed for this purpose are presented in Chart 10.6, and the reaction mechanism for a typical case is presented in Scheme 10.14 [55].

In this case, cationic polymerization is initiated by direct addition of photogenerated reactive radical cations to the appropriate monomers. Alternatively, Brønsted acids may be formed through reaction of the radical cations with hydrogen-donating constituents of the formulation, and then the initiation step involves the addition of protons to monomer molecules. The method discussed here has the advantage that virtually all kinds of radicals may be operative in the initiation process. Therefore, the polymerization can be elegantly tuned to the wavelength of the light by choosing radical sources with a suitable spectral response.

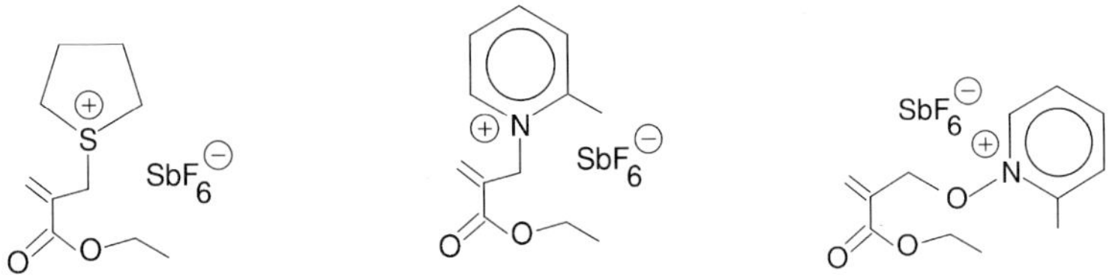

Chart 10.6 Allylic compounds employed in addition-fragmentation reactions.

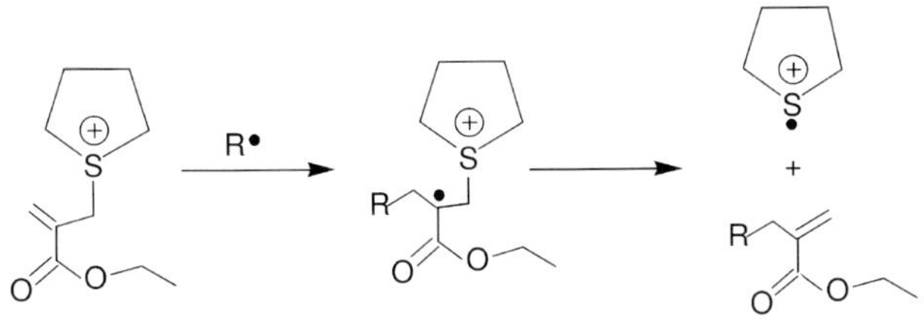

Scheme 10.14 Addition of a radical R• to the S-[2-(ethoxycarbonyl)allyl]tetrahydrothiophenium ion [55].

10.3.2
Anionic polymerization

10.3.2.1 General remarks

The possibility that photoinitiated polymerization can occur through an anionic mechanism has long been overlooked. Even today, literature reports on anionic photopolymerization are rare and there are no important commercial applications of which the author is aware. However, this situation might change, since extensive research on photoinduced base-catalyzed processes using photolatent amines has opened up new application areas [1, 3, 56].

10.3.2.2 Generation of reactive species

10.3.2.2.1 Photo-release of reactive anions

The compounds listed in Table 10.11 have been found to photoinitiate the polymerization of neat ethyl or methyl 2-cyanoacrylate, CA, that readily polymerize by an anionic mechanism. Therefore, this has been taken as evidence for the occurrence of an anionic mechanism [57–59].

The essential step in the proposed initiation mechanism is the photoinduced release of a reactive anion, which readily adds to the monomer. The polymer is then formed through the repetitive addition of CA to the growing anionic chain (see Scheme 10.15).

10 Photopolymerization

Table 10.11 Chemical structures of anionic photoinitiators [57–59].

Denotation	Chemical structure	Released anion (assumed)
Potassium Reineckate	$[Cr(NH_3)_2(NCS)_4]^\ominus \; K^\oplus$	$(NCS)^\ominus$
Platinum(II) acetylacetonate $(Pt(acac)_2)$	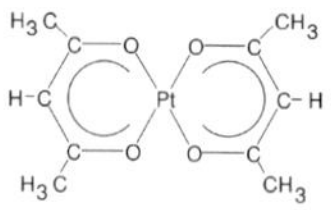 (structure of Pt(acac)$_2$)	$acac^\ominus$ [a]
Benzoylferrocene, dibenzoylferrocene	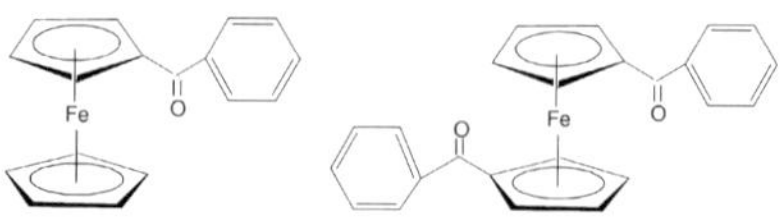 (structures of benzoylferrocene and dibenzoylferrocene)	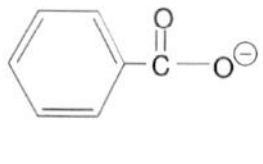 $PhCOO^\ominus$ [b]
Crystal violet leuconitrile (CVCN)	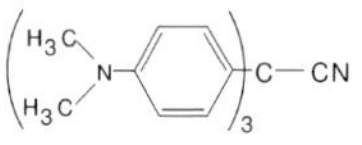 $((H_3C)_2N\text{-}C_6H_4)_3\text{C-CN}$	$CN^\ominus$
Malachite green leucohydroxide (MGOH)	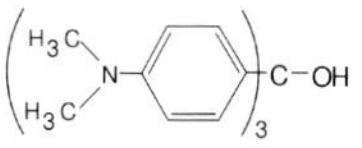 $((H_3C)_2N\text{-}C_6H_4)_3\text{C-OH}$	$HO^\ominus$

[a] acac: acetylacetonate.
[b] Forms in the presence of trace amounts of water.

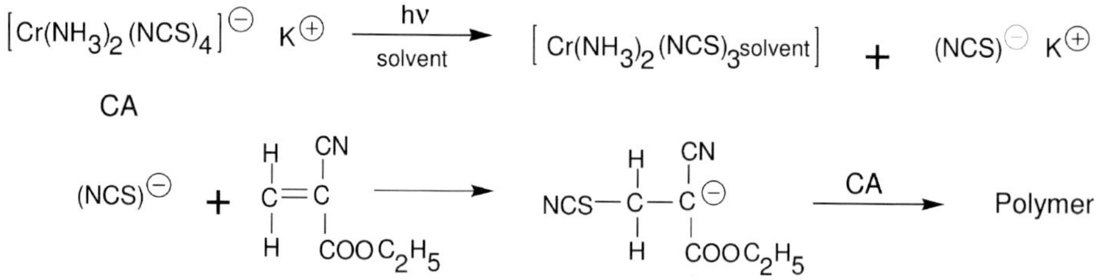

$$[Cr(NH_3)_2(NCS)_4]^\ominus \; K^\oplus \xrightarrow[\text{solvent}]{h\nu} [Cr(NH_3)_2(NCS)_3 \text{solvent}] + (NCS)^\ominus \; K^\oplus$$

CA

$$(NCS)^\ominus + \underset{H \quad COOC_2H_5}{\overset{H \quad CN}{C=C}} \longrightarrow \underset{H \quad COOC_2H_5}{\overset{H \quad CN}{NCS-C-C^\ominus}} \xrightarrow{CA} \text{Polymer}$$

Scheme 10.15 Photoinitiation of the polymerization of ethyl 2-cyanoacrylate by potassium reineckate [57].

10.3.2.2.2 Photo-production of reactive organic bases

In the context of the anionic polymerization of CA derivatives, as considered in Section 10.3.2.2.1, it is notable that the polymerization of cyanoacrylates is also photoinitiated by substituted pyridine pentacarbonyl complexes of tungsten or chromium, i.e. M(CO)$_5$L with M=Cr or W, and L=2- or 4-vinylpyridine [60]. Photo-released pyridine adds to CA, and the resulting zwitterion initiates the anionic chain polymerization (see Scheme 10.16).

Substances that release reactive bases or other reactive species upon exposure to light are often referred to as *photolatent compounds*, or, in the context of the

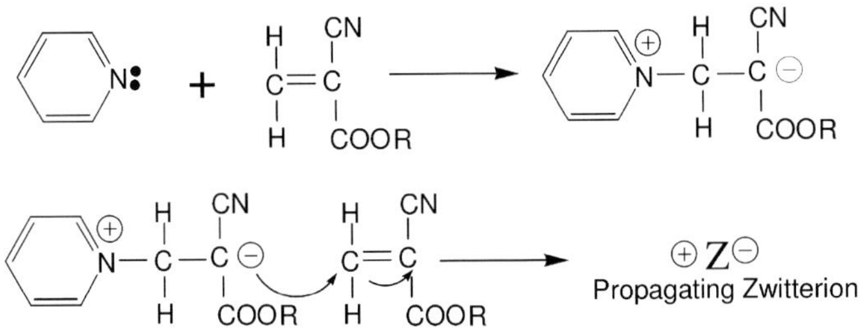

Scheme 10.16 Initiation of the polymerization of cyanoacrylate with the aid of photo-released pyridine [60].

initiation of polymerizations, as *photolatent initiators*. Actually, the photogeneration of organic bases is an important tool in inducing the polymerization of monomers of the oxirane type. Relevant research has been focused on the photogeneration of amines with the aim of developing a novel technique to cure epoxidized resins through intermolecular cross-linking [2, 3]. Strong organic bases, for instance tertiary amines or amidine bases, function as curing agents. Scheme 10.17 shows how tertiary amines act in the presence of polyols (oligomers bearing hydroxyl groups) [56]. After ring-opening is achieved by nucleophilic attack of the amine at a ring carbon, a proton is transferred from the polyol to the oxygen. The resulting alkoxide then adds to the ring carbon of another molecule and thus starts the anionic chain propagation.

Free tertiary amines can be obtained from various low molar mass compounds by irradiation with UV light. Relevant earlier work has been reviewed [3, 56]. According to more recent reports, 5-benzyl-1,5-diazabicyclo[4.3.0]nonane is a very effective photolatent initiator [1, 3]. It releases 1,5-diazabicyclo[4.3.0]-non-5-ene, DBN, a bicyclic amidine possessing a high basicity ($pK_a = 12–13$) due to the strong conjugative interaction between the two nitrogens. The suggested mechanism is depicted in Scheme 10.18.

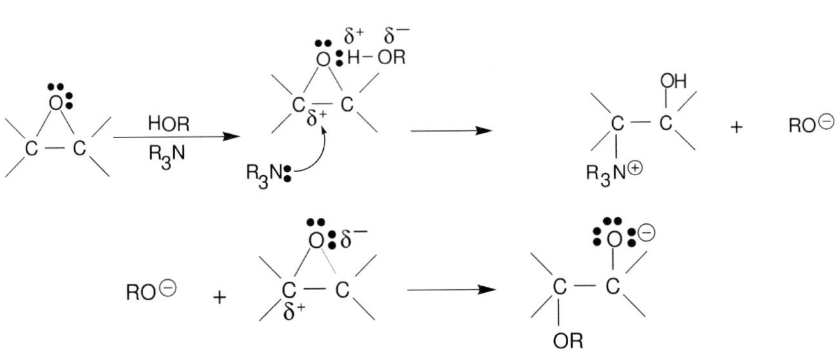

Scheme 10.17 Mechanism of the initiation of the anionic polymerization of epoxides by a tertiary amine in conjunction with a polyol [56].

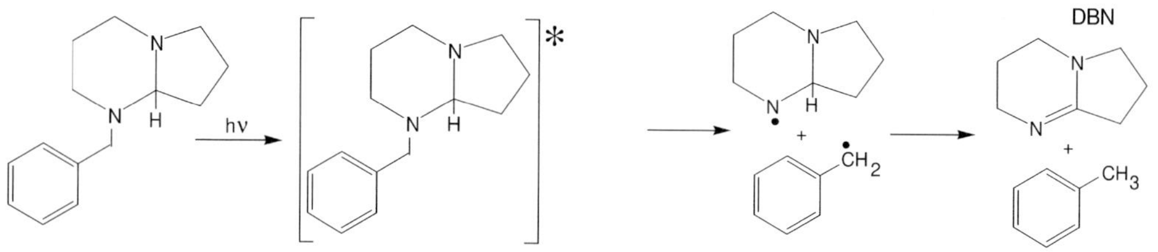

Scheme 10.18 Photoinduced release of DBN from 5-benzyl-1,5-diazabicyclo[4.3.0]nonane [1].

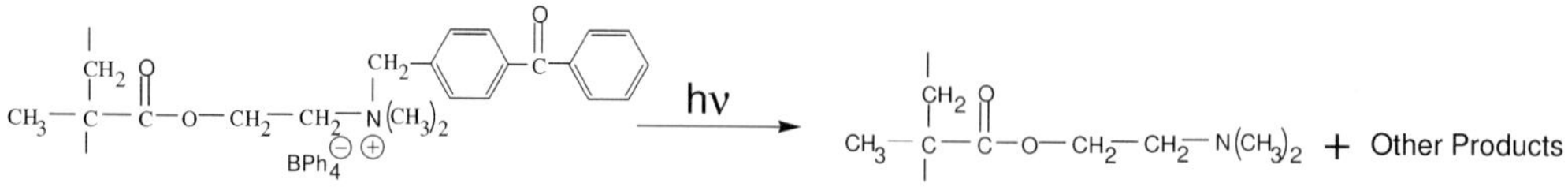

Scheme 10.19 Photoinduced generation of pendant tertiary amine groups on polymethacrylate chains [56].

Also, polymeric amines have been generated. A typical system is presented in Scheme 10.19.

Actually, in the conventional manufacture of polyurethane-based coatings, amine-catalyzed cross-linking is a widely used method. Curing of ready-to-use formulations occurs within several hours and is difficult to control. In contrast, photo-triggered curing can be performed on demand and the working window can be extended to a full day with formulations containing a photolatent compound such as the DBN-releasing initiator [1].

10.4
Topochemical polymerizations

10.4.1
General remarks

One of the most intriguing phenomena in the field of photopolymerization concerns the light-induced solid-state conversion of certain low molar mass compounds into macromolecules. Based on Schmidt's pioneering work on the dimerization of cinnamic acid and its derivatives by [2+2] photocycloaddition [61, 62], the light-induced solid-state polymerization of diacetylenes and dialkenes was discovered by Wegner [63] and Hasegawa [64], respectively. In these cases, the polymerization proceeds under crystal-lattice control. The reactivity of the starting compound and the structure of the resulting product are governed by the molecular geometry in the reactant crystal and the reaction proceeds with a minimum of atomic and molecular movement. These criteria correspond to the term *topochemical reaction*. In many cases, the topochemical polymerization proceeds homogeneously by a crystal-to-crystal transformation. Therefore, *polymer single crystals*, which are otherwise difficult to obtain, can be obtained by topochemical photopolymerization.

10.4.2
Topochemical photopolymerization of diacetylenes

The discovery of the photopolymerization of crystalline diacetylenes, such as hexa-3,5-diyne-1,6-diol and other derivatives (see Chart 10.7) [2, 30, 63], initiated scientific and technical developments extending to various fields [31–33, 35, 65, 66].

First of all, basic research concerning chemical reactions in the solid state was stimulated. As a result, various applications became feasible, since the diacetylene polymerization principle turned out to be applicable to various other organized structures, including Langmuir-Blodgett films, liposomes, vesicles, and self-assembled monolayers on metal oxide or graphite surfaces. A typical example concerns the photopolymerization of self-ordered monomolecular layers of pentacosadiynoic acid, $CH_3(CH_2)_{11}-C\equiv C-C\equiv C-(CH_2)_8COOH$, and nonacosadiynoic acid, $CH_3(CH_2)_{15}-C\equiv C-C\equiv C-(CH_2)_8COOH$, on a graphite substrate [68]. Scheme 10.20 depicts the assembly of the diacetylene molecules and the subsequent photopolymerization at 254 nm.

An exciting feature of such polymerized monolayers is the color change from blue to red that accompanies conformational changes in conjugated polydiacetylenes induced by changes in temperature or pH or by mechanical stress. This phenomenon has been exploited in the construction of direct sensing devices [70–75]. The latter consist of functionalized polydiacetylene bilayers with covalently attached receptors. Binding of biological entities (large molecules or cells) provides a mechanical stimulus. It causes conformational changes in the polydiacetylene layers (side-chain disordering and disruption of main-chain packing), resulting in a chromatic shift [67]. This method has been exploited, for example, in the direct colorimetric detection of an influenza virus [70] and of cholera toxin [71], as well as of biochemical substrates such as glucose [72].

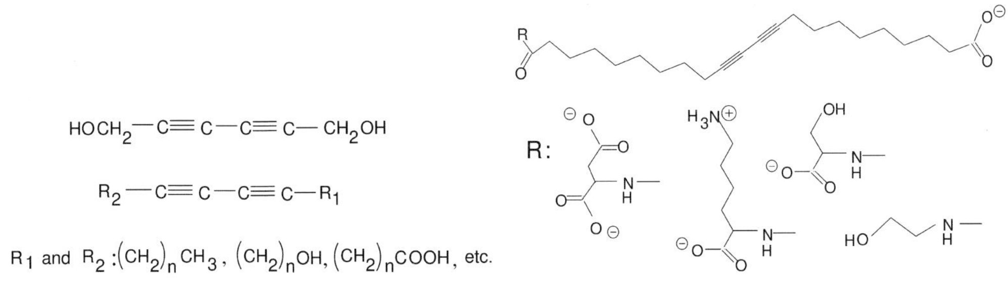

$HOCH_2-C\equiv C-C\equiv C-CH_2OH$

$R_2-C\equiv C-C\equiv C-R_1$

R_1 and R_2 : $(CH_2)_n CH_3$, $(CH_2)_n OH$, $(CH_2)_n COOH$, etc.

Chart 10.7 Typical diacetylene derivatives capable of undergoing topochemical photopolymerization. Left: classical examples [30]. Right: Self-assembling bolaamphiphilic diacetylenes [67].

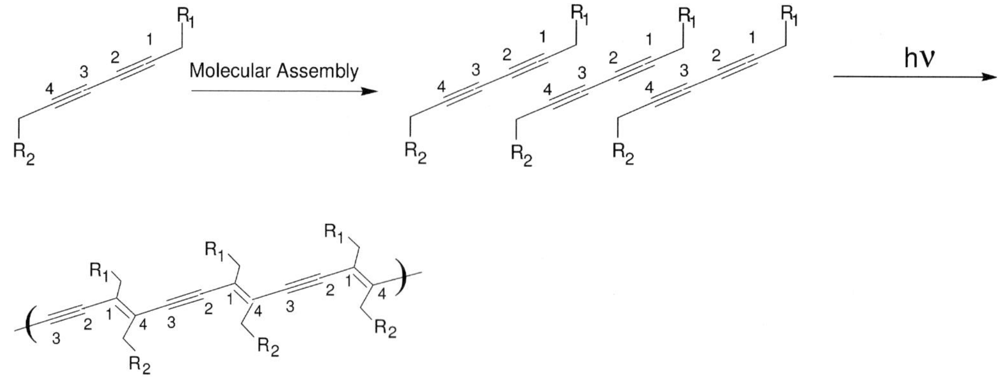

Scheme 10.20 Schematic representation of the polymerization of assembled functional diacetylenes by 1,4-addition upon exposure to UV light. R_1 and R_2 denote functionalized alkyl chains [69].

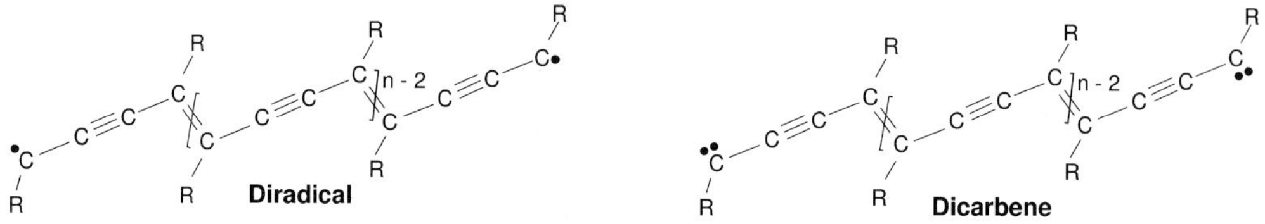

Chart 10.8 Structures of diradicals and dicarbenes involved in the topochemical photopolymerization of diacetylenes [32].

At ambient temperatures, the polymerization of diacetylenes proceeds as a chain reaction by 1,4-addition and results in alternating ene-yne polymer chains with exclusive *trans* selectivity. The quantum yield for initiation is low (ca. 0.01) [31]. Upon absorption of a photon by a diacetylene moiety of one of the molecules in the assembly or crystal, an excited diradical state with an unpaired electron at either end is generated. Subsequently, the radical sites undergo thermal addition reactions with neighboring diacetylene moieties. The resulting dimers possess reactive radical sites at their ends, which are capable of inducing chain growth [31, 34, 76]. There is experimental evidence (ESR) that dicarbenes (see Chart 10.8) are also involved in the polymerization, if chains become longer than five repeating units [31, 32].

An essential prerequisite for the topochemical polymerization of diacetylenes is a packing of the monomer molecules at a distance of $d = 4.7–5.2$ Å and a tilt angle of about 45° between the molecular axis and the packing axis [35].

10.4.3
Topochemical photopolymerization of dialkenes

The photopolymerization of diolefinic crystals was discovered in the case of the [2+2] photocyclopolymerization of 2,5-distyrylpyrazine (DSP) crystals and was named *four-center-type polymerization* (see Scheme 10.21) [36, 37].

Chart 10.9 presents four other dialkenes that are amenable to topochemical photopolymerization.

Notably, the polymerization of dialkenes proceeds as a stepwise process and not as a chain reaction. In other words, the addition of each repeating unit to the chain requires the absorption of a further photon (see Scheme 10.22).

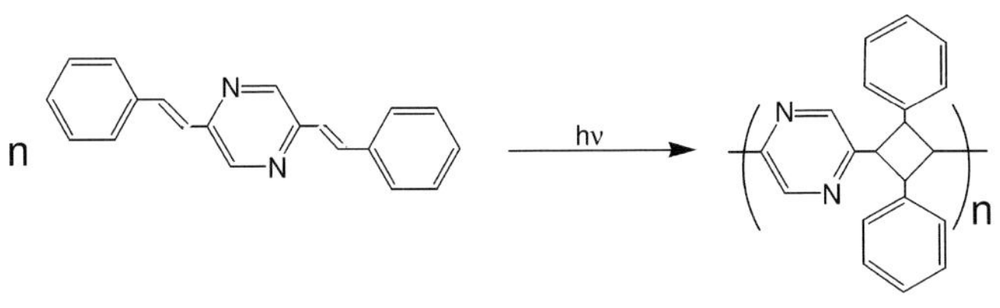

Scheme 10.21 Four-center-type photopolymerization of crystalline 2,5-distyrylpyrazine [37].

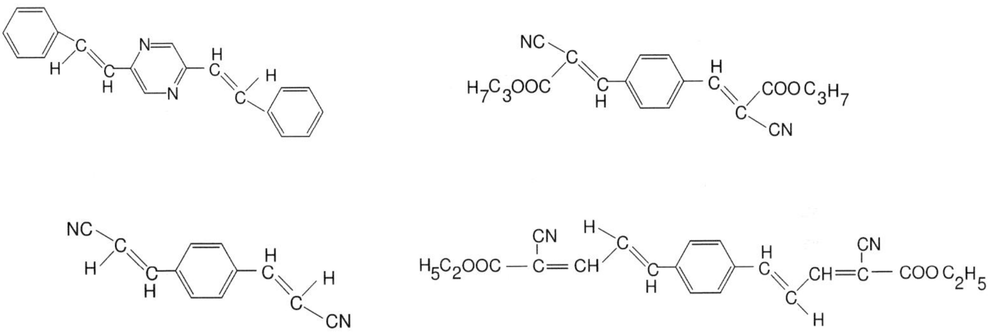

Chart 10.9 Dialkenes capable of undergoing topochemical polymerization upon exposure to UV light [37].

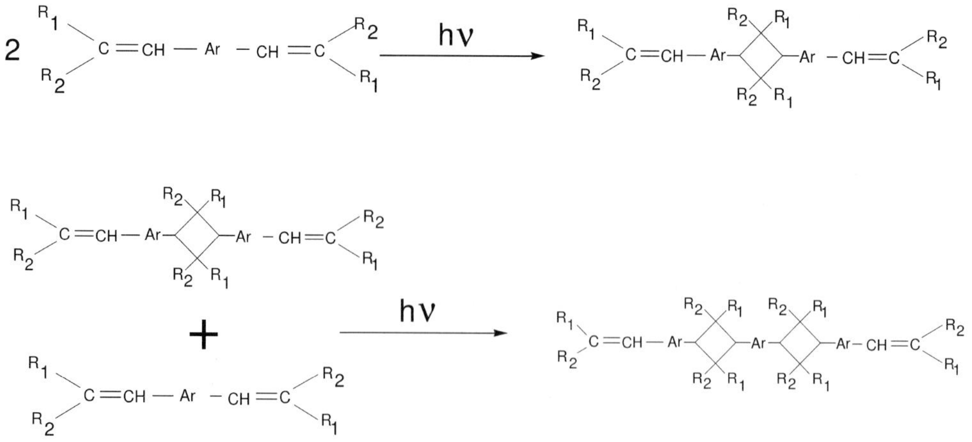

Scheme 10.22 Stepwise [2+2] photocyclopolymerization of a dialkene [36, 37].

References

1 K. Dietliker, T. Jung, J. Benkhoff, H. Kura, A. Matsumoto, H. Oka, D. Hristova, G. Gescheidt, G. Rist, *New Developments in Photoinitiators*, Macromol. Symp. 217 (2004) 77.
2 K. Dietliker, *Photoinitiators for Free Radical and Cationic Polymerization*, Vol. III, in P.K.T. Oldring (ed.), *Chemistry and Technology of UV and EB Formulations for Coatings, Inks and Paints*, SITA Technology, London (1991).
3 K. Dietliker, T. Jung, J. Benkhoff, *Photolatent Amines: New Opportunities in Radiation Curing*, Techn. Conf. Proc., UV & EB Technol. Expo & Conf., Charlotte NC, USA (2004), p. 217.
4 K. Dietliker, *A Compilation of Photoinitiators Commercially Available for UV Today*, SITA Technology, Edinburgh (2002).
5 J.V. Crivello, *Latest Developments in the Chemistry of Onium Salts*, Chapter 8 in Vol. III of [10].
6 J.V. Crivello, *The Discovery and Development of Onium Salt Cationic Photoinitiators*, J. Polym. Sci. Part A, Polym. Chem. 37 (1999) 4241.
7 J.V. Crivello, K. Dietliker, *Photoinitiators for Free Radical, Cationic and Anionic Photopolymerization*, Wiley, New York (1998).
8 I. Reetz, Y. Yagci, M.K. Mishra, *Photoinitiated Radical Vinyl Polymerization*, in M.K. Mishra, Y. Yagci (eds.), *Handbook of Radical Vinyl Polymerization*, Dekker, New York (1998).
9 J.P. Fouassier (ed.), *Photoinitiation, Photopolymerization and Photocuring. Fundamentals and Applications*, Hanser, München (1995).
10 J.P. Fouassier, J.F. Rabek (eds.), *Radiation Curing in Polymer Science and Technology*, Elsevier Applied Science, London (1993).
11 C.G. Roffey, *Photogeneration of Reactive Species for UV Curing*, Wiley, New York (1997).
12 G. Oster, N.L. Yang, *Photopolymerization of Vinyl Monomers*, Chem. Rev. 68 (1968) 125.
13 N.S. Allen (ed.), *Photopolymerization and Photoimaging, Science and Technology*, Elsevier Applied Science, London (1989).
14 N.S. Allen, *Photoinitiators for Photocuring*, in J.C. Salamone (Ed.), *Concise Polymeric Materials Encyclopedia*, CRC Press, Boca Raton, FL, USA (1999), p. 1047.
15 H.J. Timpe, S. Jokusch, K. Körner, *Dye-Sensitized Photopolymerization*, Chapter 13 in Vol. II of [10].
16 A.B. Scranton, C.N. Bowman, R.W. Pfeiffer (eds.), *Photopolymerization*, ACS

Symposium Series 673, American Chemical Society, Washington, D.C. (1996).
17 S. P. Pappas (ed.), *UV Curing, Science and Technology*, 2nd ed, Technology Marketing Corp., Stamford, CT, USA (1985).
18 S. P. Pappas (ed.), *Radiation Curing, Science and Technology*, Plenum Press, New York (1992).
19 R. S. Davidson, *Polymeric and Polymerisable Free Radical Photoinitiators*, J. Photochem. Photobiol. A Chem. 69 (1993) 263.
20 H. F. Gruber, *Photoinitiators for Free Radical Polymerization*, Prog. Polym. Sci. 17 (1993) 953.
21 C. Decker, *Photoinitiated Crosslinking Polymerization*, Prog. Polym. Sci. 21 (1996) 593.
22 T. Yamaoka, K. Naitoh, *Visible Light Photoinitiation Systems Based on Electron Transfer and Energy Transfer Processes*, in V. V. Krongauz, A. D. Trifunac (eds.), *Processes in Photoreactive Polymers*, Chapman & Hall, New York (1995).
23 D. Billy, C. Kutal, *Inorganic and Organometallic Photoinitiators*, Chapter 2 in [18].
24 A. F. Cunningham, V. Desobry, *Metal-Based Photoinitiators*, Chapter 6 of Vol. II in [10].
25 W. Schnabel, *Cationic Photopolymerization with the Aid of Pyridinium-Type Salts*, Macromol. Rapid Commun. 21 (2000) 628.
26 W. Schnabel, *Photoinitiation of Ionic Polymerizations*, Chapter 7 in N. S. Allen, M. Edge, I. R. Bellobono, E. Selli (eds.), *Current Trends in Polymer Photochemistry*, Horwood, New York (1995).
27 Y. Yagci, I. Reetz, *Externally Stimulated Initiator Systems for Cationic Polymerization*, Prog. Polym. Sci. 23 (1998) 1485.
28 R. Lazauskaite, J. V. Grazulevicius, *Cationic Photopolymerization*, Chapter 7 of Vol. 2, in H. S. Nalwa (ed.), *Handbook of Photochemistry and Photobiology*, American Scientific Publishers, Stevenson Ranch, CA, USA (2003).
29 V. Strehmel, *Epoxies, Structures, Photoinduced Cross-Linking, Network Properties, and Applications*, Vol. 2, p. 2, in H. S. Nalwa (ed.), *Handbook of Photochemistry and Photobiology*, American Scientific Publishers, Stevenson Ranch, CA, USA (2003).
30 G. Wegner, *Solid-State Polymerization Mechanisms*, Pure & Appl. Chem. 49 (1977) 443.
31 H. Sixl, *Spectroscopy of the Intermediate State of the Solid-State Polymerization Reaction in Diacetylene Crystals*, Adv. Polym. Sci. 63 (1984) 49.
32 H. Bässler, *Photopolymerization of Polydiacetylenes*, Adv. Polym. Sci. 63 (1984) 1.
33 D. Bloor, R. R. Chance (eds.), *Polydiacetylenes, Synthesis, Structures and Electronic Properties*, Nijhoff, Dordrecht (1985).
34 M. Schwörer, H. Niederwald, *Photopolymerization of Diacetylene Single Crystals*, Makromol. Chem. Suppl. 12 (1985) 61.
35 V. Enkelmann, *Structural Aspects of the Topochemical Polymerization of Diacetylenes*, Adv. Polym. Sci. 63 (1984) 91.
36 M. Hasegawa, *Photopolymerization of Diolefin Crystals*, Chem. Rev. 83 (1983) 507.
37 M. Hasegawa, *Product Control in Topochemical Photoreactions*, Chapter 10 in N. S. Allen, M. Edge, I. R. Bellobono, E. Selli (eds.), *Current Trends in Polymer Photochemistry*, Horwood, New York (1995).
38 C. E. Hoyle, J. F. Kinstle (eds.), *Radiation Curing of Polymeric Materials*, ACS Symposium Series 417, American Chemical Society, Washington, D.C. (1990).
39 G. Odian, *Principles of Photopolymerization*, Wiley, New York (1991).
40 C. Carlini, L. Angiolini, *Polymers as Free Radical Photoinitiators*, Adv. Polym. Sci. 123, Springer, Berlin (1995).
41 A. Kajiwara, Y. Konishi, Y. Morishima, W. Schnabel, K. Kuwata, M. Kamachi, Macromolecules 26 (1993) 1656.
42 T. Sumiyoshi, W. Schnabel, A. Henne, P. Lechtken, Polymer 26 (1985) 141.
43 A. Wrzynszczynski, J. Bartoszewicz, G. L. Hig, B. Marciniak, K. Paczkowski, J. Photochem. Photobiol. Chem. 155 (2003) 253.
44 R. West, A. R. Wolff, D. J. Peterson, J. Radiat. Curing 13 (1986) 35.
45 C. Peinado, A. Alonso, F. Catalina, W. Schnabel, Macromol. Chem. Phys. 201 (2000) 1156.

46 C. Badarau, Z. Y. Wang, Macromolecules 36 (2003) 6959.
47 J. Finter, M. Riedicker, O. Rohde, B. Rotzinger, Makromol. Chem. Makromol. Symp. 24 (1989) 177.
48 C. Bibaut-Renauld, D. Burget, J. P. Fouassier, C. G. Varelas, J. Thomatos, G. Tsagaropoulos, L. O. Ryrfors, O. J. Karlsson, J. Polym. Sci. Part A: Polym. Chem. 40 (2002) 371.
49 C. Dong, X. Ni, J. Macromol. Sci. Part A, Pure & Appl. Chem. A 41 (2004) 547.
50 A. J. Hoffman, G. Mills, H. Yee, M. R. Hoffmann, J. Phys. Chem. 96 (1992) 5540 and 5546.
51 I. G. Popovic, L. Katzikas, U. Müller, J. S. Velickovic, H. Weller, Macromol. Chem. Phys. 195 (1994) 889.
52 J. V. Crivello, J. Ma, F. Jiang, J. Polym. Sci.: Part A: Polym. Chem. 40 (2002) 3465.
53 H. Li, K. Ren, D. C. Neckers, Macromolecules 34 (2001) 8637.
54 J. V. Crivello, M. Jang, J. Macromol. Sci., Pure Appl. Chem. A42 (2005) 1.
55 S. Denizligil, Y. Yagci, C. M. McArdle, Polymer 36 (1995) 3093.
56 A. Mejiritski, A. M. Sarker, B. Wheaton, D. C. Neckers, Chem. Mater. 9 (1997) 1488.
57 C. Kutal, P. A. Grutsch, D. B. Yang, Macromolecules 24 (1991) 6872.
58 Y. Yamaguchi, B. J. Palmer, C. Kutal, T. Wakamatsu, D. B. Yang, Macromolecules 31 (1998) 5155.
59 V. Jarikov, D. C. Neckers, Macromolecules 33 (2000) 7761.
60 R. B. Paul, J. M. Kelly, D. C. Pepper, C. Long, Polymer 38 (1997) 2011.
61 M. D. Cohen, G. M. Schmidt, J. Chem. Soc. (1964) 1006.
62 G. M. Schmidt, Pure & Appl. Chem. 27 (1971) 647.
63 G. Wegner, Z. Naturforsch. 24B (1967) 824.
64 M. Hasegawa, Y. Susuki, J. Polym. Sci. B 5 (1967) 813.
65 V. Enkelmann, G. Wegner, K. Novak, K. B. Wagner, J. Am. Chem. Soc. 115 (1993) 1678.
66 C. Bubeck, B. Tieke, G. Wegner, Ber. Bunsenges. Phys. Chem. 86 (1982) 495.
67 J. Song, J. S. Cisar, C. R. Bertozzi, J. Am. Chem. Soc. 126 (2004) 8459.
68 Y. Okawa, M. Aono, Nature 409 (2001) 683.
69 J.-M. Kim, E.-K. Ji, S. M. Woo, H. Lee, D. J. Ahn, Adv. Mater. 15 (2003) 1118.
70 D. H. Charych, J. O. Nagy, W. Spevak, M. D. Benarski, Science 261 (1993) 585.
71 D. H. Charych, Q. Cheng, A. Reichert, G. Kuzienko, M. Stroh, J. O. Nagy, W. Spevak, R. C. Stevens, Chem. Biol. 3 (1996) 113.
72 Q. Cheng, R. C. Stevens, Adv. Mater. 9 (1997) 481.
73 K. Morigaki, T. Baumgart, A. Offenhäuser, W. Knoll, Angew. Chem. Int. Ed. 40 (2001) 172.
74 T. S. Kim, K. C. Chan, R. M. Crooks, J. Am. Chem. Soc. 119 (1997) 189.
75 Q. Huo, K. C. Russel, R. M. Leblanc, Langmuir 15 (1999) 3972.
76 W. Neumann, H. Sixl, Chem. Phys. 58 (1981) 303.

11
Technical developments related to photopolymerization

11.1
General remarks

Photopolymerization is the basis of some very important practical applications, for instance in the areas of surface coating and printing plates. In these cases, low molar mass liquid compounds are converted into rigid intermolecularly cross-linked materials that are insoluble in solvents. The relevant technological processes are denoted by the term *curing*. Detailed information is available in various books and review articles [1–15]. In contrast to thermal curing, *photocuring* can be performed at ambient temperatures with solvent-free formulations, i.e. *volatile organic compounds (VOCs)* are not released. In many cases, photocuring processes that proceed within a fraction of a second have replaced conventional thermal curing of solvent-containing formulations.

The main industrially applied photocuring processes are based on four chemical systems that are converted into three-dimensional networks upon irradiation [16]: (1) Unsaturated maleic/fumaric acid-containing polyesters (UPEs) dissolved in styrene, (2) acrylate/methacrylate systems, (3) thiol/ene systems, and (4) epoxide- or vinyl ether-containing systems. In the case of systems (1)–(3), free radical polymerizations are operative, while in case (4) cationic species are involved (see Chapter 10). Regarding thiol/ene systems, the mechanism of free radical thiol/alkene polymerization outlined in Scheme 11.1 is assumed to be operative [17]. Here, the chemistry depends upon the rate of hydrogen transfer from the thiol being competitive with the rate of alkene polymerization. By employing polyfunctional thiol compounds, very tough abrasion-resistant coatings are formed [8].

Industrially applied polymerizable formulations are composed of mixtures of mono- and multifunctional monomers and oligomers (see Table 11.1) containing a photoinitiator and, if required, also additives such as polymers (pre-polymers, resins) and pigments. Table 11.2 presents, as a typical example, the composition of a formulation applied for microfabrication (see Section 11.4).

Whether radical or cationic initiators are employed depends on the kind of mechanism (free radical or cationic, see Sections 10.2 and 10.3) according to which the monomers polymerize. Industrial applications of photocuring are extremely varied and include the coating of metals (automotive varnishes), the

Polymers and Light. Fundamentals and Technical Applications. W. Schnabel
Copyright © 2007 WILEY-VCH Verlag GmbH & Co. KGaA, Weinheim
ISBN: 978-3-527-31866-7

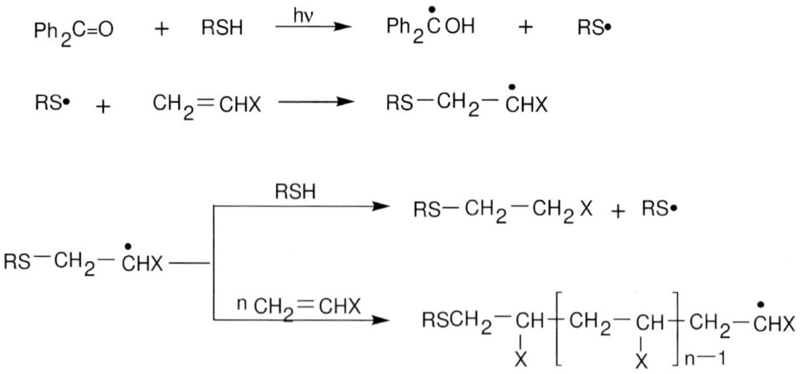

$$Ph_2C=O + RSH \xrightarrow{h\nu} Ph_2\dot{C}OH + RS^\bullet$$

$$RS^\bullet + CH_2=CHX \longrightarrow RS-CH_2-\dot{C}HX$$

$$RS-CH_2-\dot{C}HX \begin{cases} \xrightarrow{RSH} RS-CH_2-CH_2X + RS^\bullet \\ \xrightarrow{n\,CH_2=CHX} RSCH_2-CH(X)-[CH_2-CH(X)]_{n-1}-CH_2-\dot{C}HX \end{cases}$$

Scheme 11.1 Free radical thiol/alkene polymerization [8].

Table 11.1 Typical di- and trifunctional compounds used for photocuring.

Class	Chemical structure	Mode of polymerization
Trifunctional acrylates	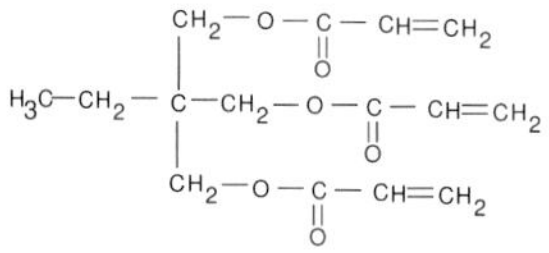 Trimethylolpropane triacrylate 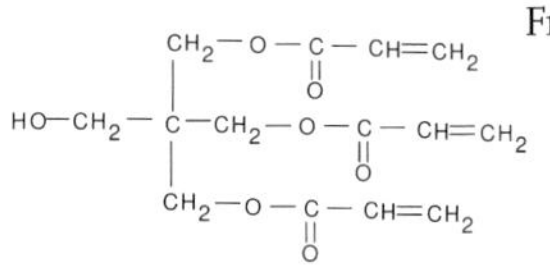 / Pentaerythritol triacrylate	Free radical
Oligomeric diacrylates	$H_2C=CH-C(O)-O-X-O-C(O)-CH=CH_2$; X: Polyester, Polyether, Polyurethane, Polysiloxane	Free radical
Thiol/Enes	$C(X-SH)_4 / H_2C=CH-X'-CH=CH_2$	Free radical
Difunctional epoxides	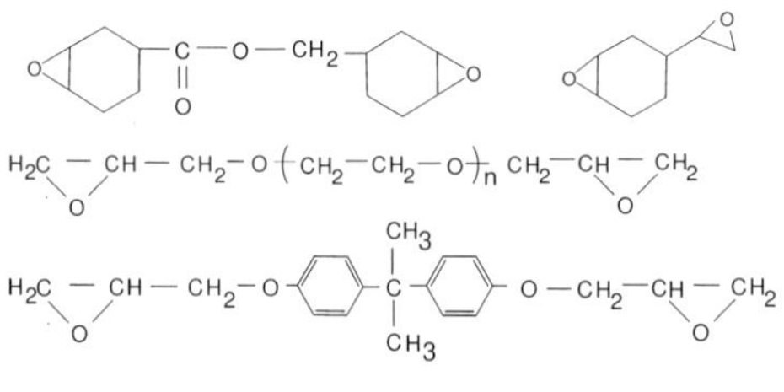 (cycloaliphatic diepoxide structures; bisphenol-A diglycidyl ether derivatives)	Cationic
Epoxidized siloxanes	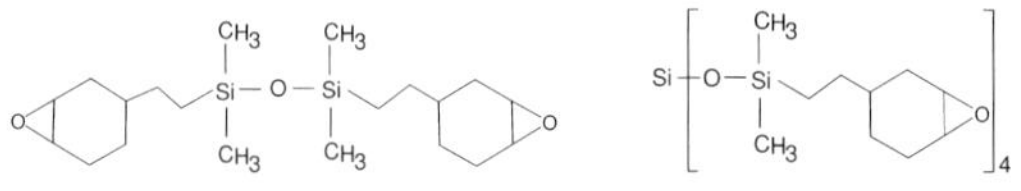 (cyclohexyl-epoxide terminated siloxanes)	Cationic
Difunctional vinyl ethers	$H_2C=CH-(O-CH_2-CH_2-O)_3-CH=CH_2$; $H_2C=CH-O-X-O-CH=CH_2$; X: Polyester, Polyether, Polyurethane, Polysiloxane	Cationic

Table 11.2 Composition of a formulation applied for microfabrication [18].

Monomer/Oligomer	wt%
Alkoxylated trifunctional acrylate	10
Tris(2-hydroxyethyl) isocyanurate triacrylate	10
Trifunctional methacrylate	11
Ethoxylated trimethylolpropane triacrylate	10
Triethyleneglycol diacrylate	11
Isobornyl acrylate	2.5
Trimethylolpropane trimethacrylate	2.75
Brominated urethane acrylate (oligomer)	7.5
Aliphatic polyester-based urethane dimethacrylate (oligomer)	8.25
Aromatic urethane acrylate (oligomer)	27

production of printed circuit boards, and the generation of 3-D models. Some of the applications are described in more detail in the following sections.

11.2
Curing of coatings, sealants, and structural adhesives

11.2.1
Free radical curing

11.2.1.1 Solvent-free formulations

UV curing of coatings was first commercially applied about four decades ago in the wood and furniture industries. It opened the door to significant savings by delivering shorter production times, improved product quality (better gloss), lower energy and equipment costs, as well as environmental friendliness because of the greatly reduced VOC emission. Today, UV curing is widely used and all sorts of substrates, including paper, plastic, and metal, are coated by employing this technique, as can be seen in Table 11.3.

Important commercial applications include clear coatings for paper, in particular overprint varnishes, as commonly applied to magazines and consumer good packaging. Paper coatings are applied at extremely high speeds, typically 5 m s^{-1}, and the coated products are immediately ready for testing or shipment. Such high-performance applications require a fast curing speed in conjunction with a conversion of reactive groups closely approaching 100%. In this context, the reactivity of the monomers and the viscosity of the formulation are of great importance. Regarding polyester acrylate-based formulations, for example, monomers bearing carbamate or oxazolidone groups (see Chart 11.1) proved to play a key role in allowing a remarkable level of performance [20]. These monomers are very reactive and ensure a slow increase in the viscosity of the formulation with conversion.

Table 11.3 Typical commercial applications of radiation-cured coatings in major industries [19].

Industrial sector	Mode of application
Furniture and construction	Hardwood flooring, PVC flooring, wood and metal furniture, particle board sealer, galvanized tubing, fencing, etc.
Electronics and telecommunications	Electrical conductor wire, printed circuit board coatings, optical fibers, magnetic media, coatings, computer disc clearcoats, coatings for metallized substrates
Printing and packaging	Inks, release coatings, overcoats for graphic art, magazine covers, coatings on beverage cans, coatings on non-food packaging, barrier coatings, DVD laminates
Automotive	Headlamps, printed dashboard components, refinishing coatings
Consumer goods	Release coatings for adhesives, pressure-sensitive adhesives, leather coatings, coatings on plastic housings (cell phones, computers, etc.), eyeglass lenses, mirror coatings

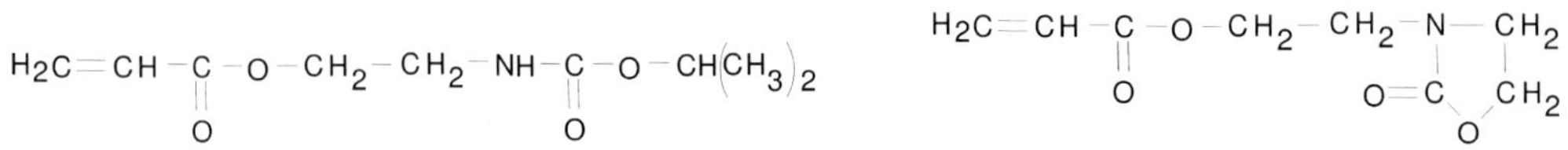

Chart 11.1 Chemical structures of acrylates containing carbamate (left) or oxazolidone groups (right) [20].

Silicones have the advantage of softness, biological inertness, good substrate wettability, and superb permeability of gases. Therefore, UV-cured silicones are suitable for various interesting product applications [21], including ophthalmic devices (hard and soft contact lenses, intraocular lens implants), gaskets, sealings, and optical fiber coatings. Photocurable formulations appropriate for the fabrication of such products contain siloxane derivatives bearing unsaturated carbon-carbon double bonds (see Chart 11.2).

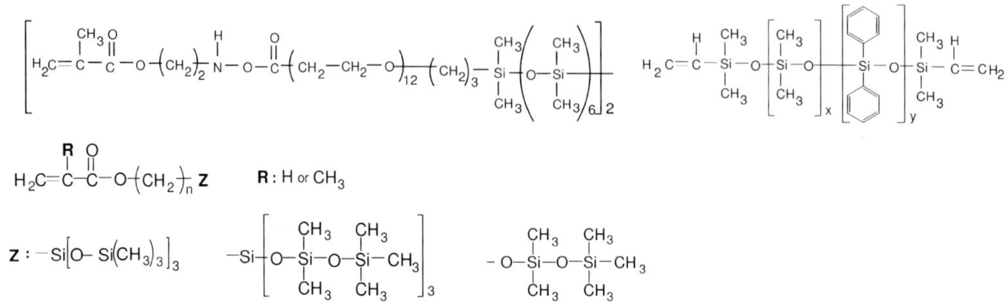

Chart 11.2 Chemical structures of typical siloxane-derived monomers [21].

11.2.1.2 Waterborn formulations

Waterborn formulations have been developed with the aim of extending the applicability of radiation curing. Representing a clear departure from the concept of solvent-free systems, waterborn formulations seem to be especially advantageous for the radiation curing of wood coatings. Formulations in the form of aqueous emulsions can be easily thinned by the addition of further water. Moreover, emulsions can be easily dispensed onto the substrate, e.g. by spraying. The resultant coatings possess good matting properties and adhere tightly to the substrate due to reduced shrinkage during curing. Naturally, the use of waterborn systems necessitates a drying step following the radiation curing process. High-frequency, near-infrared or microwave heating can be applied for this purpose [22].

11.2.2
Cationic curing

While in the early days acrylate-based systems, cured by a free radical mechanism, were overwhelmingly employed for surface coatings, nowadays epoxide-based systems, cured cationically, are also used to an increasing extent. Epoxide-based formulations yield excellent overprint varnishes on tin-free steel and aluminum for rigid packaging, especially in the production of steel food cans and aluminum beer and beverage cans. The cured films exhibit good adhesion, flexibility, and abrasion resistance, and the high production rates (up to 1600 cans per minute) are astounding [12].

Since coatings containing cycloaliphatic epoxies tend to be brittle, other compounds such as oligomeric polyols are frequently added as flexibilizing agents. Alcohols can react with the oxonium ions formed by the addition of protons to epoxide groups (see Scheme 11.2) and are thereby copolymerized with the epoxides.

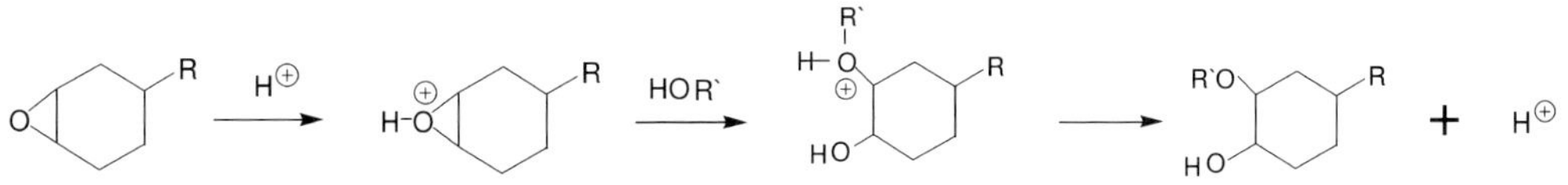

Scheme 11.2 Formation of oxonium ions by the addition of protons to epoxides and their subsequent reaction with alcohols.

When alcohols add to the growing polymer chains, protons are produced in equivalent amounts. Since these protons can, in turn, react with epoxide groups, the addition reaction represents a chain-transfer process. The use of difunctional alcohols results in an extension of the polymer chains, whilst polyfunctional alcohols contribute strongly to the formation of a three-dimensional polymer network. Chart 11.3 depicts the structure of part of such a three-dimensional network.

A major difference between cationic and free radical curing is the degree of shrinkage caused by the polymerization. Cationic ring-opening polymerization

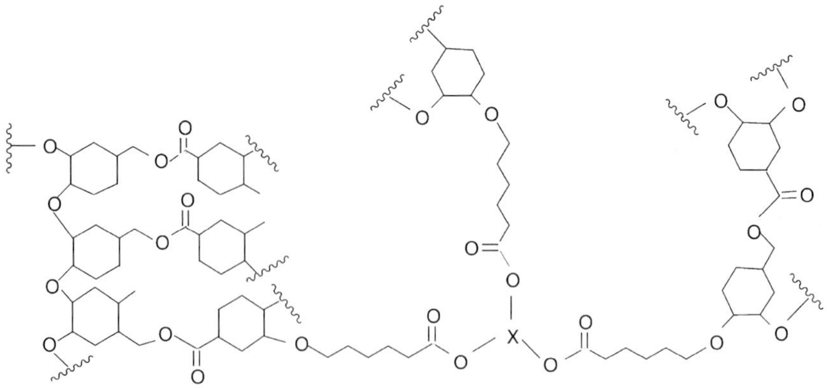

Chart 11.3 Structure of part of a network formed by the reaction of a difunctional epoxide with a trifunctional polyol.

leads to a shrinkage of 1–2%, as compared to 5–20% for radical polymerization of double bonds. A lower degree of shrinkage implies a stronger adhesion of the coating to the substrate.

11.2.3
Dual curing

Coatings protecting three-dimensional objects can be produced by *dual-curing* methods using chemical systems that are converted in two separate stages of polymerization or polycondensation [23]. Regarding the coating of three-dimensional objects, problems often arise from shadow areas that cannot be reached by the incident UV light and, therefore, remain uncured. Similar problems arise in the case of UV-curing of coatings on porous substrates such as wood and of thick pigmented coatings, where pigment particle screening prevents the penetration of light to deep-lying layers. In all of these cases, and also in the case of automotive top coatings, dual curing is successfully employed in industrial processes [24, 25]. A typical dual-curing method combining UV irradiation and thermal treatment operates with substances bearing two types of reactive functions, for example, UV-curable acrylate groups and thermally curable isocyanate groups associated with a polyol [24, 26]. Typical oligomers bearing both acrylate and isocyanate groups are shown in Chart 11.4.

First, UV irradiation initiates, with the aid of an appropriate initiator, the polymerization of acrylate groups, and then heating causes the isocyanate groups to react with hydroxyl groups. The latter reaction results in the formation of urethane linkages (see Scheme 11.3).

Polyols used for this purpose include trimethylol propane and propylene glycol. Systems containing urethane-acrylate oligomers bearing doubly-functionalized isocyanate groups are commercially available [24]. The chemical structure

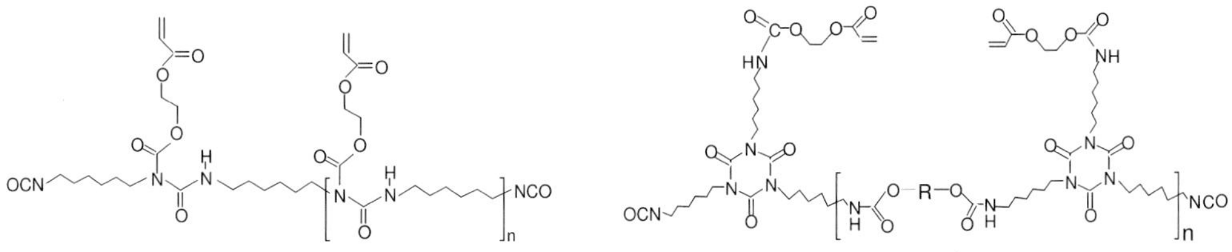

Chart 11.4 Chemical structures of oligomers bearing acrylate and isocyanate groups: isocyanato-allophane acrylate (left) and isocyanato-urethane acrylate (right) [24].

$$R_1-N=C=O \ + \ R_2OH \longrightarrow R_1-\underset{H}{\underset{|}{N}}-\underset{O}{\overset{\|}{C}}-O-R_2$$

Scheme 11.3 Formation of urethane linkages by the reaction of isocyanate with hydroxyl groups.

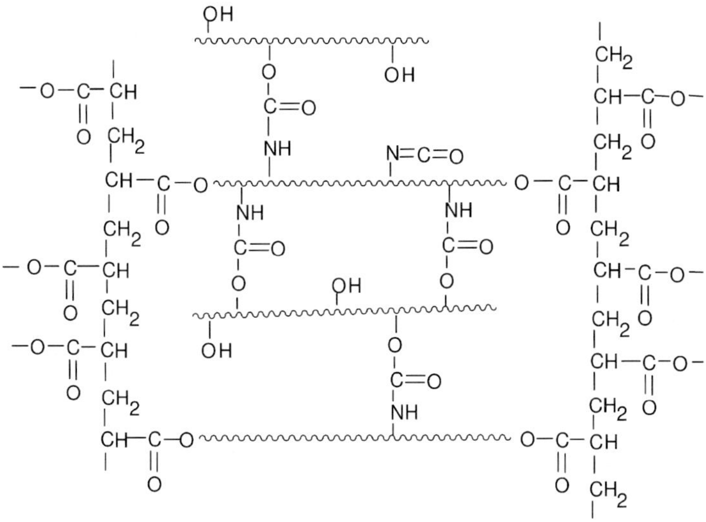

Chart 11.5 Part of a three-dimensional network formed by UV irradiation and thermal treatment of a dual-cure acrylic urethane system [27].

of a three-dimensional network formed from a dual-cure acrylic urethane system is presented in Chart 11.5.

Another mode of dual curing involves the simultaneous occurrence of free radical and cationic radiation-induced cross-linking polymerization of formulations containing appropriate initiators [20, 23, 28]. This method, which is called *hybrid curing*, leads to coatings with unique properties. A typical hybrid-cure system contains a diacrylate and a diepoxide, the former polymerizing by a free radical and the latter by a cationic mechanism. Exposure of the system to in-

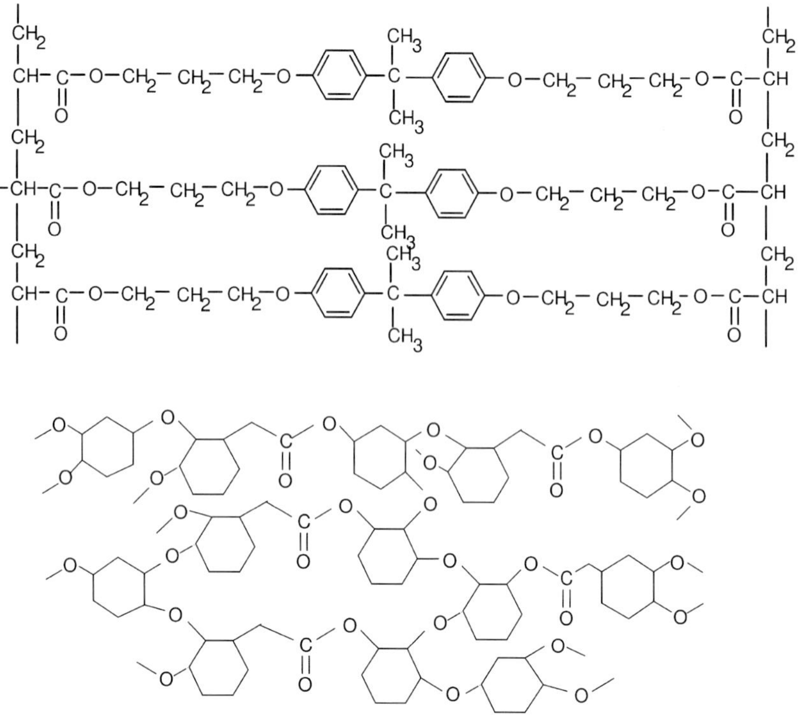

Chart 11.6 Segments of network structures formed by the radical polymerization of a diacrylate (top) and the cationic polymerization of a biscycloaliphatic diepoxide (bottom) [27].

tense UV radiation results in the formation of *interpenetrating networks* (IPNs, see Chart 11.6).

Often, IPN polymers combine the main features of the different networks. For example, elasticity and rigidity are combined in the case of interpenetrating networks formed from a vinyl ether and an acrylate, respectively.

11.3
Curing of dental preventive and restorative systems

Radiation-cured dental adhesives began replacing amalgam fillings in the early 1970s. The growth of cosmetic dentistry created new applications and, at present, dental adhesives comprise a major portion of all radiation-cured adhesives in terms of market value. Photocurable dental preventive and restorative formulations are composed of a mixture of monomeric and oligomeric esters of methacrylic and acrylic acid, a filler such as ultrafine silica, and a free-radical-type initiator system [29–36]. In the early days, curing was initiated at $\lambda \approx 360$ nm with benzoin and its derivatives or benzil ketals serving as photoinitiators. Nowadays,

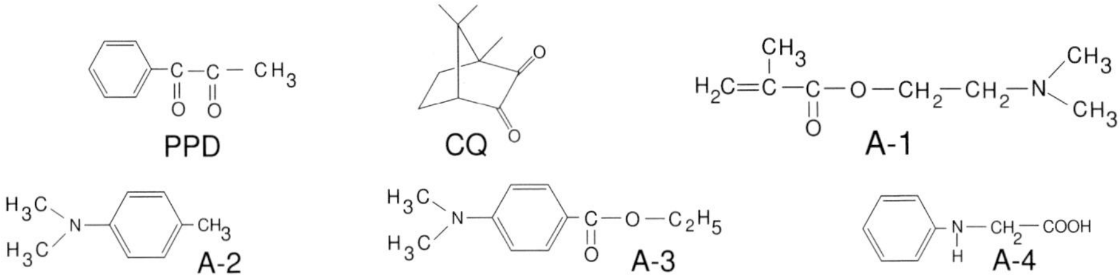

Chart 11.7 Chemical structures of diketones and amines serving as co-initiators in the curing of dental formulations.

curing is accomplished with visible light, e.g. with 488 nm light emitted by an argon-ion laser, using 1,2-diketone/amine initiator systems (see Subsection 10.2.2.4.3). The diketones include camphor quinone, CQ (1,7,7-trimethylbicyclo[2.2.1]heptane-2,3-dione), and 1-phenyl-1,2-propanedione, PPD. Appropriate amines include dimethylaminoethyl methacrylate (A-1), N,N-dimethyl-p-toluidine (A-2), p-N,N-dimethylaminobenzoic acid ethyl ester (A-3), and N-phenylglycine (A-4). The chemical structures are presented in Chart 11.7. N-Phenylglycine (A-4) is reportedly less biologically harmful than the other amines [31].

Chemical structures of typical polymerizable compounds employed for the generation of the polymeric matrix of dental formulations are presented in Chart 11.8.

Polymerized acrylate- and methacrylate-based resins are characterized by excellent aesthetics and good mechanical strength. Shortcomings include incomplete conversion, lack of durable adhesion to tooth structure, and, most importantly, polymerization shrinkage. The latter results from a volume contraction reflecting the conversion of van der Waals distances between free monomer units to the distances of covalent bonds linking these units in the polymer chain. To avoid multilayer application, this problem can be overcome by employing non-shrinking formulations containing oxaspiro monomers such as M-7 and M-8 (see Chart 11.9), together with diepoxides that undergo ring-opening polymerization initiated by cationically functioning photoinitiators upon exposure to visible light. Methacrylate-substituted oxaspiro monomers such as M-9 polymerize by a simultaneous free radical and cationic dual-photo-cure process to yield cross-linked, ring-opened structures. These aspects are discussed in a review by Antonucci et al. [30].

11.4
Stereolithography – microfabrication

Stereolithography is a technique widely adopted in industry in conjunction with *computer-aided design, CAD*, and *computer-aided manufacturing, CAM*, i.e. *micromachining* [32, 37, 38]. Stereolithography allows the fabrication of solid, plastic, three-dimensional (3-D) prototypes or models of products and devices from CAD drawings in a matter of hours. Rapid prototyping by means of stereo-

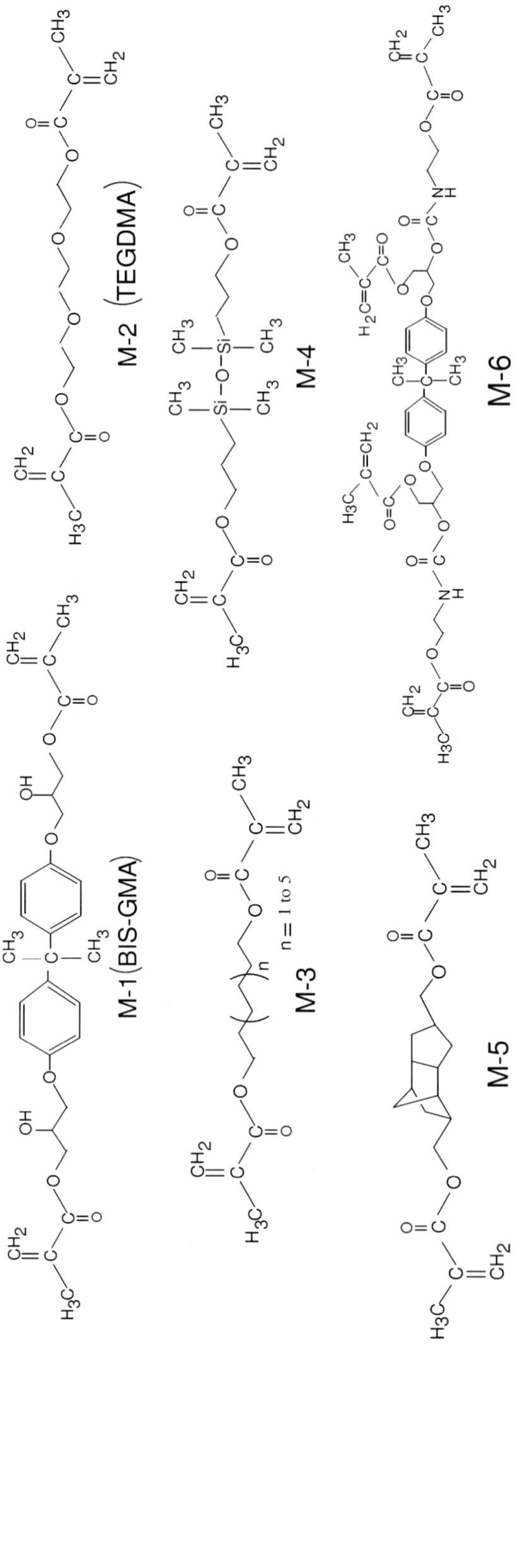

Chart 11.8 Chemical structures of typical polymerizable compounds contained in dental formulations [30].

Chart 11.9 Oxaspiro monomers used in non-shrinking dental formulations [30].

lithography is used everywhere from designing automotive and airplane parts to designing artificial hips and other replacement joints. The designer simply digitizes the plan, punches it into a computer, and gets a prototype within hours. The procedure involves hitting a photosensitive liquid contained in a vat with a laser beam. Under computer guidance, the beam outlines a shape. Wherever the light strikes the liquid, rapid polymerization occurs and thus the liquid solidifies. Since this process is restricted to a thin layer, a three-dimensional plastic model is built-up in a layer-by-layer growth procedure. This is accomplished by steadily lowering a movable table in the vat or by continuously pumping monomer into the vat from an external reservoir. Both procedures are depicted schematically in Fig. 11.1.

Rapid prototyping is an "additive" process, combining layers of plastic to create a solid object. In contrast, most machining processes (milling, drilling, grinding) are "subtractive" processes that remove material from a solid block.

Stereolithography also allows the creation of tiny parts of micrometer dimensions, including microgears that may be employed for the construction of micromachines such as micropumps and micromotors, artificial organs, surgical operating tools, etc. [39]. Also, polymeric three-dimensional photonic crystals, i.e. polymeric materials consisting of periodic microstructures, such as µm-sized rods, can

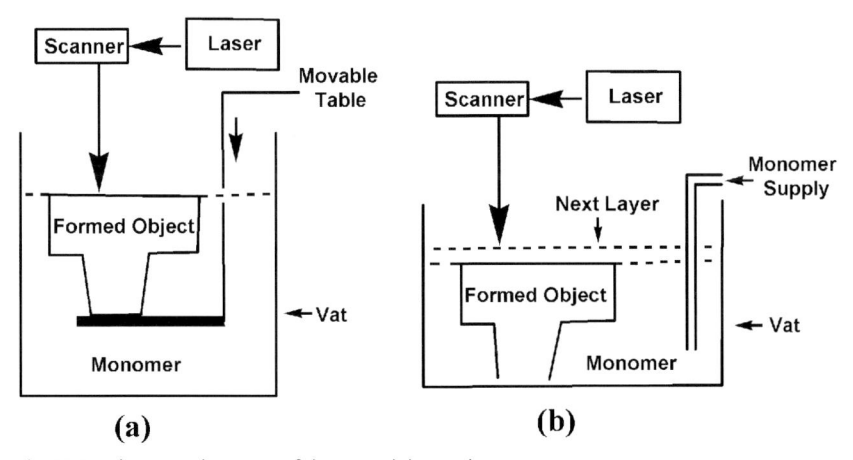

Fig. 11.1 Schematic depiction of the stereolithographic creation of solid, plastic, three-dimensional (3-D) prototypes with the aid of a movable table (a) or by monomer pumping (b).

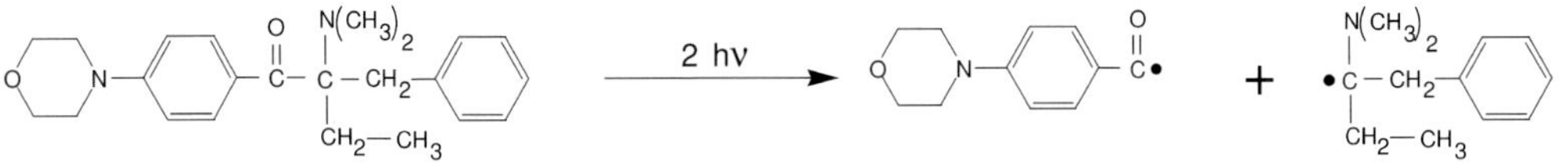

Scheme 11.4 Main pathway of the photolysis of a 4-morpholinophenyl amino ketone following two-photon absorption at $\lambda_{exc} = 600$ nm [18].

be generated by the laser microfabrication technique. Due to the presence of periodic microstructures, these materials possess photonic band gaps, i.e. wavelength regions in which propagating modes are forbidden in all directions. This offers the possibility to manipulate and control light [18, 40–44]. The development of this field pertains to *two-photon polymerization*, which relies on the simultaneous absorption of two photons by appropriate photoinitiators by way of a virtual electronic excitation state (see Section 3.3.2.3 and Fig. 3.6). In contrast to single-photon absorption, whereby the absorbed dose rate Dr_{abs} is proportional to the incident intensity I_0 ($Dr_{abs} \sim I_0$), Dr_{abs} is proportional to I_0^2 in the case of two-photon absorption ($Dr_{abs} \sim I_0^2$). This implies that photopolymerization can be confined to volumes with dimensions of the order of the wavelength of the light, as no out-of-focus absorption and thus polymerization can occur. Free radical two-photon polymerization has been performed with the aid of commercially available photoinitiators such as phosphine oxides or 2-benzyl-2-dimethylamino-1-(4-morpholinophenyl)butan-1-one, as shown in Scheme 11.4 [18, 43, 44].

Research has also been devoted to cationic two-photon photopolymerization using conventional initiator systems such as isopropylthioxanthone (ITX)/diaryl iodonium salt, with ITX serving as the photosensitizer [45, 46]. Mode-locked operated Ti:sapphire laser systems emitting femtosecond light pulses at 600, 710, or 795 nm were employed in these studies.

11.5
Printing plates

11.5.1
Introductory remarks

Printing processes use *printing plates* to transfer an image to paper or other substrates. The plates may be made of different materials. The image is applied to the printing plate by means of photomechanical, photochemical, or laser engraving processes. For printing, the plates are attached to a cylinder. Ink is applied to the image areas of the plate and transferred to the paper or, in the case of offset printing, to an intermediate cylinder and then to the paper.

During the past decades, photosensitive polymer printing plates have largely displaced the classical letterpress printing plates made of metals such as lead [47–51]. This technological revolution commenced in the 1950s [52], when the Dycryl (DuPont) and Nyloprint (BASF) letterpress plates entered the market [47]. The letterpress technique based on light-sensitive polymer printing plates is used to print newspapers, paperback books, business stationary, postage stamps, adhesive labels, and many other items. Print runs of 500 000 or more can be easily achieved. The letterpress printing plates are relief-structured, i.e. the printing areas are raised above the non-printing areas. During printing, ink dispensed on the raised areas is transferred to the substrate. Depending on the printing mode, the relief depth ranges from 0.2 to several mm. Instead of stiff printing plates, relief plates on a flexible support are employed in a special relief printing technique termed *flexography*. This technique can also be used for coarser and larger-scale work, such as in corrugated board printing. Besides letterpress printing, which is considered in the following subsections, photosensitive systems are also employed in other printing modes, such as gravure and screen printing [48, 49, 51].

11.5.2
Structure of polymer letterpress plates

As can be seen in Fig. 11.2, polymer letterpress plates consist of various layers: a protective cover layer, a photosensitive layer, an adhesion layer, and a support layer.

11.5.3
Composition of the photosensitive layer

The photosensitive layers of early plates were composed of acrylate/methacrylate and acrylated cellulose acetate mixtures. Other printing plates contained polyamides or nylon derivatives as binders. Generally, printing plates contain a mixture of reactive monomers and multifunctional oligomers (pre-polymers), polymeric binders, and photoinitiators with exceptional cure depth. The original photoinitiators were benzoin derivatives. Later, anthraquinone and other systems were used. Both free radical polymerization and cationic polymerization are applicable [49].

11.5.4
Generation of the relief structure

The printing plate, covered with the polymerizable material, often incorrectly referred to as the *photopolymer*, is irradiated through a film negative to initiate photocuring. Thereby, the areas of the photosensitive layer corresponding to the

Protective Layer	ca. 100 μm
Photosensitive Layer	30 to 6500 μm
Adhesion Layer	ca. 20 μm
Support Layer	100 to 200 μm

Fig. 11.2 Schematic depiction of the structure of a typical polymer letterpress plate. Adapted from Frass et al. [49] with permission from Wiley-VCH.

transparent regions of the negative film are polymerized and become insoluble in the developer. The relief structures generated in this way are required to possess a high cross-linking density so as to provide for sufficient hardness and heat and water resistance. Following irradiation, the plate is developed with an appropriate liquid (mostly water or alcohol/water mixtures), washed, dried, and, if necessary, re-exposed. A modern technique employs solvent-free thermal development [53]; the irradiated plate is fixed onto an internally heated drum in a processor heated to around 50 °C. At this temperature, the unexposed monomer forms a fluid that can be lifted from the plate with a fleece that is pressed against the plate. In 10–12 revolutions, a relief depth of 0.6–0.9 mm is reached, at which point the plate is ejected. Recently, printing plate fabrication techniques employing computer-to-plate digital laser exposure have been introduced, thus rendering the negative film process obsolete [54]. These techniques rely on the use of infrared lasers, particularly fiber lasers emitting at $\lambda = 1110$ nm [55]. Digital imaging of photopolymer plates requires a special plate composition. The photosensitive material, adhered to the substrate layer, is coated with a layer of carbon black only a few μm thick. The black layer is then ablated by the IR laser beam, resulting in a digital image on the surface of the plate. The printing plate is subsequently processed in much the same way as conventional photopolymer plates, by exposure to UV light, washout, drying, and finishing. Computer-to-plate printing is also accomplished with printing plates bearing a heat-sensitive mask layer containing IR absorbers. Prior to UV exposure, these plates are irradiated with a computer-guided IR laser in order to generate a mask through imagewise exposure [56].

11.6
Curing of printing inks

UV curing of inks is employed in flexographic and offset printing [57]. Besides pigments, appropriate inks typically contain unsaturated polymers based on polyacrylates and polyesters, photoinitiators, and additives [58]. The ink is cured after printing by exposing the printed items to UV light. Since the ink hardens within a fraction of a second, printing speeds of up to 300 m min^{-1} can be attained. UV-cured printing inks are superior to water-based, thermally cured inks due to their higher gloss and better fastness, i.e. abrasion resistance.

11.7
Holography

11.7.1
Principal aspects

The fact that photopolymerization can be used to record volume phase holograms is the basis of various commercial products made, for instance, by DuPont, Lucent/InPhase, and Polaroid. Therefore, the basic principles of holography are briefly described here, although other methods for writing holograms have been dealt with previously in the context of photorefractivity (Section 4.5) and photochromism (Section 5.8.2). There are various books that deal with the general area of holography [59–63].

The term holography derives from the Greek words *holos* (whole) and *graphein* (write) and denotes whole or total recording. A hologram is a two-dimensional recording but produces a three-dimensional image. Holography, invented by Gabor (Nobel Prize in 1971) [64], involves recording the complete wave field scattered by an object, that is to say, both the phase and the amplitude of the light waves diffracted by the object are recorded. This is in contrast to conventional imaging techniques, such as photography, which merely permit the recording of the intensity distribution in the original scene, and therefore, all information on the relative phases of the light waves coming from different points of the object is lost. Since recording media respond only to the light intensity, holography converts phase information into intensity variations. This is accomplished by using coherent illumination in conjunction with an interference technique. Figure 11.3 depicts schematically how a hologram is written.

Light generated by a laser simultaneously falls on the object and a mirror. The light waves diffracted from the object and those reflected by the mirror pro-

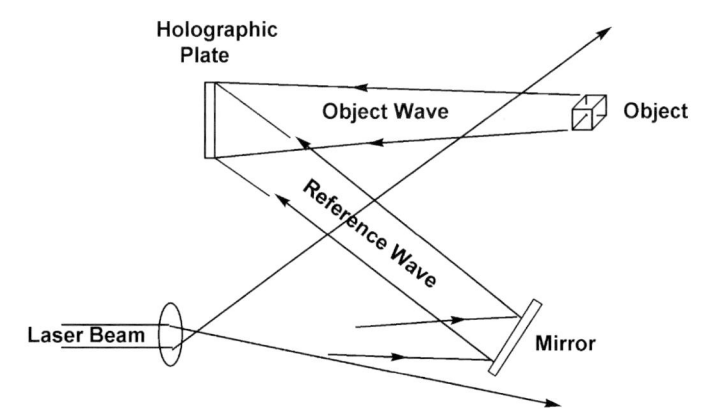

Fig. 11.3 Recording of a hologram of an object by generating an interference pattern on the detection plate.

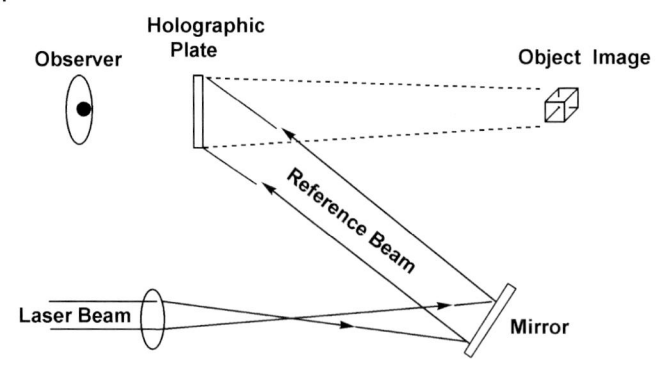

Fig. 11.4 Reconstruction of the image of an object recorded in a hologram by illuminating the detection plate with the reference light wave.

duce an interference pattern on the detection plate by generating a local refractive index modulation (phase hologram) or an absorption coefficient modulation (amplitude hologram). After processing, the image can be reconstructed by illuminating the hologram with only the reference light beam. As demonstrated in Fig. 11.4, light diffracted by the hologram appears to come from the original object.

The quality of a hologram is characterized by the efficiency factor $\eta = I/I_R$, where I and I_R are the intensities of the diffracted beam and the incident reference beam, respectively.

The term *volume holography* refers to recording plates with a thickness of up to a few millimeters. In such voluminous matrices, data storage in three dimensions is possible. This implies an enormous increase in storage capacity in comparison with other methods. If *multiplexing* techniques (see Section 12.1) are applied, thousands of holograms can be superimposed in the same plate.

Photopolymerizable systems appropriate for recording holograms are often, and sometimes also in this book, incorrectly referred to as *photopolymers*, although their essential components are monomers and not polymers. They typically comprise one or more monomers, a photoinitiator system, an inactive component (binder), and occasionally substances that serve to regulate pre-exposure shelf-life or viscosity. The resulting formulation is typically a viscous fluid, or a solid with a low glass transition temperature. For exposure, the formulation is coated onto a solid or flexible substrate or dispensed between two optically flat glass slides. Detailed information on the topic of *polymers in holography* is available in various reviews [65–72].

11.7.2
Mechanism of hologram formation

The formation of a hologram in a formulation containing polymerizable monomers is due to the generation of a refractive index grating [73]. When the holographic formulation is exposed to a light interference grating, the dispersed monomer polymerizes rapidly in the regions of high intensity, i.e., in the bright regions. Since the monomer concentration is depleted in these regions, concentration gradients are generated, which cause *component segregation*, i.e. the gradients drive the diffusion of the monomer from the dark into the depleted bright regions, where it polymerizes. Ultimately, the bright regions are characterized as areas of high concentration of newly formed polymer and the dark regions as areas of high binder concentration. Since the two materials differ in their refractive indices, a phase grating, recorded in real time, results. To increase the efficiency, the hologram may be heated for a short period to temperatures of 100–160 °C [73]. Further monomer diffusion, leading to an increased refractive index modulation, is believed to occur during the heating step. Any unreacted monomer can be finally converted by briefly exposing the plate to incoherent UV light (360–400 nm). No wet-processing is required with modern holographic formulations.

11.7.3
Multicolor holographic recording

Color holography allows the addition of life-like color to holographic images. As well as full-color display holograms, multi-wavelength holographic optical elements can also be made with the aid of color holography [74–76]. The phenomenon of *color mixing* employed in color photography is utilized to generate color holograms. By utilizing three recording wavelengths, usually red, green, and blue, which are simultaneously incident on the holographic plate, the impression of a wide variety of colors is created. Actually, the image of an object obtained from a color hologram is the superposition of the images of three holograms written with three laser beams. Typical laser wavelengths are 647 nm (red), 532 nm (green), and 476 nm (blue). If photopolymerizable formulations are employed, color holograms can be created by writing the holograms in a single holographic plate containing polymerization-initiating systems that are sufficiently sensitive at the specific wavelengths of the laser beams. Alternatively, color holograms can be created by employing multiple-layer holographic plates composed of wavelength-specific photopolymer layers (see Fig. 11.5) [75].

After recording, the plates are commonly subjected to a thermal treatment to increase the refractive index modulation and flood-exposed to UV light to fix the hologram. Wet-processing is not required [75]. For image retrieval, the holographic plate is simultaneously exposed to the three laser beams, whereby the colored image of the object is formed by the additive mixture of the individual holograms. Image retrieval with white light is possible, provided that the co-

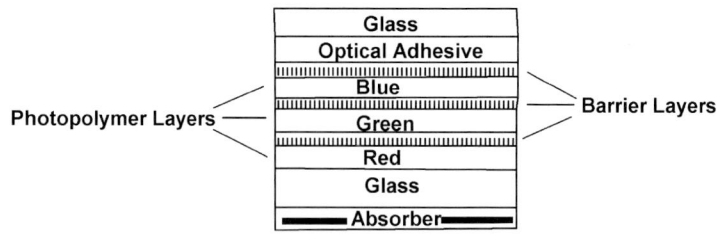

Fig. 11.5 Structure of a holographic three-layer plate employed for color holographic recording. Adapted from Trout et al. [74] with permission from the International Society for Optical Engineering (SPIE).

lored hologram was written with the reference beams incident on the reverse side of the plate.

11.7.4
Holographic materials

For many years, the most widely used holographic materials were *silver halide photographic emulsions* and *dichromated gelatin*. Upon exposure to light, gelatin layers containing a small amount of a dichromate such as $(NH_4)_2Cr_2O_7$ become progressively harder, since photochemically produced Cr^{3+} ions form localized cross-links between carboxylate groups of neighboring gelatin chains. This results in a modulation of the refractive index. The drawbacks of these materials are the need for wet-processing, high grain noise, and environmental sensitivity. During the last decades, various polymeric formulations have emerged as alternatives for practical holographic applications [72]. Although the precise compositions of relevant commercial formulations are not disclosed by the producers, it is generally agreed that in most cases acrylate- and methacrylate-based monomers are used as polymerizable components [66]. In typical holographic storage studies, the formulation comprises a difunctional acrylate oligomer, N-vinyl carbazole, and isobornyl acrylate [77]. In these cases, the polymerization proceeds by a free radical mechanism and initiator systems operating in the visible or near-IR wavelength region are employed. Multifunctional monomers are often added to the formulation so as to produce a molecular architecture that consists of a cross-linked polymer network, which improves dimensional stability and image fidelity.

Moreover, cationically polymerizable epoxide monomers capable of undergoing ring-opening polymerization (see Chart 11.10) are used in volume holographic recording [78, 79].

Actually, volume shrinkage is an important drawback regarding hologram recording based on vinyl monomer polymerization. On the other hand, no volume shrinkage or even a slight volume increase occurs upon polymerization of epoxide monomers. Therefore, in holographic formulations containing both

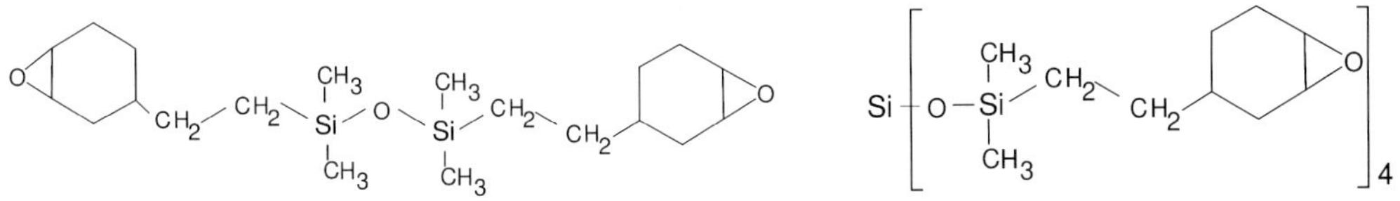

Chart 11.10 Structures of typical epoxide monomers employed in volume holography [78].

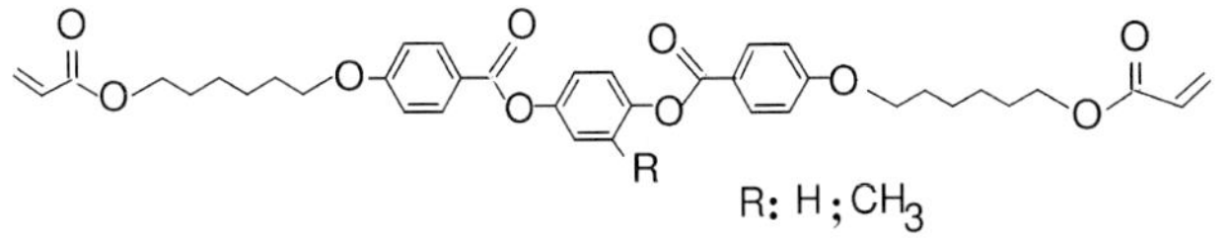

Chart 11.11 Chemical structure of a liquid-crystal-forming monomer [80].

types of monomers, volume shrinkage is largely reduced. This is especially the case if, prior to recording, a rather stable matrix is formed by in situ polymerization of the epoxide monomer. Thereby, a cross-linked network is formed in the presence of the unreacted acrylate monomer, which is ready for subsequent holographic recording [67].

Electrically switchable holograms can be generated with formulations containing a liquid-crystalline monomer. A typical example is given in Chart 11.11.

During recording, a highly cross-linked polymer is formed in the bright regions of the interference grating. Since it retains the initial order of the nematic monomer, the refractive index remains essentially unchanged. However, upon application of an electric field, the mobile monomeric regions corresponding to the dark regions are selectively reoriented, resulting in a large refractive index change. By repeatedly switching the electric field on and off, the hologram is also switched on and off [80]. Alternatively, electrically switchable holograms can be made by using formulations containing an unreactive liquid crystal and a non-liquid-crystal monomer. As the monomer diffuses from the dark regions to the bright regions to polymerize there, the liquid crystal is forced into the dark regions. There, it undergoes phase separation, appearing as droplets. The resulting so-called holographic polymer-dispersed liquid crystal (H-PDLC) can also be switched on and off by switching of the applied electric field [81–83].

11.7.5
Holographic applications

Holography has found a remarkably wide range of applications. Several companies produce photopolymer holograms for use in graphic arts, security, and goods authentication devices. Photopolymer holograms have the capability to offer bright and easily viewable displays for cell phones and other consumer electronics products, as well as unique eye-catching 3D color images that can be attached to a variety of products. Additional application fields include holographic optical elements, particle size analysis, high-resolution imaging, multiple imaging, stress analysis, and vibration studies. The importance of holography in information storage and processing is dealt with in Section 12.3.2. Actually, polymeric holographic formulations are promising materials for write-once-read-many (WORM) and read-only-memory (ROM) data storage applications, because of their good light sensitivity, good image stability, format flexibility, large dynamic range, and relatively low cost. There are various formulations that yield images directly upon exposure to light, i.e. images are developed in real time.

11.8
Light-induced synthesis of block and graft copolymers

11.8.1
Principal aspects

The copolymerization of monomers of different chemical nature often results in polymers possessing a specific combination of physical properties and is, therefore, of interest for the development of novel high-tech devices. This applies, in particular, to block and graft copolymers of the general structures indicated in Chart 11.12.

Block copolymers are composed of long chain segments of repeating units of types A or B, whereas graft copolymers are composed of chains of repeating units A, onto which side chains composed of repeating units B are grafted. Both types of copolymers can be synthesized by means of photochemical methods based on free radical or cationic mechanisms. For practical applications, cationic polymerizations are less attractive than free radical polymerizations. Therefore, only the latter will be dealt with in the following subsections.

Most of the known photochemical procedures for the synthesis of block and graft copolymers are based on the modification of already existing polymers with photolabile groups incorporated at defined positions, i.e. at the chain end, at side chains, or in the main chain (see Chart 11.13) [84].

Upon absorption of light, the photolabile groups can dissociate into pairs of free radicals capable of initiating the polymerization of a monomer present in the system (see Scheme 11.5).

Typical chromophoric groups that have been chemically incorporated into or attached to linear macromolecules for the purpose of photosynthesizing block

A∼∼∼∼A—B∼∼∼∼B   A∼∼∼∼∼∼∼∼∼∼∼∼A
                                    |
                              B∼∼∼∼∼∼B

      Block Copolymer   **Graft Copolymer**

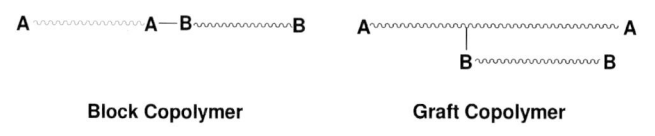

Chart 11.12 General chemical structures of block copolymers and graft copolymers consisting of monomer units A and B.

11.8 Light-induced synthesis of block and graft copolymers

$$\text{A}\sim\sim\text{A-X-Y} \quad\quad \text{A}\sim\sim\sim\underset{\underset{\text{X-Y}}{|}}{\text{A}}\sim\sim\sim\text{A} \quad\quad \text{A}\sim\sim\sim\text{X-Y}\sim\sim\sim\sim\text{A}$$

(a)   (b)   (c)

Chart 11.13 General structures of polymers bearing photolabile groups at the chain end (a), at side chains (b), or incorporated into the main chain (c).

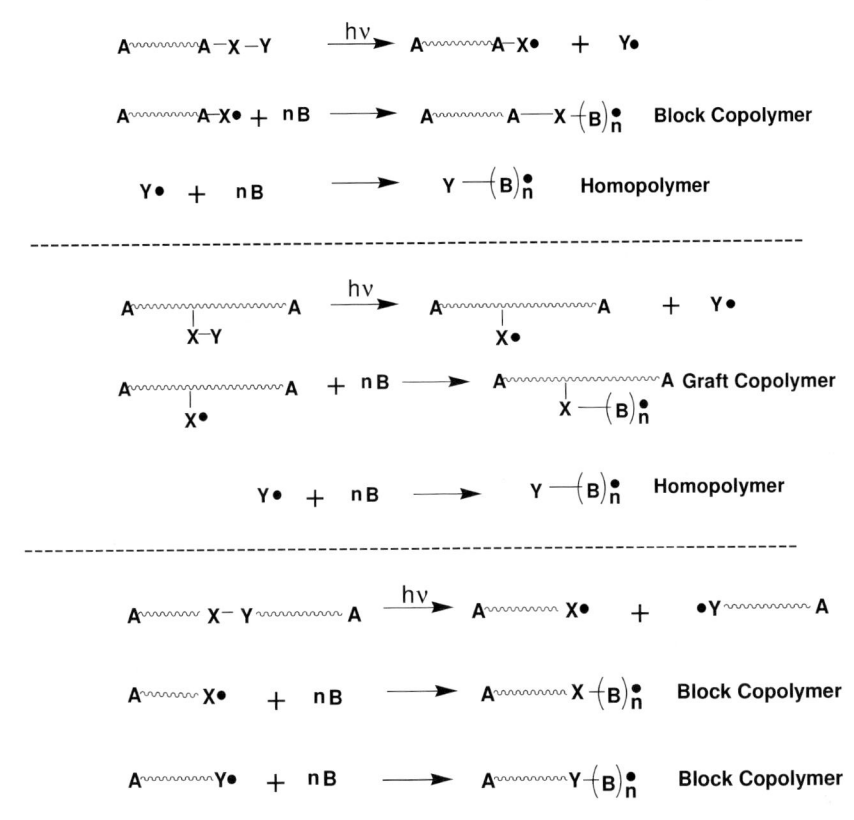

$$\text{A}\sim\sim\text{A-X-Y} \xrightarrow{h\nu} \text{A}\sim\sim\text{A-X}\bullet + \text{Y}\bullet$$

$$\text{A}\sim\sim\text{A-X}\bullet + n\text{B} \longrightarrow \text{A}\sim\sim\text{A-X}\text{-}(\text{B})_n^\bullet \quad \text{Block Copolymer}$$

$$\text{Y}\bullet + n\text{B} \longrightarrow \text{Y-}(\text{B})_n^\bullet \quad \text{Homopolymer}$$

$$\text{A}\sim\sim\sim\underset{\underset{\text{X-Y}}{|}}{\text{A}}\sim\sim\sim\text{A} \xrightarrow{h\nu} \text{A}\sim\sim\sim\underset{\underset{\text{X}\bullet}{|}}{\text{A}}\sim\sim\sim\text{A} + \text{Y}\bullet$$

$$\text{A}\sim\sim\sim\underset{\underset{\text{X}\bullet}{|}}{\text{A}}\sim\sim\sim\text{A} + n\text{B} \longrightarrow \text{A}\sim\sim\sim\underset{\underset{\text{X-}(\text{B})_n^\bullet}{|}}{\text{A}}\sim\sim\sim\text{A} \quad \text{Graft Copolymer}$$

$$\text{Y}\bullet + n\text{B} \longrightarrow \text{Y-}(\text{B})_n^\bullet \quad \text{Homopolymer}$$

$$\text{A}\sim\sim\text{X-Y}\sim\sim\sim\text{A} \xrightarrow{h\nu} \text{A}\sim\sim\text{X}\bullet + \bullet\text{Y}\sim\sim\sim\text{A}$$

$$\text{A}\sim\sim\text{X}\bullet + n\text{B} \longrightarrow \text{A}\sim\sim\text{X-}(\text{B})_n^\bullet \quad \text{Block Copolymer}$$

$$\text{A}\sim\sim\sim\text{Y}\bullet + n\text{B} \longrightarrow \text{A}\sim\sim\sim\text{Y-}(\text{B})_n^\bullet \quad \text{Block Copolymer}$$

Scheme 11.5 Formation of block and graft copolymers following the photodissociation of chromophoric groups. For the sake of simplicity, chain-termination reactions are not included.

or graft copolymers are compiled in Table 11.4 (see also Chart 10.2 in Section 10.2.2.3). Macromolecules bearing photolabile groups are occasionally also termed *macroinitiators* [85].

Apart from the photoreactions of dithiocarbamate groups (last entry in Table 11.4), no details on the radical-generating photoreactions referred to in Table 11.4 are given here. These can be found in [84–86]. Dithiocarbamate groups play a special role with regard to the photoinitiation of polymerizations. This is

Table 11.4 Photolabile groups, chemically incorporated into linear polymers at in-chain, lateral or terminal positions, giving rise to the formation of reactive free radicals [84].

Photolabile groups [a]	Free radicals
Carbonyl groups	$R_1-\overset{\bullet}{C}H_2$ + $\overset{\bullet}{C}(=O)-R_2$
Keto oxime ester groups $R_1-CH_2-\overset{O}{\underset{}{C}}-\underset{R_2}{C}=N-O-\overset{O}{\underset{}{C}}-R_3$	$R_1-CH_2-\overset{O}{\underset{}{C}}-\underset{R_2}{C}=N^\bullet$ + CO_2 + $^\bullet R_3$
Benzoin methyl ether groups $R_1-C_6H_4-C(=O)-CH(OCH_3)-C_6H_4-R_2$	$R_1-C_6H_4-C(=O)^\bullet$ + $^\bullet CH(OCH_3)-C_6H_4-R_2$
N-Nitroso groups $R_1-C(=O)-N(NO)-CH_2-R_2$	$R_1-C(=O)-O^\bullet$ + N_2 + $^\bullet CH_2-R_2$
Disulfide groups $R_1-S-S-C_6H_4-R_2$	$R_1-S^\bullet$ + $^\bullet S-C_6H_4-R_2$
Phenyl sulfide groups $R_1-CH_2-S-C_6H_5$	$R_1-\overset{\bullet}{C}H_2$ + $^\bullet S-C_6H_5$
Dithiocarbamate groups $R_1-CH_2-C(=S)-S-N(C_2H_5)_2$	$R_1-\overset{\bullet}{C}H_2$ + $^\bullet S-N(C_2H_5)_2$ (with C=S)

[a] R_1 denotes a macromolecular substituent.

due to the fact that the sulfur-centered radical is much less reactive than the carbon-centered radical and, hence, does not react with vinyl monomers but rather acts as a terminator of growing macroradicals. Thus, polymerizations initiated by the photolysis of polymeric dithiocarbamates result in macromolecules possessing the original end groups (see Scheme 11.6). Initiators behaving in this way were termed *iniferters* by Otsu, as an acronym for *initiator-transfer-agent-terminator* [87].

Block and graft copolymerization can also be initiated in indirect modes. Here, light is absorbed by independent initiator molecules that are present in the reaction system but are not incorporated into the polymer. Reactive species formed in this way interact with the polymer so as to generate free radical sites

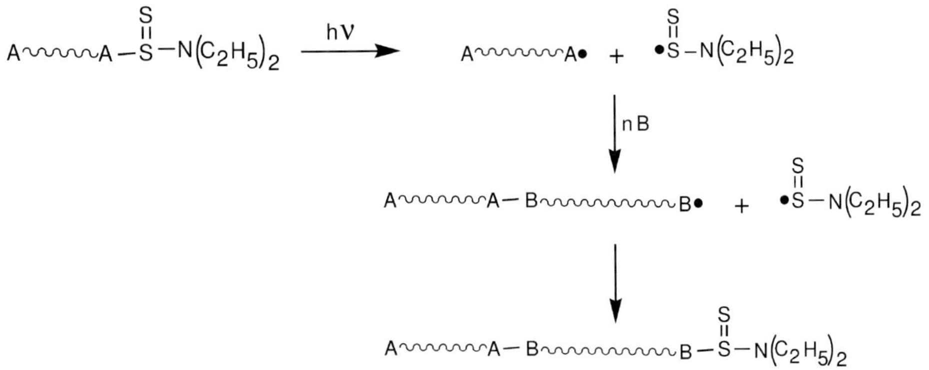

11.8 Light-induced synthesis of block and graft copolymers

$$A\sim\sim A-\overset{S}{\underset{\|}{S}}-N(C_2H_5)_2 \xrightarrow{h\nu} A\sim\sim A\bullet + \bullet S-\overset{S}{\underset{\|}{}}-N(C_2H_5)_2$$

$$\downarrow nB$$

$$A\sim\sim A-B\sim\sim\sim B\bullet + \bullet S-\overset{S}{\underset{\|}{}}-N(C_2H_5)_2$$

$$\downarrow$$

$$A\sim\sim A-B\sim\sim\sim B-\overset{S}{\underset{\|}{S}}-N(C_2H_5)_2$$

Scheme 11.6 Formation of a diblock copolymer with the aid of a photoiniferter.

on the latter that are capable of reacting with monomer molecules. Such systems are presented in Table 11.5. Of general importance is the system based on hydrogen abstraction from the trunk polymer by excited aromatic carbonyl groups.

The methods described above commonly do not lead to pure products. Instead, mixtures composed of starting material and copolymer are obtained. Moreover, homopolymer is produced if one of the free radicals released from the initiator radical pair is of low molar mass (see Scheme 11.5). These are serious drawbacks for practical applications regarding the production of novel polymeric materials based on block copolymers. However, there is important technical potential with respect to *photografting of surfaces of polymeric articles*. Having been widely explored by many investigators during the last decades [88, 89], this

Table 11.5 Indirect generation of free radical sites at lateral or terminal positions of linear polymers.

Precursor reaction	Attack of polymer	Product free radicals
$Mn_2(CO)_{10} \xrightarrow{h\nu} 2\,Mn(CO)_5$	$\sim\sim\sim\underset{CX_3}{\overset{\|}{\text{C}}}\sim\sim + Mn(CO)_5$	$\sim\sim\underset{\bullet CX_2}{\overset{\|}{\text{C}}}\sim\sim$
	$\sim\sim\sim CX_3 + Mn(CO)_5$	$\sim\sim\sim \bullet CX_2$
	X: Cl, Br	
$\diagdown_{C=O} \xrightarrow{h\nu} \diagdown_{C=O*}$ a)	$\diagdown_{C=O*} + \underset{NR_2}{\overset{\|}{CH_2}}$	$\diagdown_{\bullet C-OH} + \underset{NR_2}{\overset{\|}{\bullet CH}}$
$\diagdown_{C=O} \xrightarrow{h\nu} \diagdown_{C=O*}$ a)	$\diagdown_{C=O*} + \sim\sim\underset{H}{\overset{\|}{}}\sim\sim$	$\diagdown_{\bullet C-OH} + \sim\sim\bullet\sim\sim$

a) Refers to aromatic carbonyl compounds such as benzophenone or anthraquinone.

11.8.2
Surface modification by photografting

Photografting can change the surface properties of polymeric articles. For example, photografting can impart hydrophilicity to hydrophobic surfaces of polyalkenes and bring about antifogging, antistatic, and antistaining properties and improvements in dyeability, adhesiveness, printability, and biocompatibility. Photografting competes with other techniques of surface modification, including corona discharge, plasma treatment, chemical oxidation, and coating. Photografting has the advantage over these methods that a large variety of property changes can be imparted to plastic articles by grafting monomers of quite different chemical nature onto the same polymer. Surface grafting can also be accomplished with high-energy radiation of low penetration depth, including electron beam radiation and soft X-rays. Photografting is advantageous over high-energy radiation grafting in that it is virtually restricted to a very thin surface layer and in that it can be applied with rather little effort with respect to the radiation sources. Polyalkenes and other polymers that are produced industrially in large quantities lack chromophoric groups capable of absorbing UV light emitted from commonly available light sources. To circumvent this problem, procedures based on the adsorption of monomers and initiators by pre-soaking have been

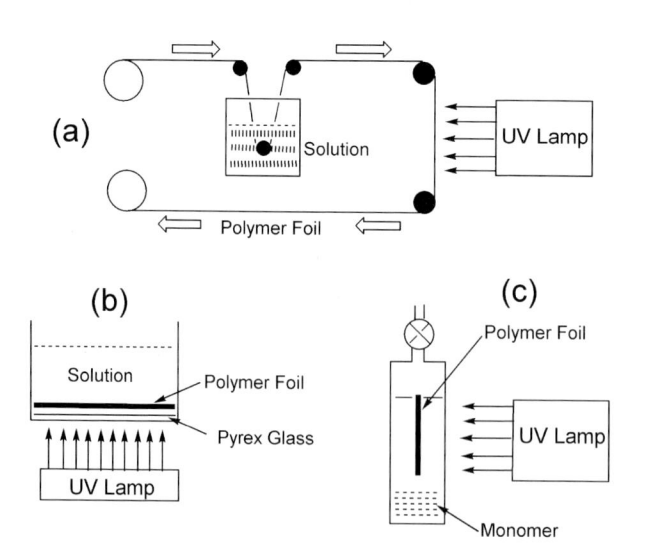

Fig. 11.6 Schematic depiction of surface photografting processes: (a) continuous grafting [91], (b) immersion grafting [97], (c) vapor-phase grafting. Adapted from Ogiwara et al. [98] with permission from John Wiley & Sons, Inc.

Table 11.6 Surface grafting of monomers; recent investigations.

Substrate	Monomers	Remarks	Refs.
Low-density polyethylene	Acrylic acid, acrylamide, vinyl pyridine, glycidyl acrylate	Enhanced hydrophilicity and dye adsorption, adhesion to different substrates	[91, 92]
Low-density polyethylene Polypropylene	Acrylic acid, hydroxypropyl acrylate	Enhanced hydrophilicity, wettability	[100]
Low-density polyethylene Polypropylene	Maleic anhydride, vinyl acetate/maleic anhydride	Enhanced hydrophilicity	[101] [102]
Polyurethane	Methacrylic acid	Enhanced hydrophilicity, enhanced biological cell compatibility	[103]
Ultrafine inorganic particles (silica, titania)	Acrylic acid, acrylamide, acrylonitrile, styrene	Grafted materials give stable dispersions in appropriate liquids	[104] [105]

elaborated. For example, acrylic acid, acrylamide, vinylpyridine, or glycidyl acrylate can be grafted onto low-density polyethylene or linear polyesters in layers ranging from 2 to 8 nm in a continuous process using benzophenone as a hydrogen abstraction-type initiator. As shown in Fig. 11.6a, the polymer foil is drawn from a roll through a solution of initiator and monomer to a reaction chamber for irradiation at $\lambda \approx 250$ nm and is subsequently reeled up [91]. Figure 11.6 also depicts batch processes, i.e. immersion grafting (Fig. 11.6b) and vapor-phase grafting (Fig. 11.6c). In the latter case, the initiator-coated polymer is irradiated in an atmosphere of the monomer.

The aim within the frame of this book is not to survey the plethora of publications devoted to surface photografting. Typical work published in recent years is compiled in Table 11.6, which demonstrates that the enhancement of hydrophilicity and wettability of hydrophobic polymers and the improvement of adhesion of polymers to various substrates are still major research topics (see also [99]). Moreover, the grafting of ultrafine inorganic particles, such as nanosized silica and titania, with vinyl monomers is an attractive subject. Relevant earlier work on surface photografting has been reviewed by Yaĝci and Schnabel [84].

References

1 C.G. Roffey, *Photopolymerization of Surface Coatings*, Wiley, Chichester (1982).
2 S.P. Pappas (ed.), *UV Curing: Science and Technology*, Vols. I and II, Technology Marketing Corp., Norwalk, CT, USA (1978) and (1985).
3 S.P. Pappas (ed.), *Radiation Curing, Science and Technology*, Plenum Press, New York (1992).

4 C. E. Hoyle, J. F. Kinstle (eds.), *Radiation Curing of Polymeric Materials*, ACS Symposium Series 417, American Chemical Society, Washington, D.C. (1990).

5 P. K. T. Oldring (ed.), *Chemistry and Technology of UV and EB Formulations for Coatings, Inks and Paints*, Volumes I–IV, SITA Technology, London (1991).

6 H. Böttcher (ed.), *Technical Applications of Photochemistry*, Deutscher Verlag für Grundstoffindustrie, Leipzig (1991).

7 J. P. Fouassier, J. F. Rabek (eds.), *Radiation Curing in Polymer Science and Technology*, Elsevier Applied Science, London (1993).

8 N. S. Allen, M. Edge, *UV and Electron Beam Curable Pre-Polymers and Diluent Monomers. Classification, Preparation and Properties*, in Vol. I of [7], p. 225.

9 R. Mehnert, A. Pincus, I. Janorski, R. Stowe, A. Berejka, *UV and EB Technology and Equipment*, SITA Technology, London (1998).

10 (a) G. Webster, *Prepolymers & Reactive Diluents*, SITA Technology, London (1998); (b) G. Webster, G. Bradley, C. Lowe, *A Compilation of Oligomers and Monomers Commercially Available for UV Today*, SITA Technology, London (2001).

11 C. Lowe, G. Webster, S. Kessel, I. McDonald, *Formulation*, SITA Technology, London (1999).

12 J. V. Crivello, *The Discovery and Development of Onium Salt Cationic Photoinitiators*, J. Polym. Sci. Part A: Polym. Chem. 37 (1999) 4241.

13 R. S. Davidson, *Radiation Curing*, Rapra Technology, Shawbury, Shrewsbury (2001).

14 K. Dietliker, *A Compilation of Photoinitiators Commercially Available for UV Today*, SITA Technology, London (2002).

15 D. C. Neckers, W. Jager, *Photoinitiation for Polymerisation*, SITA Technology, London (1999).

16 P. Dufour, *State-of-the-Art and Trends in the Radiation Curing Market*, in Vol. I of [7], p. 1.

17 A. F. Jacobine, *Thiol-Ene Photopolymers*, Chapter 7 in Vol. III of [7].

18 L. H. Nguyen, M. Straub, M. Gu, Adv. Funct. Mater. 15 (2005) 209.

19 Information of Radiation Curing Center, SpecialChem S.A. {http://www.specialchem4coatings.com/tc/radiation%2Dcuring}.

20 C. Decker, *Photoinitiated Curing of Multifunctional Monomers*, Acta Polym. 45 (1994) 333.

21 A. F. Jacobine, S. T. Nakos, *Polymerizable Silicone Monomers, Oligomers, and Resins*, Chapter 5 of [3].

22 H. H. Bankowsky, P. Enenkel, M. Lokai, K. Menzel, Paint & Coatings Ind., April (2000).

23 S. Peeters, *Overview of Dual-Cure and Hybrid-Cure Systems in Radiation Curing*, Chapter 6 in Vol. III of [7].

24 E. Beck, B. Bruchmann, Proc. XXVII FATIPEC Congress, Aix-en-Provence (2004), Vol. I, p. 41.

25 J. Ortmeier, Proc. Conf. Automotive Circle International, Bad Nauheim (2003), p. 54.

26 K. Studer, C. Decker, E. Beck, R. Schwalm, Prog. Org. Coat. 48 (2003) 101.

27 C. Decker, F. Masson, R. Schwalm, Macromol. Mater. Eng. 288 (2003) 17.

28 C. Decker, *Photoinitiated Crosslinking Polymerization*, Prog. Polym. Sci. 21 (1996) 593.

29 M. Braden, R. L. Clarke, J. Nicholson, S. Parker, *Polymeric Dental Materials*, Springer, Berlin (1997).

30 J. M. Antonucci, J. W. Stansbury, *Molecularly Designed Dental Polymers*, in R. Arshady (ed.), *Desk Reference of Functional Polymers, Synthesis and Applications*, American Chemical Society, Washington, D.C. (1997), p. 719.

31 Z. Kucybala, M. Pietrzak, J. Paczkowski, L. A. Lindén, J. F. Rabek, Polymer 37 (1996) 4585.

32 J. Jakubiak, X. Allonas, J. P. Fouassier, A. Sionkowska, E. Andrzejewska, L. A. Lindén, J. F. Rabek, Polymer 44 (2003) 5219.

33 Y. J. Park, H. H. Chae, H. R. Rawls, Dental Mater. 15 (1999) 120.

34 J.-F. Roulet, *Degradation of Dental Polymers*, Karger, Basel (1987).

35 L. A. Lindén, *Photocuring of Polymeric Dental Materials and Plastic Composite Resins*, Chapter 13 in Vol. IV of [7].

36 L.A. Lindén, Proceed. RadTech Europe Mediterranean Conference, RadTech Europe, Fribourg, Switzerland (1993), p. 557.
37 J. Jakubiak, J.F. Rabek, Poliymery 45 (2000) 759 and 46 (2001) 5219.
38 P.F. Jacobs (ed.), *Rapid Prototyping and Technology: Fundamentals of Stereolithography*, Society of Manufacturing Engineers, Dearborn (1992).
39 K. Yamaguchi, T. Nakamoto, *Manufacturing of Unidirectional Whisker Reinforced Plastic Microstructures*, in N. Ikawa, T. Kishinami, F. Kimura (eds.), *Rapid Product Developments*, Chapman & Hall, London (1997), p. 159.
40 V. Mizeikis, H.B. Sun, A. Marcinkevičius, J. Nishii, S. Matsuo, S. Juodkazis, H. Misawa, J. Photochem. Photobiol. A: Chem. 145 (2001) 41.
41 H.B. Sun, S. Matsuo, H. Misawa, Appl. Phys. Lett. 74 (1999) 786.
42 (a) T.-C. Lin, S.-J. Chung, K.-S. Kim, X. Wang, G.S. He, J. Swiatkiewicz, H.E. Pudavar, P.N. Prasad, *Organics and Polymers with High Two-Photon Activities and Their Applications*, in K.S. Lee (ed.), *Polymers for Photonics Applications II*, Springer, Berlin (2002). (b) B. Strehmel, V. Strehmel, *Two-Photon Physical, Organic, and Polymer Chemistry: Theory, Techniques, Chromophore Design, and Applications*, in D.C. Neckers, W.S. Jenks, T. Wolff (eds.), *Advances in Photochemistry*. 29, Wiley-Interscience, New York (2007) p. 111.
43 R.A. Borisov, G.N. Dorojkina, N.I. Koroteev, V.M. Kozenkov, S.A. Magnitskii, D.V. Malakov, A.V. Tarasishin, A.M. Zheltikov, Appl. Phys. B 67 (1998) 765.
44 K.D. Belfield, X. Ren, E.W. van Stryland, D.V. Hagan, V. Dobikovsky, E.J. Miesak, J. Am. Chem. Soc. 122 (2000) 1217.
45 Y. Boiko, J.M. Costa, M. Wang, S. Esener, Proc. SPIE 4279 (2001) 212.
46 S.M. Kuebler, B.H. Cumpson, S. Ananthavel, S. Barlow, J.E. Ehrlich, L.L. Etsine, A.A. Heikal, D. McCord-Maughon, J. Qin, H. Rocel, M. Rumi, S.R. Marder, J.W. Perry, Proc. SPIE 3937 (2000) 97.
47 A. Reiser, *Photoreactive Polymers. The Science and Technology of Resists*, Wiley, New York (1989).
48 J. Bendig, H.-J. Timpe, *Photostructuring*, in H. Böttcher (ed.), *Technical Applications of Photochemistry*, Deutscher Verlag für Grundstoffindustrie, Leipzig (1991), p. 172.
49 W. Frass, H. Hoffmann, B. Bronstert, *Imaging Technology*, in B. Elvers, S. Hawkins, M. Ravenscroft, G. Schulz (eds.), *Ullmann's Encyclopedia of Industrial Chemistry*, VCH, Weinheim (1989), Vol. A 13, p. 629.
50 H.R. Ragin, *Radiation-Curable Coatings with Emphasis on the Graphic Arts*, Chapter 7 of [3].
51 H. Vollmann, W. Frass, D. Mohr, Angew. Makromol. Chem. 145/146 (1986) 411.
52 L. Plambeck, U.S. Patent 2,760,863 (1956).
53 The Seybold Report on Publishing Systems 30 (2000) 5.
54 Newsletter for the Flexographic Pre-Press Platemakers Association {http://fppa.net/newsletter/summer03/members.htm}
55 Flexo and Gravure International, March (2001).
56 Y. Ichii, S. Ichikawa, S. Tanaka, Jap. Patent (Toray Industries), JP 2005326442 (2005).
57 R. Kübler, *Printing Inks*, in B. Elvers, S. Hawkins, W. Russey, G. Schulz (eds.), *Ullmann's Encyclopedia of Industrial Chemistry*, VCH, Weinheim (1993), Vol. A 22, p. 143.
58 A.J. Bean, *Radiation Curing of Printing Inks*, Chapter 8 of [3].
59 H.M. Smith, *Principles of Holography*, Wiley, New York (1969).
60 H.J. Caulfield (ed.), *Handbook of Optical Holography*, Academic Press, New York (1979).
61 N. Abramson, *The Making and Evaluation of Optical Holograms*, Academic Press, London (1981).
62 H.J. Coufal, D. Psaltis, G.T. Sincerbox (eds.), *Holographic Data Storage*, Springer, Berlin (2000).
63 P. Hariharan, *Optical Holography, Principles, Techniques, and Applications*, 2nd Edition, Cambridge University Press, Cambridge (1996).
64 D. Gabor, Nature 161 (1948) 777.

65 S. Reich, *Photodielectric Polymers for Holography*, Angew. Chem. Int. Ed. 16 (1977) 441.

66 R. T. Ingwall, D. Waldmann, *Photopolymer Systems*, in [62], p. 171.

67 I. Dhar, M. G. Schnoes, H. E. Katz, A. Hale, M. T. Schilling, A. L. Harris, *Photopolymers for Digital Holographic Data Storage*, in [62], p. 199.

68 M. B. Sponsler, (a) *Photochemical and Photophysical Processes in the Design of Holographic Recording Materials*, Spectrum 13 (2000) 7; (b) *Hologram Switching and Erasing Strategies with Liquid Crystals*, in V. Ramamurthy, K. S. Schanze (eds.), *Optical Sensors and Switches*, Marcel Dekker, New York (2001), p. 363.

69 D. J. Lougnot, *Photopolymers and Holography*, Chapter 3 in Vol. III of [7].

70 D. J. Lougnot, *Self-Processing Photopolymer Materials for Holographic Recording*, Crit. Rev. Opt. Sci. Technol. CR 63 (1996) 190.

71 M. L. Schilling, V. L. Colvin, I. Dhar, A. L. Harris, F. C. Schilling, H. E. Katz, T. Wysocki, A. L. Hale, L. L. Blyer, C. Boyd, *Acrylate Oligomer-Based Photopolymers for Optical Storage Applications*, Chem. Mater. 11 (1999) 247.

72 L. V. Natarajan, R. L. Sutherland, V. Tondiglia, T. J. Bunning, W. W. Adams, *Photopolymer Materials, Development of Holographic Gratings*, in J. C. Salamone (ed.), *Concise Polymeric Materials Encyclopedia*, CRC Press, Boca Raton, Florida, USA (1999), p. 1052.

73 S. H. Stevenson, M. L. Armstrong, P. J. O'Connor, D. F. Tipton, Proc. SPIE 2333 (1994) 60.

74 T. J. Trout, W. J. Gambogi, S. H. Stevenson, Proc. SPIE 2577 (1995) 94.

75 W. J. Gambogi, W. K. Smothers, K. W. Steijn, Proc. SPIE 2405 (1995) 62.

76 K. W. Steijn, Proc. SPIE 2688 (1996) 123.

77 V. L. Colvin, R. G. Larson, A. L. Harris, M. L. Schilling, J. Appl. Phys. 81 (1997) 5913.

78 D. A. Waldman, R. T. Ingwall, P. K. Dal, M. G. Horner, E. S. Kolb, H.-Y. S. Li, R. A. Minns, H. G. Schild, Proc. SPIE 2689 (1996) 127; U.S. Patent 5 759 721 (1998).

79 V. Strehmel, *Epoxies: Structures, Photoinduced Crosslinking, Network Properties and Applications*, in H. S. Nalwa (ed.), *Handbook of Photochemistry and Photobiology*, Vol. 2, American Scientific Publ., Stevenson Ranch, CA, USA (2003), p. 102.

80 J. Zhang, C. A. Carlen, S. Palmer, M. B. Sponsler, J. Am. Chem. Soc. 116 (1994) 7055.

81 R. T. Pogue, R. L. Sutherland, M. G. Schmitt, L. V. Natarajan, V. P. Tondiglia, T. J. Bunning, Appl. Spectrosc. 54 (2000) 12A.

82 C. C. Bowley, G. P. Crawford, Appl. Phys. Lett. 76 (2000) 2235.

83 D. Duca, A. V. Sukhov, C. Umeton, Liq. Cryst. 26 (1999) 931.

84 Y. Yaĝci, W. Schnabel, *Light-Induced Synthesis of Block and Graft Copolymers*, Prog. Polym. Sci. 15 (1990) 551.

85 I. Reetz, Y. Yaĝci, M. K. Mishra, *Photoinitiated Radical Vinyl Polymerization*, in M. K. Mishra, Y. Yaĝci (eds.), *Handbook of Radical Vinyl Polymerization*, Dekker, New York (1998), p. 149.

86 A. E. Muftuoglu, M. A. Tasdelen, Y. Yaĝci, *Photoinduced Synthesis of Block Copolymers*, Chapter 29 in J. P. Fouassier (ed.), *Photochemistry and UV Curing: New Trends*, Research Signpost, Trivandrum, India (2006).

87 (a) T. Otsu, M. Yoshida, Makromol. Chem. Rapid Commun. 3 (1982) 127. (b) A. Šebenik, *Living Free-Radical Block Copolymerization Using Thio-Inifterters*, Prog. Polym. Sci. 23 (1998) 875.

88 J. C. Arthur, *Photoinitiated Grafting of Monomers onto Cellulose Substrates*, in N. S. Allen (ed.), *Developments in Polymer Photochemistry – 1*, Appl. Science Publ., London (1980), p. 69.

89 J. C. Arthur, *Photografting of Monomers onto Synthetic Polymer Substrates*, in N. S. Allen (ed.), *Developments in Polymer Photochemistry – 2*, Appl. Science Publ., London (1981), p. 39.

90 K. L. Mittal (ed.), *Polymer Surface Modification: Relevance to Adhesion*, VSP, Utrecht (1996).

91 B. Rånby, *Surface Photografting onto Polymers – A New Method in Adhesion Control*, in Part 3 of [90].

92 B. Rånby, *Surface Modification and Lamination of Polymers by Photografting*, Int. J. Adhesion and Adhesives 19 (1999) 337.

93 B. Rånby, *Photoinitiated Modification of Synthetic Polymers: Photocrosslinking and Surface Photografting*, in N.S. Allen, M. Edge, I.R. Bellobono, E. Selli (eds.), *Current Trends in Polymer Photochemistry*, Ellis Horwood, New York (1995), Chapter 2.
94 M.J. Swanson, G.W. Oppermann, *Photochemical Surface Modification. Photografting of Polymers for Improved Adhesion*, in Part 3 of [90].
95 J.P. Bilz, C.B. Lottle (eds.), *Fundamental and Applied Aspects of Chemically Modified Surfaces*, The Royal Chemical Society, London (1999).
96 P.A. Dworjanyn, J.L. Garnett, *Role of Grafting in UV- and EB-Curing Reactions*, Chapter 6 of Vol. I of [7].
97 S. Tazuke, M. Matoba, H. Kimura, T. Okada, in C.E. Carraher Jr., M. Tsuda (eds.), *Modification of Polymers*, ACS Symp. Series 121, Washington, D.C. (1980).
98 Y. Ogiwara, M. Kanda, M. Takumi, H. Kubota, J. Polym. Sci. Lett. Ed. 19 (1981) 457.
99 N.S. Allen, *Polymer Photochemistry*, Photochem. 34 (2003) 197.
100 K. Zahouilly, Techn. Conf. Proceed. RadTech 1 (2002) 1079.
101 J. Deng, W. Yang, J. Appl. Polym. Sci. 97 (2005) 2230.
102 J. Deng, W. Yang, J. Appl. Polym. Sci. 95 (2005) 903.
103 Y. Zhu, C. Gao, J. Guan, J. Chen, J. Biomed. Mater. Res. 67 A (2003) 1367.
104 M. Satoh, K. Shirai, H. Saitoh, T. Yamauchi, N. Tsubokawa, J. Polym. Sci. Part A: Polym. Chem. 43 (2005) 600.
105 N. Tsubokawa, Y. Shirai, H. Tsuchida, S. Handa, J. Polym. Sci. Part A: Polym. Chem. 32 (1994) 2327.

Part IV
Miscellaneous technical developments

12
Polymers in optical memories

12.1
General aspects

The revolutionary development in computer technology during the last decades has been inextricably linked with the elaboration of novel data storage methods and the invention of relevant devices. Impetus for innovations in the data storage field has also come from the steadily increasing demand for larger storage capacity in the disparate fields of scientific research, industrial production, and daily life entertainment [1]. At present, optical storage techniques reliant on polymeric recording media play a prominent role. Actually, polymers are being used in various ways, not only as disk substrates but also as surfacing/subbing layers for the substrate, protective and antistatic overcoatings, etc.

The history of modern storage media commenced with magnetic memories, which proved very reliable in terms of stability and recording/reading speed. When they could no longer meet capacity requirements, a new optical storage system consisting of a drive unit and a storage medium in rotating disk form,

Table 12.1 Characteristics of single-sided, single-layer 12 cm disks [2-4].

Disk Format	d [a] (mm)	TP [b] (µm)	λ [c] (nm)	NA [d]	C_{St} [e] (GB)	r_{trans} [f] (Mb s^{-1})
Compact Disk (CD)	1.2	1.6	780	0.45	0.65	0.1
Digital Versatile Disk (DVD)	1.2	0.74	650	0.60	4.7	11
HD-DVD [g]	1.2	0.40	405	0.65	15	36
Blu-ray Disk (BD) [h]	1.2	0.32	405	0.85	25	36

a) Substrate thickness.
b) Track pitch.
c) Laser wavelength.
d) Numerical aperture of objective lenses.
e) Storage capacity, 1 Byte (B) = 8 bits (b).
f) Data transfer rate.
g) High Definition DVD, developed by Toshiba and NEC within the DVD Forum.
h) Developed by Blu-ray Disc Association.

Polymers and Light. Fundamentals and Technical Applications. W. Schnabel
Copyright © 2007 WILEY-VCH Verlag GmbH & Co. KGaA, Weinheim
ISBN: 978-3-527-31866-7

the compact disk, CD (storage capacity: 650 MB), was invented. Then, following the constantly increasing demand for larger storage capacity, the digital versatile disk, DVD (storage capacity: 4.7 GB), was developed. At present, disks having a storage capacity of about 25 GB, manufactured with the aid of advanced techniques, are poised to enter the market. The characteristics of single-sided, single-layer disks are listed in Table 12.1. Because of the given limit in information storage of these optical media, novel storage systems emerging from a hybrid technology (magneto-optical disks, MO) or developed on the basis of solid immersion techniques or volume holography can be foreseen.

12.2
Current optical data storage systems

12.2.1
Compact disk (CD) and digital versatile disk (DVD)

Since its release in 1982, the compact disk has taken the world by storm, and billions of CDs have been manufactured [5, 6]. Most of them are of the read-only memory (ROM) type, made from transparent polycarbonate (see Chart 12.1) and providing almost perfect resolution.

In the cases of both CD-ROM and DVD, the information is binary coded bit-wise in the form of pits and lands (see Fig. 12.1). Commencing at the inside, spirally arranged tracks of pits and lands are engraved into the disk.

Standard stamper-injection molding is the most commonly used method for manufacturing compact disks [2, 3]. It comprises various steps, which are depicted schematically in Fig. 12.2. First, a plane glass substrate is coated with a photoreactive layer, which is patterned with a pit/land structure by an appropriate technique such as photolithography (see Section 9.1). In the latter case, the disk is rotated at a constant linear velocity while being exposed along a spiral path from the inside to the outer edge to a laser beam, e.g. of an Ar laser emitting 442 nm light. Since the exposure is intermittent, subsequent development results in a pit/land structure of the tracks. The master disk obtained in this way is then electroformed to create a stamper for use in an injection-molding process. Disks generated in this way are first coated with a thin reflective metal layer (typically Al) and then with two layers, a protective acrylic layer and a label layer, both of which are cured using UV light (see Section 11.2). Finally, the disks, having a total thickness of 1.2 mm, are packaged in jewel-boxes for ship-

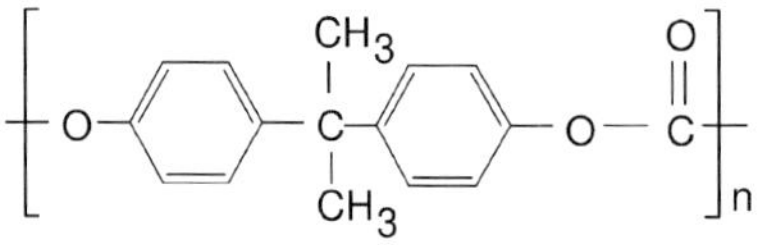

Chart 12.1 Chemical structure of polycarbonate used for compact disks.

12.2 Current optical data storage systems

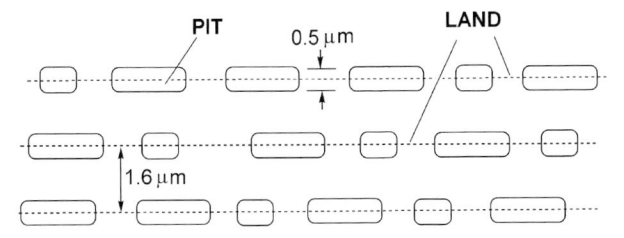

Fig. 12.1 Pit/land structure of tracks of compact disks.

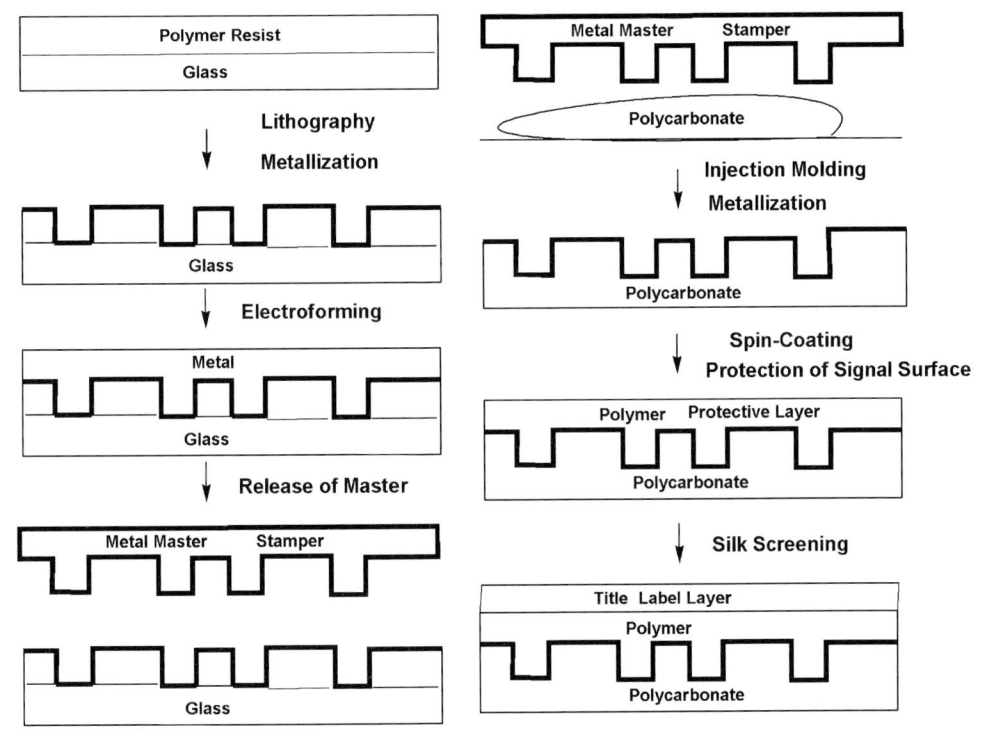

Fig. 12.2 Schematic depiction of the commonly used method for the manufacture of compact disks.

ping. DVDs are also fabricated by injection molding. In this case, two 0.6 mm thick disks, one of them containing the recording layer, are glued together [7].

During reading, coherent laser light shone onto the tracks is reflected by the metal in the case of lands and is scattered in the case of pits, which corresponds to the photocell-aided recording of 0 or 1, respectively. The maximum disk storage capacity is set by the resolving power, i.e. the size and the packing density of the pits. This is limited by the wavelength of the laser light, since the focus of the laser beam used for writing and reading cannot be smaller than the wavelength λ. In the case of optical systems operated with conventional lenses, the diameter δ of the laser spot at the recording medium is given by Eq. (12-1). It can be seen that δ is proportional to λ/NA, where k is a constant and NA is the numerical aperture of the objective lens.

$$\delta = k\frac{\lambda}{NA} \tag{12-1}$$

Past strategies for increasing the storage capacity of optical disks were based on a reduction of λ and an increase in NA, as can be seen from Table 12.1. In principle, a reduction in the spot size can be achieved with the aid of solid immersion lenses. This as yet not practically exploited technique, operating with a hemispherical or a Weierstrass superspherical lens placed near the recording medium (< 100 nm), yields a reduced spot size, δ, as is evident from Eqs. (12-2) and (12-3), respectively, where n denotes the refractive index of the lens [8].

$$\text{Hemispherical lens:} \quad \delta = k\frac{\lambda}{nNA} \tag{12-2}$$

$$\text{Weierstrass superspherical lens:} \quad \delta = k\frac{\lambda}{n^2 NA} \tag{12-3}$$

In addition to read-only systems, there are recordable (write-once/read-many, CD-R) and rewritable CD formats (CD-RW), which will not be treated here. Current recordable storage systems are based on laser-induced pit formation in organic dye films or a laser-induced amorphous-to-crystalline phase change in an inorganic alloy film. Current rewritable optical recording methods involve phase change recording and magneto-optical (MO) recording. The latter is based on switching the magnetization direction of perpendicularly magnetized domains in a magnetic film [9].

12.2.2
Blue-ray disks

As blue diode lasers became available on a large scale [10], a new generation of storage disks with further increased storage capacity was developed by the Blu-ray Disc Association and by Toshiba and NEC within the DVD Forum [4]. As can be seen in Table 12.1, a HD-DVD holds 15 GB and BDs hold 25 GB (single-layer DB) or 50 GB (dual-layer BD). Figure 12.3 shows, as a typical example, the cross-section of a novel disk type having a triple-layer structure: one BD layer and a dual DVD layer to be read by a blue and a red laser, respectively.

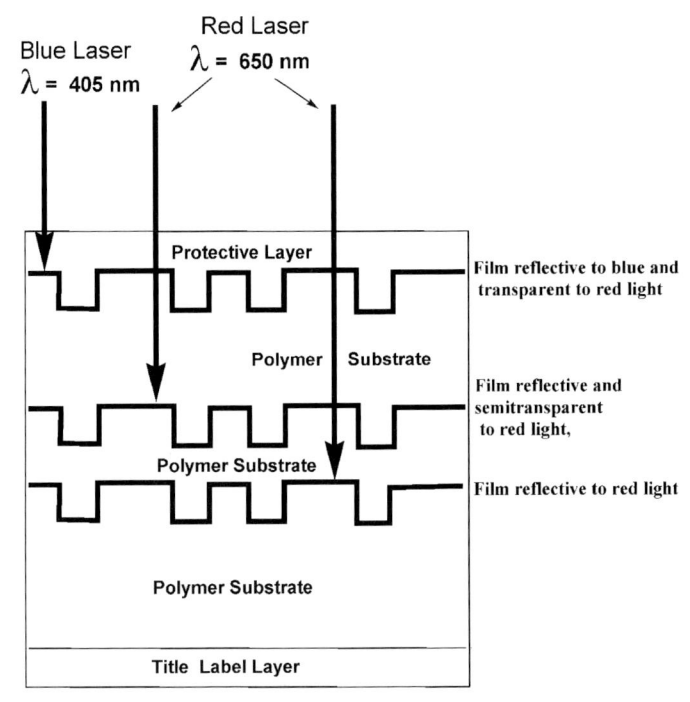

Fig. 12.3 Schematic depiction of the cross-section of a Blu-ray DVD ROM disk having a triple-layer structure. BD single layer: 25 GB, DVD dual layer: 8.5 GB.

12.3
Future optical data storage systems

12.3.1
General aspects

Considering the currently applied optical and magnetic recording methods, there are physical limitations to a further increase in storage capacity. Near-field optical recording with an expected recording density of more than 1 Tb in^{-2} (ca. 19 GB cm^{-2}), thus exceeding that of Blue-ray disks by about two orders of magnitude, might be a method to overcome these limitations. Here, the data bits are written and read by using an optical near field, generated near a nanometer-scale object. In this case, the size of the optical spot can be reduced to less than 1 nm, because it is not limited by light diffraction [8]. Pioneering near-field recording experiments with chromophoric compounds embedded in a polymeric matrix yielded recording marks with a diameter smaller than 100 nm [11]. Although the desired high recording density is realized in this way, practical application is hampered by rather slow data transfer rates, which are on the level of the storage systems in current use.

An interesting, non-optical technique developed in another attempt to achieve larger data storage capacities relates to an atomic microscope-based data storage technique operating with very thin polymer films. With this technique, 30–40 nm-sized bit indentations with a similar pitch size are made by a single cantilever in thin polymer films, typically a 50 nm poly(methyl methacrylate) thin film, resulting in a storage density of 8–10 GB cm^{-2} [12]. While this new technique is also unlikely to lead to products on the market in the near future, holography seems to be more promising. The principle of holography has been outlined in Section 11.7, and the applicability of holography as an optical storage method has been alluded to elsewhere (see Sections 3.5.2, 4.5, and 5.8.2). Therefore, in this chapter, mainly application-related aspects are discussed.

12.3.2
Volume holography

12.3.2.1 Storage mechanism

Holography offers the potential for data storage, since a large number of holograms can be superimposed in one volume element of an appropriate matrix [13–15]. Bit recording in three dimensions implies an enormous increase in storage capacity in comparison to the techniques described in Section 12.2, since multiple pages of data can be stored in the same volume of holographic material. The storage mechanism is based on the generation of light-induced local changes in the refractive index (phase hologram) or in the absorption coefficient (amplitude hologram). As outlined in Section 11.7, interference patterns are generated upon superimposing the light beam carrying the information with a reference beam. Read-out is achieved with the aid of the reference beam. A schematic set-up for recording phase holograms on an appropriate holographic plate is depicted in Fig. 12.4.

A large storage capacity corresponding to a density ranging up to 0.6 GB cm^{-2} (ca. 60 GB per 12 cm disk) is feasible if thousands of holograms are superim-

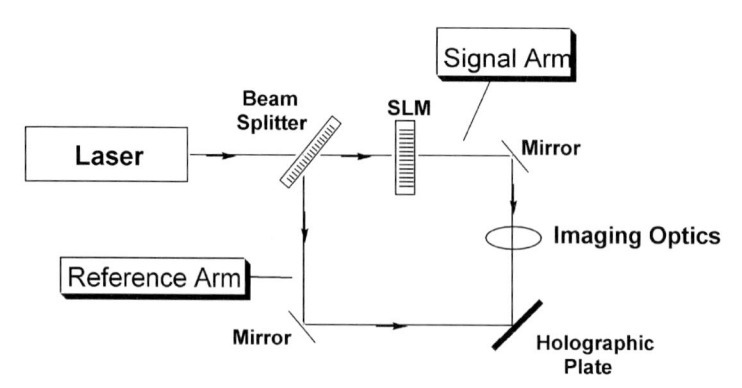

Fig. 12.4 Schematic depiction of a set-up for recording holograms. SLM: Spatial light modulator.

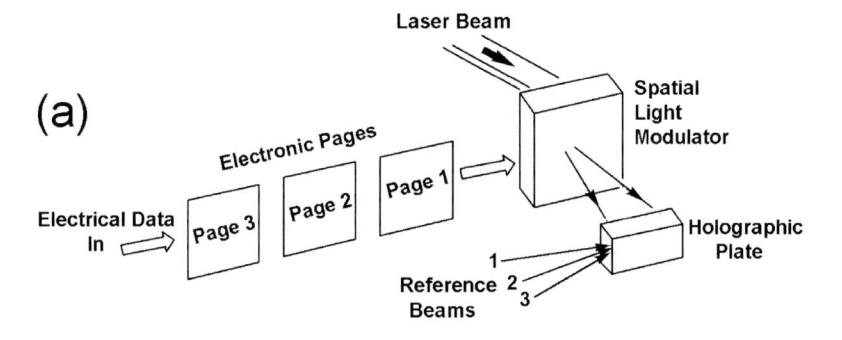

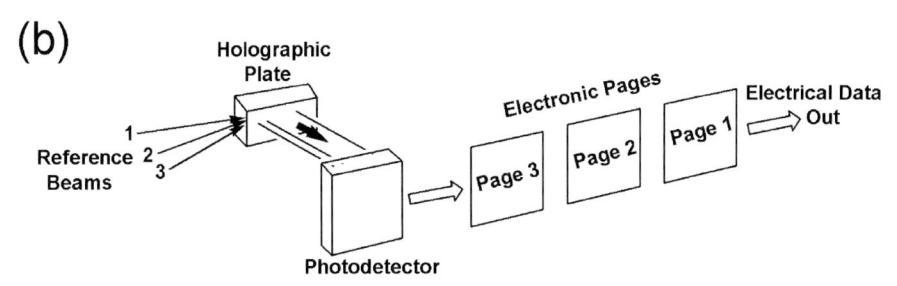

Fig. 12.5 Volume holography in conjunction with angular multiplexing. Set-ups for the recording of digital data (a) and the retrieval of stored data (b). Adapted from Sincerbox [17] with permission from Springer.

posed in the same disk. This can be achieved by means of *multiplexing*, i.e. by addressing individual high-density data pages to holographic plates by changing the angle, wavelength, or phase code of the reference beam [16]. Figure 12.5 shows a set-up operating on the basis of angular multiplexing, i.e. by varying the angle between the writing and reference beams.

The great success of ROM disks (CD and DVD) relies on the availability of inexpensive methods to mass-produce copies of recorded disks. In this context, a method to replicate holographic disks containing page-formatted data with the aid of a replicator operating with ten reference beams is noteworthy [18].

12.3.2.2 Storage materials

Holographic storage materials appropriate for commercial application have to fulfil various requirements, the most important of which are as follows: high storage density (>1 GB cm^{-2}), fast writing time (ms), high sensitivity (mW), long memory (years), fast access time (µs), and reversibility ($>10^4$ cycles) for write/erase systems [19]. In this context, three categories of materials have been

Table 12.2 Light-sensitive materials suitable for volume holography.

Polymeric systems	Inorganic crystals [a]	Inorganic glasses
Photopolymerizable systems Photorefractive systems Photochromic systems (Photoaddressable polymers)	$LiNiO_3$, $KNiO_3$, $LiTaO_3$, $BaTiO_3$, $Sr_xBa_{1-x}Nb_2O_6$ [b], $Bi_{12}TiO_{20}$	Chalcogenide glasses containing group VI elements such as As_2S_3, As_2Se_3

a) For the recording of holograms, crystals are doped with Fe, Cr, Cu, Mg or Zn.
b) x: varying from 0 to 1.

found appropriate for volume holography [20]: inorganic crystals [21, 22], inorganic glasses [23–25], and photopolymer systems [26–28] (see Table 12.2).

Of the polymeric systems, the photopolymerizable systems (commonly referred to as photopolymers) show the most promise (see Section 11.7). At present, InPhase Technologies and Aprilis Inc. are reported to commercialize ROM products with storage densities of 1.2 and 1.9 GB cm^{-2} and negligible shrinkage during writing [20]. The read/write speed is said to be comparable to that of an optical disk. A competitor in the race to the market is Polight Technologies Ltd., who are commercializing products based on inorganic glasses. In this case, the recording media are rewritable, since the light-induced refractive index changes are reversible. However, chalcogenide glasses are much less sensitive compared to polymerizable systems, because the latter exhibit an amplification mechanism based on a chain reaction, i.e. each absorbed photon induces the polymerization of a large number of molecules. On the other hand, there is no shrinkage problem in the case of inorganic glasses, which, moreover, have a much smaller thermal expansion coefficient than polymers. The latter is of importance when the temperature soars in disk drives. Photorefractive crystals, which were the subject of much attention for a while, do not compete with the other materials with regard to the commercialization of a product. This is mainly due to the fact that the light used to read holograms also erases them. Therefore, in inorganic crystals, holograms have to be fixed after writing by heating. Another drawback in this case is the low photosensitivity [20]. Finally, the so-called photoaddressable polymers, PAPs, were considered as potential candidates for data storing materials. For instance, PAP systems consisting of linear polymers bearing pendant liquid-crystalline side chains and azobenzene chromophores (see Sections 3.5.2 and 5.8.2) seemed to be very suitable for recording volume phase holograms [21]. However, even under favorable illumination conditions, the writing time of holograms was found to be of the order of 100 ms (for some systems of the order of several seconds). This writing speed is at least one order of magnitude too long for technical applications.

12.3.3
Photo-induced surface relief storing

A novel optical recording method based on large-scale, light-driven mass transport in films of azobenzene polymers has been proposed. As outlined in Section 5.6, the phenomenon of light-induced mass transport is due to the photoisomerization of azobenzene groups. It can be utilized to inscribe narrow relief structures in the surfaces of appropriate polymer films by using light of the requisite wavelength. The relief structures can be erased and rewritten. On this basis, a novel technique for high-density optical data storage has been developed [29]. Since data can be stored at a recording density of up to 10^8 B cm^{-2} by combining angular and depth gradation, this method has potential for practical application.

References

1. D. Day, M. Gu, A. Smallridge, *Review of Optical Data Storage*, in P. Boffi, D. Piccini, M. C. Ubaldi (eds.), *Infrared Holography for Optical Communications*, Springer, Berlin (2003), p. 1.
2. E. A. LeMaster, *Compact Disc Manufacturing* {http://www.ee.washington.edu/conselec/W94/edward/edward.htm}, (1994).
3. K. J. Kuhn, *Audio Compact Disk – An Introduction* {http://www.ee.washington.edu/conselec/CE/kuhn/cdaudio/95x6.htm}, (1994).
4. Blu-ray FAQ {http://www.blu-ray.com.faq}.
5. K. C. Pohlmann, *The CD ROM Handbook*, A-R Editions, Madison (1992).
6. C. Sherman, *The Compact Disc Handbook*, Intertext Publications, New York (1988).
7. S. Watson {http://electronics.howstuffworks.comb/blu-ray3.htm}, (2004).
8. T. Matsumoto, *Near-Field Optical Head Technology for High Density, Near-Field Optical Recording*, in M. Ohtsu (ed.), Progress in Nano-Electro-Optics III, Springer Series in Optical Sciences, Berlin, 96 (2005) 93.
9. H. J. Borg, R. van Woudenberg, *Trends in Optical Recording*, J. Magnetism Magnet. Mater. 193 (1999) 519.
10. S. Nakamura, S. Fasol, *The Blue Diode Laser*, Springer, Berlin (1997).
11. M. Irie, *High-Density Optical Memory and Ultrafine Photofabrication*, in S. Kawata, M. Ohtsu, M. Irie (eds.), *Nano-Optics*, Springer Series in Optical Sciences, Berlin, 84 (2002) 137.
12. P. Vettiger, M. Despont, U. Dürig, M. Lantz, H. E. Rothuizen, K. G. Binnig, *AFM-Based Mass Storage – The Millipede Concept*, in R. Waser (ed.), *Nanoelectronics and Information Technology*, Wiley-VCH, Weinheim (2005), p. 685.
13. V. A. Barachevsky, *Organic Storage Media for Holographic Optical Memory: State of the Art and Future*, Optical Memory and Neural Networks 9 (2000) 251, and Proc. SPIE 4149 (2000) 205.
14. H. J. Coufal, D. Psaltis, G. T. Sincerbox (eds.), *Holographic Data Storage*, Springer, Berlin (2000).
15. R. M. Shelby, *Materials for Holographic Digital Data Storage*, Proc. SPIE 4659 (2002) 344.
16. G. Barbastathis, D. Psaltis, *Volume Holographic Multiplexing Methods*, in [14], p. 21.
17. G. T. Sincerbox, *History and Physical Principles*, in [14], p. 3.
18. F. Mok, G. Zhou, D. Psaltis, *Holographic Read-Only Memory*, in [14], p. 399.

19 L. Lucchetti, F. Simoni, *Soft Materials for Optical Data Storage*, Rivista del Nuovo Cimento 23 (2000) 1.
20 N. Anscombe, *Holographic Data Storage: When Will it Happen?*, Photonics Spectra, June (2003) 54.
21 M. Imlau, T. Bieringer, S.G. Odoulov, T. Woike, *Holographic Data Storage*, in R. Waser (ed.), *Nanoelectronics and Information Technology*, Wiley-VCH, Weinheim (2005), p. 657.
22 K. Buse, E. Krätzig, *Inorganic Photorefractive Materials*, in [14], p. 113.
23 V.I. Minko, I.Z. Indutniy, P.E. Shepeliavyi, P.M. Litvin, J. Optoelectron. Adv. Mater. 7 (2005) 1429.
24 A. Feigel, Z. Kotler, B. Sfez, A. Arsh, M. Klebanov, V. Lyubin, Appl. Phys. Lett. 13 (2000) 3221.
25 S. Ramachandran, S.G. Bishop, J.P. Guo, D.J. Bradley, Photon. Technol. Lett., IEEE, 8 (1996) 1041.
26 R.T. Ingwall, D. Waldmann, *Photopolymer Systems*, in [14], p. 171.
27 I. Dhar, M.G. Schnoes, H.E. Katz, A. Hale, M.L. Schilling, A.L. Harris, *Photopolymers for Digital Holographic Data Storage*, in [14], p. 199.
28 S. Kawata, Y. Kawata, *Three-Dimensional Optical Data Storage Using Photochromic Materials*, Chem. Rev. 100 (2000) 1777.
29 T. Fukuda, *Rewritable High-Density Optical Recording on Azobenzene Polymer Thin Films*, Opt. Rev. 12 (2005) 126.

13
Polymeric photosensors

13.1
General aspects

The increasing desire to detect analytes (components of mixtures of compounds) in situ and in real time, and to monitor continuously the chemical changes in industrial and biological processes, has given impetus to interesting developments in the field of chemical sensors, also referred to as chemosensors [1–15]. Chemosensing can be accomplished by measuring a chemical or physical property of either a particular analyte or of a chemical transducer interacting with a particular analyte. For practical applications, the latter type of chemical sensor is most important. Prominent in this context are highly fluorescent conjugated polymers that possess a large number of receptor sites for analytes, in fact one receptor site per repeating unit. Non-covalent binding of an analyte results in a shift of the maximum of the emission spectrum or causes quenching or enhancement of the fluorescence intensity. A somewhat different type of chemosensor comprises molecules, in some cases supramolecules, that recognize and signal the presence of analytes on the basis of a *3R scheme* – "recognize, relay, and report", which is schematically depicted in Fig. 13.1. The sensor system consists of a receptor site and a reporter site, which are commonly covalently linked. A non-covalent recognition event at the receptor site is communicated to the reporter site, which produces a measurable signal. Energy transfer, electron transfer, a conformational change in the molecular structure, or a combination of these processes constitutes the relay mechanism. Commonly, chemosensor

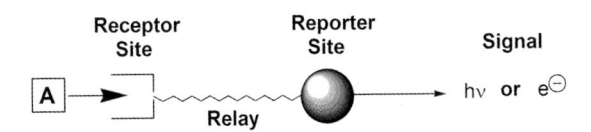

Fig. 13.1 Schematic depiction of chemical sensor action. An optical or electrical signal reports the non-covalent binding of an analyte to the receptor site.

Polymers and Light. Fundamentals and Technical Applications. W. Schnabel
Copyright © 2007 WILEY-VCH Verlag GmbH & Co. KGaA, Weinheim
ISBN: 978-3-527-31866-7

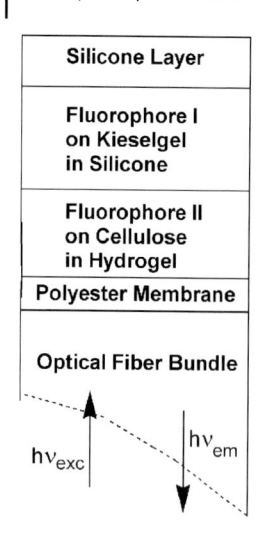

Fig. 13.2 Structure of an optode for the detection of molecular oxygen and carbon dioxide. Fluorophore I (O_2): tris(2,2'-bipyridyl)ruthenium(II) dichloride; fluorophore II (CO_2): 1-hydroxypyrene-3,6,8-trisulfonate. Adapted from Baldini et al. [4] with permission from Springer.

systems operating according to the 3R scheme consist of sensor molecules or groups that are physically admixed or covalently linked to a polymer matrix.

The magnitude of the signal generated by the sensor is normally proportional to the concentration of the analyte. Regarding practical applications, optical chemosensors that monitor changes in fluorescence intensity, or to a lesser extent in optical absorption, are much more prevalent as compared to chemosensors that monitor changes in electrical conductivity or electrical current.

In many cases, optical chemosensor devices consist of a probe, called an *optode*, in which modulation of the optical signal takes place, and an optical link connecting the probe to the instrumentation. The main parts of the latter are a light source, a photodetector, and an electronic signal-processing unit. A schematic depiction of a typical optode is shown in Fig. 13.2. This optode operates with the aid of two fluorophores that undergo a change in fluorescent light emission in the presence of O_2 or CO_2. Fluorophore I is admixed and fluorophore II is chemically linked to the polymer.

In conclusion, polymers play a versatile role in the field of chemosensors. Most interestingly, certain polymers can actively serve as sensors. This pertains to certain strongly fluorescent conjugated polymers, as pointed out above, and to polymers employed as cladding for optical fibers in evanescent wave-based sensors. Moreover, polymers are widely used as supports for transducers, which are either admixed or chemically linked to the polymer matrices. Typical examples are given in the following sections.

13.2
Polymers as active chemical sensors

13.2.1
Conjugated polymers

Conjugated polymers are powerful fluorescent materials, which makes them suitable for applications as chemical sensors. Chart 13.1 presents the structures of some typical polymers that are applicable for the detection of analytes at low concentrations. These polymers include poly(p-phenylene ethynylene), PPE; poly(p-phenylene vinylene), PPV; polyacetylene; and polyfluorene. Those polymers bearing ionizable pendant groups are water-soluble polyelectrolytes.

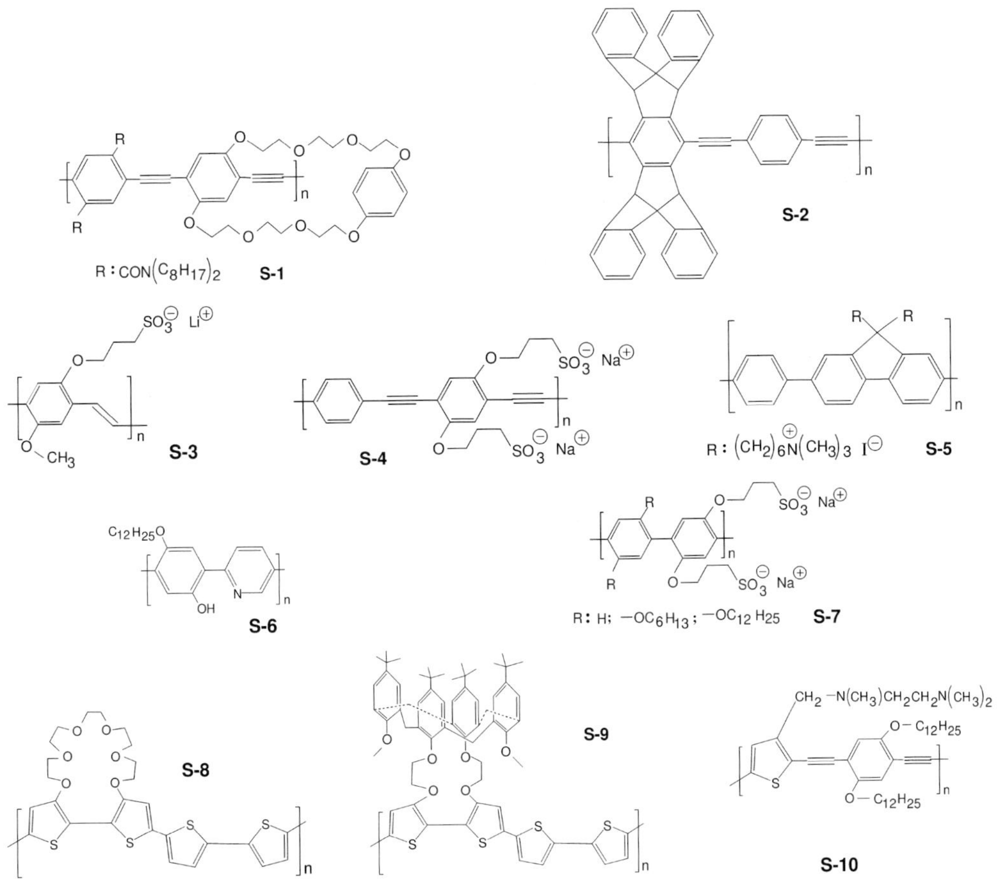

Chart 13.1 Chemical structures of typical conjugated polymers used as chemical sensors for organic compounds.

The sensing ability of conjugated polymers relies on the fact that non-covalent binding of extremely small amounts of analytes can quench their fluorescence. This phenomenon, referred to as *superquenching*, is due to the pronounced delocalization of excitons formed in conjugated polymers upon light absorption. Owing to this delocalization, excitons can rapidly travel along the polymer chain to quenching sites. This mode of action is referred to as *fluorescence turn-off sensing*. On the other hand, fluorescence *turn-on sensing* is observed when an analyte is capable of selectively detaching a quencher previously non-covalently linked to the polymer. Examples of both mechanisms are described in the following subsections.

13.2.1.1 Turn-off fluorescence detection

Conjugated polymer based chemosensors operating in the fluorescence turn-off mode are used to quickly detect trace amounts of certain organic substances in the gas phase or in solution. This is important in areas such as forensics, or the packaging and distribution of food, etc. An interesting example relates to the fast detection of 2,4,6-trinitrotoluene vapor (see Chart 13.2) [16]. TNT is present in about 80% of the 120 million landmines that are buried in over 70 countries [17]. A TNT sensor is based on a PPE polymer functionalized with pentiptycene groups (S-2 in Chart 13.1). An industrially developed portable landmine detector operating in this way is reported to detect femtogram quantities of TNT in one second, thus performing better than a TNT sniffer dog [18].

13.2.1.2 Turn-on fluorescence detection

Chemical sensors based on the turn-on fluorescence mode are used to selectively detect certain proteins and carbohydrates [12]. Moreover, the activity of protease enzymes playing important roles in regulating biological systems, such as thrombin (blood coagulation) or caspace (apoptosis), can be detected in this way. Scheme 13.1 illustrates how avidin, a glycoprotein of molar mass 6.6×10^4 g mol^{-1} that is present in raw egg white, is detected with the aid of an anionic PPV polymer to which cationic biotin-tethered viologen is linked by electrostatic interaction. The adduct does not fluoresce. Upon addition of avidin, however, the fluorescence is restored, since the biotin group is bound very tightly within the active site of avidin [19].

Another example is related to enzyme activity. Scheme 13.2 demonstrates how turn-on fluorescence can be used to monitor protease activity [20, 21].

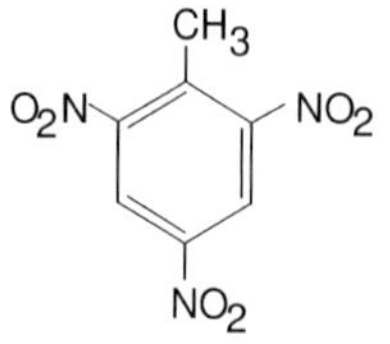

Chart 13.2 Chemical structure of 2,4,6-trinitrotoluene, TNT, an explosive constituent of landmines.

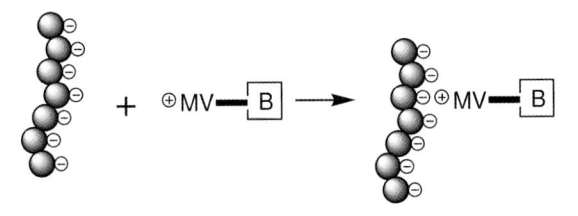

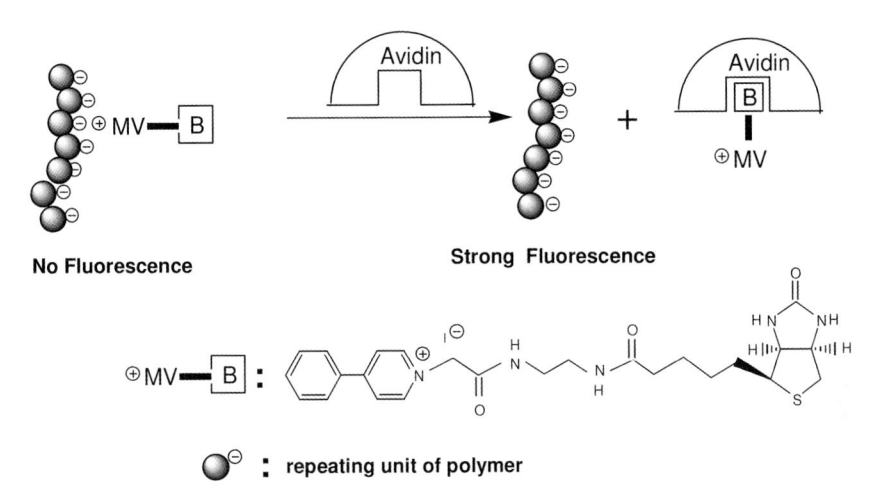

Scheme 13.1 Detection of avidin by turn-on fluorescence. Adapted from Chen et al. [19] with permission from the National Academy of Sciences USA.

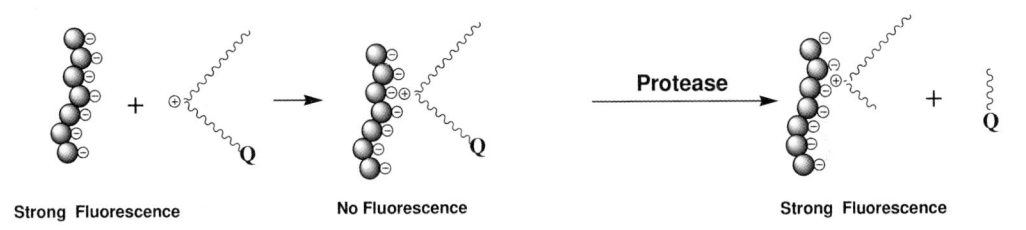

Scheme 13.2 Detection of protease by turn-on fluorescence. Adapted from Kumaraswani et al. [21] with permission from the National Academy of Sciences USA.

Here, a protein functionalized with a quencher, Q, is linked to the polymer by electrostatic interaction so that initially fluorescence is quenched. When added protease cleaves a specific bond in the peptide chain, the quencher is released into solution and fluorescence is restored.

13.2.1.3 ssDNA base sequence detection

Conjugated polymers also permit the detection of DNA hybridization (pairing of complementary DNA single strands, ssDNAs) and thus act as *ssDNA sequence sensors* [22]. These sensors comprise an aqueous solution containing CP, a cationic conjugated polymer (e.g., S-5 in Chart 13.1), and ssDNA-FL, a single-stranded DNA with a known base sequence and labeled with a chromophore such as fluorescein, FL. CP and ssDNA do not interact. Irradiation with light of relatively short wavelength that is not absorbed by FL causes the fluorescence of CP. Upon addition of an ssDNA with a specific base sequence, complementary to that of the probe ssDNA-FL, hybridization occurs. The double-strand thus formed becomes electrostatically linked to CP, thus allowing energy transfer from electronically excited CP* to FL (see Scheme 13.3). The characteristic fluorescence of the FL groups generated in this way signals hybridization. The FL fluorescence is not observed upon the addition of non-complementary ssDNA. Relative to the CP* emission, the FL emission spectrum is shifted to the long-wavelength region and can therefore be reliably detected. Recent research on strand-specific DNA detection with cationic conjugated polymers has been concerned with their incorporation into DNA chips and microarrays [23, 24].

13.2.1.4 Sensors for metal ions

As the recognition of possible effects of metal ions is of paramount importance with regard to human health, considerable effort has been directed towards the development of suitable chemosensors [14, 15, 25, 26]. Interesting work in this field concerns sensors based on regiospecific polythiophenes with substituted crown-ether macrocycles, such as S-8 in Chart 13.1. Depending on the ring size of the macrocycle substituent, these polymers display selectivity for specific alkali metal cations. Accommodation of ions causes a substantial blue shift of the maximum of the emission spectrum. Similarly, calix[4]arene-substituted poly-(phenylene bithiophene)s exhibit selectivity towards certain metal ions. For example, S-9 in Chart 13.1 selectively binds sodium ions, as indicated by a blue shift of the maximum of the emission spectrum [26]. Certain conjugated polymers bearing pendant amino groups are capable of selectively binding divalent cations such as Ca^{2+}, Zn^{2+}, and Hg^{2+} in aqueous solution [27]. This applies, for example, to polymer S-10 in Chart 13.1, which bears pendant N,N,N'-trimethyl-ethylenediamino groups. Chelation of the cation results in a pronounced increase in the fluorescence intensity, in particular in the case of Hg^{2+}. The augmented light emission may be rationalized in terms of the 3R scheme (see Sec-

$$CP^* + FL \longrightarrow CP + FL^*$$

$$FL^* \longrightarrow FL + h\nu$$

Scheme 13.3 Energy transfer from an electronically excited conjugated polymer to fluorescein.

tion 13.1), with photoinduced electron transfer, PET, as the relay mechanism. Rapid intramolecular electron transfer from the nonbonding electron pair at the N atom of the receptor site to the excited reporter site quenches the fluorescence in the absence of the analyte. Cation binding prevents PET.

13.2.1.5 Image sensors

Large-area (15×15 cm), full-color image sensors can be made on the basis of photoinduced charge generation in conjugated polymers (see Chapter 2) [28, 29]. Figure 13.3 shows the structure of a thin-film sandwich device in the metal/polymer/ITO configuration.

In typical work of Yu et al. [29], the arrays were fabricated on ITO glass substrates. The ITO glass layer was patterned by photolithography into perpendicular rows of electrode strips (width: 450 µm, spacing: 185 µm). The polymer film, a blend of poly(3-octyl thiophene) and fullerene PCBM[6,6] (see Section 6.3), was spin-cast onto the substrate.

Such microfabricated array devices are suitable for linear or two-dimensional (2D) digital optical cameras. In principle, they may also be actively used as electroluminescent devices.

13.2.2
Optical fiber sensors

Besides acting as wave guides in sensor devices (see Fig. 13.2), optical fibers play an important role as actively functioning sensing elements in *evanescent field absorption sensors*. In this case, part of the fiber cladding is replaced by a modified, solvent-repellent polymer, which, when inserted into a solution, is capable of selectively adsorbing specific analytes [4]. The working principle of evanescent field absorption sensors is based on the interaction of the analyte with the evanescent field generated when light passes through the core of an optical fiber. The light travels down the core as a result of numerous total internal reflections at the core–cladding interface. Optical interference occurs between parallel wavefronts during the succession of skips along the core, resulting in a standing wave and an electromagnetic evanescent field that penetrates the core–cladding interface. In other words, some of the radiation at the core–cladding interface penetrates a certain distance into the cladding. The depth of penetration, d_p, is defined as

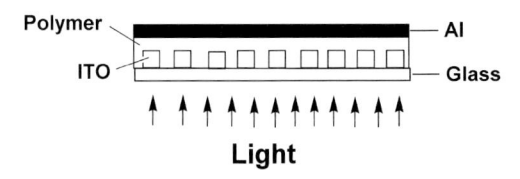

Fig. 13.3 Structure of a large image sensor device operated with a polythiophene/fullerene blend. Adapted from Yu et al. [29] with permission from Wiley-VCH.

the distance into the cladding over which the evanescent field is reduced to 1/e of its interface value; d_p can be calculated according to Eq. (13-1):

$$d_p = \frac{\lambda}{2\pi\sqrt{n_1^2 \sin^2 \theta - n_2^2}} \quad (13\text{-}1)$$

where λ is the wavelength of light propagating down the fiber; n_1 and n_2 are the refractive indices of the core and the surrounding cladding, respectively; and θ is the angle of incidence at the core–cladding interface. Typical values of d_p are of the order of the light wavelength λ. The strength of the evanescent field is reduced if it interacts with absorbing species. The penetrating light is then absorbed, and the intensity of the light passing through the fiber is attenuated. This reduction in intensity can be measured and related to the chromophore concentration at the core interface. Fiber evanescent field absorption (FEFA) spectroscopy offers advantages over conventional absorption spectroscopy using cuvettes, i.e. the effective absorption path length can be made very small and the technique can be applied to strongly absorbing chromophores. Moreover, due to the low value of d_p, FEFA is insensitive to scattering particles, thus permitting light absorption measurements in turbid water [30]. The FEFA technique is quite versatile; measurements in aqueous solutions can be readily performed with optical fibers made of poly(methyl methacrylate), PMMA, after complete removal of the cladding over the length that is to be immersed in the solution. In this case, the solution behaves as cladding and the evanescent field penetrates into the liquid [31]. The sensing sensitivity can be increased by coiling the fiber, e.g. to a length of 1.5 m, on a Teflon support of radius 1.5 cm. Coupling of a coiled polysiloxane-cladded fiber with a near-infrared spectrometer, operated in the 1.0–2.2 μm range, permits the recognition of organic compounds in mixtures such as chloroform in carbon tetrachloride or toluene in cyclohexane [32].

13.2.3
Displacement sensors

The working principle of displacement sensors is the swelling and shrinking of polymer beads, located at the end of polymer fibers, as a function of analyte concentration. Variations in the bead volume due to changes in analyte concentration alter the intensity of probe light guided through the bead to a reflector. Typical optode types operating in this way are listed in Table 13.1. Owing to the fragility of the beads, there are problems related to the reproducibility and the durability of these sensors [4].

Table 13.1 Displacement sensor systems based on reversible swelling.

Analyte	Polymer	Refs.
Protons in water (pH)	Polystyrene bearing amino groups	[33]
Ions (ionic strength)	Sulfonated polystyrene, sulfonated dextran	[34]
Water in organic liquids	Polystyrene bearing quaternary ammonium groups	[35]
Hydrocarbons in water	Poly(methyl trifluoropropyl siloxane), poly(dimethyl siloxane), poly(styrene-co-butyl methacrylate)	[36, 37]

13.3
Polymers as transducer supports

A large number of optodes developed for the selective detection of inorganic anions and cations, so-called ion-selective optodes (see Table 13.2), consist of polymer membranes that contain transducers. The latter are mostly physically admixed, but in some cases they are covalently bound to the polymer matrix. Most of these optodes [7, 8] are based on poly(vinyl chloride), plasticized with DOS, BBPA, DOP, o-NPOE or other plasticizers (see Chart 13.3). Typically, membranes are composed of 33 wt.% PVC, 66 wt% plasticizer, and 1 wt% ionophore (analyte-complexing agent) and lipophilic salt (ion-exchanger). Other polymers occasionally employed in hydrophobic optodes include polysiloxanes and poly(vi-

Table 13.2 Typical optode-detectable analytes [7]

Analyte class	Analytes
Inorganic cations	H^+, Li^+, Na^+, K^+, Mg^{2+}, Ca^{2+}, Ag^+, Zn^{2+}, Hg^{2+}, Pb^{2+}, NH_4^+
Inorganic anions	CO_3^{2-}, SCN^-, NO_2^-, Cl^-, I^-
Organic cations	Ammonium ions of 1-phenylethylamine, octylamine
Organic anions	Salicylate, guanosine triphosphate, heparin
Neutral analytes	H_2O, NH_3, SO_2, O_2, ethanol

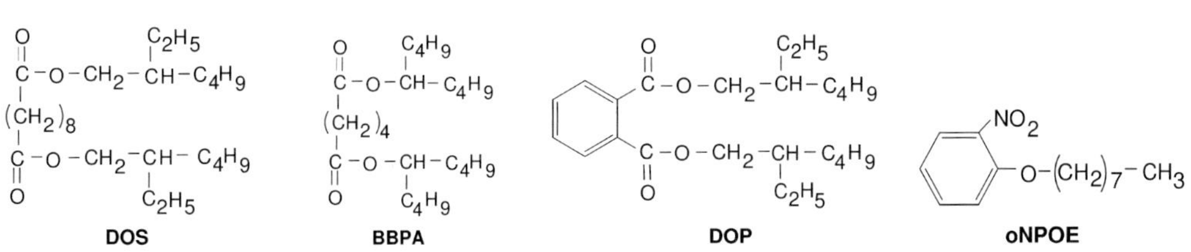

Chart 13.3 Plasticizers used in PVC-based optodes.
DOS: dioctyl sebacate, BBPA: bis(1-butylpentyl) adipate,
DOP: dioctyl phthalate, o-NPOE: o-nitrophenyl octyl ether.

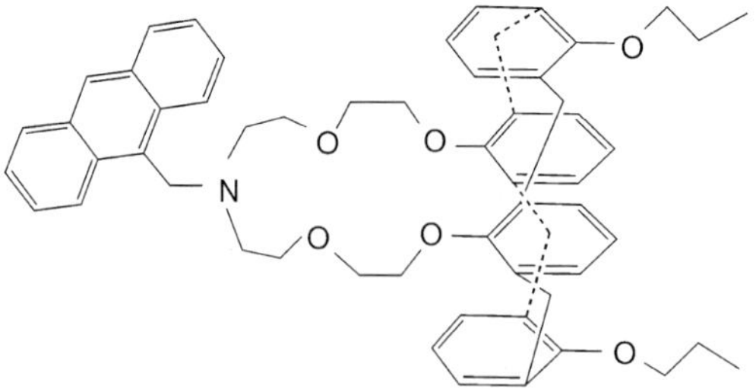

Chart 13.4 Chemical structure of N-(9-methyl-anthracene)-25,27-bis(1-propyloxy) calix[4]arene azacrown-5, used as a selective potassium ion sensor [39].

nylidene chloride). Polyacrylamide or other hydrogel-forming polymers are used in the case of hydrophilic membrane-based optodes.

Many of the optodes referred to here employ sensors operating on the basis of the 3R scheme (see Section 13.1), the relay mechanism being photoinduced electron transfer, PET. Due to their applicability in various chemical and biological processes, they have received much attention in recent years [1, 7, 8, 10]. Of note in this context are sensors that become fluorescent upon complexation of an analyte because the binding of the analyte within the sensor prevents the PET that suppresses fluoresence in the absence of the analyte [38]. Anthryl aza-crown-calix[4]arene, a K^+-selective sensor (see Chart 13.4), exhibits such behavior. It selectively binds potassium ions, and this triggers a substantial increase in anthryl fluorescence through disruption of the PET quenching process [9, 39].

References

1 V. Ramamurthy, K. S. Schanze (eds.), *Optical Sensors and Switches*, Marcel Dekker, New York (2001).
2 Y. Osada, D. E. Rossi (eds.), *Polymer Sensors and Actuators. Macromolecular Systems – Material Approach*, Springer, Berlin (2000).
3 J. Wackerly, *Conjugated Polymers as Fluorescence-Based Chemical Sensors* {www.scs.uiuc.edu/chem/gradprogram/chem435/fall04/06_Wackerly_Abstract.pdt}.
4 F. Baldini, S. Bracci, *Polymers for Optical Fiber Sensors*, Chapter 3 of [2], p. 91.
5 B. R. Eggins, *Chemical Sensors and Biosensors*, Wiley, Chichester (2002).
6 A. Mulchandani, O. A. Sadik (eds.), *Chemical and Biological Sensors for Environmental Monitoring*, ACS Symposium Series 762, American Chemical Society, Washington, D.C. (2000).
7 E. Bakker, P. Bühlmann, E. Pretsch, *Carrier-Based Ion-Selective Electrodes and Bulk Optodes. 1. General Characteristics*, Chem. Rev. 97 (1997) 3083.
8 P. Bühlmann, E. Pretsch, E. Bakker, *Carrier-Based Ion-Selective Electrodes and Bulk Optodes. 2. Ionophores for Potentiometric and Optical Sensors*, Chem. Rev. 98 (1998) 1593.
9 J. B. Benco, H. A. Nienaber, W. G. McGimpsey, *Optical Sensors for Blood Analytes*, The Spectrum 14 (2002) 1.
10 A. P. de Silva, H. Q. N. Gunaratne, T. Gunnlaugsson, A. J. M. Huxley, C. P. McCoy, J. T. Rademacher, T. E. Rice, *Signaling Recognition Events with Fluorescent Sensors and Switches*, Chem. Rev. 97 (1997) 1515.

11 C. M. Rudzinski, D. G. Nocera, *Buckets of Light*, Chapter 1 of [1].
12 D. Whitten, R. Jones, T. Bergstedt, D. McBranch, L. Chen, P. Heeger, *From Superquenching to Biodetection: Building Sensors Based on Fluorescent Polyelectrolytes*, Chapter 4 of [1].
13 T. Ishii, M. Kaneko, *Photoluminescent Polymers for Chemical Sensors*, in R. Arshady (ed.), *Desk Reference of Functional Polymers. Syntheses and Applications*, American Chemical Society, Washington, D.C. (1997), Chapter 4.3.
14 L. Dai, P. Soundarrajan, T. Kim, *Sensors and Sensor Arrays Based on Conjugated Polymers and Carbon Nanotubes*, Pure Appl. Chem. 74 (2002) 1753.
15 T. M. Swager, *The Molecular Wire Approach to Sensory Signal Amplification*, Acc. Chem. Res. 31 (1998) 201.
16 J.-S. Yang, T. M. Swager, J. Am. Chem. Soc. 120 (1998) 5321 and 11864.
17 J. Yinon, Anal. Chem. (2003) 99A.
18 M. La Grone, C. Cumming, M. Fisher, M. Fox, S. Jacob, D. Reust, M. Rockley, E. Towers, Proc. SPIE 4038 (2000) 553.
19 L. Chen, D. W. McBranch, H.-L. Helgeson, R. Wudl, D. Whitten, Proc. Natl. Acad. Sci. USA 96 (1999) 12287.
20 M. R. Pinto, K. S. Schanze, Proc. Natl. Acad. Sci. USA 101 (2004) 7505.
21 S. Kumaraswamy, T. Bergstedt, X. Shi, F. Rininsland, S. Kushon, W. Xia, K. Ley, K. Achyuthan, D. W. McBranch, D. Whitten, Proc. Natl. Acad. Sci. USA 101 (2004) 7511.
22 B. S. Gaylord, A. J. Heeger, G. C. Bazan, J. Am. Chem. Soc. 125 (2003) 896.
23 B. Liu, G. C. Bazan, Proc. Natl. Acad. Sci. USA 102 (2005) 589.
24 H. Xu, H. Wu, F. Huang, S. Song, W. Li, Y. Cao, C. Fan, Nucl. Acid Res. 33 (2005) e83.
25 J. Li, Y. Lu, J. Am. Chem. Soc. 122 (2000) 10466.
26 K. B. Crawford, M. B. Goldfinger, T. M. Swager, J. Am. Chem. Soc. 120 (1998) 5178.
27 L.-J. Fan, Y. Zhang, W. E. Jones, Jr., Macromolecules 38 (2005) 2844.
28 D. Pede, E. Smela, T. Johansson, M. Johansson, O. Inganäs, Adv. Mater. 10 (1998) 233.
29 G. Yu, J. Wang, J. McElvain, A. J. Heeger, Adv. Mater. 10 (1998) 1431.
30 D. W. Lamb, Y. Bunganaen, J. Louis, G. A. Woolsey, R. Oliver, G. White, Marine and Freshwater Research 55 (2004) 533.
31 P. G. Leye, M. Boerkamp, A. Ernest, D. W. Lamb, J. Phys.: Conf. Series 15 (2005) 262.
32 M. D. Degrandpre, L. W. Burgess, Appl. Spectrosc. 44 (1990) 273.
33 Z. Shakhsher, R. W. Seitz, Anal. Chem. 66 (1994) 1731.
34 M. F. McCurley, R. W. Seitz, Anal. Chim. Acta 249 (1991) 373.
35 M. Bai, R. W. Seitz, Talanta 41 (1994) 993.
36 G. Kraus, A. Brecht, V. Vasic, G. Gauglitz, Fresen. J. Anal. Chem. 348 (1994) 598.
37 G. Gauglitz, A. Brecht, G. Kraus, W. Nahm, Sensor Actuat. B 11 (1993) 21.
38 H. F. Ji, R. Dabestani, G. M. Brown, J. Am. Chem. Soc. 122 (2000) 9306.
39 J. B. Benco, H. A. Nienaber, K. Dennen, W. G. McGimpsey, J. Photochem. Photobiol. A: Chem. 152 (2002) 33.

14
Polymeric photocatalysts

14.1
General aspects

Photocatalysts are substances that initiate chemical reactions under the influence of light without being consumed during the process. Although the field of photocatalysts is largely dominated by inorganic substances such as titanium dioxide [1-4], polymers also have roles to play, in particular as catalyst-supporting materials. However, there are also some interesting developments concerning special polymers that function as active photocatalysts. These developments pertain not only to certain conjugated polymers, but also to polymers bearing pendant aromatic groups. In general, a photocatalytic process commences with the absorption of photons by the catalyst. Subsequent chemical alterations in the surrounding substrate molecules are the result of interactions with relatively long-lived excited states or electrically charged species formed in the catalyst. Typical polymeric photocatalysts and mechanistic aspects are presented in the following subsections.

14.2
Polymers as active photocatalysts

14.2.1
Conjugated polymers

It has been shown in Chapters 2 and 6 that conjugated polymers are quite versatile with regard to practical applications. For example, they play an outstanding role in the fields of organic light-emitting diodes and photovoltaic devices (see Sections 6.2 and 6.3, respectively). Here, their photocatalytic capability is highlighted by referring to the fixation of carbon dioxide, CO_2, a process of quite general importance, since methods of fixation of carbon dioxide are

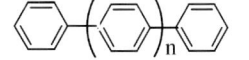

Chart 14.1 Chemical structure of PPP.

Polymers and Light. Fundamentals and Technical Applications. W. Schnabel
Copyright © 2007 WILEY-VCH Verlag GmbH & Co. KGaA, Weinheim
ISBN: 978-3-527-31866-7

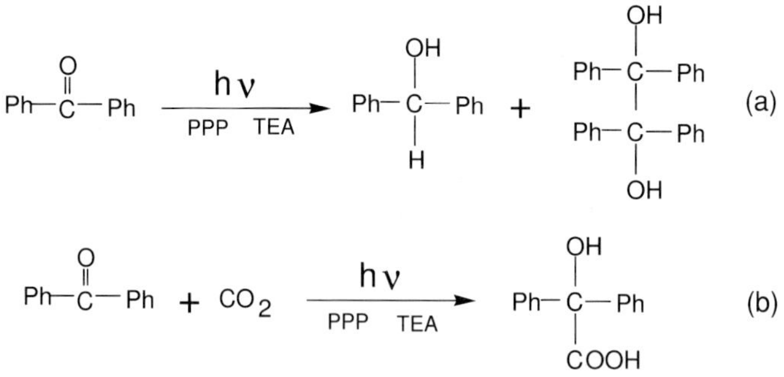

14 Polymeric photocatalysts

Scheme 14.1 PPP-catalyzed photoreactions of benzophenone in the absence (a) and in the presence of CO_2 [5].

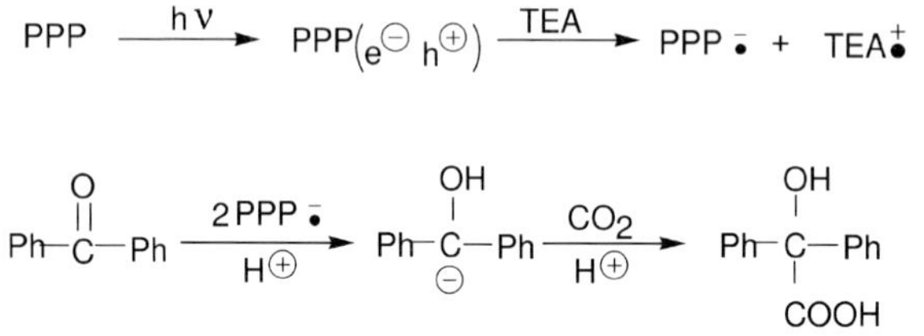

Scheme 14.2 Simplified reaction mechanism of the PPP-catalyzed photofixation of CO_2 in benzophenone [5].

needed to prevent the uncontrolled release of this greenhouse gas into the atmosphere [5]. The process reported here operates with a solution of benzophenone and triethylamine, TEA, in dimethylformamide containing dispersed poly(p-phenylene), PPP, the structure of which is shown in Chart 14.1. Upon exposure to visible light ($\lambda > 400$ nm), PPP catalyzes the photoreduction of benzophenone yielding benzhydrol and benzopinacol (Scheme 14.1a). If the system is saturated with CO_2, diphenylglycolic acid is formed, i.e. CO_2 is fixed (Scheme 14.1b).

The somewhat simplified reaction mechanism shown in Scheme 14.2 is based on the photogeneration of electron/hole pairs in PPP. While the holes react with triethylamine present in the system, the electrons remain in the polymer as delocalized anion radicals. They react with benzophenone to form the diphenylcarbinol anion, and the latter eventually reacts with CO_2. The CO_2 fixation is strongly enhanced by the presence of tetraethylammonium chloride. The soft onium cations are thought to stabilize the diphenylcarbinol anion, the precursor of the final product.

14.2.2
Linear polymers bearing pendant aromatic groups

This type of reaction has been pioneered by Guillet et al. using poly(sodium styrene sulfonate-co-2-vinylnaphthalene), a copolymer consisting, in this case, of about equal parts of the respective monomers (see Chart 14.2) [6].

In aqueous solution, this copolymer adopts a pseudo-micellar conformation, i.e. the macromolecules form hydrophobic microdomains capable of solubilizing organic compounds that are sparingly soluble in water. Table 14.1 presents typical systems explored in this work.

The reaction mechanism depends on the system and may be based on energy or electron transfer between the naphthalene moieties of the copolymer and the substrate molecule. In the case of oxidations, singlet oxygen, generated by energy transfer from the naphthalene moiety to $^3O_2$, may be involved. Typical reaction mechanisms are presented in Schemes 14.3 and 14.4.

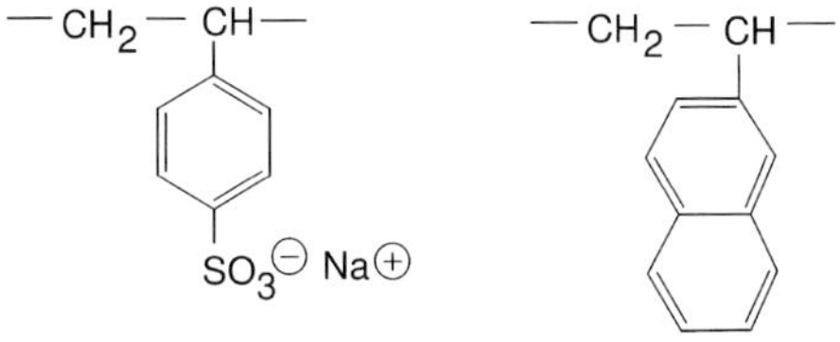

Chart 14.2 Chemical structures of the base units of poly(sodium styrene sulfonate-co-2-vinylnaphthalene).

Table 14.1 Reactions photocatalyzed by poly(sodium styrene sulfonate-co-2-vinylnaphthalene) in aqueous solution under solar irradiation [6].

Process	Products	Reaction mechanism
Oxidation of cyanide, CN^-	NCO^-	Electron transfer
Oxidation of styrene 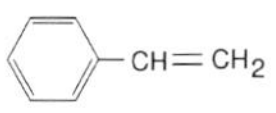	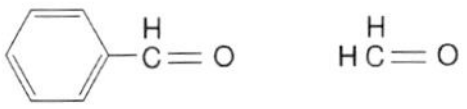 H-C=O, HC=O	Singlet oxygen reaction
Photodechlorination of hexachlorobiphenyl 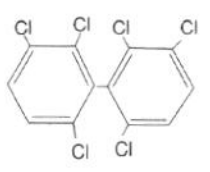	 HCl	Electron transfer
Photosynthesis of previtamin D_3 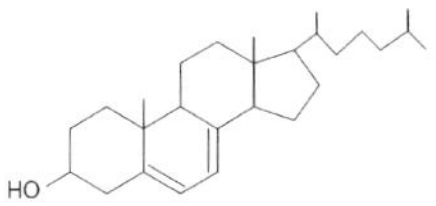	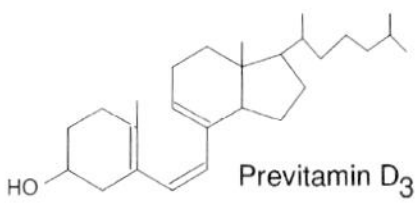 Previtamin D_3	Isomerization of 7-dehydrocholesterol

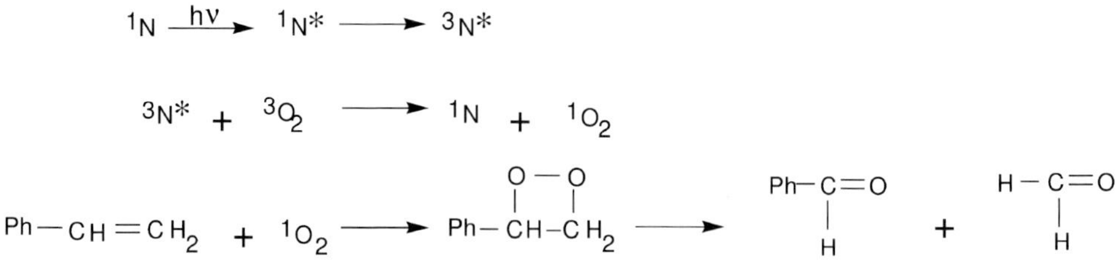

$$^1N \xrightarrow{h\nu} {^1N^*} \longrightarrow {^3N^*}$$

$$^3N^* + {^3O_2} \longrightarrow {^1N} + {^1O_2}$$

$$Ph-CH=CH_2 + {^1O_2} \longrightarrow Ph-\underset{|}{CH}-\underset{|}{CH_2} \longrightarrow \underset{|}{Ph-C=O} + \underset{|}{H-C=O}$$
$$\phantom{Ph-CH=CH_2 + {^1O_2} \longrightarrow Ph-}\overset{O-O}{} H H$$

Scheme 14.3 Singlet oxygen-mediated oxidation of styrene photocatalyzed by poly(sodium styrene sulfonate-co-2-vinylnaphthalene); N denotes the naphthalene moiety of the copolymer and Ph the phenyl group of styrene [7].

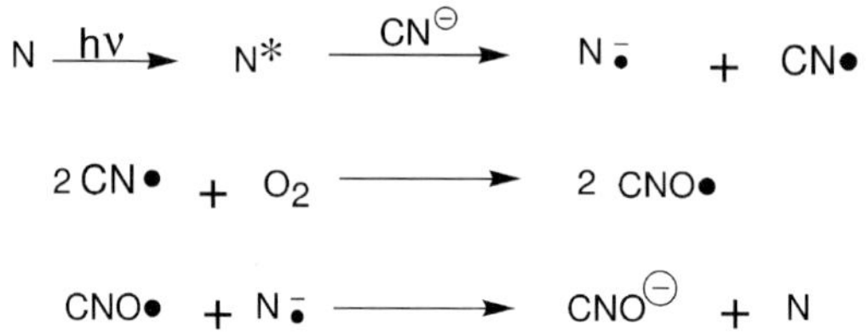

$$N \xrightarrow{h\nu} N^* \xrightarrow{CN^\ominus} N^{\bar{\bullet}} + CN\bullet$$

$$2\,CN\bullet + O_2 \longrightarrow 2\,CNO\bullet$$

$$CNO\bullet + N^{\bar{\bullet}} \longrightarrow CNO^\ominus + N$$

Scheme 14.4 Oxidation of cyanide ions photocatalyzed by poly(sodium styrene sulfonate-co-2-vinylnaphthalene); N denotes the naphthalene moiety contained in the copolymer as a pendant group [8].

14.3
Polymers as supports for inorganic photocatalysts

Certain inorganic materials can be employed as photocatalysts for the synthesis or degradation of compounds in heterogeneous systems. Relevant devices contain, for example, films incorporating immobilized photocatalyst particles. Typically, titania, TiO_2, is used for the treatment of water contaminated with chemical pollutants and/or bacteria [9]. The contaminants are oxidized by reactive species, i.e. hydroxyl and superoxide radicals, generated by reaction of electron/hole pairs with O_2 and water adsorbed at the particle surface. Electron/hole pairs are formed when UV light ($\lambda < 400$ nm) is absorbed by titania (see Scheme 14.5).

Titania is especially suitable as a photocatalyst, because it is highly catalytically active, yet chemically and biologically inert, photostable, and cheap. The photocatalytic efficiency of inorganic particles depends strongly on their specific surface area and their accessibility, since only substrate molecules in close contact with the particle surface can undergo chemical alterations. Both requirements, i.e. large surface area and accessibility, can be very well fulfilled by using nanoparticles embedded in polymer films of high porosity, as has been demonstrated in the case of titania [10, 11]. For example, photocatalytic porous films containing nanocrystalline anatase, the active TiO_2 modification, have been prepared on polycarbonate and poly(methyl methacrylate) substrates [10]. In an-

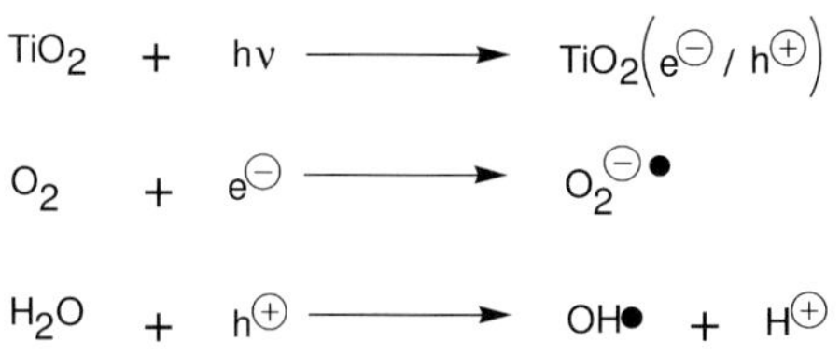

Scheme 14.5 Photogeneration of oxidizing species upon irradiation of titania with UV light.

$$TiO_2 + h\nu \longrightarrow TiO_2(e^{\ominus}/h^{\oplus})$$

$$O_2 + e^{\ominus} \longrightarrow O_2^{\ominus}\bullet$$

$$H_2O + h^{\oplus} \longrightarrow OH\bullet + H^{\oplus}$$

other case, photocatalytic films consisting of layers of cationic poly(allylamine hydrochloride), anionic poly(acrylic acid) (see Chart 14.3), and positively charged TiO$_2$ nanoparticles were fabricated in a layer-by-layer self-assembling method [11]. Besides the fact that polymer films are flexible, the advantages of using polymer-supported catalysts for the synthesis or degradation of compounds include reagent stability, suitability for automation, ease of work, and reduced contamination in the final product.

The performance of polymer-coated TiO$_2$ particles in an aqueous environment is also noteworthy. The presence of Nafion adlayers (see Chart 14.3) ensures that the surface charge on the TiO$_2$ particles is highly negative over the entire pH range. As a consequence, the photocatalytic degradation, PCD, of cationic substrates is enhanced, while that of anionic or neutral substrates is not significantly retarded [12]. In contrast, the efficiency and rate of PCD are much more pH-dependent in the case of naked TiO$_2$ particles, which are positively charged at low pH and negatively charged at high pH, due to the presence of $\equiv$TiOH$_2^+$ and $\equiv$TiO$^-$ groups, respectively.

From a survey of the patent literature, it is inferred that industrial research and development is focused to a significant extent on polymer-supported photocatalysts. While most of the numerous patents deal with titania, a few are devoted to other materials, such as ruthenium complexes or iridium oxide. Novel applications concerning the deodorization of air in automobiles with the aid of polytetrafluoroethylene-supported photocatalysts are noteworthy [13, 14].

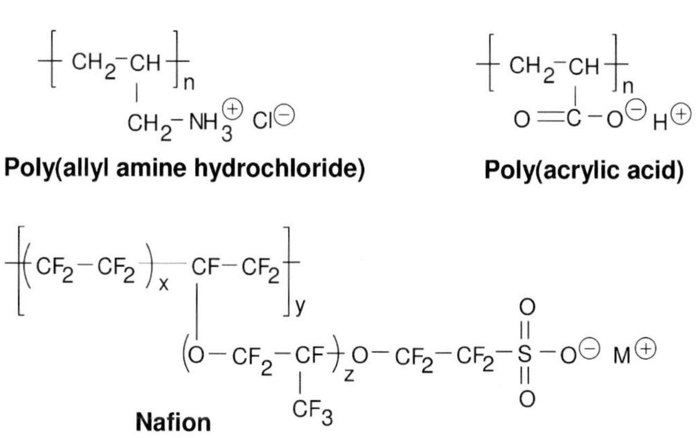

Poly(allyl amine hydrochloride)

Poly(acrylic acid)

Nafion

Chart 14.3 Polymers employed as supports for inorganic photocatalysts.

References

1 J. M. Herrmann, Catalysis Today 53 (1999) 115.
2 M. R. Hoffmann, S. T. Martin, W. Choi, D. W. Bahnemann, Chem. Rev. 95 (1995) 69.
3 D. F. Olis, H. Al-Ekabi (eds.), *Photocatalytic Purification and Treatment of Water and Air*, Elsevier, Amsterdam (1993).
4 N. Serpone, E. Pelizetti (eds.), *Photocatalysis: Fundamentals and Applications*, Wiley, New York (1989).
5 Y. Wada, T. Ogata, K. Hiranaga, H. Yasuda, T. Kitamura, K. Murakoshi, S. Yanagida, J. Chem. Soc. Perkin Trans. 2 (1998) 1999.
6 J. E. Guillet, *Biomimetic Polymer Catalysts for Important Photochemical Reactions*, Can. Chem. News 52 (2000) 16.
7 M. Nowakowska, J. E. Guillet, Macromolecules 24 (1991) 474.
8 M. Nowakowska, N. A. D. Burke, J. E. Guillet, Chemosphere 39 (1999) 2249.
9 J. M. C. Robertson, P. K. J. Robertson, L. A. Lawton, J. Photochem. Photobiol. A: Chem. 175 (2005) 51.
10 M. Langlet, A. Kim, M. Audier, J. M. Herrmann, J. Sol-Gel Sci. Tech. 25 (2002) 223.
11 T.-H. Kim, B.-H. Sohn, Appl. Surf. Sci. 201 (2002) 109.
12 H. Park, W. Choi, J. Phys. Chem. B 109 (2005) 11667.
13 K. Yamamoto, K. Sakaguchi, J. Asano, Patent JP 2000300984 (2001).
14 T. Hiyori, T. Domoto, Patent JP 2000296168 (2001).

Subject Index

a

absorbance (extinction, optical density) 7
absorption of light 5, 14
– photoinduced absorption 41
– T-T absorption 41
acetophenones
– type I free radical photoinitiators 278
acrylate- and methacrylate-based monomers
– volume holography 324
acrylonitrile/butadiene/styrene (ABS) co-polymer
– photodegradation 199
O-acyl-α-oximo ketones
– type I free radical photoinitiators 278
acylphosphine oxides
– type I free radical photoinitiators 278
acylphosphonates
– type I free radical photoinitiators 278
N-alkoxy pyridinium and isoquinolinium salts
– cationic photoinitiators 290
Alzheimer's disease 224
amine-catalyzed cross-linking
– photo-triggered curing 298
– polyurethane-based coatings 298
amines 315
– curing of dental formulations 315
amino ethers
– reaction with alkyl peroxyl or acyl peroxyl radicals 264
amplified spontaneous emission 44
angular multiplexing
– volume holography 345
anionic polymerization
– photo-production of reactive organic bases
– – amidine bases 297
– – tertiary amines 297
– photo-release of reactive anions 296
anisotropic contraction 131

anisotropy
– generation by trans-cis-trans isomerization 124
antenna effect 17
anthraquinones
– type II free radical photoinitiators 280
anthryl aza-crown-calix[4]arene
– potassium ion sensor 357
antioxidants
– radical scavengers 257
apoptosis 223
– turn-on fluorescence detection 352
applications of NLO polymers
– optical limiters 100
– phase conjugation 100
– transphasor, the optical transistor 100
aromatic amino acids
– phenylalanine (Phe) 209
– tryptophan (Trp) 209
– tyrosine (Tyr) 209
aromatic ketones
– water-soluble 280
aspect ratio 236
atomic force microscopy (AFM)
– detection of surface gratings 133
autoacceleration 199
automotive accessories 310
– photocured coatings 310
automotive applications
– polymer optical fibers 169
autooxidation
– polymers 199, 200
autoretardation 199
avidin
– turn-on fluorescence detection 353
azobenzene compounds
– isomerization quantum yields 125
azobenzene groups 115
– in polyamides 117

Polymers and Light. Fundamentals and Technical Applications. W. Schnabel
Copyright © 2007 WILEY-VCH Verlag GmbH & Co. KGaA, Weinheim
ISBN: 978-3-527-31866-7

– – conformational change   119
– in polyimides   125, 135
– in polymer films   123
– in polymers   116
– in polypeptides   119
azobenzene-modified polymers
– surface gratings   133

b

BD
– blu-ray disk   339
benzoylferrocene
– anionic photoinitiators   296
benzyl ketals
– type I free radical photoinitiators   278
benzoin and benzoin ethers
– type I free radical photoinitiators   278
benzophenone derivatives
– type II free radical photoinitiators   280
benzotriazoles
– UV absorbers   258
bioluminescence   207
biopolymer structures   208
bipolarons   55
birefringence   73, 124
– light-induced   123
birefringent modulator   96, 97
bisazides   189
– poly(cis-isoprene)   188
– photo-cross-linking of linear polymers   188
bisphenol A polycarbonate   68
blepharismins
– photosensors   211
block copolymers   326
– formation   327
– structures   327
blood coagulation
– turn-on fluorescence detection   352
blue diode lasers   342
blue-ray disks
– storage capacity   342
blu-ray disk   339
bond cleavage   177
bond dissociation energies   177
bovine serum albumin
– optical absorption   209
Bragg condition   160
Bragg reflector   160
Bragg wavelength filters   96
Brønsted (protonic) acids
– photogeneration   240

business stationary
– polymer printing plates   319

c

CAD
– computer-aided design   315
cadmium sulfide, CdS
– inorganic photoinitiators   286
CaF_2
– lens material at   157 nm   246
calf thymus DNA
– optical absorption   209
CAM
– computer-aided manufacturing   315
cancer   211
– photochemotherapy   223
cans
– aluminum beer and beverage cans
– – photocured coatings   311
– food cans
– – photocured coatings   311
carbamate containing acrylates
– photocured coatings   310
carbohydrates
– turn-on fluorescence detection   352
carbonyl groups
– photoreactions   182
carotenoids
– photoreceptors   209, 210
cationic polymerization   288
– chemical structures of monomers   289
CD
– compact disk   339
CD-ROM   340
cellulose   208
– photoreactions   221
chain breakers
– radical scavengers   257
chain polymerization   275
chain reactions
– dehydrochlorination of PVC   197
– photo-oxidation of polymers   201
– polymerization   275
– – of diacetylenes   300
– topochemical   300
chain terminators   262
– radical scavengers   257
chalcogenide glasses   346
charge carriers
– bipolarons   54
– dissociation of excitons   56
– drift mobility   60
– generation   55

- polarons 54, 55
- quantum yields 57, 58
- radical cations 55
- transport 60
- transport in amorphous polymers 64
-- disorder concept 64
-- hopping mechanism 64
charge-coupled device (CCD) 41
charge generation layers
- xerography 145
charge hopping 52
charge-transfer molecules 88
charge-transport layers
- xerography 146
chemical amplification resists 239
chemical sensor action
- schematic depiction 349
chemosensing 349
chirality
- enantioselective induction 32
chiral molecules 23
chlorophylls 217
- photoreceptors 211
chromophores
- electro-optically active 98
chromophoric groups 6, 177
circadian rhythm 217
circular birefringence 24
circular dichroism 24, 25
- circular dichroism spectroscopy 25
circular dichroism (CD) spectra
- polypeptide structures 120
circular dichroism spectroscopy
- characterization of the chirality 32
- nucleic acids 32
- polypeptides 32
- proteins 32
- spectra of PMBET 34
- spectra of polyisocyanate, PICS 34
cis-trans isomerization 54
Claisen rearrangement 242
claddings of optical fibers
- polymers 170
clear coatings for paper 309
cleavage of chemical bonds
- polystyrene 178
- poly(methyl methacrylate) 178
CO_2 fixation 362
coatings
- radiation-cured
-- commercial applications 310
coil helix transition
- in poly(L-glutamic acids) 119

co-initiators
- type II free radical photoinitiators 280
collagen 214
- thermal denaturation 31
color holograms
- holography 323
color mixing
- holography 323
command surfaces 127
compact disk 339
- manufacture 341
- storage capacity 340
computer-aided design, CAD 315
computer-aided manufacturing, CAM 315
computer-assisted design, CAD
- photoinitiators for visible light 281
computer chip fabrication 236
conjugated polymers 156
- absorption spectra 12
- chemical sensors 351
- chemosensors 349
- exciton model 12
- laser materials 157
- photocatalysts 361
construction
- photocured coatings 310
consumer goods
- photocured coatings 310
contact lenses 310
contact printing
- photolithography 232
copper wire cables 168
copying machines
- xerography 143
cornea reprofiling and sculpting 254
Cotton effect 24
cross-linking
- [2+2] cycloaddition
-- poly(vinyl cinnamate) 185
- cleavage of phenolic OH groups 192
- cycloaddition of C=C bonds
-- poly(vinyl cinnamate) 184
- intermolecular cross-links 183
- mechanism 183
- photoacid-catalzyed
-- epoxide groups 242
- photogenerated reactive species 188
- photopolymerization 186
- polymerization of reactive moieties in pendant groups 186
- quantum yields 194
- thick polymer films 184
- triplet nitrene 190

cryptochromes
- photoreceptors 209
crystal violet leuconitrile (CVCN)
- anionic photoinitiators 296
curing 307
- cationic curing 311
- dual curing 312
- free radical curing 309
- of inks 320
cyanide ions
- photocatalyzed oxidation 364
[2+2] cycloaddition 185
- DNA, dimeric photoproducts 212
cycloaliphatic structures
- in random copolymers 244
cystine bridges
- rupture 216
cytochromes 217
cytoskeleton 223

d

3D color images
- holography 325
data transfer rate 339
degenerate four-wave mixing (DFWM) 86
dendritic polymers 19 f.
dental formulations
- curing 315
- photocurable formulations 314
- polymerizable compounds 316
dental preventive and restorative systems
- photocuring 314
deodorization of air 365
deoxyribonucleic acid (DNA) 208
- photoreactions 211
- thermal denaturation 31
depletion
- of stabilizers 267
desktop printing
- xerography 143
detrimental degradation
- of unstabilized commercial polymeric products 182
Dexter mechanism 15
diacetylenes 299
- bolaamphiphilic diacetylenes 300
- polymerization 300
- topochemical photopolymerization 300
dialkenes
- stepwise [2+2] photocyclopolymerization 302
diarylethenes 114

diazonium salts
- cationic photoinitiators 290
dibenzoylferrocene
- anionic photoinitiators 296
dicarbenes
- diacetylene polymerization 301
dichromated gelatine 324
digital optical cameras
- image sensors 355
digital versatile disks 339
- storage capacity 340
diglycol diallylcarbonate resin
- POFs 169
diketones
- curing of dental formulations 315
1,2-diketones (benzils and camphorquinone)
- type II free radical photoinitiators 280
diphenyliodonium salts
- photolysis 241
dipole moment 6
- aligning of permanent dipole moments 78
- electric field dependence 74
- hyperpolarizabilities β and γ 74
- linear polarizability 74
diradicals
- diacetylene polymerization 301
displacement sensors 357
- swelling and shrinking of polymer beads 356
displays for cell phones
- holography 325
dissolution inhibitor 236
distributed Bragg reflector, DBR 159
distributed Bragg reflector device 161
2,5-distyrylpyrazine
- four-center-type photopolymerization 301
- four-center-type polymerization 303
disulfide bridges
- proteins 215
DNA 207, 209
- dimeric photoproducts 212
- photodimers 213
- repair of dimer lesions 213
- sequence-selective photocleavage 226
- strand cleavage 226
DNA lesions 212
DNA photolyases 219
DNA strands
- sequence-specific cleavage 227
dopants
- dinitrobenzene 69

– fullerene, C_{60} 69
– in photoconducting polymeric systems 50
– isopropylcarbazole (ICP) 67
– phenylcarbazole (PhC) 67
– tetracyanoquinone (TCNQ) 69
– trinitrofluorenone (TNF) 68
– triphenylamine (TPA) 67
doped polymers 49
– dopants 67
– hole mobility 67
– photoconductivity 66, 68
– quantum yields of charge carrier generation 67
– temperature dependence of the hole mobility 68
DRAM
– dynamic random access memory 234
dual-cure acrylic urethane system 313
dual curing
– coatings protecting three-dimensional objects 312
– method combining UV irradiation and thermal treatment 312
– oligomers bearing acrylate and isocyanate groups 312
dual-layer photoreceptors 145
– charge generation layer 144
– charge transport layer 144
dual-layer systems
– xerography 143
DVD
– digital versatile disk 339
Dycryl
– letterpress plates 318
dye/co-initiator systems
– photoinitiators for visible light 281
dye-sensitized free radical polymerization
– co-initiators 286
dynamic random access memory (DRAM) 234

e
EFISH method 79
– electric field-induced second harmonic generation 79
elastin 214
electrical-to-optical signal transducers 96
electroluminescence
– polymer-based 148
– quantum yields 152
electroluminescence spectra

– oriented substituted poly(p-phenylene) 155
electron/hole pairs 53
– dissociation 55
– organic solarcells 165
– PPP 362
electronics
– photocured coatings 310
electron-spin resonance (ESR) 54
electron transition 9
electro-optic (EO) phenomena 73 ff.
electrophotography – xerography
– photoreceptors 143
ellipticity
– mean residue weight ellipticity 25
– molar ellipticity 25
enantiomers 23
energy migration 16, 17
energy quenchers 257
– light stabilizers 260
energy quenching 177
energy transfer 14, 17, 38
– Dexter mechanism 15
– Förster mechanism 15
– long-range interaction 15
– short-range interaction 15
enzymes
– inactivation 215
EO (electro-optic) materials 73
EO modulators 73
epoxide monomers
– volume holography 324
epoxide/polyol formulations
– photocured coatings 312
epoxides
– photo-cross-linking
– – stereolithography 186
– – surface coating 186
– – volume holography 186
ESCAP
– Environmentally Stable Chemical Amplification Positive Photoresist 241
Escherichia coli
– resurrection of UV-killed 219
ESIPT 260, 268
– excited-state intramolecular proton transfer 259
ethylene propylene diene copolymers (EPDM elastomers)
– photo-cross-linking 191
EUV
– extreme ultraviolet radiation ($\lambda = 13$ nm) 234

evanescent field absorption sensors
- optical fiber sensors   355
excimer emission   17
excimers   16
excited molecules   10
- annihilation   16
- deactivation by chemical reactions   21
- excimers   16
- intermolecular deactivation   14
- intramolecular deactivation   13
exciton concept   52
exciton model   12
excitons   52, 152
- CT excitons   53
- dissociation   56
- emission   56
- Frenkel excitons   53
- organic solar cells   165
- Wannier excitons   53
exposure characteristic curves   238
extinction coefficient   7, 9, 11

f

fatigue resistance
- photochromic systems
-- diarylethenes   137
-- fulgides   137
femtosecond spectroscopy   43
Fermi level   51
ferrocenium salts
- cationic photoinitiators   290
- photoinitiators   283
fiber evanescent field absorption (FEFA) spectroscopy   356
fiber-optic sensors   169
fiber-optic systems
- high-bandwidth   168
fiber-to-the-home systems   169
fibroin (silk)   214
films
- Langmuir-Blodgett (LB) film   22
flash photolysis   39
flavin adenine dinucleotide, FAD   219
flavins
- photoreceptors   210
flexographic printing   320
fluorescence   10, 13, 14
- depolarization   28, 29
fluorescence turn-off sensing   352
fluorescence turn-on sensing   352
fluorine-containing polymers
- F_2 *(157 nm) lithography*   245
Förster mechanism   15

four-center-type photopolymerization   301
four-center-type polymerization   303
Fourier-transform infrared (FTIR) spectra   36
Fowler-Nordheim (FN) tunnelling
- OLEDs   151
Franck-Condon factor   6
free radical polymerization
- two-photon absorption   99
free-radical-promoted cationic polymerization   293
free radicals   178
- generation   182
Frenkel excitons   53
- in polysilanes   53
fulgides   114
fulgimides   114
fullerene derivatives
- organic solar cells   166
furniture
- photocured coatings   310

g

gaskets   310
gel dose, D_{gel}
- cross-linking   195
geminate electron/hole pairs   57
generation of light   146
glow discharge   70
graded-index polymer optical fibers   170
graft copolymers   326, 327

h

halogenated compounds
- type I free radical photoinitiators   278
HALSs
- hindered amine light stabilizers   262
HASs
- chemical structures   263
- hindered amine stabilizers   262
- oxidation of   264
HD-DVD   342
- high definition DVD   339
hemispherical lens   342
heterolytic bond cleavage   113
hole mobility   62, 63
- electric field dependence   66
- temperature dependence   66
holograms
- electrically switchable   325
- reconstruction of the image   322
- recording   321
- set-up for recording   344

hologram formation
- mechanism 323
holographic disks
- replication 345
holographic imaging
- photorefractive materials 111
- time-gated holographic imaging 111
holographic materials 324
holographic plate 344
holographic storage materials
- volume holography 345
holographic three-layer plate
- color holographic recording 324
holography
- applications 325
- volume phase holograms
-- photopolymerization 321
host/guest systems 156
HRS (hyper-Rayleigh scattering) method 79
hybrid curing
- dual curing
-- simultaneous free radical and cationic cross-linking polymerization 313
hydrogen abstraction 182
hydrogen bonds
- destruction 31
hydroperoxide decomposers
- alkyl and aryl phosphites 265
- chemical structures 265
- dialkyl dithiocarbamates 265
- dithioalkyl propionates 265
- dithiophosphates 265
hydroperoxide groups 200
- generation of hydroxyl radicals 221
hydroxyalkylphenones
- type I free radical photoinitiators 278
o-hydroxybenzophenones
- UV absorbers 258
hydroxyl radicals 180
hyperchromicity 31
hyperpolarizability 74, 77
- electric field-induced second harmonic generation, EFISH 79
- hyper-Rayleigh scattering, HRS 79
hyper-Rayleigh scattering, HRS 80
hypochromicity 31

i

image sensors
- full-color sensors 355
immunoglobulins
- segmental motions 29

impurity chromophores
- carbonyl groups 180
- charge-transfer complexes 181
- commercial polymer formulations 180
- conjugated double bonds 181
- double bonds 181
- hydroperoxide groups 180
- in commercial polyalkenes and poly(vinyl chloride)s 181
- metal ions 181
- polynuclear aromatics 181
index of refraction 74
- electric field dependence 78
influenza virus
- colorimetric detection
-- polydiacetylene 300
information density 231
information storage
- holography 326
infrared (IR) spectroscopy
- analysis and identification of polymers 35
iniferters
- initiator-transfer-agent-terminators 328
initiation techniques
- electrochemical initiation 275
- high-energy radiation initiation 275
- photoinitiation 275
- thermochemical initiation 275
injection of charges 150
inorganic particles
- surface grafting 331
inorganic photocatalysts 364
intermolecular cross-linking 178
interpenetrating networks
- hybrid curing 313
intraocular lens implants 310
iodonium salts
- cationic photoinitiators 290
- photolysis 291
IPN
- interpenetrating networks 313
IPN polymers 314
IR spectra of polymers 36
IR spectrometers 36

j

Jablonski diagram 10

k

keratin (wool) 214
Kerr effect 73
α-keto coumarins

– type II free radical photoinitiators 280
Kleinman symmetry 77

l

α-lactalbumin 215
Lambert-Beer law 7
Langmuir-Blodgett (LB) film 22
laser ablation 248
– dopant-enhanced 250
– generation of periodic nanostructures in polymer surfaces 256
– keratectomy 253
– molecular mechanism 250
– multi-photon absorption 250
– plasma thrusters 256
– plume 250
– polymers designed for 251
– synthesis of organic compounds 252
laser direct imaging, LDI
– photoinitiators for visible light 281
lasers 156
lasing mechanism
– Boltzmann equilibrium of states 158
– population inversion 158
– stimulated emission 158
lasing threshold 159
LDMS
– laser ablation 254
– laser desorption mass spectrometry 254
LED (light emitting diode)
– multilayer polymer LED 149
– single-layer polymer LED 149
letterpress plates
– structure 319
letterpress printing plates 318
light attenuation
– in POFs 169
light-driven mass transport 347
light-emitting diodes, LEDs 147
light-harvesting
– in multiporphyrin arrays 21
light sources
– extreme ultraviolet (EUV) sources 234
– Hg discharge lamps 234
– lasers 234
light stabilizers
– bifunctional and trifunctional stabilizers 266
– energy quenchers 257
– radical scavengers 257
– UV absorbers 257
lignins
– formation of quinoid structures 222

– optical absorption spectra 208
– phenoxyl radicals 222
– photoreactions 221
– wood 207
linear electro-optic effect (Pockels effect) 78
linear polarizability 74
liquid-crystal displays (LCDs)
– polarized backlights 38
liquid-crystalline copolymers
– forgery-proof storage systems 139
liquid-crystalline polymers
– alignment 127
– amplified photoalignment 126
– birefringence 125
– command surfaces 127
– image storage 127
– optical dichroism 125
– photochromic amplification effect 127
– trans-cis-trans isomerization of azobenzene groups 126
liquid immersion lithography
– photolithography 234
lithium niobate 99
lithographic process 232
lithography
– imprinting lithography 235
– maskless lithography 235
– photolithography 231
local area networks (LANs)
– polymer optical fibers 169
luminance–voltage characteristic
– polymer LED 150
luminescence 37
– excimer emission 16
– fluorescence 13
– monomer emission 16
– phosphorescence 13
luminophores 28
lysozyme
– thermal denaturation 30

m

Mach-Zehnder (MZ) interferometer 96, 97
macroinitiators 327
macromolecular photoinitiators 279, 282
macromolecules
– photochromic transformations
– – aggregation 117
– – coil contraction 117
– – coil expansion and contraction 116
– – precipitation 117
macroradicals 190, 199

– generation 191
magneto-optical disk 340
magneto-optical (MO) recording 342
main-chain cleavage
– quantum yields 194
main-chain scission 178
malachite green leucohydroxide (MGOH)
– anionic photoinitiators 296
MALDI
– laser ablation 254
– matrix-assisted laser desorption/ionization 254
MALDI mass spectra 255
mask
– photolithography 232
maskless lithography
– ion-beam lithography 235
mass transport
– light-induced 132
mechanical energy
– by light energy conversion 130
mechanical machining
– laser ablation 248
media-oriented system transport (MOST) devices
– polymer optical fibers 169
membranes
– photochromic transformations
– – control of physical properties 122
– photoresponsive behaviour 122
metal-based photoinitiators 283
metal ions
– detection by chemosensors 354
microcavity
– vertical cavity lasing device 160
microfabrication 246, 309, 315
microlithography 231
micromachining 248, 315
– photomicrolithography 247
microring laser 161
microstructures
– high aspect ratio 247
mirror
– conventional mirror 86
– phase conjugate mirror 86
molecular orbitals 7, 8, 9
molecular wires 63
monomer emission 17
monomers
– surface grafting 331
4-morpholinophenyl amino ketone
– two-photon absorption
– – photolysis 318

multicolour holographic recording
– holography 323
multiplexing
– holography 322
– volume holography 345

n

nafion
– polymer support for inorganic photocatalysts 365
nanofabrication 246
naphthodianthrones
– photosensors 211
near-field optical recording
– recording density 343
newspapers
– polymer printing plates 319
nickel chelates
– light stabilizers 261
nitrene
– singlet nitrene
– – reactions 189
– triplet nitrene
– – reactions 189
o-nitrobenzyl ester photo-rearrangement
– nitronic acid 203
nitronic acid 203
nitroxyl (aminoxyl) radicals > N–O 262
nitroxyl radicals 264
– photolysis 268
– reaction with polymers 265
nonacosadiynoic acid 299
non-conjugated polymers
– absorption of light 10
– absorption spectra 11
nonlinear optical materials
– applications of NLO polymers 100
– – optical data storage 99
– – telecommunications 96
– second-order NLO materials 87, 89
– third-order NLO materials 88
nonlinear optical phenomena 73 ff.
– second-order phenomena 79
– third-order phenomena 82
nonlinear optical (NLO) properties
– second-order optical nonlinearity 77
– third-order optical nonlinearity 77
Norrish reactions 268
Norrish type I and II processes 260
Norrish type I reaction 21, 182, 183
Norrish type II reaction 182, 183
Novolak resists 236, 237
nucleases see photochemical nucleases 227

nucleic acids 207
numerical aperture 233, 339
– of objective lenses 342
Nyloprint
– letterpress plates 318

o

offset printing 320
OLED (organic light emitting diode) displays 147
OLEDs
– injection-limited conduction 151
– polarized light 154
– structure of a two-layer OLED 151
– transport-limited conduction 151
– white-light 155
– – Pt-containing compounds 156
oligopeptides
– for optical storage 139
Onsager theory
– quantum yield of charge carrier generation 57
optical absorption 9
optical activity 23
optical data storage 99
– photochromic systems
– – diarylethenes 137
– – fulgides 137
– – liquid-crystalline copolymers 138
– – liquid-crystalline polyesters 138
optical dichroism
– light-induced 123
optical fiber cables 168
optical fiber coatings 310
optical fibers 167
– information networks 168
– step-index optical fibers 168
optical fiber sensors
– evanescent field absorption sensors 355
optical limiters
– applications of NLO polymers 100
optical memories 339
optical near field recording 343
optical phase conjugation (OPC) 86
optical recording materials 126
optical resonators
– feedback structures
– – flat microdisks 159
– – microrings 159
– – microspheres 159
optical rotary dispersion (ORD) 24
optical storage techniques
– blu-ray disk 339

– compact disk 339
– digital versatile disk 339
– high definition DVD 339
– light-driven mass transport 347
– near-field recording 343
optical transistor 100
optical waveguides 167
optodes
– detection of molecular oxygen and carbon dioxide 350
– ion-selective optodes 357
– polymer transducer supports 357
organic light-emitting diodes, OLEDs 147
organometallic initiators
– photoinitiators for visible light 281
orientation of polymers
– electric field-induced 75
oscillator strength 7
overprint varnishes 309
– aluminum 311
– tin-free steel 311
oxaspiro monomers
– non-shrinking dental formulations 317
– – curing of dental formulations 315
oxazolidone containing acrylates
– photocured coatings 310
oxidation
– polymers 199
oxyl radicals 200
– reactions 201

p

packaging
– photocured coatings 310
paperback books
– polymer printing plates 319
paper coatings 309
Paterno-Büchi-type reaction 213
– DNA, dimeric photoproducts 212
PBOCSt
– acidolysis 241
– poly(t-butoxycarbonyl oxystyrene) 240
pentacosadiynoic acid 299
pericyclic reactions (electrocyclizations) 113
peroxyl radicals 264
– reactions 201
phase conjugate mirror 86, 87
phase conjugation
– applications of NLO polymers 100
phase controllers 96
phase holograms
– recording 344

phenacyl anilinium salts
- cationic photoinitiators 290
phenylalanine (Phe) 209
phenylglyoxylates
- type I free radical photoinitiators 278
phenyl salicylates
- UV absorbers 258
S-phenyl thiobenzoates
- type I free radical photoinitiators 278
phosphonium salts
- cationic photoinitiators 290
phosphorescence 10, 13, 14
photoacid generators 243
photoaddressable polymers 346
photoalignment
- liquid-crystalline compounds 128
- of liquid-crystal molecules 23
- of liquid-crystal systems 126
- surface-assisted 129
photocatalysts 361
- inorganic materials 364
photocatalytic polymer films 365
photochemical nucleases 228
photochemical reactions
- amplification effects 178
- polymers 178
photochromic compounds 114
photochromic eyewear
- photochromic lenses 136
photochromic lenses
- indolinospironaphthoxazines 136
- pyridobenzoxazines 136
photochromic systems 346
photochromic transformations 114
- activation of second-order NLO properties 134
- conformational changes in linear polymers 115
- data storage 137
- heterolytic bond cleavage 113
- pericyclic reactions (electrocyclizations) 113
- photoalignment of liquid-crystal systems 126
- photochromic lenses 136
- photocontrol of enzymatic activity 123
- photoinduced anisotropy (PIA) 123
- photomechanical effects 130
- trans-cis (E/Z) isomerization 113
photochromism 113
photoconductive polymers 49
- produced by glow discharge 70
- produced by heat 69

- produced by high-energy radiation 69
- produced by plasma polymerization 70
photoconductivity 49 ff.
- electron conduction 61
- hole conduction 61
photocontrol of enzymatic activity 123
photo-cross-linking
- bisazides 188
- co-polypeptide 185
- intermolecular cross-links 183
- mechanism 183
- simultaneous cross-linking and main-chain cleavage 193
- thin films
-- photolithographic processes 184
photocured coatings
- waterborn formulations 311
photocuring see also curing
- di- and trifunctional compounds 308
- industrial applications 307
- polymerizable formulations 307
[2+2] photocycloaddition 299
photodegradation
- polymers 196
photodynamic therapy, PDT
- cancer 223
- sensitizers 224
photo-Fries rearrangement 260
- aromatic esters, amides, urethanes 202
- polycarbonates 203
photogeneration of charge carriers 50
- dissociation of excitons 56
photografting 330
photoinduced absorption (PIA) 42
photoinduced anisotropy (PIA) 123
photo-induced surface relief storing
- recording density 347
photoinitiation of cationic polymerizations
- direct photolysis of the initiator 289
- sensitized photolysis of the initiator 292
photoinitiation of free radical polymerizations 276, 277
photoinitiation of ionic polymerizations
- anionic polymerization 296
- cationic polymerization 288
- free radical-mediated generation of cations
-- addition-fragmentation reactions 295
-- oxidation of radicals 293
photoinitiators 275
- anionic photoinitiators 296
- cationic photoinitiators 290
- dye/co-initiator systems 284

- free radical polymerizations
-- type I initiators 276
-- type II initiators 276
- inorganic photoinitiators 286
- metal-based initiators 283
- photoinitiators for visible light 281
- quinones and 1,2-diketones 285
- type I free radical photoinitiators 276, 278
- type II free radical photoinitiators 279, 280
photoionization
- tryptophan 215
- tyrosine 215
photolatent compounds 297
photolatent initiators 297, 298
photolithography 231
- maskless lithography 235
- phase-shifting transmission masks 234
- projection optical lithography 233
- soft lithography 246
- zone-plate array lithography, ZPAL 235
photoluminescence
photolysis
- poly(methyl methacrylate) 179
- polystyrene 179, 180
photo-mask production
- electron-beam lithography 235
- ion-beam lithography 235
photomechanical effects 130, 131, 132
- in hairy-rod type poly(glutamate)s 134
- in monolayers 134
photomorphogenic control functions
- of photoreceptor proteins 219
photon harvesting 16
- role of anthracene groups 18
- role of naphthalene groups 18
photonic crystals
- polymeric materials consisting of periodic microstructures 317
photopolymerizable systems 346
photopolymerization 275
- epoxides 186
photopolymers
- holography 322
photoreactivation 220
- of organisms 219
photorearrangements 204
photoreceptor action
- in biological processes 217
photoreceptor proteins 210
- regulatory action 217
- transformation modes of chromophores 217

photoreceptors 143, 208
photoreceptors, dual layer 143
photorefractive formulations 105
- polymers 106
photorefractive (PR) effect 103
- applications
-- dynamic holographic interferometry 110
-- holographic storage 110
-- real-time processing 110
- diffraction efficiency
-- four-wave mixing technique 109
- evidence for PR effect
-- two-beam coupling experiments 108
- mechanism 104
photorefractive systems 346
photorefractivity 103
- orientational photorefractivity 107
photo-release of reactive anions 296
photosensitizers
- nucleic acid/protein cross-linking 225
photosensors 349
photosynthesis 207
photovoltaic (PV) cells
- classical PV cells
-- CdTe 162
-- CuInSe$_2$ 162
-- GaAs 162
-- silicon 162
- polymeric solar cells 163
phytochrome kinase 218
phytochrome interacting factor 218
phytochromes 217
- interdomain signal transmission 218
- photoreceptors 209, 210
phytochromobilin
- photocycle 218
PICUP (photo-induced cross-linking of unmodified proteins) 223
pigments
- light absorbers 257
pinacol rearrangement 242
pit/land structure
- compact disks 340, 341
planar waveguides
- polymeric 170
plants
- photomorphogenic processes 211
plasma thrusters
- laser ablation 256
platinum(II) acetyl-acetonate (Pt(acac)$_2$)
- anionic photoinitiators 296
plume

– laser ablation   250
Pockels effect   73
Pockels tensor   78
POFs (polymer opticals fibers)   168 ff.
polarization   75
– electric field dependence   74
polarized electroluminescence
– background illumination of liquid-crystal displays   154
polarized light
– absorption   22
– circularly polarized light   23, 28
– creation of anisotropy   23
– degree of polarization   26
– elliptically polarized light   24
– emission   22, 26
– fluorescence   26
– generation of anisotropy   124
– generation of birefringence   124
– linearly polarized light   22, 23
polarizing excitonic energy transfer, EET   38
poling
– electro-optical poling   93
– Langmuir-Blodgett (LB) technique   93
– optical poling   93
– self-assembly techniques   93
poly(4-acetoxy styrene)
– photo-rearrangement   204
polyacetylene
– chemical sensors   351
poly(acrylic acid)
– polymer support for inorganic photocatalysts   365
polyacrylonitrile
– cross-linking   195
– main-chain cleavage   195
poly(allyl amine hydrochloride)
– polymer support for inorganic photocatalysts   365
polyaniline   51
poly[bis(2-naphthoxy)phosphazene], P2NP   69
polycarbonates
– compact disks   340
– photo-rearrangement   204
– POFs   169
poly(cis-isoprene)   189
– photolithography   236
polydiacetylenes
– color change   300
poly(dialkyl fluorine)   51
poly(2,5-di-isopentyloxy-p-phenylene), DPOPP   23

polyester acrylate-based formulations
– coatings   309
polyester with pendant azobenzene groups
– holographically recorded gratings   138
polyethylene
– surface grafting   331
polyfluorene
– chemical sensors   351
poly(glutamic acids)
– coil helix transition   121
poly(L-glutamic acids)   119
– modified   120
poly(4-hydroxystyrene)
– photo-cross-linking   192
polyimides
– laser ablation   249
– resists   237
polyisocyanates   33
– CD spectra   34
poly(L-lysine)
– CD spectra   26
– circular dichroism   25
polymer fibers
– information networks   168
polymer films
– anisotropic contraction   131
– chain alignment   154
– light-induced dimensional alterations   131
– light-induced mass transport   132
– photoinduced anisotropy (PIA)   123
– surface relief gratings   132
polymeric light sources   146
polymeric materials   248
polymer lasers
– conjugated polymers   156
– electrically pumped   162
– host/guest systems   156
polymer LEDs
– hole and electron transport materials   153
– luminance–voltage characteristic   150
polymer optical fibers (POFs)   169
polymer optical waveguides   167
polymers
– light-emitting diodes   148
polymer single crystals
– topochemical photopolymerization   299
polymers bearing pendant aromatic groups
– photocatalysts   363
polymers in holography   322
polymer transducer supports
– polyacrylamide   357

- polysiloxanes 357
- poly(vinylidene chloride) 357
- PVC 357
poly(methyl methacrylate) 11, 37
- POFs 169
poly(methyl vinyl ketone) 11
poly(phenyl vinylene)s
- BuEH-PPV 45
- spectral narrowing 45
poly(phenylene vinylene)s
- MEH-DSB 43
- MEH-PPV 43
poly(1,4-phenylene vinylene) 12
poly(p-phenylene)s 32, 51, 362
- ladder-type 51
- m-LPPP 51, 55
poly(p-phenylene ethynylene), PPE
- chemical sensors 351
poly(p-phenylene vinylene), PPV 51
- chemical sensors 351
- DOO-PPV 12
- light-emitting diodes 147
- MEH-PPV 27
- PMCYHPV 12
- PPFPV 12
- PPV 12
- solar cells 164
poly(phenyl vinyl ketone) 11, 42
polypropylene
- surface grafting 331
polysaccharides 207, 208
polysilanes
- main-chain cleavage 198
- photodegradation 198
polysilylene 51, 57
- main-chain cleavage 198
- photodegradation 198
poly(sodium styrene sulfonate-co-2-vinyl-naphthalene)
- photocatalyst 363
polystyrene 11
- excimer formation 17
- POFs 169
- segmental motions 29
poly(thiophene)s 28, 51
- CD spectrum 33
- PDMBT 32
- PMBET 33
- solar cells 164
polyurethane
- surface grafting 331
polyurethane-based coatings 298
poly(uridylic acid)

- intra-chain hydrogen abstraction 227
poly(vinyl acetate) 11
poly(N-vinyl carbazole) 51, 53, 54
poly(vinyl chloride)
- dehydrochlorination 197
- discoloration 196, 197
- photodegradation 196
poly(vinyl cinnamate) 23
positive resists 239
potassium
- anionic photoinitiators 296
potassium ion sensor 358
PPP
- active photocatalyst 361
- poly(p-phenylene) 361
printing
- photocured coatings 310
printing inks
- curing 320
printing plates 318
- composition of the photosensitive layer 319
- generation of the relief structure 319
projection optical lithography 233
protease activity 353
- turn-on fluorescence detection 352
protein-nucleic acid assemblies
- photochemical cross-linking 223
protein-protein assemblies
- photochemical cross-linking 223
proteins 207, 209
- cross-linking 216
- denaturation 214
- photoreactions 214
- rotational correlation 29
- turn-on fluorescence detection 352
proximity printing
- photolithography 232
pterins
- photoreceptors 210
PTBVB
- poly(t-butyl-p-vinyl benzoate) 240

q
quantum yield of photodecomposition
- [2+2] cycloreversion 213
- purines 212
- pyrimidines 212
quantum yields
- cross-linking 194
- electroluminescence 152
- initiation of diacetylene polymerization 301

- main-chain cleavage 194
- of charge generation 145
- photoproducts of selected polymers 195
quinones and 1,2-diketones
- chemical structures 288

r
3R scheme
- chemosensing 349
radiant flux of light, same as intensity 7
radical combination 180
radical disproportionation 180
radical scavengers 257, 262
read-only memory (ROM) 340
rearrangements
- o-nitrobenzyl ester rearrangement 202
- photo-Fries rearrangement 202
refractive index
- complex refractive index 76
- electric field-induced changes 74
- imaginary part 76
regioregularity
- poly(3-hexylthiophene), P3HT 62
Rehm–Weller equation 285, 293
reineckate
- anionic photoinitiators 296
repair of lesions
- DNA photolyases 219
resists
- ArF (193 nm) lithography 242
- chemical amplification resists
- computer chip fabrication 236
- F_2 *(157 nm) lithography* 245
- negative resists 238
- photolithography 232
- positive resists 238
- sensitivity 238
-- of deep UV resists 240
Richardson-Schottky (RS) thermionic emission
- OLEDs 151
ROM
- read-only memory 326, 340
rotational correlation times
- proteins 29
rotational diffusion constant 29
ruby laser 74
Russel mechanism
- combination of peroxyl radicals 200

s
sacrificial consumption
- of stabilizers 267
sealings 310
second harmonic generation (SHG) 74, 76, 82
- photochromic activation 134
second-order NLO materials 87
- alignment of AπD moieties 92
- commercially available NLO polymers 92
- electric field-induced alignment (poling) 92
- guest-host systems 89
- NLO polymers 89, 91
- orientation techniques 92
- poled polymer films 91
- poling 93
second-order NLO properties
- light-induced generation 135
second-order optical nonlinearity 77
self-focusing/defocusing 84
sensitizers
- photochemotherapy of cancer cells 224
shrinkage
- curing of dental formulations 315
SIA International Roadmap 231
signal modulators 96
silica
- surface grafting 331
silicones
- UV-cured 310
silver halide photographic emulsions 324
silyl benzyl ethers
- cationic photoinitiators 290
singlet-oxygen
- formation 202
- reactions with unsaturated polymers 202
skin cancer 211
soft lithographic process 246
solar cells
- donor/acceptor heterojunctions 166
-- CN-PPV 165
-- MEH-PPV 165
- flat-heterojunction organic solar cells 165
- performance characteristics 167
- phase-separated polymer blends 165
- p-n homojunction crystalline silicon solar cells 163
- semiconducting polymers 164
solid immersion lenses

- hemispherical lenses   342
- Weierstrass superspherical lenses   342
solid immersion techniques   340
solitons
- negatively charged soliton   54
- neutral soliton   54
- positively charged soliton   54
spectral narrowing   44
spectroscopy
- time-resolved   38
spectrum
- optical absorption   209
spirooxazines   114
spiropyran groups
- in polypeptides   119
spiropyrans   114
ssDNA (single-strand DNA)
- base sequence detection   354
stabilization
- light stabilizers   257
stabilization of polymers
- by energy transfer   15
stabilizers see also light stabilizers
- hydroperoxide decomposers   265
- packages   266
- sacrificial consumption and depletion   267
stencils
- photolithography   232
step-index polymer optical fibers   170
stepwise [2+2] photocyclopolymerization   302
stepwise processes   303
stereolithography   315, 317
storage capacity   339, 340
- blue-ray disks   342
- HD-DVDs   342
- volume holography   344
storage materials
- chalcogenide glasses   346
- inorganic crystals   346
- photoaddressable polymers, PAPs   346
- photopolymerizable systems   346
- photopolymers   346
- photorefractive crystals   346
- volume holography   346
storage systems
- forgery-proof   139
streak camera   41
stress proteins
- light-induced formation   216
sulfonium salts
- cationic photoinitiators   290

sulfonyloxy ketones
- cationic photoinitiators   290
superquenching   352
surface grafting   331
surface modification
- photografting   330
surface relief gratings   132
susceptibility
- linear electro-optic (EO) effect   81
- second harmonic generation, SHG   81
susceptibility tensors   74 f., 78
synchrotron radiation   246

t

telecommunications
- photocured coatings   310
terephthalophenones
- type II free radical photoinitiators   280
tertiary amines   298
- initiators of anionic polymerizations   297
thioanthrenium salts
- cationic photoinitiators   290
thioxanthone derivatives
- type II free radical photoinitiators   280
third harmonic generation, THG   78, 83
third-order NLO materials
- conjugated compounds   88
- polyacetylenes   88
- polydiacetylenes   88, 93
- poly(phenylene vinylene)s   93
- poly(p-phenylene)s   93
- polysilanes   88
- polythiophenes   93
- susceptibilities   94, 95
- trans-polyacetylenes   93
third-order optical nonlinearity   77
third-order phenomena   82
- degenerate four-wave mixing   83
- electric field-induced second harmonic generation   83
- optical Kerr gate   83
- third harmonic generation   83
- two-photon absorption   83
- Z-scan   83
third-order susceptibilities   83
threshold fluence   248
time-of-flight (TOF) method
- poly(methyl phenyl silylene)   62
- determination of the mobility   60
time-resolved optical absorption measurements   39 f.
- flash photolysis   39
- Nd^{3+}:YAG laser   39

- ruby laser 39
- Ti:sapphire laser 39
time-resolved spectroscopy 38, 55
- amplified spontaneous emission 44
- fluorescence 44
- luminescence 44
- optical absorption 41
- spectral narrowing 44 f.
TiO(F_4-Pc):TTA
- dual-layer system
-- xerography 145
titania
- photocatalyst 364
- polymer-coated TiO_2 particles 365
- surface grafting 331
titanium dioxide, TiO_2
- generation of reactive free radicals 288
- inorganic photoinitiators 286
titanocenes
- photoinitiators 283
TMP
- 2,2-6,6-tetramethylpiperidine 262
TMPO
- piperidinoxyl radical 264
- reaction with alkyl radicals 264
TNT
- 2,4,6-trinitrotoluene 352
TNT sensor 352
topochemical photopolymerization of diacetylenes 299
topochemical polymerizations 299
track pitch 339
trans-cis (E/Z) isomerization 113
trans-coniferyl alcohol
- lignins 207
trans-p-coumaryl alcohol
- lignins 207
transphasor, the optical transistor
- applications of NLO polymers 100
trans-polyacetylene 51
trans-sinapyl alcohol
- lignins 207
triallyl cyanurate, TAC
- cross-linking enhancer 191
triaryl cyclopropenium salts
- cationic photoinitiators 290
triarylmethanes 114
1,3,5-triazines
- UV absorbers 258
trinitrofluorenone 54
triphenylmethyl cations
- photogeneration 118
triphenylsulfonium salts

- photolysis 241
triplet-triplet absorption 42
tris(8-oxyquinolato)-aluminum, Alq_3
- electron conduction 61
tryptophan (Trp) 209
tungsten hexacarbonyl, $W(CO)_6$
- photoiniator
-- photo-cross-linking 187
two-photon absorption (TPA) 85
two-photon polymerization 318
type I free radical photoinitiators
- chemical structures 278
type II free radical photoinitiators
- bimolecular reactions 279
- chemical structures 280
tyrosine (Tyr) 209

u
UV absorbers 257, 258
UVAs
- UV absorbers 258
UV/Vis spectroscopy 30

v
VOC emission 309
VOCs
- volatile organic compounds 307
volume holography 340
- holography 322
- storage materials 345, 346
- storage mechanism 344
volume shrinkage
- volume holography 324

w
Wannier excitons 53
water-soluble aromatic ketones
- type II free radical photoinitiators 280
Weierstrass superspherical lens 342
wood
- darkening 222
- photoreactions 221
- yellowing 222
wool tendering 214
WORM
- write-once-read-many 326

x

xanthopsins
– photoreceptors 210
xerographic discharge method 58
– quantum yields for charge carriers 59
xerography
– charge-generation systems
– – pigment particles of dyes 145

y

yellow proteins
– photoreceptors 210
yellowing
– wood 222

z

Z-scan experiment 84, 85

Related Titles

Elias, H.-G.

Macromolecules

Volume 4: Applications of Polymers

2008

ISBN 978-3-527-31175-0

Parmon, V., Vorontsov, A., Kozlov, D., Smirniotis, P.

Photocatalysis

Catalysts, Kinetics and Reactors

2008

ISBN 978-3-527-31784-4

Elias, H.-G.

Macromolecules

Volume 3: Physical Structures and Properties

2007

ISBN 978-3-527-31174-3

Elias, H.-G.

Macromolecules

Volume 2: Industrial Polymers and Syntheses

2007

ISBN 978-3-527-31173-6

Elias, H.-G.

Macromolecules

Volume 1: Chemical Structures and Syntheses

2005

ISBN 978-3-527-31172-9

Kemmere, M. F., Meyer, T. (eds.)

Supercritical Carbon Dioxide

in Polymer Reaction Engineering

2005

ISBN 978-3-527-31092-0

Meyer, T., Keurentjes, J. (eds.)

Handbook of Polymer Reaction Engineering

2005

ISBN 978-3-527-31014-2

Xanthos, M. (ed.)

Functional Fillers for Plastics

2005

ISBN 978-3-527-31054-8

Advincula, R. C., Brittain, W. J., Caster, K. C., Rühe, J. (eds.)

Polymer Brushes

Synthesis, Characterization, Applications

2004

ISBN 978-3-527-31033-3

Elias, H.-G.

An Introduction to Plastics

2003

ISBN 978-3-527-29602-6